FIFTH EDITION

Technical Communication

SITUATIONS AND STRATEGIES

Mike Markel

Boise State University

St. Martin's Press

New York

EXECUTIVE EDITOR: Nancy Perry
MANAGER, PUBLISHING SERVICES: Emily Berleth
ASSOCIATE EDITOR, PUBLISHING SERVICES: Meryl Gross
PROJECT MANAGEMENT: Books By Design, Inc.
PRODUCTION SUPERVISOR: Scott Lavelle
TEXT DESIGN: Books By Design, Inc.
COVER DESIGN: Patricia McFadden
COVER ART: © 1996 Bill Milne Studio

Library of Congress Catalog Card Number: 97–66314

Manufactured in the United States of America

3 2 1 0 9 8
f e d c

For information, write:
St. Martin's Press, Inc.
175 Fifth Avenue
New York, NY 10010

ISBN: 0–312–17087–4

ACKNOWLEDGMENTS
Acknowledgments and copyrights are on page xx, which constitutes an extension of the copyright page.

Brief Contents

Contents   *iv*
Preface   *xv*
Acknowledgments   *xx*

PART ONE.   The Technical Communication Environment

Chapter 1.   Introduction to Technical Communication *1*
Chapter 2.   Understanding Ethical and Legal Considerations *21*
Chapter 3.   Writing Collaboratively *48*
Chapter 4.   Using Electronic Communication Tools *67*

PART TWO.   Planning the Document

Chapter 5.   Analyzing Your Audience *100*
Chapter 6.   Determining Your Purpose and Strategy *129*
Chapter 7.   Researching Your Subject *142*

PART THREE.   Drafting the Document

Chapter 8.   Developing the Argument *189*
Chapter 9.   Drafting Definitions and Descriptions *235*
Chapter 10.   Drafting the Front and Back Matter *268*

PART FOUR.   Integrating Design Elements and Graphics

Chapter 11.   Integrating Design Elements *308*
Chapter 12.   Integrating Graphics *344*

PART FIVE.   Revising the Document

Chapter 13.   Revising for Coherence *399*
Chapter 14.   Revising for Sentence Effectiveness *424*

PART SIX.   Writing Proposals and Reports

Chapter 15.   Writing Proposals *459*
Chapter 16.   Writing Memos and Informal Reports *491*
Chapter 17.   Writing Formal Reports *519*

PART SEVEN.   Preparing Other Common Types of Applications

Chapter 18.   Preparing Job-Application Materials *556*
Chapter 19.   Writing Letters *595*
Chapter 20.   Writing Instructions and Manuals *623*
Chapter 21.   Making Oral Presentations *659*

APPENDICES

Appendix A.   Handbook *683*
Appendix B.   Guidelines for Speakers of English as a Second Language *707*
Appendix C.   Commonly Misused Words *718*
Appendix D.   Selected Bibliography *720*

Index *725*

Contents

Preface   *xv*
Acknowledgments   *xx*

PART ONE. The Technical Communication Environment

CHAPTER 1. Introduction to Technical Communication   1

What Is Technical Communication?   *2*
Who Creates Technical Communication?   *2*
The Evolution of Technical Communication   *3*
The Role of Technical Communication in Business and Industry   *4*
Technical Communication and Your Career   *4*
Characteristics of Technical Communication   *6*
   Addresses Particular Readers   *7*   •   Helps Readers Solve Problems   *8*   •
   Furthers an Organization's Goals   *8*   •   Is Created Collaboratively   *9*   •
   Uses Design to Increase Readability   *9*   •   Involves Words and
   Graphics   *10*   •   Involves High-Tech Tools   *10*
Measures of Excellence in Technical Communication   *11*
   Honesty   *11*   •   Clarity   *13*   •   Accuracy   *13*   •
   Comprehensiveness   *14*   •   Accessibility   *14*   •   Conciseness   *14*   •
   Professional Appearance   *15*   •   Correctness   *15*
CASE: **Teaching Technical Communication across Campus**   *16*

Exercises   *16*
   COLLABORATIVE EXERCISE   *16*   •   WEB EXERCISE   *16*

CHAPTER 2. Understanding Ethical and Legal Considerations   21

A Brief Introduction to Ethics   *22*
Your Ethical Obligations   *25*
   Obligations to Your Employer   *25*   •   Obligations to the Public   *27*   •
   Obligations to the Environment   *29*
Your Legal Obligations   *29*
   Copyright Law   *29*   •   Trademark Law   *32*   •   Contract Law   *34*   •
   Liability Law   *35*
Codes of Conduct   *37*

Whistleblowing   *40*

Guidelines for Ethical Communication   *42*
Tell the Truth   *42*   •   Do Not Mislead   *42*

CASES: Irritating Insulation   *43*   •   Mysterious Internet Code   *44*

Exercises   *45*
COLLABORATIVE EXERCISES   *46*   •   WEB EXERCISE   *46*

CHAPTER 3. Writing Collaboratively   48

Patterns of Collaboration   *49*

Advantages and Disadvantages of Collaboration   *51*
Advantages of Collaboration   *51*   •   Disadvantages of Collaboration   *52*

Organizing and Maintaining a Collaborative Group   *53*

Reducing Conflict   *57*

Improving Listening Skills   *59*

Gender and Collaboration   *60*

Multiculturalism and Collaboration   *62*

CASE: The Reluctant Collaborator   *63*

Writer's Checklist   *64*

Exercises   *64*
COLLABORATIVE EXERCISES   *65*   •   WEB EXERCISE   *65*

CHAPTER 4. Using Electronic Communication Tools   67

Drafting Tools on the Word Processor   *69*
The Outline Feature   *69*   •   Templates and Styles   *69*   •   Automated
Hypertext Linking and Web-Conversion Tools   *72*

Revision Tools on the Word Processor   *72*
Spell Checkers   *72*   •   Grammar Checkers   *73*   •   Thesaurus Software   *73*

Working Collaboratively Using Electronic Tools   *74*
Electronic Mail   *74*   •   Groupware   *77*   •   Other Electronic Tools   *80*

Using Web Pages   *81*
Understanding Hypertext   *82*   •   How to Use Web Pages   *82*   •
Ethics, Copyright Law, and the Web   *86*

Creating Web Pages   *88*
Analyze Your Audience and Purpose   *90*   •   Gather or Create Your Text
and Other Information   *90*   •   Create a Site Design   *90*   •   Create a Page
Design   *92*   •   Convert the Text and Graphics   *93*   •   Revise and Test the
Information   *94*   •   Get the Files to an Internet Server   *95*   •   Register the
Site with the Search Engines   *95*

The Future of the Web   *96*

CASE: Creating a Web Site   96

Writer's Checklist   97

Exercises   97

    COLLABORATIVE EXERCISES   *98*   •   WEB EXERCISES   *99*

PART TWO. Planning the Document

CHAPTER 5. Analyzing Your Audience   100

Understanding the Writing Situation   *101*

Primary and Secondary Audiences   *101*

Basic Categories of Readers   *102*
    The Expert   *103*   •   The Technician   *104*   •   The Manager   *104*   •
    The General Reader   *108*

Individual Characteristics of Readers   *110*
    Education   *110*   •   Professional Experience   *111*   •   Job
    Responsibilities   *112*   •   Personal Characteristics   *112*   •   Personal
    Preferences   *112*   •   Attitudes toward You and Your Subject   *113*   •
    Cultural Characteristics   *114*   •   Document Use   *114*   •   Physical
    Environment   *115*

The Audience Profile Sheet   *115*

Accommodating the Multiple Audience   *118*

Understanding Multiculturalism   *118*
    Categories of Cultural Patterns   *119*   •   Writing to Multicultural
    Audiences   *124*

CASE: Analyzing an Audience   125

Writer's Checklist   *126*

Exercises   *127*

    COLLABORATIVE EXERCISE   *127*   •   WEB EXERCISE   *128*

CHAPTER 6. Determining Your Purpose and Strategy   129

Determining Your Purpose   *130*

Determining Your Strategy   *131*
    Ethical Constraints   *131*   •   Legal Constraints   *131*   •
    Informational Constraints   *132*   •   Personnel Constraints   *132*   •
    Financial Constraints   *133*   •   Time Constraints   *133*   •   Format and
    Tone Constraints   *134*

Planning Your Tactics   *134*

Proposing a Strategy   *136*
    Proposing a Small Project   *137*   •   Proposing a Large Project   *137*

CASE: Making a Q-and-A Sheet   138

Writer's Checklist   140

Exercises   140

COLLABORATIVE EXERCISE   140 • WEB EXERCISE   140

CHAPTER 7. Researching Your Subject   142

The Role of Research in the Problem-Solving Process   143

Characteristics of Good Information   144

Choosing a Topic   145

Generating Ideas about the Topic   146
Asking Journalistic Questions   147 • Brainstorming   148 •
Freewriting   148 • Talking   149 • Sketching   150

Conducting Secondary Research   152
Understanding the Research Media   153 • Using Basic
Research Tools   156 • Researching Government Information   160 •
Understanding the Information   161 • Evaluating the Information   169 •
Evaluating Internet Sources   170

Conducting Primary Research   171
Inspections   171 • Experiments   171 • Field Research   172 •
Interviews   173 • Letters of Inquiry   176 • Questionnaires   176

An Example of How to Plan a Research Strategy   180
Deciding on a Research Sequence   182 • Carrying Out Secondary
Research   182 • Planning Primary Research   184

CASE: Creating a Research Guidebook for Your Major   185

Writer's Checklist   186

Exercises   186

COLLABORATIVE EXERCISE   187 • WEB EXERCISE   187

PART THREE. **Drafting the Document**

CHAPTER 8. Developing the Argument   189

A Brief Review of Outlining   190
Grouping Similar Items   190 • Ordering the Items in the Groups   191 •
Organizing the Groups   191 • Avoiding Common Logical Problems   191 •
Choosing an Outline Format   192

Guidelines for Drafting   193

How to Develop an Argument   195
Understanding Your Audience's Goals   196 • Understanding the Nature of
Persuasive Evidence   196 • Presenting Yourself Effectively   197 •
Understanding Political Realities   198

Basic Patterns of Organizing Arguments   *198*
Chronological   *199*   •   Spatial   *200*   •   General to Specific   *203*   •
More Important to Less Important   *206*   •   Comparison
and Contrast   *207*   •   Classification and Partition   *211*   •
Problem-Methods-Solution   *217*   •   Cause and Effect   *221*

Introducing and Concluding the Body   *224*
Introducing the Body   *224*   •   Concluding the Body   *227*

CASE: Introducing a Document   *229*

Writer's Checklist   *230*

Exercises   231
COLLABORATIVE EXERCISE   *233*   •   WEB EXERCISE   *234*

CHAPTER 9. Drafting Definitions and Descriptions   235

Understanding the Role of Definitions   *236*

Analyzing the Writing Situation for Definitions   *237*

Determining the Kind of Definition to Write   *238*
Drafting Parenthetical Definitions   *238*   •   Drafting Sentence
Definitions   *239*   •   Drafting Extended Definitions   *240*

Deciding Where to Place the Definition   *246*

Drafting Descriptions   *248*

Understanding the Role of Descriptions   *249*

Analyzing the Writing Situation for Descriptions   *249*

Guidelines for Writing Descriptions   *250*
Clearly Indicate the Nature and Scope of the Description   *250*   •   Introduce
the Description Clearly   *251*   •   Provide Appropriate Detail   *251*   •
Conclude the Description   *253*

Analyzing Some Descriptions   *255*

CASE: Annotating a Patent Application   *261*

Writer's Checklist   *262*

Exercises   262
COLLABORATIVE EXERCISE   *267*   •   WEB EXERCISE   *267*

CHAPTER 10. Drafting the Front and Back Matter   268

Using the Word Processor   *269*

Writing the Front Matter   *270*
The Letter of Transmittal   *270*   •   The Cover   *271*   •   The Title
Page   *271*   •   The Abstract   *274*   •   The Table of Contents   *276*   •
The List of Illustrations   *279*   •   The Executive Summary   *279*

APA and MLA Documentation of Sources   *283*
APA Style   *284*   •   MLA Style   *289*

Writing the Back Matter   *295*
   The Glossary and List of Symbols   *296*   •   References   *297*   •
   Appendices   *297*   •   The Index   *298*
CASE: Comparing Online Indexes   *304*
Writer's Checklist   *304*

Exercises   305
   COLLABORATIVE EXERCISE   *307*   •   WEB EXERCISE   *307*

PART FOUR. **Integrating Design Elements and Graphics**

CHAPTER 11. Integrating Design Elements   308

Goals of Document Design and Page Design   *309*

Planning the Design of the Document and the Page   *311*
   Analyzing Your Audience's Needs and Expectations   *311*   •
   Determining Your Resources   *311*

Designing the Document   *312*
   Size   *312*   •   Paper   *314*   •   Bindings   *315*   •   Accessing Tools   *316*

Designing the Page   *319*
   Principles of Effective Page Design   *320*   •   Learning Theory and Its Relation
   to Page Design   *321*   •   Page Grids, White Space, and Columns   *323*   •
   Typography   *330*   •   Designing Titles and Headings   *337*   •   Other Design
   Techniques   *338*
CASES: The Brochure with the Inconsistent Design   *340*
       Updating an Old Design   *340*
Writer's Checklist   *340*

Exercises   340
   COLLABORATIVE EXERCISE   *340*   •   WEB EXERCISE   *343*

CHAPTER 12. Integrating Graphics   344

Characteristics of an Effective Graphic   *347*

Planning, Creating, and Revising the Graphics   *350*
   Planning the Graphics   *350*   •   Creating the Graphics   *351*   •   Revising the
   Graphics   *355*

Using Color Effectively   *356*

Choosing the Appropriate Kind of Graphic   *361*
   Illustrating Numerical Values   *361*   •   Illustrating Logical
   Relationships   *376*   •   Illustrating Instructions and Process
   Descriptions   *378*   •   Illustrating Visual and Spatial Characteristics   *381*

Showing Motion in Your Graphics   *387*

Creating Graphics for International Readers   *388*

Understanding Graphics Software *391*
Spreadsheet Business Graphics *391* • Paint Programs and
Draw Programs *391*

CASE: Adding Graphics to a Web Site *394*

Writer's Checklist *394*

Exercises 394
COLLABORATIVE EXERCISE *397* • WEB EXERCISE *397*

PART FIVE. Revising the Document

CHAPTER 13. Revising for Coherence   399

Understanding the Given/New Strategy *400*

Techniques Used in Revising *401*
Revising by Yourself *401* • Obtaining Help from Someone Else *403* •
Testing Readers as They Use the Document *404*

Revising the Whole Document for Coherence *405*

Revising Titles for Coherence *406*

Revising Headings for Coherence *407*

Revising Lists for Coherence *408*

Revising Paragraphs for Coherence *410*
Structure Paragraphs Clearly *411* • Keep Paragraphs to a Manageable
Length *413* • Use Coherence Devices within and between Paragraphs *415* •
Turning a "Writer's Paragraph" into a "Reader's Paragraph" *418*

CASE: Writing Guidelines about Coherence *419*

Writer's Checklist *421*

Exercises 421
COLLABORATIVE EXERCISE *422* • WEB EXERCISE *423*

CHAPTER 14. Revising for Sentence Effectiveness   424

Structuring Effective Sentences *425*
Use Lists *425* • Put New and Important Information at the End of
the Sentence *430* • Choose an Appropriate Sentence Length *430* •
Focus on the "Real" Subject *432* • Focus on the "Real" Verb *433* •
Express Parallel Elements in Parallel Structures *434* • Use Modifiers
Effectively *435*

Choosing an Appropriate Level of Formality *437*

Choosing the Right Words and Phrases *438*
Be Clear and Specific *438* • Be Concise *444* • Use Inoffensive
Language *448*

Understanding Simplified English for Nonnative Speakers   *449*

Understanding Structured Writing   *450*

CASE: Revising a Draft for Sentence Effectiveness   *453*

Writer's Checklist   *454*

Exercises   454

COLLABORATIVE EXERCISE   *458*   •   WEB EXERCISE   *458*

PART SIX. **Writing Proposals and Reports**

CHAPTER 15. Writing Proposals   459

The Logistics of Proposals   *460*
External and Internal Proposals   *460*   •   Solicited and Unsolicited
Proposals   *461*

The "Deliverables" of Proposals   *463*
Research Proposals   *463*   •   Goods-and-Services Proposals   *464*

Persuasion and Proposals   *464*
Understanding the Readers' Needs   *465*   •   Describing What You Plan
to Do   *466*   •   Demonstrating Your Professionalism   *467*

Writing a Proposal   *468*
The Role of Resource Planning   *468*   •   The Role of Collaboration   *468*

The Structure of the Proposal   *469*
Summary   *470*   •   Introduction   *471*   •   Proposed Program   *472*   •
Qualifications and Experience   *473*   •   Budget   *476*   •   Appendices   *476*

The Format of the Proposal   *481*

Sample Internal Proposal   *481*

CASES: Ethics and Proposals   *488*
           Selecting Project-Management Software   *488*

Writer's Checklist   *488*

Exercises   489

COLLABORATIVE EXERCISE   *489*   •   WEB EXERCISE   *489*

CHAPTER 16. Writing Memos and Informal Reports   491

Guidelines for Informal Reports   *492*

Formats for Informal Reports   *493*
Memos   *494*   •   Other Formats   *499*

Writing Directives   *499*

Writing Trip Reports   *500*

Writing Field and Lab Reports   *502*

Writing Progress and Status Reports   *502*
Organizing Progress and Status Reports   *505*   •   Concluding Progress and Status Reports   *506*

Writing Meeting Minutes   *506*
Recording the Logistical Details of the Meeting   *512*   •   Providing an Accurate Record of the Meeting   *512*   •   Reflecting Positively on the Participants and the Organization   *512*

CASE: Amending a Proposal   514

Writer's Checklist   *514*

Exercises   *515*
COLLABORATIVE EXERCISE   *517*   •   WEB EXERCISE   *518*

CHAPTER 17. Writing Formal Reports   519

Understanding the Types of Formal Reports   *520*
Informational Reports   *520*   •   Analytical Reports   *521*   •   Recommendation Reports   *522*

A Problem-Solving Model for Preparing Formal Reports   *524*
Analyze Your Audience   *524*   •   Analyze Your Purpose   *525*   •   Identify the Questions That Need to Be Answered   *526*   •   Carry Out Appropriate Research   *527*   •   Draw Conclusions from the Results   *527*   •   Formulate Recommendations Based on the Conclusions   *527*

A Problem-Solving Model for Preparing Feasibility Reports   *528*
Identify the Problem or Opportunity   *529*   •   Establish Criteria for Responding to the Problem or Opportunity   *530*   •   Determine the Options   *530*   •   Study Each Option according to the Criteria   *531*   •   Draw Conclusions about Each Option   *532*   •   Formulate Recommendations Based on the Conclusions   *533*

Organizing the Body of the Formal Report   *533*
Introduction   *534*   •   Methods   *535*   •   Results   *536*   •   Conclusions   *537*   •   Recommendations   *537*

Sample Formal Report   *542*

CASE: Writing a Recommendation Report   553

Writer's Checklist   *554*

Exercises   *554*
COLLABORATIVE EXERCISE   *554*   •   WEB EXERCISE   *555*

PART SEVEN. **Preparing Other Common Types of Applications**

CHAPTER 18. Preparing Job-Application Materials   556

Planning the Job Search   *557*

Understanding Seven Ways to Look for a Position   *558*

Writing Résumés   *561*
  Appearance of the Résumé   *562* • Content of the Résumé   *563* •
  Elements of the Chronological Résumé   *564* • Elements of the Analytical
  Résumé   *571*

Writing Electronic Résumés   *576*
  Content of an Electronic Résumé   *577* • Format of an Electronic
  Résumé   *577*

Writing Job-Application Letters   *580*
  Content of the Job-Application Letter   *580* • Elements of the
  Job-Application Letter   *580*

Writing Follow-Up Letters   *585*

CASE: Updating Career-Center Materials   *590*

Writer's Checklist   *590*

Exercises   *591*

  COLLABORATIVE EXERCISE   *592* • WEB EXERCISES   *593*

CHAPTER 19. Writing Letters   595

Projecting the "You Attitude"   *596*

Avoiding Letter Clichés   *598*

Understanding the Elements of the Letter   *599*
  Heading   *600* • Inside Address   *600* • Attention Line   *601* • Subject
  Line   *601* • Salutation   *601* • Body   *602* • Complimentary
  Close   *602* • Signature   *602* • Enclosure Line   *602* • Copy Line   *603*

Learning the Format of the Letter   *603*

Understanding Common Types of Letters   *603*
  Order Letters   *603* • Inquiry Letters   *605* • Responses to Inquiries   *607* •
  Sales Letters   *607* • Claim Letters   *610* • Adjustment Letters   *612*

CASE: Dangerous Wrenches   *616*

Writer's Checklist   *617*

Exercises   *617*

  COLLABORATIVE EXERCISE   *620* • WEB EXERCISE   *622*

CHAPTER 20. Writing Instructions and Manuals   623

Understanding the Role of Instructions and Manuals   *624*

Analyzing Your Audience   *625*

Planning for Safety   *625*
  Writing Safety Information   *626* • Designing Safety Information   *627* •
  Placing Safety Information in the Appropriate Locations   *628*

Drafting Effective Instructions   *628*
  Drafting General Introductions   *628* • Drafting Step-by-Step
  Instructions   *630* • Drafting Conclusions   *631*

Drafting Effective Manuals   *633*
Drafting Front Matter   *637*   •   Drafting the Body   *638*   •   Drafting Back
Matter   *644*   •   Drafting Revisions of Manuals   *644*

Writing Instructions and Manuals for Nonnative Speakers of English   *646*

Hypertext   *647*
Advantages of Hypertext   *648*   •   Disadvantages of Hypertext   *649*

Usability Testing of Instructions and Manuals   *650*
The Goals of Usability Testing   *651*   •   Basic Concepts of Usability
Testing   *651*   •   Preparing for a Usability Test   *653*   •   Conducting a
Usability Test   *654*   •   Interpreting and Reporting the Data from a
Usability Test   *655*

CASE: Writing a Set of Instructions   *656*

Writer's Checklist   *656*

Exercises   *657*
COLLABORATIVE EXERCISE   *657*   •   WEB EXERCISE   *657*

CHAPTER 21. Making Oral Presentations   *659*

Understanding the Role of Oral Presentations   *660*

Preparing an Oral Presentation   *662*
Assessing the Speaking Situation   *663*   •   Choosing Effective Language   *664*   •
Preparing an Outline or Note Cards   *666*   •   Preparing Presentation
Graphics   *668*   •   Rehearsing the Presentation   *672*

Giving the Oral Presentation   *675*
Dealing with Nerves   *675*   •   Using Your Voice Effectively   *676*   •   Using Your
Body Effectively   *677*

Answering Questions after the Presentation   *678*

Sample Evaluation Form   *679*

CASE: Publicizing the School Newspaper   *680*

Speaker's Checklist   *680*

Exercises   *681*
COLLABORATIVE EXERCISES   *681*   •   WEB EXERCISE   *681*

APPENDICES

Appendix A. Handbook   *683*

Appendix B. Guidelines for Speakers of English as a Second Language   *707*

Appendix C. Commonly Misused Words   *718*

Appendix D. Selected Bibliography   *720*

Index   *725*

Preface

The principles of good writing have not changed much in the fourteen years since *Technical Communication: Situations and Strategies* was first published in 1984, but almost everything else about technical communication has. The Fifth Edition of this text reflects the many exciting changes that have occurred in technical communication—and in the ways to teach it—in the last few years, most especially the greatly expanded importance of the Internet and electronic communication tools. Every chapter has been revised and updated, and new examples, exercises, and cases have been added throughout.

New Chapter and New Elements

- *A new chapter on electronic communication tools.* More than ever, students need to know how to function in an electronic environment. Although high-tech tools are discussed in virtually every chapter, the Fifth Edition includes a new chapter on electronic communication tools (Chapter 4). Topics in this chapter include drafting tools such as the outline feature, templates, styles, and automated hypertext linking, as well as groupware tools. The chapter describes how to use browsers and search engines to find information on the Web. In addition, the chapter explains the concepts used in creating effective Web sites.

- *Cases in every chapter.* The Fifth Edition includes at least one case in each chapter. These cases provide realistic scenarios that test students' problem-solving communication skills. Some cases are meant to be completed individually, others collaboratively. Most of the cases can be completed either way. Some of the cases are self-contained in that all the information students need is provided in the case. Other cases call for students to conduct research through the Web first.

- *Collaborative exercises in every chapter.* The Fifth Edition includes at least one collaborative exercise in each chapter. These collaborative exercises, indicated by ⚎, not only call for students to work together cooperatively, but also provide an opportunity for students to analyze the successes and challenges of the collaborative process.

- *Web-based exercises in every chapter.* The Fifth Edition includes at least one Web-based exercise in each chapter. These exercises, indicated by ▣, help students become comfortable with the research medium that is certain to become more and more important each year. In addition, the Web-based exercises greatly expand the amount of technical communication to which your students are exposed.

Expanded Coverage

- *Research methods.* The discussion of research methods includes expanded coverage of using electronic media. In addition, the research chapter (Chapter 7) now shows the step-by-step process used by a group of students in planning a research strategy for a formal report.
- *Ethics and the Web.* The discussion of the Web in Chapter 4 has been expanded to address the evolving legal and ethical issues related to the use of electronic sources.
- *Feasibility studies.* The expanded coverage of feasibility studies (Chapter 17) now includes a problem-solving model that will help students understand how to perform this important task.
- *Electronic job searches.* The discussion of job-application materials (Chapter 18) now includes thorough treatment of how the Internet has changed the process of searching for a job, including coverage of job boards, Web sites, and electronic résumés.
- *Citing electronic sources.* The discussions of APA and MLA documentation styles (Chapter 10) now include formats for common electronic sources.

New Organization

The Fifth Edition of *Technical Communication* has been completely reorganized to help students use electronic tools more effectively, to understand the research process better, and to plan, research, and write reports more easily.

A new chapter—Chapter 4, "Using Electronic Communication Tools"—has been added to Part One, "The Technical Communication Environment." Along with an introductory chapter and chapters on ethical and legal issues and on collaboration, Part One now provides the background students need to work constructively and ethically with other people, using modern communication tools.

Part Two, "Planning the Document," describes the planning process used for both individual and collaborative projects, then presents techniques for finding and using information. Chapter 5, "Analyzing Your Audience," discusses different kinds of audiences, including multicultural readers. Chapter 6, "Determining Your Purpose and Strategy," explains how to work within constraints (such as time and money) and how to propose a strategy to a supervisor. Chapter 7, "Researching Your Subject," describes how to perform both primary and secondary research. Concluding that chapter is the description of how a group of students planned and carried out a research strategy combining a number of different kinds of research.

Part Three, "Drafting the Document," concentrates on the skills and tasks involved in drafting, beginning with the basic rhetorical patterns used in an argument (Chapter 8), then focusing on definitions and descriptions (Chapter 9), and front and back matter (Chapter 10).

Part Four, "Integrating Design Elements and Graphics," discusses the important graphical elements of technical communication. Although many instructors will want to cover design (Chapter 11) and graphics (Chapter 12) after they have covered planning and drafting, others will want to treat these subjects earlier. For this reason, I have made both chapters as flexible as possible.

Part Five, "Revising the Document," consists of two chapters that cover strategies for making technical communication clearer and easier to follow. Chapter 13, "Revising for Coherence," describes techniques for revising the document as a whole, down to the paragraph level. Chapter 14, "Revising for Sentence Effectiveness," focuses on sentence-level concerns, including such topics as sentence emphasis, formality, euphemisms, stating the obvious, meaningless modifiers, unnecessary prepositions, writing about people with disabilities, Simplified English, and structured writing. Although these issues are the focus of revising, some instructors may wish to cover them as part of the drafting phase.

Part Six, "Writing Proposals and Reports," has been reconceived and reorganized to improve its clarity and comprehensiveness. Chapter 15, "Writing Proposals," begins the part. Chapter 16, "Writing Memos and Informal Reports," covers a number of informal reporting situations, including progress and status reports, trip reports, and meeting minutes, and discusses the different formats used in these contexts. Chapter 17, "Writing Formal Reports," focuses on the different kinds of formal reports, including informational, analytical, and recommendations reports. This chapter includes the new problem-solving model for feasibility studies.

Part Seven, "Preparing Other Common Types of Applications," begins with Chapter 18, "Preparing Job-Application Materials." This chapter covers the use of the Internet as a means of both discovering available positions and applying for positions. Chapter 19, "Writing Letters," discusses this still-important application. Chapter 20, "Writing Instructions and Manuals," focuses on an increasingly important application for technical communicators and technical professionals alike. Chapter 21, "Making Oral Presentations," relates oral presentations to written documents, emphasizing how to construct a clear and compelling argument.

The Web Site

As with previous editions, the Fifth Edition includes a print version of the *Instructor's Resource Manual.* One of the most important changes to the Fifth Edition, however, is the new Web site, which includes a number of elements of the *Instructor's Resource Manual,* plus new features:

- A note to new teachers of technical communication.
- A general description of the text.
- Sample syllabi.
- A chapter-by-chapter guide to the text.
- A note on using computers in the classroom.
- Links to resources of value to students in performing research.
- Additional exercises not included in the text.
- Transparency masters that instructors can modify to meet their own needs.
- Forms that students will use: an audience-profile form, collaborative-writing evaluation forms, and an oral-presentation evaluation form. Students can download these forms and fill them out on their word processors.
- New cases every month.

You can find the site at <http://www.smpcollege.com/technical>.

Acknowledgments

All the examples in the book—from single sentences to complete documents—are real. Some were written by my students at Drexel University and Boise State University. Some were written by engineers, scientists, health-care providers, and businesspersons with whom I have worked as a consultant for over twenty years. Because much of the information in these documents is proprietary, I have silently changed brand names and other identifying information. I sincerely thank these dozens of individuals—students and professionals alike—who have graciously allowed me to reprint their writing. They have been my best teachers.

The Fifth Edition of *Technical Communication: Situation and Strategies* has benefited greatly from the perceptive observations and helpful suggestions of my fellow instructors throughout the country. Some completed extensive questionnaires about the previous edition; others reviewed the current edition in its draft form. I thank Christiana Birchak, University of Houston–Downtown; Jeanne Blum, Indiana University–Purdue University, Fort Wayne; David Bourdeaux, Nicholls State University; Kevin Conway, Santa Clara University; Virgil Cook, Virginia Polytechnic Institute and State University; David Hailey, Utah State University; Michael Hassett, Boise State University; Sharon Humphries, Montgomery

College; Sharon Irvin, Florida Institute of Technology; Johndan Johnson-Eiolola, Purdue University; Susan Lang, Southern Illinois University–Carbondale; Diane Maldonaldo, Community College of Allegheny County–South Campus; Mary Massirer, Baylor University; Harriet Napierkowski, University of Colorado—Colorado Springs; Matthew Novak, California Polytechnic State University–San Luis Obispo; Carolyn Plumb, University of Washington; Arnold Preussner, Truman State University; Sara Putzell-Shoemaker, Georgia Institute of Technology; Mark Rollins, Ohio University; Stuart Selber, Texas Tech University; and Donald Swanson, Wright State University.

I would like to acknowledge one other reader. Kevin S. Wilson of Boise State University, an extraordinary editor, has helped me clarify my thinking and simplify my writing in this edition, as he did in the previous edition. The book is much improved due to his efforts, and I thank him.

I have been fortunate, too, to work with a superb team at St. Martin's Press, led by Barbara Heinssen, Nancy Perry, Susan Cottenden, Emily Berleth, and Natalie Hart. For me, St. Martin's continues to exemplify the highest standards of professionalism in publishing. They have been demanding but endlessly encouraging and helpful. I hope they realize the value of their contributions to this book.

I want to thank, too, my colleagues at Boise State University—Chaman Sahni, Phil Eastman, and Daryl Jones—who have created a congenial, productive atmosphere for teaching and studying technical communication. In addition, I have benefited greatly from working with my colleagues in the technical communication program at Boise State: John Battalio and Michael Hassett.

My greatest debt, however, is to my wife, Rita, who over the course of many months and, now, five editions, has helped me say what I mean.

A Final Word

I am more aware than ever before of how much I learn from my students, my fellow instructors, and my colleagues in industry and academia. If you have comments or suggestions for making this a better book, please get in touch with me at the Department of English at Boise State University, Boise, ID 83725. My phone number is (208) 385-3088, or you can send me an e-mail from the Web site: <http://www.smpcollege.com/technical>. I hope to hear from you.

Mike Markel

Acknowledgments

CHAPTER 1

Introduction to Technical Communication

What Is Technical Communication? 2

Who Creates Technical Communication? 2

The Evolution of Technical Communication 3

The Role of Technical Communication in Business
and Industry 4

Technical Communication and Your Career 4

Characteristics of Technical Communication 6

    Addresses Particular Readers 7
    Helps Readers Solve Problems 8
    Furthers an Organization's Goals 8
    Is Created Collaboratively 9
    Uses Design to Increase Readability 9
    Involves Words and Graphics 10
    Involves High-Tech Tools 10

Measures of Excellence in Technical Communication 11

    Honesty 11
    Clarity 13
    Accuracy 13
    Comprehensiveness 14
    Accessibility 14
    Conciseness 14
    Professional Appearance 15
    Correctness 15

Case: Teaching Technical Communication across Campus 16

Exercises 16

     Collaborative Exercise 16
    Web Exercise 16

What Is Technical Communication?

Technical communication is the process of creating, designing, and transmitting technical information so that people can understand it easily and use it safely, effectively, and efficiently. Most technical communication is written by people working in or for organizations. Technical communication is read by people who need to carry out procedures and solve problems.

Much of what you read every day—textbooks, phone books, cookbooks, procedures manuals at the office, environmental impact statements, journal articles, the owner's manual for your car—is technical communication. The words and graphics of such documents are meant to be practical, that is, to communicate a body of factual information that will help an audience understand a subject or carry out a task. For example, an introductory biology text helps students understand the fundamentals of plant and animal biology and perform basic experiments. A user's manual for a software program describes how to use the program effectively.

Who Creates Technical Communication?

Most technical communication is created by one of two different categories of people:

- *Technical professionals.* Engineers, scientists, businesspeople, and other technically trained individuals are sometimes surprised to learn how much writing they do on the job. According to one survey, technical professionals can expect to devote at least one-fifth of their time to writing (Anderson, 1985). Engineers can expect to spend 40 percent of their time writing (Beer & McMurrey, 1997). And the percentage of time that technical professionals devote to communication increases as they advance: supervisors spend 40 percent of their time reading and writing; managers, 50 percent (MIT, 1984). A technical professional is a writer.
- *Technical communicators.* The job of a technical communicator is to write documents, including manuals, proposals, reports, sales literature, letters, journal articles, and speeches. Many technical communicators still call themselves *technical writers* (or *tech writers*), even though the term *technical communicator* more accurately reflects the increasing importance of graphics and the use of other media, such as online documentation. Other terms you will see include *learning product developer, information developer, documentation writer, information architect,* and *information engineer.* In small organizations, one technical communica-

tor might be responsible for all aspects of creating the document, from drafting through production. In large organizations, a technical communicator might be a specialist, working full-time as a writer, editor, graphic artist, designer, or production specialist.

Although most technical communicators work as employees in organizations, many are independent contractors—sometimes called *freelancers*—who work on individual projects for organizations. These projects can last a few weeks, a few months, or a few years; the independent contractor can work at home, at the client's facility, or both.

The Evolution of Technical Communication

The earliest samples of writing are technical communication: invoices, receipts, and deeds. The English poet Geoffrey Chaucer (ca. 1342–1400) wrote some technical communication about the astrolabe, a navigational instrument. *On the Origin of Species,* Charles Darwin's famous work on evolution, is another example of technical communication. But technical communication didn't become a field until around the time of World War II, when the military needed people to write user's manuals and maintenance manuals for hardware and weapons systems. In the past twenty years, the number of technical communicators has grown tremendously, largely because of the growth of the computer industry and related high-technology fields.

Before the 1970s, technical communication was often an afterthought. A computer manufacturer would market large systems without any instruction manuals at all. The company would send out technicians to install the system, and while they were there they tried to explain how to use it. Naturally, this wasn't a very effective or efficient way to educate the purchaser.

As more companies entered the richly profitable field, manufacturers realized that customers were frustrated when the product came without a user's manual, or when the manual arrived six months after the system, or when the manual was useless because of its poor quality. To keep their customers, high-tech companies started to pay more attention to their *documentation,* as these manuals are called.

In many high-tech companies today, technical communicators work closely with design engineers and the legal and marketing staffs in creating a new product. The emphasis today is on "user-friendly" products, and no product is friendly if the user can't figure out how to use it. Because technical communicators are valued members of professional organizations, their salaries and prestige have grown substantially. At many companies, such as IBM, technical communicators receive the same salary as hardware and software engineers.

The Role of Technical Communication in Business and Industry

The working world depends on written communication. Virtually every action taken in a modern organization is documented in writing (whether on paper or online). Here are a few examples of writing within an organization:

- Memos and electronic mail (e-mail) messages are written to request information and communicate new policies.
- Travel reports are written after technical professionals return from attending a conference, inspecting a site, or teaching a training session.
- Detailed step-by-step instructions and policy statements are written when new processes and procedures are introduced in an organization.
- When a major project is being contemplated within an organization, a proposal is written to persuade management that the project would be not only feasible but also in the best interests of the organization.
- Once the project is approved and under way, progress reports are written periodically to inform the supervisors and managers about the current status of the project and about any unexpected developments.
- When the project has ended, a completion report is written to document the work and enable the organization to implement any recommendations.

In addition to in-house writing, every organization must communicate with other organizations and the public. Following are some examples:

- Inquiry letters, sales letters, goodwill letters, and claim and adjustment letters are written to customers, clients, and suppliers.
- Sales and marketing literature is written for distribution to potential customers.
- Web sites are created to provide product information and marketing and to solicit job applications.
- Research reports are written for external organizations.
- Articles are written for trade and professional journals.

The world of business and industry is a world of communication.

Technical Communication and Your Career

If you are taking a technical-communication course now, it was probably created to meet the needs of the working world. You undoubtedly have seen articles in newspapers and magazines about the importance of writing and other communication skills in a professional's career.

The first step in obtaining a professional-level position, in fact, is to write an application letter and a résumé. Your writing will help determine whether an organization decides to interview you. At the interview, your oral communication will be evaluated along with your other qualifications.

Once you start work, you will write e-mail, memos, letters, marketing materials, and short reports, and you might be asked to contribute to larger projects, such as proposals and Web sites. During your first few months on the job, your supervisors will be looking at both your technical abilities and your communication skills.

A survey of engineering managers found that nearly two-thirds felt that technical communication was "extremely important" as a management tool (Spretnak, 1982). Richard C. Levine, an engineering manager with Bell Northern Research, has this to say about his employees' ability to write clearly:

> . . . when I see a report or memo that has repeated errors I immediately question the ability and dedication of the person who wrote it. Why didn't they take the time to do it right? Most of the successful engineers I know write clear, well-organized memos and reports. Engineers who can't write well are definitely held back from career advancement (qtd. in Beer & McMurrey, 1997, pp. 4–5).

Evidence that the working world values good communication skills is easy to find. Just look at the job ads in newspapers and professional journals:

From an organization that manufactures medical instruments:

> Senior Design Assurance Engineer. Duties include performing electronic/mechanical product, component, and material qualifications. Requires 3 years medical device industry experience, spreadsheet/word-processing abilities, and excellent written/oral communication skills. BSEE or biology degree preferred.

From a corporation that supplies agricultural products:

> Compensation Administrator. The successful candidate will administer the company-wide design, development, implementation, and monitoring of the compensation program. Minimum qualifications include 4–5 years of compensation analysis and salary administration experience. Experience writing job descriptions, developing salary structures, conducting and participating in salary surveys. Broad-based background in the general human resource function with specific knowledge of federal and state regulations affecting compensation systems. Demonstrated ability to operate a personal computer and spreadsheet software. Microsoft Excel preferred.

From a corporation that manufactures custom-molded plastic components:

> Custom injection molder has need for a process engineer. Manufacturing knowledge and experience with statistical process control required. Experience with injection molding and ASQC certification a plus. Excellent writing and speaking skills required, as well as word processing and spreadsheet skills.

Notice that these job ads mention not only communication skills but also computer skills.

The steady growth in continuing-education courses in technical and business communication also reflects this need for effective communicators. Organizations

INTERVIEW: NURSING PROFESSOR ANNE PAYNE

on the importance of communication for nurses

The most important thing is the ability to synthesize information and to record it and communicate it to other individuals succinctly and accurately. I'm talking about any sort of electronic or written medical record. We have to be able to communicate our conclusions and do it rapidly and to the appropriate individuals. We have to be accurate, but we also have to be brief; if you write three pages' worth, then nobody's going to read it.

on the importance of electronic tools for nursing students

Right now St. Alphonsus Hospital is almost completely computerized; the nursing students will walk around with the notebook computer just like the nurses that they're working with. Our nursing students have to be comfortable with the technology.

on the kinds of writing that nurses do

Unit managers have to manage budgets, defend budgets, make a budget proposal. They have to do performance appraisals of other personnel. They have to be able to create a case for discussion in a team, maybe even have to communicate between agencies about a client's condition and make recommendations and referrals. It's very important for them to be able to analyze that information and get it across quickly, clearly, and with as much brevity as possible.

on the need to communicate with patients and families of patients

In the future, more and more of this will be in a written form; right now a lot of it is oral communication. But either way, nurses have to be able to gather appropriate information from a patient or from his or her family. If the patient is a child, the nurse must teach someone else how to care for the child. To do this, the nurse has to be able to communicate how to take care of someone at home.

on the role of communication in the nursing curriculum

I have wonderful faculty who have really taken it upon themselves to do a lot of writing across the curriculum. Right from the beginning, students are doing those micro themes and journals and a lot of different projects that make them express themselves correctly and completely in writing, and we also have students doing presentations.

on the importance of communication skills for a nurse entering the job market

One of the things that we do is have someone come over from the placement center; students work on writing a résumé and presenting themselves in writing and in interview situations. They have to learn to present themselves professionally. Written and oral communication skills can have a large impact on whether you're hired for a job.

Anne Payne, Ph.D., is interim Dean of the College of Health Science at Boise State University and a nurse practitioner.

are paying to send their professionals back to the classroom to improve their technical-communication skills.

The facts of corporate life today are simple. If you can communicate well, you are valuable; if you cannot, you are much less valuable.

Characteristics of Technical Communication

This section discusses seven characteristics of technical communication, as shown in Figure 1.1.

FIGURE 1.1

Characteristics of
Technical
Communication

**Addresses
Particular Readers**

Perhaps the most significant characteristic of technical communication is that it addresses particular readers. Although a journalist might have a profile of a typical reader in mind, you start with a much more specific profile when you create technical documents.

Sometimes you know the reader. For instance, if you are planning to write a proposal to your supervisor, you will think about that person's background, job responsibilities, history with the organization, attitudes toward your idea, and so forth. You will consider these factors as you decide what kind of document to write, how to structure it, what kinds of paragraphs to create, and what types of sentences and words to use. If you are writing to several different people with various backgrounds and needs, you might structure the document so that they can locate and understand the information they seek.

Sometimes you do not know the readers personally, such as when you write a brochure describing one of your company's products, but you create a profile. For example, your readers are police officers who are responsible for capital purchases. Even though you don't know their gender, age, or other personal characteristics, at least you know that if they are reading the brochure, they share a police background and the responsibility for approving capital purchases.

Even though technical communication is addressed to particular readers, sometimes it is read by people who were never intended to be the audience. Other workers at your organization—especially managers and executives—might read sensitive documents; in addition, members of the public or the press might see documents that were never addressed to them. For these reasons, you must be careful that all your writing reflects the highest standards of professionalism.

Helps Readers Solve Problems

Technical communication is meant not to express the writer's creativity or to entertain readers, but to help readers learn something or do something.

For instance, you look at your college's catalog to figure out the procedures for registration, to understand the requirements for a particular major, and to learn what is covered in a particular course. You read the owner's manual for your car to find out whether the manufacturer recommends that you rotate the tires front-to-back or diagonally. You read your company's employee-benefits manual to help you decide which benefits package you should select. You read technical communication because you need the information to help you analyze a situation and decide how to proceed.

People read technical communication to help them solve problems; people also write and distribute technical communication to help solve problems. If there is a parking problem on your campus, university officials educate the college community about the parking regulations and recommend alternative means of transportation. If one of your company's new products is not selling well, the company creates marketing materials and places ads in trade publications and on its Web site.

Furthers an Organization's Goals

Technical communication is created by people working within or for an organization to further its goals. Consider, for example, a department in a state government that oversees the state's programs in vocational education. Every activity undertaken by that department involves technical communication.

The department submits an annual report to the state legislature. This report covers each program offered by the department, indicating what need it was intended to address, who delivered it, who enrolled in it, where and when it was offered, how much it cost, and how much money it generated. The report also explains the successes and failures of each program and offers recommendations on how to make it more effective next time.

The department produces a vast quantity of technical information for the public as well. This information includes flyers, brochures, pamphlets, and even radio and television advertisements to publicize its offerings. In addition, the department produces materials used in the courses themselves, including written course texts and workbooks, and Internet-based instructional material.

All the activities that support these services are technical communication. Staff members conduct interviews, distribute surveys, and perform database research to determine what programs should be offered; they make oral presentations to potential students at high schools and factories; they retain, train, and evaluate instructors; they write and produce publicity materials and monitor their effectiveness; they coordinate the programs; and they write letters to their students.

Is Created Collaboratively

You will often work alone in writing e-mail, memos, and letters, but you will work as part of a team in writing most of the larger, more sophisticated documents. Collaboration can take many forms, from having a colleague read a two-page memo you have written to working with a team of a dozen technical professionals and technical communicators to produce a 200-page catalog.

Collaboration is common in technical communication because no one person has all the information, skills, and time to put together a large document. Writers, editors, designers, production specialists, lawyers, and subject-matter experts—the various technical professionals—work together to create a better document than any one of them could have made working alone.

Much of the time spent on projects in organizations is devoted to communicating with collaborators. Large projects often begin with brainstorming sessions, at which the participants contribute ideas about the content, organization, and style of the document and try to reach consensus. Often, the different team members go off to work on different portions of the project, meeting periodically for briefings. Team members write progress reports and, at the end of a project, a completion report.

Because of the collaborative nature of technical communication, interpersonal skills are essential. You have to be able to listen to people with other views, express yourself clearly and diplomatically, and compromise.

Collaboration is discussed in more detail in Chapter 3.

Uses Design to Increase Readability

Technical communicators use design features to make their documents more effective. Design features have three basic purposes:

- *To make the document attractive.* If the document looks attractive and professional, the reader is more likely to read it and to form a positive impression of it and of you. Therefore, you are more likely to accomplish your purpose in writing.

- *To help the reader navigate the document.* Because a technical document can be long and complicated, and because most readers want to read only parts of it, design features help readers see where they are and get where they want to be. For Web sites, effective navigation is essential.
- *To help the reader understand the document.* Design features help the reader see the organization of the document. For instance, if all first-level headings have one design and all second-level headings another, the reader will recognize this pattern and be better able to understand the discussion. You might also choose to display safety warnings in a color and size different from the rest of the text.

Design is discussed in Chapter 11.

Involves Words and Graphics Virtually every technical document is a combination of words and graphics. In technical communication, graphics help the writer perform five main functions:

- communicate and reinforce difficult concepts
- communicate how-to instructions and descriptions of objects and processes
- communicate large amounts of quantifiable data
- communicate with nonnative speakers of English
- make the document more interesting and appealing to readers

Graphics are discussed in Chapter 12.

Involves High-Tech Tools Technical communication is produced on high-tech tools. The personal computer—the essential tool, along with the printer—is used in every phase of the production of documents. Everyone uses word-processing software; graphics software and desktop-publishing software are common as well. The personal computer is also the basic tool for using e-mail and the Internet.

The evolution from paper to online information is occurring not only on the input end (using tools to make the document) but also on the output end (using tools to publish it). The explosive growth of the Web is one obvious example. And the development of CD-ROM, with its massive storage capabilities, has made it practical to deliver large quantities of information—including text, video, and animation—in an inexpensive medium.

As information technology develops, becoming more powerful, easier to use, and less expensive, technical communicators and technical professionals alike are continuously upgrading their skills. We are all lifelong learners now.

Figure 1.2 shows a page from a user's manual for a computer printer (*LaserJet*, 1993, pp. 5–7). This page displays the characteristics of technical communication discussed in this section.

Printing Transparencies

The page is part of the user's manual, which is addressed to purchasers of this printer.

The page helps readers perform a particular task: print transparencies.

The manual was written collaboratively by employees at Hewlett-Packard and its subcontractors.

Design features include:

- integrated words and graphics, with each graphic keyed to a numbered instruction
- clear diagrams and icons
- rules (lines) and boxes
- plenty of white space

In addition, the writing is simple and clear, addressed to the reader's needs.

According to a note at the beginning, the manual "was created using text formatting software on an HP Vectra personal computer."

FIGURE 1.2

An Example of Technical Communication

Selecting the Correct Media

CAUTION

To avoid damaging your printer, use only transparencies designed for use with laser printers. See Appendix B for detailed specifications.

Setting Up the Printer

First, select "manual feed" or "single sheet feed" through your software and send your job to print. Then set up the printer:

1 Set the paper path knob to the down position. This allows the transparencies to feed straight out the rear of the printer, reducing film curl.

2 Gently insert the transparency film into the manual feed slot until you feel some resistance on the upper left corner. The printer grabs the end of the transparency sheet.

3 Let the other end of the transparency hang down from the manual feed slot and rest on the table. The printer then prints the page.

4 After printing the transparencies, reset the paper path knob to the up position.

> **To Avoid Transparency Curl**
>
> To avoid transparency curl, grab the transparency as it comes out of the printer and set it on a flat surface to cool.

Measures of Excellence in Technical Communication

Figure 1.3 on page 12 shows eight basic measures of excellence of technical communication.

Honesty The most important measure of excellence in technical communication is that it is honest. You have to tell the truth and not mislead the reader.

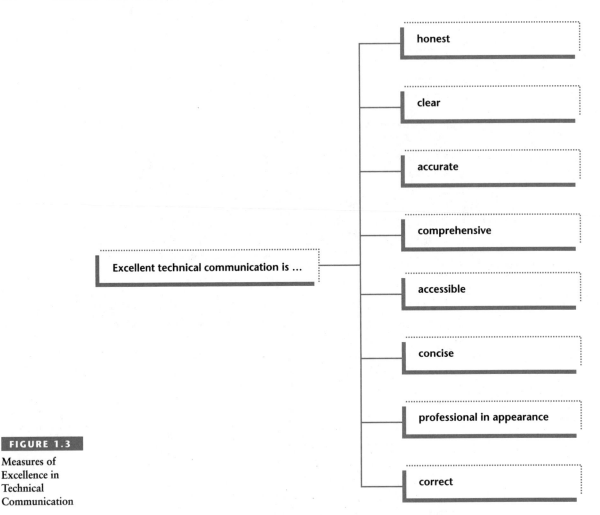

FIGURE 1.3

Measures of
Excellence in
Technical
Communication

There are three reasons to be honest as a communicator:

- *It is the right thing to do.* Technical communication is not about using words and pictures to mislead or lie to people; technical communication is about helping people understand how to make wise choices as they use the artifacts of a high-tech culture.
- *Readers can get hurt if you are dishonest.* If you lie or mislead in writing about medication dosages, people could become seriously ill or could die. If you fail to warn readers about the safety hazards involved in operating a piece of equipment, they could be seriously injured or killed.
- *You and your organization could get into serious legal trouble if you are dishonest.* If a plaintiff can convince a judge or jury that your document failed to provide honest, appropriate information, and that this failure

led to a substantial injury or loss, you and your organization are likely to have to pay millions of dollars.

Chapter 2 discusses the ethical and legal aspects of technical communication.

Clarity An important characteristic of effective technical communication is clarity. That is, your goal is to produce a document that conveys a single meaning that the reader can understand easily.

The following directive, written by the British navy (*Technical Communication*, 1990), is an example of what you *don't* want to do:

> It is necessary for technical reasons that these warheads should be stored upside down, that is, with the top at the bottom and the bottom at the top. In order that there may be no doubt as to which is the top and which is the bottom, for storage purposes, it will be seen that the bottom of each warhead has been labeled with the word TOP.

Technical communication must be clear for two reasons:

- *Unclear technical communication can be dangerous.* A carelessly drafted building code tempts contractors to save money by using inferior materials or techniques. The 1986 space-shuttle tragedy might have been prevented had the officials responsible for deciding whether to launch received clear reports of the safety risks involved that day.

- *Unclear technical communication is expensive.* The cost of a typical business letter today—counting the writer's time and the cost of equipment, stationery, and postage—is more than ten dollars. In addition, every time an unclear letter is sent out, the reader has to write or phone to ask for clarification. The letter often must be rewritten and sent again to confirm the correct information. On a larger scale, clarity in technical communication is essential because of the collaborative nature of most projects. One employee investigates one aspect of the problem, while other employees work on other aspects of it. Unclear communication can jeopardize the entire project.

Accuracy Inaccurate writing causes as many problems as unclear writing. In one sense, accuracy is a simple concept: you must record your facts carefully. If you mean to write *4,000,* don't write *40,000.* The slightest inaccuracy will at least confuse and annoy your readers. A major inaccuracy, naturally, can be dangerous and expensive.

In another sense, however, accuracy is a question of ethics. Technical communication must be as objective and free of bias as you can make it. If readers suspect that you are slanting information—by overstating the significance of a particular fact or by omitting an important point—they have every right to doubt the validity of the entire document. Technical communication must be effective by virtue of its clarity and organization, but it also must be reasonable, fair, and honest.

Comprehensiveness   A good technical document is comprehensive; that is, it provides all the information that readers will need. It describes the background so that readers who are unfamiliar with the subject can understand it. It contains sufficient detail so that readers can follow the discussion and carry out any tasks called for. It refers clearly to supporting materials or includes them as attachments.

Comprehensiveness is crucial because the people who will act on the document need a complete, self-contained discussion in order to apply the information safely, effectively, and efficiently. In addition, they need an official company record of the project, from its inception to its completion.

Here are two examples of the importance of being comprehensive:

- A scientific article reporting on an experiment comparing the reaction of a new strain of bacterium to two different compounds will not be considered for publication unless the writer has fully described the methods used in the experiment. Because other scientists should be able to replicate the researcher's methods, every detail must be provided, including the names of the companies from which the researcher obtained all the materials.

- A report recommending that a company open a plant in a new location will be analyzed in detail before management commits itself to such an expensive and important project. The team charged with studying the report will need all the details. If the recommendations are implemented, the company will need a single, complete record in case changes have to be made several months or years later.

Accessibility   *Accessibility* refers to the ease with which readers can locate the information they seek. Most technical documents are made up of small, relatively independent sections. Some readers are interested in only one or two sections; others might want to read more. Because few people will pick up the document and start reading from the first page all the way through, your job is to make the various parts of the document accessible. That is, readers should not be forced to flip through the pages to find the appropriate section.

Chapter 10 describes the major elements that make documents accessible, such as tables of contents and indexes. Chapter 11 discusses how to use design elements to improve accessibility.

Conciseness   To be useful, technical communication must be concise. The longer the document, the more difficult it is to use, for the obvious reason that it takes more of the reader's time. In a sense, conciseness works against clarity and comprehensiveness; for a technical explanation to be absolutely clear, it must describe every aspect of the subject in great detail. To balance the claims of clarity, conciseness, and comprehensiveness, you must make the document just long enough to be clear—given the audience, purpose, and subject—but not a word longer. You can shorten most writing by 10 to 20 percent simply by eliminating unnecessary phrases, choosing short words rather than long ones, and using economical grammatical forms.

The battle for concise writing, however, is often more a matter of psychology than of grammar. Some writers produce long documents to show their readers that they are trying hard. But if the document needs to be short to fulfill its purpose, your job is to figure out how to convey a lot of information in a little space. That is one of the challenges of writing technical communication.

Professional Appearance

Documents start to communicate before anyone reads the first word. If the document looks neat and professional, readers will form a positive impression of both the content and the authors.

Your documents should adhere to the format standards that apply in your organization or your professional field. In addition, your documents should be well designed and neatly printed. For example, a letter should follow one of the traditional letter formats and have generous margins. It should be balanced both vertically and horizontally on the page.

Correctness

Good technical communication correctly observes the conventions of grammar, punctuation, spelling, and usage.

Many of the rules of correctness are clearly important. If you mean to write "The three inspectors—Bill, Harry, and I—attended the session" but you use commas instead of dashes, your readers might think six people (not three) attended. If you write "While feeding on the algae, the researchers captured the fish," some of your readers might have a little trouble following you—at least for a few moments as they consider the image of researchers feeding on algae.

Most of the rules, however, make a difference primarily because readers will judge your thinking on how your writing looks and sounds. Carelessness and grammar errors make your readers doubt the accuracy of your information, or at least lose their concentration. You will still be communicating, but the message won't be the one you had intended. As a result, the document will not achieve its purpose, and readers could well judge you incompetent and unprofessional.

Technical communication is meant to fulfill a mission: to convey information to a particular audience or to persuade that audience to accept a particular point of view. To accomplish these goals, it must be honest, clear, accurate, comprehensive, accessible, concise, professional in appearance, and correct.

The rest of this book describes ways to help you say what you want to say.

CASE

Teaching Technical Communication across Campus

Form groups according to major. The faculty in your major department have decided to do a thorough review of its curriculum, and they have decided to involve students. The department chair has asked your group to determine the extent to which the courses in your major prepare you to communicate effectively. Here are some questions you might focus on:

- What do the professors in the department think is the best approach to helping students learn how to communicate? How do they incorporate technical communication into their courses?
- To what extent do the course descriptions in your department describe communication skills that will be part of a professional's set of skills?
- To what extent do the course syllabi for required courses describe the class activities and assignments that call for technical-communication skills?
- What approach is used at other similar colleges and universities?

Write a memo to the chair of your major department, describing and evaluating your department's attempt to teach technical communication. What aspects do you think are effective? How might the department do more to teach students the importance of technical communication to the professional's career? See Chapter 16 for a discussion of memos.

EXERCISES

The following exercises call for you to write memos. See Chapter 16 for a detailed discussion of memos.

1. Locate an owner's manual for a consumer product, such as a coffeemaker, bicycle, or hair dryer. In a memo to your instructor, describe and evaluate the manual. To what extent does it meet the measures of excellence discussed in this chapter? In what ways does it fall short of these measures? Write your response in a memo to your instructor. Submit a photocopy of the document (or a representative portion of it) with your memo.

2. Study the job ads in your field from a large newspaper, Web site, or professional journal. To what extent do they suggest that the employers value technical-communication skills? What percentage of the ads explicitly mention communication skills? What kinds of skills are mentioned most often? Are the ads themselves examples of effective technical communication? Write your response in a memo to your instructor.

3. Evaluate the example of technical communication on page 17, an excerpt from a description of medical plans available to state employees. In a memo to your

instructor, explain the strengths and weaknesses of the excerpt by referring to the measures of excellence described in this chapter.

Collaborative Exercise

4. Form small groups to study the excerpt from the Concord Coalition brochure, which appears on pages 18–19. Meet to discuss the characteristics of technical communication that you see in this excerpt. How effective is this excerpt? What changes would you make to improve it? Present your ideas in a memo to your instructor.

Web Exercise

5. Locate a document on the Web that you believe to be an example of technical communication. Describe the aspects of the document that show the characteristics of technical communication discussed in this chapter. Then evaluate the effectiveness of the document using the strategy described in exercise 1. Write your response in a memo to your instructor. Submit a printout of the document (or a representative portion of it) with your assignment.

The former Lincoln National/Peak Health HMO plan which was purchased by Idaho Preferred Healthcare in September, 1992, **WILL NOT** be offered this year. **Current members of the Peak Health plan will AUTOMATICALLY BE ENROLLED IN THE IDAHO PREFERRED HEALTHCARE PLAN unless they choose to enroll in another plan during Open Enrollment.**

If you would like to add yourself/dependents to medical coverage or change medical plans you must complete the appropriate enrollment application form and return it to your human resources/payroll office by May 14. **Late applications will not be accepted.**

Although there is no Open Enrollment for dental coverage at this time, if you are adding dependents to medical coverage, and **HAVE NOT PREVIOUSLY** declined dependent dental coverage, your dependents will automatically be added to dental coverage unless you complete a Dental Declination Form.

If you are moving from one medical plan to another, you and your dependents **currently covered by a dental plan** will be enrolled in the dental plan which corresponds to your new medical plan. For instance, if you are enrolled in Blue Shield Module 1/Dental Plan 1 and you enroll in Blue Shield Module 2, 3 or one of the HMO plans you will be covered under Dental Plan 2. If you are enrolled in Blue Shield Module 2, 3 or one of the HMO plans and you enroll in Blue Shield Module 1 you will be covered under Dental Plan 1.

The waiting periods for pre-existing and/or chronic conditions, and certain other conditions specifically listed under the Blue Shield, Group Health Northwest and Idaho Preferred Healthcare **MEDICAL** plans will be waived for all members **CHANGING** plans during Open Enrollment WHO ARE NOT CURRENTLY SUBJECT TO A WAITING PERIOD.

Any members WHO ARE SUBJECT TO THE WAITING PERIODS in their current **MEDICAL** plan WILL BE SUBJECT TO THE WAITING PERIODS in the plan they are enrolled in as of July 1. However, any months of coverage under a member's current medical plan will be credited toward the waiting periods in the plan he/she is enrolled in as of July 1.

All members changing from Dental Plan 1 to Dental Plan 2 **WILL BE SUBJECT** to the waiting periods and limitations in Dental Plan 2.

ALL NEW ENROLLEES, NEW HIRES, AND NEWLY ACQUIRED DEPENDENTS (EXCEPT NEWBORNS) WILL BE SUBJECT TO THE WAITING PERIODS IN EACH MEDICAL/DENTAL PLAN.

If you are adding dependents to medical coverage, or are changing plans, be sure to list all members of your family to be covered, not just the new enrollees. Members not listed on your new enrollment application form will not be covered, even if they were previously enrolled.

ALL CHANGES YOU MAKE WILL BE EFFECTIVE JULY 1, 1993. NECESSARY ENROLLMENT APPLICATION FORMS, HMO ENROLLMENT PACKETS AND DENTAL DECLINATION FORMS CAN BE OBTAINED FROM YOUR HUMAN RESOURCES/PAYROLL OFFICE.

Consult your Employee Group Insurance Handbook and HMO enrollment packets to compare the services of the HMOs and the Blue Shield modules.

FY94 MONTHLY EMPLOYEE PAID MEDICAL/DENTAL PLAN RATES

	Blue Shield Module 1/ Dental Plan 1	Blue Shield Module 2/ Dental Plan 2	Blue Shield Module 3/ Dental Plan 2
Single	$ 6.00	$ 6.00	$ 24.65
2-Party with Dependent Dental	34.90	34.90	72.00
2-Party **without** Dependent Dental	30.93	30.93	59.63
Family with Dependent Dental	60.80	60.80	113.55
Family **without** Dependent Dental	50.22	50.22	83.90
*Split with Dependent Dental	6.00	6.00	40.10
*Split **without** Dependent Dental	6.00	6.00	29.33

	Group Health Northwest/ Dental Plan 2	Idaho Preferred Healthcare/ Dental Plan 2
Single	$ 35.40	$ 44.98
2-Party with Dependent Dental	93.51	112.65
2-Party **without** Dependent Dental	81.14	100.28
Family with Dependent Dental	148.50	179.61
Family **without** Dependent Dental	118.85	149.96
*Split with Dependent Dental	64.29	85.83
*Split **without** Dependent Dental	53.52	75.06

*A Split membership applies when both spouses in a family, with dependent children covered, are employed by the state and are both enrolled in the **same** plan. One spouse enrolls as a single, the spouse covering the children enrolls as a split.

—————— The Problem ——————
A TOWERING MOUNTAIN OF DEBT

America is at war—an economic war. Our nation's wealth is being drained, drop by drop, by federal budget deficits. Each year, the government spends $200 billion more than it has. These deficits are never paid off. They are added to the huge mountain of debt that now stands at over $4.5 trillion.

This debt mortgages our future to pay for current consumption. We're squandering our fiscal integrity and threatening America's prosperity. Without quick action to eliminate budget deficits, today's adults will be the first generation in American history to leave the nation worse off than it was when they came of age.

The Facts

◆ The federal government has run a budget deficit every year since 1969.

◆ Between 1980 and 1992, the national debt tripled from **$1 trillion** to **$4 trillion.** If unchecked, the debt will increase by **$1 trillion** more by 1996.

◆ Annual interest on the federal debt equals the income taxes paid by all those living west of the Mississippi River.

Why Worry about the Deficit?

Budget deficits strangle economic growth and hold down our standard of living. If the U.S. economy had grown as fast in the last 20 years as it did in the 1950s and 1960s, the median family income today would be **$50,000** instead of **$35,000.** In part, that's because large deficits use up the savings that would otherwise be available for private investment. Without that investment, it is harder for the economy to grow and provide more and better jobs.

Deficits cause cuts in important programs. Interest on the debt is the third largest program in the budget. Interest payments squeeze out programs we care about. This year, we'll spend more on interest than on agriculture, crime, education, environment, medical research, science and technology, and transportation — **combined.**

Deficits rob future generations. Instead of leaving a stronger, more competitive economy for our children, we will leave them ever-rising interest payments and stagnant economic growth. This is not the American Dream.

———— The Solution ————
JOIN THE CONCORD COALITION

The Concord Coalition is the citizens' voice against federal budget deficits. Even though most of the nation's elected officials say they want to eliminate the huge deficits that sap our economy and darken our future, too often they listen to the wishes of special interests when deciding budget issues.

The Concord Coalition is a nonpartisan grassroots movement to eliminate budget deficits. We are educating and mobilizing citizens and members of Congress to do something before it is too late.

How Will We Achieve Our Goals?

Large and vocal membership. Coalition chapters are building a broad base of participation through membership drives, highly visible meetings and activities, and local media attention. Our goal is to build large chapters across the country that provide a countervailing voice against special interest groups.

Education. State and local chapters take economic issues back from the economists and explain them in terms that everyone can understand. Coalition chapters host teach-ins, discussion groups, and budget-balancing exercises. A scorecard showing how senators and representatives voted on deficit issues is published annually.

Empowerment. Local and state Concord chapters hold well-publicized activities such as town meetings, teach-ins, deficit reduction exercises, candidates' forums, Bring Home Your Congressman meetings, and other educational events.

Local volunteers help fundraise, organize events, put together newsletters, arrange phone trees, write letters to members of Congress, staff information booths, and network with local media.

No matter how you wish to get involved, the most important thing is to **join** The Concord Coalition and add your voice to the rising chorus of Americans demanding an end to federal deficits.

To find out the name of your state coodinator, call 202-737-1077.

What Will I Receive as a Concord Coalition Member?

- ◆ Automatic membership in your local chapter.
- ◆ Quarterly newsletter on budget developments and the activities of Coalition chapters nationwide.
- ◆ Invitation to annual national policy forum.
- ◆ Invitations to local candidates' forums, educational meetings, and other activities.
- ◆ Opportunities to order issue papers, booklets, and other educational materials.
- ◆ An effective way to make your voice heard.

REFERENCES

Anderson, P. (1985). What survey research tells us about writing at work. In L. Odell & D. Goswami (Eds.), *Writing in nonacademic settings.* New York: Guilford.

Beer, D., & McMurrey, D. (1997). *A guide to writing as an engineer.* New York: Wiley.

Concord Coalition. (1994). The Concord Coalition [Brochure]. Washington, DC: The Concord Coalition.

LaserJet 4L user's manual. (1993). Boise, ID: Hewlett-Packard Company.

MIT Industrial Liaison Program. (1984). *Communication skills: A top priority for engineers and scientists* (Report No. 32929).

Spretnak, C. M. (1982). A survey of the frequency and importance of technical communication in an engineering career. *The Technical Writing Teacher, 9*(3), 133–136.

Technical Communication. (1990). *37*(4), 385.

CHAPTER
2

Understanding Ethical and Legal Considerations

A Brief Introduction to Ethics 22

Your Ethical Obligations 25
  Obligations to Your Employer 25
  Obligations to the Public 27
  Obligations to the Environment 29

Your Legal Obligations 29
  Copyright Law 29
  Trademark Law 32
  Contract Law 34
  Liability Law 35

Codes of Conduct 37

Whistleblowing 40

Guidelines for Ethical Communication 42
  Tell the Truth 42
  Do Not Mislead 42

Cases: Irritating Insulation 43
          Mysterious Internet Code 44

Exercises 45
  Collaborative Exercises 46
  Web Exercise 46

A power company in a big American city faced a problem recently: its obsolete generating equipment produced illegal levels of air pollution. Unfortunately, the company didn't have the millions of dollars needed to replace the old equipment or even to add scrubbers that would lower the pollution to legal levels. The Environmental Protection Agency, fully aware of the pollution levels, routinely fined the utility—$250, four times a year. Why were the penalties so low? Because the courts recognized that older power companies would not be able to meet modern pollution standards and continue to supply power to the city, as well as employ thousands of people. The utility industry and the lawmakers in effect had worked out a plan to replace the older technology with newer equipment over a period of years, thus enabling the companies to stay in business.

What does this have to do with you? As a technical communicator or a technical professional, you will likely be involved with problems like this—problems that have an ethical dimension. Everyone knows why pollution is undesirable, but getting rid of it, or lowering it to acceptable levels, is often difficult and expensive. In business, government, and industry, people work every day to solve ethical problems in ways that address the needs of different constituencies: in this case, the home owners and businesspeople who need the power, the company employees who need paychecks, and everyone who breathes the city's air.

As a professional, you will have to make decisions about how to treat clients, customers, and other organizations; how your organization will deal with the government regulatory bodies that oversee your industry; and how the actions of your organization will affect the environment. All these decisions will be communicated in your writing and speaking.

This chapter provides an introduction to professional ethics. Throughout the rest of this book, you will also read other discussions of ethics. The chapter on proposals, for example, offers a discussion of some of the ethical issues involved in writing proposals.

This chapter briefly introduces ethics and explains the standards commonly used to decide ethical questions. Then it discusses legal considerations for communicators: copyrights, trademarks, contract law, and liability. Next, the chapter discusses codes of conduct and whistleblowing. Finally, the chapter provides a set of guidelines for communicating technical information ethically.

A Brief Introduction to Ethics

According to Claudia Huff and Saul Carliner (personal communication, June 7, 1993), 35 percent of the software used in the United States is pirated, that is, copied illegally. In Spain, the figure is 86 percent; in Thailand, 99 percent. In 1995, U.S. manufacturers of business software lost almost $3 billion (Aversa, 1996, p. 8B).

A recent survey of two thousand faculty members and two thousand doctoral students in the United States found that 22 percent of the faculty know of department members who have overlooked sloppy use of data, over 30 percent report cases of inappropriate credit given for authorship of articles, and 40 percent know of colleagues who have used college resources for personal business (Swazey, Louis, & Anderson, 1994). Other widespread types of unethical conduct reported in the survey include violations of safety and animal-research guidelines, sexual harassment of students, discrimination, and cheating by students.

Computer-monitoring software tells an employer how long a networked computer was used, what files the user accessed, whether any files were modified or deleted, how quickly the user worked, and how long the user was away from the desk. This software is used to monitor some 26 million American workers; the pay of 10 million workers is based on computer monitoring (Forester & Morrison, 1994, p. 157). Does the use of this software invade the worker's privacy and degrade the work environment intolerably, or does it provide a fair, objective way to measure productivity and reward good performance? Does the company fulfill its ethical obligation by informing workers that it is using, or might use, the software? Questions such as these are being debated across the country.

Virtually every day you see an article in a newspaper or journal about ethics. But what exactly is ethics? Some people equate ethical conduct and legal conduct; if an act is legal, it is ethical. Most people, however, believe that ethical standards are more demanding than legal standards. It is perfectly legal, for example, to try to sell an expensive life insurance policy to an impoverished elderly person who has no dependents and therefore no need for such a policy. Yet many people would consider the attempt unethical.

When businesspeople were asked to give their definition of ethics, half of them answered, "What my feelings tell me is right." One-quarter of the respondents replied, "What is in accord with my religious beliefs." And most of the rest said, "What conforms to 'the golden rule'" (Baumhart, 1968, pp. 11–12). Although philosophers cannot agree on what constitutes ethical conduct, most would agree to the following definition: *Ethics is the study of the principles of conduct that apply to an individual or a group.*

What are the basic principles used to think through an ethical problem? Ethicist Manuel G. Velasquez (1992) outlines three kinds of moral standards useful in confronting ethical dilemmas:

- *Rights.* The standard of rights concerns the basic needs and welfare of particular individuals. If a company has agreed to provide continuing employment to its workers, the standard of rights requires that the company either keep the plant open or provide adequate job training and placement services.

- *Justice.* The standard of justice concerns how the positive and negative effects of an action or policy can be distributed fairly among a group. For example, it would seem reasonable that the expense of maintaining a highway be borne primarily by people who use that highway. However, since everyone benefits from the highway, it is just that general funds also be used for maintenance.

- *Utility.* The standard of utility concerns the effects, both positive and negative, of a particular action or policy on the general public. If, for example, a company is considering closing a plant, the standard of utility requires that the company's leaders consider not only their savings from shutting down the plant but also the financial hardships on the unemployed workers and the economic impact on the rest of the community.

Although it is best to think about the implications of any serious act in terms of all three standards, often they conflict. For instance, from the point of view of utility, no-fault car insurance laws—under which people are prohibited from suing for damages below a certain dollar amount—are good because they reduce the number of nuisance suits that clog up the court system and increase everyone's insurance costs. However, from the point of view of justice, these laws are unfair because they force innocent drivers to pay for the repairs.

In conflicts among the three standards, the standard of rights is generally considered the most important, and the standard of justice the second most important. Simply ranking the three standards cannot solve all ethical problems, however. If an action or policy were to greatly affect the general public, the standard of utility—the "least important" standard—might outweigh the standard of rights. For instance, if the power company has to cross your property to repair a transformer on a utility pole, the standard of utility (the need to get power to all the people affected by the problem) takes precedence over the standard of rights (your right to private property). Of course, the power company is obligated to respect your rights, insofar as possible, by explaining what it wants to do, trying to accommodate your schedule, and repairing any resulting damage to your property.

Ethical problems are difficult to resolve precisely because no rules exist to determine when one standard outweighs another. In the example of the transformer, how many customers have to be deprived of their power before the power company is morally entitled to violate the property owner's right of privacy? Is it a thousand? Five hundred? Ten? Only one?

Most people do not debate the conflict among rights, justice, and utility when they confront a serious ethical dilemma; instead, they simply do what they think is right. Perhaps this is good news. However, the depth of ethical thinking varies dramatically from one person to another, and the consequences of superficial ethical thinking can be profound. For these reasons, ethicists have described a general set of principles that can help people organize their thinking about the role of ethics within an organizational context. These principles form a web of rights and obligations that connect an employee, an organization, and the world in which the organization is situated.

For example, in exchange for their labor, employees enjoy three basic rights:

- fair wages
- safe and healthy working conditions
- due process in the handling of such matters as promotions, salary increases, and firing

Although there is still serious debate about employee rights, such as the freedom from surreptitious surveillance and unreasonable searches in the case of drug investigations, the question almost always concerns the extent of employees' rights, not the existence of the basic rights themselves. For instance, ethicists disagree about whether hiring undercover investigators to discover drug users at a job site is an unwarranted intrusion on the employees' rights, but there is no debate about the right of exemption from unwarranted intrusion.

Your Ethical Obligations

In addition to enjoying rights, an employee assumes obligations, which can form a clear and reasonable framework for discussing the ethics of technical communication. The following discussion outlines three sets of obligations:

- to your employer
- to the public
- to the environment

Obligations to Your Employer

You will be hired to further your employer's legitimate aims and to refrain from any activities that run counter to those aims. Specifically, you will have the four obligations shown in Figure 2.1 on page 26.

- *Competence and diligence. Competence* refers to your skills; you should have the training and experience to do the job adequately. *Diligence* simply means hard work.
- *Honesty and candor.* You should not steal from your employer. Stealing involves such obvious practices as embezzlement as well as the common occurrences of "borrowing" office supplies and padding expense accounts. *Candor* means truthfulness; report problems to your employer that might threaten the quality or safety of the organization's product or service. If, for instance, you have learned that a chemical your company wants to make could pollute the drinking water, you are obligated to inform your supervisor, even though the news may be unwelcome.

FIGURE 2.1

Obligations to
Your Employer

Another kind of problem involving honesty and candor concerns what Sigma Xi, the Scientific Research Society, calls trimming, cooking, and forging (*Honor,* 1986, p. 11). *Trimming* is the smoothing of irregularities to make the data look extremely accurate and precise. *Cooking* is retaining only those results that fit the theory and discarding the others. And *forging* is inventing some or all of the research data, and even reporting experiments that were never performed. In carrying out research, an employee might feel some pressure to report only positive, statistically significant findings but must resist the pressure.

- *Confidentiality.* You should not divulge company business outside of the company. If a competitor knew that your company is planning to introduce a new product, it might introduce its own version of that product, robbing you of your competitive advantage. Many other kinds of privileged information—such as quality-control problems, personnel matters, relocation or expansion plans, and financial restructuring—also could be used against the company. A well-known problem of confidentiality involves *insider information;* an employee who knows about a development that will increase the value of the company's stock buys the stock before the information is made public, thus reaping an unfair (and illegal) profit.

- *Loyalty.* You should act in the employer's interest, not in your own. Therefore, it is unethical to invest heavily in a competitor's stock, because that could jeopardize your objectivity and judgment. For the same reason, it is unethical to accept bribes or kickbacks. It is unethical to devote considerable time to moonlighting (performing an outside job, such as private consulting) because the outside job could lead to a conflict of interest and because the heavy workload could make you less productive in your primary position.

Obligations to the Public

Every organization that offers products or provides services is obligated to treat its customers fairly. As a representative of an organization, and especially as an employee communicating technical information, you will frequently confront ethical questions.

In general, an organization is acting ethically if its product or service is both *safe* and *effective*. It must not injure or harm the consumer, and it must fulfill its promised function. However, these commonsense principles provide little guidance in dealing with the complicated ethical problems that arise routinely.

Every year, some 34 million Americans are hurt in accidents related to consumer items; 28,000 of these people die, and 100,000 are permanently disabled. The annual cost of these injuries is $12 billion (Velasquez, 1992, p. 271). Even more commonplace, of course, are product and service failures: the items are difficult to assemble or operate, they don't do what they are supposed to do, they break down, or they require more expensive maintenance than indicated in the product information.

Who is responsible for the injuries and product failures: the company that produces the product or service, or the consumer who purchases it? In individual cases, blame is sometimes easy enough to fix. A person who operates a chain saw without reading the safety warnings and without seeking any instruction in how to use it is surely to blame for any injuries caused by the normal operation of the saw. But a manufacturer who knows that the chain on the saw is liable to break when used under certain circumstances, and who failed to remedy this problem or warn the consumer, is responsible for any resulting accidents.

However, such cases do not outline a rational theory that can help people understand how to act ethically in fulfilling their obligations to the public. There is no clear consensus on this issue. Today three main theories describe a company's obligations to the public, as shown in Figure 2.2 on page 28.

- *The contract theory.* The contract theory holds that a person who buys a product or service is entering into a contract with the manufacturer. The manufacturer and, by implication, the employee representing the manufacturer have four main obligations:

  – to ensure that the product or service complies with the contract in several respects: it does what its advertisements say it does, it operates a certain period of time before needing service or maintenance, and it is at least as safe as the product information states and the advertising suggests

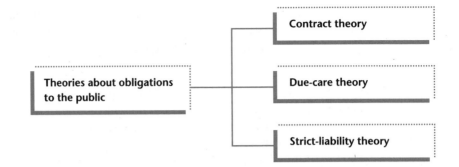

FIGURE 2.2

Theories about
Obligations to
the Public

– to disclose all pertinent information about the product or service, so
the potential consumer can make an informed decision about whether
to purchase it
– to avoid misrepresenting the product or service
– to avoid coercion, like that used by an unethical funeral director who
takes advantage of a consumer's grief to sell an expensive funeral

Critics of the contract theory argue that the typical consumer cannot
understand the product as well as the manufacturer does, and that consumer ignorance invalidates the "contract."

- *The due-care theory.* The due-care theory places somewhat more responsibility on the manufacturer than the contract theory does. This theory
holds that the manufacturer knows more about the product or service
than the consumer does and therefore has a greater responsibility to
make sure it complies with all manufacturer's claims and is safe. Therefore, in designing, manufacturing, testing, and communicating about a
product, the manufacturer has to make sure the product will be safe and
effective when used according to the instructions. However, the manufacturer is not liable when something goes wrong that it could not foresee
or prevent.

Critics of the due-care theory argue that because it is almost impossible to determine whether the manufacturer has in fact exercised due care,
the theory offers little of practical value.

- *The strict-liability theory.* The strict-liability theory, which goes one step
further than the due-care theory, holds that the manufacturer is at fault
when any injury or harm occurs from any product defect, even if it has
exercised due care and could not possibly have predicted or prevented
the failure or injury. This theory assumes that the only way to assess
blame is to hold the manufacturer guilty and thereby force it to assume
all liability costs. In so doing, society subsidizes the product when the
manufacturer builds these costs into the price of the product.

Critics of the strict-liability theory hold that, although it might be practical, it is unjust, because no person or organization should be held liable for something that is not its fault.

Obligations to the Environment

One of the most important lessons we have learned in the last two decades is that we are polluting and depleting our limited natural resources at an unacceptably high rate. Our overreliance on fossil fuels not only deprives future generations of their use but also causes possibly irreversible pollution problems, such as global warming. Everyone—government, business, and individuals—must work to preserve the ecosystem to ensure the survival not only of our own species but also of the other species with which we share the planet.

But what does this have to do with you? In your daily work, you probably do not cause pollution or deplete the environment in any extraordinary way. Yet because of the nature of your work, you will often know how your organization's actions affect the environment. For example, if you work for a manufacturing company, you might be aware of the environmental effects of making or using your company's products. Or you might contribute to the creation of an environmental impact statement.

Communicators should treat every actual or potential occurrence of environmental damage seriously. We should alert our supervisors to the situation and work with them to try to reduce the damage. The difficulty, of course, is that protecting the environment is expensive. Clean fuels cost more than dirty ones. Disposing of hazardous waste properly costs more (in the short run) than merely dumping it. Organizations that want to cut costs may be tempted to cut corners on environmental protection.

Your Legal Obligations

This chapter mentioned earlier that most people distinguish between ethical and legal obligations, and that they feel that ethical obligations are more comprehensive and more important. Still, the two are related very clearly in our culture. We attempt to use the law to enforce many of our ethical guidelines. Through the legal system, we try to punish people and organizations that violate our most fundamental ethical beliefs.

As a person involved in technical communication, you should know the basics of the four different bodies of law shown in Figure 2.3 on page 30.

Copyright Law

The Copyright Act of 1976 provides legal protection to the author of any work—such as printed material, software, photographs—whether it is published or unpublished, and whether the author is an individual or a corporation. The author

FIGURE 2.3

Bodies of Law Related
to Technical
Communication

is entitled to profit from the sale and distribution of the work in exchange for the obligation to make the work accessible.

Fair Use and Competitive Pressures

The concept by which work is made accessible is called *fair use*. According to Kozak (1993, p. 2), fair use gives you the right to use material, without getting permission, for "purposes such as criticism, comment, news reporting, teaching . . . , scholarship, or research." Unfortunately, fair use is not a specific legal term but a convention that is interpreted differently in different situations.

Kozak (1993) points out that courts consider four factors in determining whether a use of material is "fair use":

- *The purpose and character of the use, focusing on whether the use is for profit.* Profit-making organizations are scrutinized more carefully.
- *The nature and purpose of the copyrighted work.* When information is essential to the public good, such as medical information, the concept of fair use is applied more liberally.
- *The amount and substantiality of the portion of the work used.* A 200-word passage would be a small portion of a book but a large portion of a 500-word brochure. Although you will frequently see guidelines stating that 400 words are the maximum you may use, the courts have not always ruled in favor of that number.
- *The effect of the use on the potential market for the copyrighted work.* Your use of the work cannot hurt the author's potential to profit from the original work. This stipulation is deliberately vague.

Fair use does not apply to graphics: you must obtain written permission to use any graphics.

Although writers usually adhere to fair-use guidelines in reports and journal articles, the situation is different in the case of manuals and product information. For these applications, unfortunately, some companies have been known to take whole sections of another company's product information or manual, make minor changes, and publish it themselves. This kind of theft is made simpler when the victim's document is on the Web, for in almost all cases the source code—the digital text—is easy to download. These practices are unethical, of course, even if they have been done so skillfully that they are technically legal.

The difficulty in thinking about the ethics of this problem is that the difference between stealing and learning from your competitors can be subtle. Words are protected by copyright, but ideas are not. For instance, hundreds of companies produce personal computers, all of which are sold with user's guides. Rare is the manufacturer who doesn't study the competitors' user's guides to see how a feature or topic is described. Inevitably, a good idea spreads from one document to another. If one manual contains a particularly useful kind of troubleshooting guide, soon many others contain similar ones. Even though this process of imitation tends to produce a dull uniformity, it can improve the overall quality of the documents.

Many organizations treat internal and external documents differently.

- *Internal documents.* If your organization asks you to update an internal procedures manual, it probably expects you to use any material from the existing manual, even if you cannot determine the original author. This is perfectly legal under the concept known as *work for hire.* Anything written on the job by an employee being paid by your organization is the property of the organization. Any revisions you make will also be the property of the organization.

- *External documents.* Many organizations credit the authors of the external documents they publish, such as external manuals or articles. Many user's guides, as well as many software programs, acknowledge the authors and illustrators.

  However, the question of what constitutes authorship can be complicated because most large technical documents are produced collaboratively, with several persons contributing text, another doing the graphics, a third reviewing for technical accuracy, and a fourth reviewing for legal concerns. Problems are compounded when a document is revised and parts of original text or graphics are combined with new material.

Guidelines for Dealing with Copyright Questions

To deal with copyright questions, follow these five guidelines:

- *Abide by the fair-use concept.* The fair-use concept protects everyone; try not to use excessive amounts of another source's work.

- *Seek permission.* Write to the source, seeking permission to use the work. Describe accurately what portion of the work you wish to use and what kind of publication or other work you wish to use it in. The source is likely to charge you for permission.

- *Cite your sources accurately.* Not only is citing your sources an ethical matter, it strengthens your writing by showing the reader the range of your research. In addition, it protects you in case some of the facts or interpretations in your work turn out to be inaccurate: you have not claimed that you were solely responsible for them.

- *Discuss authorship questions openly.* The best way to determine the authorship of a document is to discuss the issue openly with everyone who contributed to it. Some persons might deserve to be listed as authors; others, only credited in an acknowledgment section.

- *Seek legal counsel if you have questions.* Copyright questions can be very complicated. There are questions of authorship, especially when you are dealing with electronic documents found on the Internet. Exactly who created a document, or whether the document has been altered without the author's knowledge or permission, is sometimes difficult to determine. There are also questions related to permissions, particularly in the case of a multimedia product such as a laser disc, which can require dozens of permissions for a minute of running time. Work with your organization's legal counsel. Not only can legal counsel answer many of your questions, but the mere fact that you sought counsel is evidence of good faith, an important factor should the copyright question ever have to be resolved in court.

For a more detailed discussion of plagiarism, see the discussion of references in Chapter 10.

Trademark Law   Companies use *trademarks* and *registered trademarks* to ensure that the public recognizes the name or logo of their products; therefore, you need to know the definitions of these two legal terms. The following discussion is based on excellent articles by Branscomb (1993) and Jim Grey (personal communication, January 31, 1994).

- A *trademark* is a word, phrase, name, or symbol that is identified with a company. The company uses the ™ symbol after the product name to signify that it is claiming the design or device as a trademark. For instance, Adobe Systems Incorporated claims PageMaker™ as a trademark. However, claiming a trademark does not ensure significant legal rights.

- A *registered trademark* refers to a word, phrase, name, or symbol that has been registered with the U.S. Patent and Trademark Office. The company uses the ® symbol after the product name to indicate that the name is a registered trademark. Registering a trademark, a process that can take years, ensures much more legal protection in the United States, as well as protection in other nations. The application for registration gives other companies an opportunity to challenge trademark ownership. For instance, Mead Data Central, owner of the database service Lexis®, sued Toyota after it introduced the Lexus® automobile, although it is hard to imagine someone confusing a database service with a Toyota. (Both Cadillac Pet Foods, Inc., and General Motors have registered the Cadillac trademark, apparently confident that consumers can distinguish a can of dog food from a luxury automobile.)

As a communicator, you are responsible for using the trademark and registered trademark symbols accurately when you refer to the items. Unfortunately, doing so is not always easy, because, as Branscomb (1993, p. 98) points out, companies often inappropriately claim trademark status for their products. He recommends that you or your legal staff consult sources such as the following for the most accurate and up-to-date information:

Official Gazette of the United States Patent and Trademark Office: a weekly periodical published by the U.S. Patent and Trademark Office.

Trademark Register of the United States: an annual directory published by the U.S. Patent and Trademark Office. This directory is also available online.

To protect your own company's trademarks, Jim Grey (personal communication, January 31, 1994) recommends the following four guidelines:

- *Distinguish your trademarks from other material.* Use boldface, italics, different size type, or a different color to distinguish the trademarked term.

- *Use the trademark symbol.* At least once in each document—preferably, the first time—use the appropriate symbol after the name or logo, followed by an asterisk. At the bottom of the page, include a statement such as the following: "*COKE is a registered trademark of the Coca-Cola Company."

- *Use the trademarked item as an adjective, not a noun or verb.* Trademarks can get confused with the generic term they refer to; the people at Xerox spend a lot of money running ads explaining that you cannot "xerox" anything, even on a Xerox® photocopier; you can only photocopy something. Therefore, use the trademarked item along with the generic term: LaserJet® printer.

- *Do not use the plural form or the possessive form of the term.* Doing so reduces the uniqueness of the item and encourages the public to think of the term as generic.

DOES NOT PROTECT TRADEMARK:	take some Kodacolors®
PROTECTS TRADEMARK:	take some photographs using Kodacolor® film
DOES NOT PROTECT TRADEMARK:	Kodacolor's® fine quality
PROTECTS TRADEMARK:	the fine quality of Kodacolor® film

Contract Law   Contract law deals with agreements between two parties. In most cases, disputes concern whether a product lives up to the manufacturer's claims. These claims are communicated as express warranties or implied warranties.

An *express warranty* is an explicit statement in a piece of product documentation that the product has a particular feature or can perform a particular function. An *implied warranty* is a warranty that is not written explicitly but rather inferred by the purchaser.

What exactly does *implied* mean? If a user's manual for a printer shows sophisticated graphics contained on sample pages, the company is implying that the printer can print those graphics—and the purchaser has a right to infer that the printer can—even if the manual never explicitly makes that claim. Implied warranties also occur in more casual communication, such as memos or letters to customers or potential customers. Therefore, it is important to be careful about what you write when you describe your company's product.

To protect themselves, most companies include a disclaimer with their product information. A *disclaimer* is a statement that the company hopes will limit its liability for the product. Whether the disclaimer in fact *does* limit the company's liability is determined in court on a case-by-case basis. If the court holds that the disclaimer is fair and reasonable, the company wins; if not, it loses. Manuals usually include disclaimers such as the following:

> Every effort has been made to supply complete and accurate information. However, Ottorino Publishers assumes no responsibility for the information contained herein or its use.

Software disclaimers often contain statements such as the following:

> Ottorino Software warranties this product to be free of defects at the time of purchase and agrees to replace the product at no cost if it is found to be defective under normal conditions of use, for a period of ninety (90) days from purchase. However, Ottorino Software assumes no responsibility for incidental or consequential damages.

Disclaimers such as this are often parodied, as in the following example (quoted in Forester & Morrison, 1990, p. 119):

> We don't claim Interactive EasyFlow is good for anything—if you think it is, great, but it's up to you to decide. If Interactive EasyFlow doesn't work: tough. If you lose a million because Interactive EasyFlow messes up, it's you that's out the million, not us. If you don't like this disclaimer: tough. We reserve the right to do the absolute minimum provided by law, up to and including nothing.
>
> This is basically the same disclaimer that comes with all software packages, but ours is in plain English and theirs is in legalese.
>
> We didn't really want to include a disclaimer at all, but our lawyers insisted.

Or:

> Omniverse Software warrants the diskettes to be of black color and of square shape under normal use for a period of ninety (90) days from the date of purchase.

These parodies are funny, but they make a serious point: many companies try as hard as they can to avoid taking responsibility for the quality of their products.

Liability Law   Liability law is a serious subject. A product-liability action is "a lawsuit for personal injury, death, property damage, or financial loss caused by a defective product" (Helyar, 1992, p. 126). In the years 1974–1992, the number of product-liability cases increased 800 percent. Liability is a concern for communicators because courts routinely decide that manufacturers have a legal obligation to provide adequate operating instructions and to warn consumers about the risks of using their products. In other words, product documentation in the form of instructions and manuals is treated just like any other component of the product itself.

Helyar (1992) summarizes the communicator's obligations, as reflected in recent court rulings. Helyar offers ten guidelines for abiding by liability laws:

- *Understand the product and its likely users.* Before writing the product information, perform research to understand the product thoroughly and to learn everything possible about the users and how they will use the product.

- *Describe the product's functions and limitations.* Describe the product's function to enable potential users to determine whether it is the appropriate product to buy. You must also describe the product's shortcomings. In one case, the court found the manufacturer liable for not stating that its electric smoke alarm does not work during a power outage.

- *Instruct users on all aspects of ownership.* The instructions should cover the following seven topics:

  – assembly
  – installation
  – use and storage
  – testing
  – maintenance
  – first aid and emergencies
  – disposal

- *Use appropriate words and graphics in the instructions.* The text must use common terms, simple sentences, and brief paragraphs; it must be structured logically; and it must contain explicit directions. The graphics must be unified and easy to understand; where appropriate, they should show people performing tasks.

  The words and graphics must be appropriate for the intended users in terms of their education, mechanical ability, manual dexterity, and intelligence. For products to be used by children or nonnative speakers of English, the instructions should include graphics.

- *Warn about the risks of using or misusing the product.* The instructions must warn not only about the obvious dangers of using the product (such as electrical shock or chemical poisoning) but also about less obvious dangers. For example, an automobile manufacturer was found guilty for not having warned that parking a car on grass, leaves, or other combustible material can cause a fire.

  The instructions should also describe the cause, extent, or seriousness of the danger. One judge noted in a ruling that a sign warning "Keep off the grass" is inadequate if there are deadly snakes in the grass.

  For particularly dangerous products, the instructions should include the following information:

  – the nature of the danger
  – the uses or misuses that can lead to the danger
  – the extent of the harm caused by the uses or misuses
  – a clear statement warning of that harm
  – clearly written steps on how to use the product safely

  In warning about danger, the instructions should use mandatory language, such as *must* and *shall,* rather than *might, could,* or *should.* The instructions should also use the terms *warning* and *caution* appropriately. See Chapter 20 for a discussion of *danger, warning,* and *caution.*

- *Include warnings along with assertions of safety.* When users read in the product information that a product is safe, they tend to pay less attention to warnings. Therefore, the product information must include detailed warnings to balance the claims of safety.

- *Make the directions and warnings conspicuous.* As discussed in Chapter 20, warnings and safety information must be

  – large
  – easily visible
  – in the appropriate location
  – durable

- *Make sure the instructions comply with applicable company standards and any applicable local, state, or federal statutes.*

- *Test the product and instructions to make sure they are accurate and easy to understand.* For more information on usability testing, see Chapter 20.

- *Make sure the users get the information.* Obviously, the instructions must accompany the product. But if the manufacturer discovers a problem later, it is obligated to get in touch with the users, using direct mail if possible or newspaper advertising if not. Automobile recall notices are examples of how manufacturers contact their users.

Codes of Conduct

Virtually all professional organizations, as well as three-quarters of the country's 1,500 largest corporations, have codes of conduct: written statements of the ethical standards employees are to follow (Brockmann & Rook, 1989, p. 91).

Codes of conduct vary greatly from organization to organization. Many are brief, such as the Society for Technical Communication's *Ethical Guidelines for Technical Communicators,* reprinted in Figure 2.4. An effective code should have three major characteristics:

- It should protect the public's interest rather than the interests of the members of the organization or profession. For instance, it should condemn unsafe building practices but not condemn advertising, because advertising increases competition and therefore lowers prices.

STC ETHICAL GUIDELINES FOR TECHNICAL COMMUNICATORS

As technical communicators, we observe the following ethical guidelines in our professional activities. Their purpose is to help us maintain ethical practices.

Legality
We observe the laws and regulations governing our professional activities in the workplace. We meet the terms and obligations of contracts that we undertake. We ensure that all terms of our contractual agreements are consistent with the STC Ethical Guidelines.

Honesty
We seek to promote the public good in our activites. To the best of our ability, we provide truthful and accurate communications. We dedicate ourselves to conciseness, clarity, coherence, and creativity, striving to address the needs of those who use our products. We alert our clients and employers when we believe material is ambiguous. Before using another person's work, we obtain permission. In cases where individuals are credited, we attribute authorship only to those who have made an original, substantive contribution. We do not perform work outside our job scope during hours compensated by clients or employers, except with their permission; nor do we use their facilities, equipment, or supplies without their approval. When we advertise our services, we do so truthfully.

FIGURE 2.4

Ethical Guidelines
for Technical
Communicators

(Continued)

(Figure 2.4 continued)

Confidentiality

Respecting the confidentiality of our clients, employers, and professional organizations, we disclose business-sensitive information only with their consent or when legally required. We acquire releases from clients and employers before including their business-sensitive information in our portfolios or before using such material for a different client or employer or for demo purposes.

Quality

With the goal of producing high-quality work, we negotiate realistic, candid agreement on the schedule, budget, and deliverables with clients and employers in the initial project planning stage. When working on the project, we fulfill our negotiated roles in a timely, responsible manner and meet the stated expectations.

Fairness

We respect cultural variety and other aspects of diversity in our clients, employers, development teams, and audiences. We serve the business interests of our clients and employers, as long as such loyalty does not require us to violate the public good. We avoid conflicts of interest in the fulfillment of our professional responsibilities and activities. If we are aware of a conflict of interest, we disclose it to those concerned and obtain their approval before proceeding.

Professionalism

We seek candid evaluations of our professional performance from clients and employers. We also provide candid evaluations of communication products and services. We advance the technical communication profession through our integrity, standards, and performance.

Used with permission of the Society for Technical Communication, Arlington, Virginia.

- It should be specific. If the code merely states that people must not steal, it offers little of value. If the code does not address the ethical offenses most likely to occur in the company or profession, such as bribery in companies that do a lot of business overseas, it is ineffective.
- It must be enforceable. If it does not stipulate penalties, up to and including dismissal from the company or expulsion from the profession, it is ineffective.

Do codes of conduct really encourage ethical behavior? Sometimes, yes. James Burke, chairman and CEO of Johnson & Johnson, has been praised for his deci-

INTERVIEW: ATTORNEY BRENT GARDNER

on Hewlett-Packard's long-term commitment to ethics

The founders of the company set an expectation that employees act with the highest standards of ethics. They felt it was good business: if you act ethically with your customers there will be a bond of trust established and they will come back.

on Hewlett-Packard's approach to ethics training for its employees

Every employee gets a copy of our Standards of Business Conduct, and then there is training on a periodic basis, beginning when the employee first starts at HP. We also have prepared a training tape that can be used by any supervisor. The latest edition of the Standards is simpler and more straightforward than past editions.

on Hewlett-Packard's other efforts to maintain high ethical standards

We have internal auditors who review all of the entities of the company to see whether there are any standards-of-conduct violations. Also, HP has a post office box where anyone concerned with an ethical issue can anonymously raise that issue with the company. Every manager is responsible for seeing that his or her employees follow the standards that we have established. Our efforts have been quite effective.

on the major areas covered by the Standards of Business Conduct

We cover dealings by the employee with the company, conflicts of interest, gifts to or from customers and suppliers, legal issues in dealing with customers and competitors and suppliers, and dealing with the federal government.

on the major impediments to ethical behavior in the workplace

One problem is a general lack of understanding of what ethical standards are expected. Few employees have studied ethics in school or given the subject much thought. This means that they often just adopt the standards of those with whom they work closely. Another problem is that economic pressures sometimes cause people to do things they think make business sense but which are really shortsighted.

on how ethics should be taught in colleges and universities

My oldest son graduated a couple of years ago in computer science; one of his required classes was on the ethics of computer science. I think that is a good idea. Ethics is actually a legitimate issue for every course of study. I would include ethics in teaching technical communication.

Brent Gardner is assistant general counsel for Hewlett-Packard Company in Boise, Idaho.

sion to withdraw all Tylenol products from store shelves in 1982 in the wake of tampering that cost seven lives. Burke stated at the time that withdrawing the product—a move that cost the company $100 million—was the only option because the company's code begins with this sentence: "We believe our first responsibility is to the doctors, nurses, and patients, to mothers, and to all others who use our products and services." He expressed surprise that people would find his decision unusual (De George, 1995, p. 4).

Often, however, it is difficult to tell whether codes encourage ethical behavior, for no statistics exist on how many unethical acts were prevented because people were inspired or frightened by a code of conduct. Many ethicists and officers in professional societies are skeptical. Because codes must be flexible enough to cover a wide variety of situations, they tend to be so vague that whether a person in fact violated one of their principles is virtually impossible to say.

A study conducted by the American Association for the Advancement of Science (Chalk, 1980) found that whereas most professional organizations in the sciences and engineering have codes that provide for hearing cases of unethical conduct, relatively few such cases have ever been brought before the organizations. Many societies reported, for example, that only three or four allegations were ever lodged against individuals, and most of the societies have never taken disciplinary action—censure or expulsion—against any of their members for ethical violations. And a survey about how chemical engineers make decisions about ethical conflicts reveals that of the 4,318 respondents, fewer than a half dozen even mentioned codes of conduct in describing their thinking (Bryan, 1992, p. 81).

If codes of conduct are not often systematically enforced, do they have any real value? Critics point out that almost no organization is willing to support someone who brings a charge. An accuser within an organization is likely to face overt or covert punishment; thus self-interest compels many people to remain silent, regardless of the high-minded statements in the organization's code of conduct. Few professional organizations have ever come to the financial aid of an accuser who has lost a job because of a justified allegation.

For this reason, codes of conduct sometimes seem to be public-relations tools intended to persuade employees, organization members, the general public, and the government that the organization polices its own members.

Ethicist Jack N. Behrman finds the greatest value in merely writing a code: an organization thereby clarifies its own position on ethical behavior (1988, p. 156). Of course, distributing the code might also foster an increased awareness of ethical issues, in itself a positive result.

Whistleblowing

Whistleblowing is the practice of going public with information about serious unethical conduct within the organization. For example, an engineer is blowing the whistle in telling a government regulatory agency or a newspaper that quality-control tests on a product the company sells have been faked.

In deciding whether to blow the whistle, you must choose between loyalty to your employer on the one hand and to your own standards of ethical behavior on the other. Some people believe that an employee owes complete loyalty to the employer. James M. Roche, former president of General Motors, has written: "Some of the enemies of business now encourage an employee to be *disloyal* to the enterprise. They want to create disharmony, and pry into the proprietary interests of the business. However this is labeled—industrial espionage, whistleblowing, or professional responsibility—it is another tactic for spreading disunity and creating conflict" (Beauchamp & Bowie, 1993, p. 259).

Yet most people believe that workers should not be asked to steal or lie or take actions that could physically harm others. Where does loyalty to the employer end and the employee's right to take action begin? And what should an employee do before blowing the whistle?

Ethicist Manuel Velasquez (1992, p. 403) argues that whistleblowing is justified if four conditions are satisfied:

- There is clear and compelling evidence that the organization is doing something that is hurting or will hurt other parties.
- The employee has made a serious but unsuccessful attempt to prevent the wrongdoing by going through channels internally.
- External whistleblowing is reasonably certain to prevent or stop the wrongdoing.
- The wrongdoing is serious enough to warrant the injuries that the whistleblowing will probably cause for the employee, his or her family, and any other parties.

But saying that you are *justified* in blowing the whistle is not the same as saying that you are *obligated* to do so. Velasquez suggests you are obligated to blow the whistle when you are justified and when at least one of the following two conditions is satisfied:

- Your position or professional responsibility calls for you to prevent the wrong. For instance, if you are a professional engineer, your professional code of ethics stipulates that you are to prevent certain kinds of wrongdoing, such as the building of unsafe structures.
- The wrongdoing is sufficiently serious for society, a group, or an individual. Wrongdoing that involves health, safety, or basic economic welfare falls into this category.

A number of organizations have instituted procedures that encourage employees to bring ethical questions to management rather than blow the whistle. Among the more common means that companies use are anonymous questionnaires and ombudspersons. An *ombudsperson* is an employee whose job includes bringing ethical grievances to management's attention. An ombudsperson who feels that management has not responded satisfactorily to the situation is empowered to report the information freely.

Most companies still have no formal procedures for handling serious ethical questions, however, and whistleblowing thus remains risky. Although the federal government and about half the states have laws intended to protect whistleblowers, these laws are not highly effective. It is simply too easy for the organization to penalize the whistleblower—subtly or unsubtly—through negative performance appraisals, transfers to undesirable locations, or isolation within the company. For this reason, many people feel that an employee who has unsuccessfully tried every method of alerting management to a serious ethical problem would be wise simply to resign rather than face the professional risks of whistle-blowing. Of course, resigning quietly is much less likely to force the organization to remedy the situation.

As many ethicists say, doing the ethical thing does not always advance a person's career.

Guidelines for Ethical Communication

Ethical problems are particularly common in two different kinds of writing you will do:

- *Proposals.* You might be asked to exaggerate or lie about your organization's past accomplishments, pad the résumés of the project personnel, or list as project personnel workers who in fact will not be contributing to the project.
- *Product information.* This category includes everything from descriptions in sales catalogs to specification sheets and operating instructions and manuals.

As Bryan (1992) points out, most ethical dilemmas in any kind of product information arise because you are doing two things at once: describing and advertising. These two functions are not only different, but often in conflict. (Ironically, most advertisers maintain they are merely providing product information, even though most ads are light on facts.)

The two guidelines presented here cannot solve all the ethical problems and dilemmas you will face as a working professional. However, they can help you prevent many kinds of ethical problems.

- Tell the truth.
- Do not mislead.

Tell the Truth   Perhaps the simplest and strongest guideline is to tell the facts honestly and accurately. Employees report that sometimes they are asked to lie. Obviously, lying—knowingly providing inaccurate information—is unethical. If your company's own tests of the electric screwdriver it produces show that a charge lasts 50 minutes, but the competitor's model lasts 75 minutes, you might be pressured by a supervisor to simply lie: to say that your product's charge lasts 75 minutes. Your responsibility is to resist this pressure, by going over the supervisor's head, if necessary.

Do Not Mislead   Providing misleading information is a little more complicated than lying, but ethically it amounts to the same thing. A misleading statement or graphic enables or even encourages the reader to believe false information. For instance, a product-information sheet for a set of skis is misleading if the accompanying photograph shows bindings and poles that are not included with the skis. Some stores advertise computer systems at low prices, with the small print stating "monitor optional." The monitor isn't optional if you actually want to use the computer.

Information can mislead the reader in a number of other ways:

- *Scare tactics.* A document can wrongly suggest that a product—or a particular brand of the product—is necessary. For example, a flashlight manufacturer misleads in suggesting that only its own brand of batteries will power the flashlight.

- *Euphemisms.* If you need to fire someone, use a word such as *fire* or *release,* not *alternative employment facilitation.* See Chapter 14 for more information on euphemisms.

- *Clichés.* Communicators sometimes use such terms as *user-friendly, ergonomically designed,* and *state of the art* to make the product sound better than it is. If you make a judgment about the product, back it up with accurate, specific information. Do not write "We carried out extensive market research" if all you did was make a few phone calls.

- *Ignoring negative features.* For instance, if an information sheet for a portable compact-disc player suggests that it can be used by joggers without mentioning that the bouncing will probably make it skip, the information sheet is misleading.

- *Legalistic constructions.* Avoid phrases that say one thing but actually mean another. For instance, it is unethical to write "The 3000X was designed to operate in extreme temperatures, from −40 degrees to 120 degrees Fahrenheit" if in fact the unit cannot reliably operate in those temperatures. The fact that the statement might actually be accurate—the unit was *designed* to operate in those temperatures—doesn't make it any less misleading.

- *Failing to acknowledge assistance from others.* Don't suggest that you did all the work yourself if you didn't. Cite your sources and your collaborators accurately and graciously.

CASES

1. Irritating Insulation

The town of Acton, Ohio (population 6,500), like many other small communities in the Rust Belt, has suffered economically during the past decade. Much of its infrastructure is old and in need of repair, and the town has a shrinking tax base. Young people routinely leave the area after high school in search of better jobs.

The main employer in Acton is Diversified Construction Materials, which employs more than 1,000 people from Acton and surrounding communities. Like Acton, Diversified has known better times. Its products are known for their high quality, but foreign and domestic manufacturers who have moved their production facilities to developing countries are undercutting Diversified's prices and gaining market share.

However, the research and development department at Diversified has just formulated a new type of blown insulation that the company thinks will perform as well as fiberglass but beat its price. This new substance promises to be a major part of Diversified's highly regarded insulation products. A number of retailers have ordered large quantities of the new insulation based on Diversified's exhibits at trade shows and some preliminary advertisements in industrial catalogs.

As the head technical writer at Diversified, Susan Taggert oversees the creation of all product information for the insulation. As she normally does in such cases, she gathers all the documentation from R&D and any other materials available in the company, which she will study before mapping out a strategy.

About one week into the project, Taggert discovers from laboratory notebooks that three of the seven technicians participating in the project experienced abnormally high rates of absence from work during the four months they spent developing the insulation. One of the three technicians requested to be transferred from the project at the end of the first month. His request was granted.

Taggert calls Diversified's personnel department and learns that all three of the technicians complained of the same condition—bronchial irritation of varying degrees of severity—but that the irritation ceased two to three days after the last exposure to the insulation. Apparently some compound in the insulation, which the company physician could not immediately identify, affected some of the technicians who worked closely on it.

Taggert goes to the vice president of operations, Bill Mondale, who is in charge of introducing all new products. Taggert presents her information to Mondale and suggests that the company find out what is causing the bronchial irritation before it ships any of the product. Although the irritation does not appear to be serious, there are no data on the potential effects of long-term exposure to the insulation when used in houses or offices.

Mondale points out Diversified's tight deadline; delivery is scheduled in less than two weeks. Determining the cause of the irritation could take weeks or months and cost many thousands of dollars. Taggert points out the financial risks involved in selling a product that poses a health hazard. Mondale responds that it is a risk the company will have to take, and adds that the product is in compliance with all applicable federal guidelines. The company has staked its reputation—and its third-quarter profits—on the insulation. He directs Taggert to proceed with the product literature as quickly as possible and not to spend any more time worrying about the health hazard.

In a memo to your instructor, respond to the following questions: What should Susan Taggert do? Why is her recommended course of action preferable to other courses of action? Explain carefully the ethical implications of the recommended course of action. See Chapter 16 for a discussion of memos.

2. Mysterious Internet Code

John Boorman manages the technical publications department at Santa Barbara Equipment, a company that manufactures equipment used in the semiconductor industry. John supervises four technical communicators who handle the firm's internal and external communication.

Santa Barbara Equipment is only three years old. It was founded by Eugene Froom, who invented the two major products the company produces. Froom has molded the company in his image. He doesn't delegate authority, and he has no patience with standard business practices. After three years, the company still does not have a policies-and-procedures manual for new employees, nor is there a code of conduct. If Mr. Froom thinks you are very bright, he will hire you, despite—or even because of—your lack of business experience and sophistication.

The newest member of the technical publications department is Lynn Stone, who was hired largely because of her computer skills. She knows more about computers than anyone else in her department, and in her free time she loves to "surf the Net," as she describes it.

Lynn is contributing to the department's latest project: online documentation to accompany Santa Barbara's new product. Her job is to help the other technical communicators put the information online. A week before the documentation is to be pressed onto CDs, John Boorman learns through a conversation with Lynn that the software she used in coding the information was downloaded off the Internet. The problem is that one of the executable files of this software must be included with Santa Barbara's product. Concerned about copyright questions, he asks her to find out everything she can about the software.

Lynn learns that the software was made by a company called Visionary Software, which has no idea how it got onto the Internet. After checking the coding, Visionary's president tells Lynn that the software has been altered. He also tells her that he will refer the matter to Visionary's legal counsel.

Lynn reports this information to John Boorman, who communicates it to Eugene Froom. Froom is furious.

In a memo to your instructor, respond to the following questions: What should John Boorman do? What is his obligation regarding Lynn's conduct? What is his responsibility to Eugene Froom, the company president? To Visionary Software? To Santa Barbara's customers? In writing your response, be sure to explain your reasoning and evaluate the ethical and practical implications of each recommendation you make. See Chapter 16 for a discussion of memos.

EXERCISES

The following exercises call for you to write memos. See Chapter 16 for a discussion of memos.

1. Find a newspaper or magazine ad that you feel contains untrue or misleading information. Write a memo to your instructor describing the ad and analyzing the unethical techniques. How might the information have been presented more honestly? Include a photocopy of the ad with your memo.

2. Find a newspaper or magazine article that you feel contains untrue or misleading information. Write a memo to your instructor describing the article and analyzing the unethical techniques. How might the information have been presented more honestly? Include a photocopy of the article with your memo.

3. It is late April and you need a summer job. In a local newspaper you see an ad for a potential job. The only problem is that the ad specifically mentions that

"This is a continuing, full-time position." You know that you will be returning to college in the fall. Is it ethical for you to apply for the job without mentioning this fact? Why or why not? If you feel it is not ethical to withhold the information that you plan to return to college in the fall, is there any way you can apply ethically? Respond in a brief memo to your instructor.

4. Many foods are labeled as "fat free," yet they contain substances such as monogycerides, which are treated by the body just like more common forms of fat. Is it ethical to use the phrase "fat free" in such cases? Why or why not? Respond in a brief memo to your instructor.

5. Research your college or university's code of conduct for students, and write a memo to your instructor describing and evaluating it. Consider such questions as the following: How long is the code? How comprehensive? Does it provide detailed guidelines or merely general statements? Where does the code

appear? From your experience, does it appear to be widely publicized, enforced, and adhered to? Are there sources of information on campus that could provide information on how the code is applied?

Collaborative Exercises

6. Form small groups according to major. Study the code of conduct of a company or organization in your community.
   - One group member could study the code, describing how effectively it states the ideals of the organization, describes proper and improper behavior and practices for employees, and spells out penalties.
   - Another group member could interview the officer who oversees the use of the code in the organization. Who wrote the code? What were the circumstances that led the organization to write it? Is it based on another organization's code? Does this officer of the organization believe the code is effective? Why or why not?
   - A third group member could secure the code of one of the professional groups in the organization's field (search for the code on the Web, as described in Chapter 4). For example, if the local organization produces electronic equipment, a professional group would be the Institute for Electrical and Electronics Engineers, Inc. To what extent does the code of the local organization reflect the principles and ideals of the professional group's code?

   As a team, write a memo to your instructor presenting your findings. Attach the local organization's code to your memo.

7. Form small groups and research an event that involved ethical matters, such as the Chernobyl nuclear accident, the Exxon *Valdez* oil spill, the *Challenger* tragedy, or an incident in which a news organization manipulated a film, video, or photograph. Each member of the group might investigate one of the following aspects of the case:
   - Study the incident itself. What happened? What factors led to the event, and what were the results immediately after the event?
   - Analyze the event. What were the ethical issues involved, and in what way or ways does the incident represent unethical conduct?

   - Analyze the long-term consequences of the event for the different stakeholders. In the Exxon *Valdez* spill, for instance, what have been the consequences for the environment, for the people who earned their livings from the natural resources, for Exxon, and for the general public?

   Write a memo to your instructor presenting your findings.

Web Exercise

8. Find two Web sites dealing with ethics issues related to the use of computers. (A good place to start the search would be the World Wide Web Ethics Center for Engineering and Society at http://web.mit.edu/ethics/www/home.html.) Evaluate each site in terms of the following questions:
   - How detailed and comprehensive is the site?
   - Does the site attempt to provide a balanced selection of materials, or does it seem to be more of an advocacy group whose purpose is to communicate a particular perspective?
   - How effective do you think the site is in helping people understand how to think about ethics and computers?

   Write a memo to your instructor presenting your findings. Attach printouts of representative sections from each site.

REFERENCES

Aversa, J. (1996, Dec. 19). Software piracy costs firms billions. *Idaho Statesman*, 8B.

Baumhart, R. (1968). *An honest profit: What businessmen say about ethics in business.* New York: Holt, Rinehart and Winston.

Beauchamp, T. L., & Bowie, N. E. (1993). *Ethical theory and business* (4th ed.). Upper Saddle River, NJ: Prentice-Hall.

Behrman, J. N. (1988). *Essays on ethics in business and the professions.* Upper Saddle River, NJ: Prentice-Hall.

Branscomb, E. S. (1993). Trademarks: Caveat scriptor. *Technical Communication, 40*(1), 97–99.

Brockmann, R. J., & Rook, F. (Eds.). (1989). *Technical communication and ethics.* Washington, DC: Society for Technical Communication.

Bryan, J. (1992). Down the slippery slope: Ethics and the technical writer as marketer. *Technical Communication Quarterly, 1*(1), 73–88.

Chalk, R. (1980). *AAAS professional ethics project: Professional ethics activities in the scientific and engineering societies.* Washington, DC: American Association for the Advancement of Science.

De George, R. T. (1995). *Business ethics* (4th ed.). Upper Saddle River, NJ: Prentice-Hall.

Forester, T., & Morrison, P. (1994). *Computer ethics: Cautionary tales and ethical dilemmas in computing.* (2nd ed.). Cambridge, MA: MIT Press.

Helyar, P. S. (1992). Products liability: Meeting legal standards for adequate instructions. *Journal of Technical Writing and Communication, 22*(2), 125–147.

Honor in science. (1986). New Haven: Sigma Xi. The Scientific Research Society.

Kozak, E. M. (1993). Copyright and fair use: A legal tug of war. *Editorial Eye, 16*(3), 2–3.

Swazey, J. P., Louis, K. S., & Anderson, M.S. (1994). The ethical training of graduate students requires serious and continuing attention. *Chronicle of Higher Education, 40*(27), B1–B2.

Velasquez, M. G. (1992). *Business ethics: Concepts and cases* (3rd ed.). Upper Saddle River, NJ: Prentice-Hall.

Writing Collaboratively

Patterns of Collaboration 49

Advantages and Disadvantages of Collaboration 51

  Advantages of Collaboration *51*
  Disadvantages of Collaboration *52*

Organizing and Maintaining a Collaborative Group 53

Reducing Conflict 57

Improving Listening Skills 59

Gender and Collaboration 60

Multiculturalism and Collaboration 62

Case: The Reluctant Collaborator 63

Writer's Checklist 64

Exercises 64

    Collaborative Exercises *65*
    Web Exercise *65*

You have probably noticed that many professors require collaborative writing in their courses. The popularity of collaboration on campus reflects a similar trend in the working world. This chapter focuses on the techniques of collaborative writing.

The chapter begins by describing different patterns of collaboration and the strengths and weaknesses of the technique. Next, the chapter discusses reducing conflict, improving listening skills, and understanding the role of gender differences and multiculturalism in collaboration.

In one sense, virtually every document is collaborative, for every writer uses information created, discovered, or interpreted by someone else. However, a useful description (Allen, Atkinson, Morgan, Moore, & Snow, 1987) is that collaborative writing

- leads to the production of a document
- involves substantive interaction among group members
- involves shared decision-making power over and responsibility for the document

Collaborative writing is common in organizations, and as documents and the techniques used to produce them become more complex, the amount of collaboration is likely to continue to increase. One survey found that 73.5 percent of 200 college-educated businesspeople collaborate to produce about a quarter of their documents (Faigley & Miller, 1982, p. 567). Another survey found that 87 percent of 520 professionals collaborate at least some of the time (Ede & Lunsford, 1990, p. 20). A third study of more than 400 professionals found that they often write collaboratively (Couture & Rymer, 1989, p. 78).

What kinds of documents do people write collaboratively? Everything from brief notes to books. However, three kinds of documents that appear to be written collaboratively most often are proposals, memos, and reports (Ede & Lunsford, 1990, p. 63), and Web sites are almost always created collaboratively.

Collaboration exists in highly structured bureaucracies and more flexible organizations. Some organizations follow the *division-of-labor model,* in which the manager assigns roles, and then employees work separately and submit their products to the manager. Other organizations follow the *integrative model,* in which people work together cooperatively, sometimes forming and reforming their own groups (Killingsworth & Jones, 1989).

Patterns of Collaboration

Regardless of whether the organization uses the division-of-labor model or the integrative model, there are three different patterns of collaboration, as shown in Figure 3.1 on page 50:

- *Collaboration based on job specialty.* Perhaps the most common pattern is to divide the work according to the expertise of the group members.

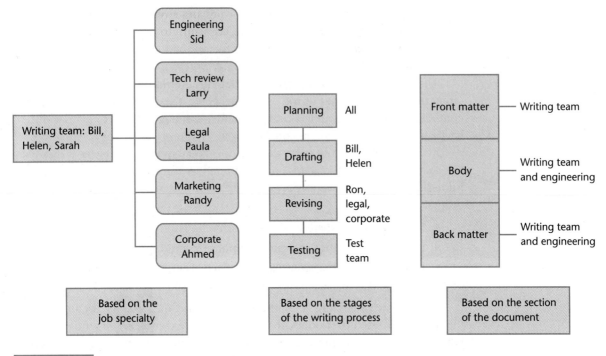

Patterns of Collaboration

For instance, an engineer is the subject-matter expert in charge of contributing all the technical information; a graphic artist is responsible for the graphics and the design; a technical communicator drafts the document.

- *Collaboration based on the stages of the writing process.* In many organizations, a great deal of collaboration occurs during the *prewriting* stage. Different technical professionals and technical communicators come together to plan a lengthy document, such as a manual. They share ideas about the content, organization, and style of the document, and they establish a production schedule and an evaluation program.

  Collaboration during the *drafting* stage is much less common, largely because working together is much more time consuming than working individually. However, advances in groupware are making it more convenient to draft collaboratively. *Groupware* is computer software that lets several people at different locations work on the same document at the same time. (See Chapter 4 for a discussion of groupware.)

Collaboration during the *revising* phase is very common. Often the people who helped launch the project return to review the draft and make suggestions.

- *Collaboration based on the section of the document.* Sometimes one person takes responsibility for one section of the document, another person does another section, and so forth. This pattern is common for large projects that have discrete sections. For example, the various parts of proposals—which have standard components, such as the project description and the credentials section—are often written by different people.

Advantages and Disadvantages of Collaboration

Ede and Lunsford's study (1990, p. 50) found that 58 percent of writers considered collaborative writing to be very productive or productive, whereas 42 percent found it not very productive or not at all productive. As a technique for creating documents, collaboration has both strengths and weaknesses.

Advantages of Collaboration

Writers who collaborate can create a better document and improve the functioning of an organization:

- *Collaboration draws on a greater knowledge base.* Two heads are better than one, and three are better yet. The more people involved in planning and writing the document, the larger the knowledge base. This principle translates into a document that can be more comprehensive and more accurate than the product of a single author.
- *Collaboration draws on a greater skills base.* No one person can be an expert manager, writer, editor, graphic artist, and production person.
- *Collaboration provides a better sense of how the audience will read the document.* Working collaboratively involves having different people read drafts. Each person who reads and comments on a draft acts as an audience, raising questions and suggesting improvements that one person writing alone would not likely generate.
- *Collaboration enables people to share the responsibility for the document.* No one person should have to shoulder the responsibility for a 400-page manual or proposal.
- *Collaboration improves communication among employees.* Collaboration breaks down physical, organizational, and emotional barriers. People who work together for a long time get to know one another and learn about their jobs, responsibilities, and frustrations. A shared goal—putting together a professional-quality document that will reflect

positively on everyone involved—can enable people to work together more productively.

- *Collaboration improves the socialization of new employees.* Collaboration helps acclimate new employees to an organization. They get to know people from different parts and different levels of the company. Working collaboratively helps new employees learn the procedures used in the organization: how to get things done, which people to see, what forms to fill out, and so forth. In addition, collaboration teaches the values of the company, such as the importance of shared goals and the willingness to work hard and sacrifice for an important initiative.

Disadvantages of Collaboration

Six principal problems are associated with collaboration:

- *Collaboration takes more time than individual writing.* It takes longer to create a document collaboratively than individually simply because of the need for the collaborators to communicate.
- *Collaboration can lead to groupthink.* When members of a group place a higher value on getting along with each other than on thinking clearly and critically about the subject, *groupthink* can be the result. Groupthink can lead to an inferior document because no one wants to cause a scene by asking tough questions about the group's work.
- *Collaboration can yield a disunified document.* The document can end up disunified in content or style. For example, sections written by two people can contradict each other or contain unintended duplication. Or one section can refer to another section that never gets written. And the more people involved in the collaboration, the greater the variation of style, in everything from design to spelling. To prevent this sort of disunity, writers must plan and edit the document with extreme care.
- *Collaboration can lead to inequitable workloads.* No matter how hard the project leader tries to prevent it, some people will end up working more than others.
- *Collaboration can reduce the collaborators' sense of pride in the document.* If the project calls for many people to contribute relatively minor effort, they might miss the sense of satisfaction that writers generally feel. They can start to think of themselves as very small cogs in a very large wheel; as a result, the quality of the document can suffer.
- *Collaboration can lead to interpersonal conflict.* People can disagree about the best way to create the document or about the document itself. Such disagreements can poison the working relationship not only during the project but long after it is complete. Techniques for reducing conflict are discussed in more detail later in this chapter.

Organizing and Maintaining a Collaborative Group

Although collaborative groups take many shapes and function in many different ways, the following seven guidelines will help you set up and maintain a successful group:

- *Define the group's task.* In some groups, the first and most important guideline is to define your task. Your group members must agree on what you are supposed to accomplish. You have to reach a consensus, for instance, that your task is to revise your company's policies-and-procedures manual, and that it has to be finished by April 10. If there are other requirements, such as that the page count not exceed 300, you should articulate them as well.

- *Choose a group leader.* In some groups, the first task is to choose a group leader. This person serves as the link between the group and management. In addition, the group leader is responsible for keeping the group on track, leading the meetings, and coordinating the communication among group members.

  Sometimes group leaders are not chosen by the group members but are assigned by management. For instance, companies that bid for contracts often have project managers whose job is to organize and supervise proposal-writing teams.

- *Define tasks for the group members.* In most groups, work is divided among the group members. As shown earlier in Figure 3.1, the division of labor can follow different patterns.

- *Establish working procedures.* To prevent problems later, establish working procedures to cover meetings, communication, and conflict resolution. Members need answers to the following questions—in writing, if possible—before the work of the group proceeds:

  – When do we meet?
  – Where do we meet?
  – What procedures will we follow in the meetings?
  – How are we to communicate with other members of the group, including the group leader?
  – How often are we to communicate?
  – What access will we have to people outside the group and to other resources?
  – What procedures are we to follow in resolving conflict?

- *Establish a work schedule.* The first thing you did was define your task, focusing on the "deliverable": the document you are to produce. If you know the date on which it is due, you have to work backward and

Notice that sometimes milestones are presented in reverse chronological order; the delivery-date milestone, for instance, comes first. In other forms, items are presented in normal chronological order.

The form includes spaces to list the person responsible for each of the milestones and progress reports, as well as spaces to state the progress of each milestone and progress report.

Name of Project:
Principal Reader:
Other Readers:
Group Members:

Milestones	Responsible Member	Status	Date
Deliver Document			
Proofread Document			
Send Document to Print Shop			
Complete Revision			
Review Draft Elements			
Assemble Draft			
Establish Tasks			

Progress Reports	Responsible Member	Status	Date
Progress Report 3			
Progress Report 2			
Progress Report 1			

Meetings	Agenda	Location	Date	Time
Meeting 3				
Meeting 2				
Meeting 1				

Notes

FIGURE 3.2

Work-Schedule Form

create a schedule for the entire project: for a document to be shipped on April 10, it might have to be at shipping by April 5 and at the printer by March 19. Each of these dates is called a *milestone*. Figure 3.2 shows a work-schedule form that can be used to assist in planning the schedule.

- *Create evaluation materials.* Group members have a right to know how they will be evaluated for their participation in the project. In some hierarchical organizations, the project manager evaluates the group members without communicating to them either the criteria or the results of the evaluation. If it is possible to make the criteria explicit, however, do so, for a more open process creates a better atmosphere and increases the chances that the group members will perform satisfactorily. Figure 3.3 is a sample evaluation form that can be used by each group member to evaluate the contributions of other members of the group.

Your name_____   Date_____

Title of the project_____

Instructions

Use this form to evaluate the other members of your group. Write the name of each group member in one of the columns, then assign a grade of 0 to 10 (0 being the lowest grade, 10 the highest) for each group member for each criterion. Then total the grades for each member. Because each group member has different strengths and weaknesses, the scores you write will differ. On the back of this sheet, write any comments you wish to make.

Criterion	Group Members			
1. Regularly attends meetings	1._____	1._____	1._____	1._____
2. Is prepared at the meetings	2._____	2._____	2._____	2._____
3. Meets deadlines	3._____	3._____	3._____	3._____
4. Contributes good ideas in meetings	4._____	4._____	4._____	4._____
5. Contributes ideas diplomatically	5._____	5._____	5._____	5._____
6. Submits good work	6._____	6._____	6._____	6._____
7. Listens to other members	7._____	7._____	7._____	7._____
8. Shows respect for other members	8._____	8._____	8._____	8._____
9. Helps to reduce conflict	9._____	9._____	9._____	9._____
10. Your overall assessment of this person's contribution	10._____	10._____	10._____	10._____
Total Points				

FIGURE 3.3

Evaluation Form

Some organizations and college instructors also use self-evaluation to help people assess their own performance in group projects. Figure 3.4 on page 56 is a self-evaluation form.

- *Involve outsiders.* Groups benefit from the involvement of outsiders for the same reasons that individuals benefit from groups: the more

Your name _____ Date _____
Title of the project _____

Instructions

On this form, record and evaluate your activities on this project. In the Log section, record the activities you performed as an individual; then record the activities you performed as part of the group. For all activities, record the date and the number of hours you spent. In the Evaluation section, write two brief statements: one about the aspects of your contribution to the group that you think were successful, and one about the aspects that you want to improve.

Log

Individual Activities   Date   Number of Hours

_____
_____
_____
_____
_____
_____
_____
_____

Activities as Part of Group   Date   Number of Hours

_____
_____
_____
_____
_____
_____
_____

Evaluation

Aspects of My Participation That Were Successful

_____
_____
_____
_____
_____

Aspects of My Participation That I Want to Improve

_____
_____
_____
_____
_____

FIGURE 3.4

Self-Evaluation Form

perspectives the group has, the more successful the project will be. For this reason, groups often call in persons or organizations to assess the project periodically. For example, a group commonly submits its project plan to management for review and suggestions before beginning the project. Then, periodically, the group consults with outside subject-matter experts. At the end of the project, the group usually presents its work to subject-matter experts and management for review and approval.

Of course, you should also consult with other members of your group as much as possible. Before you articulate an idea in a group meeting, try it out on as many group members as you can privately; you will see their reactions to the idea and be able to refine it.

Reducing Conflict

Because collaborating on an important project is stressful, it can lead to interpersonal conflict. People can get frustrated and angry with each other because of personality clashes or disputes about the process of collaborating or about the substance of the project itself. But for the project to succeed, the group members must be able to work together productively. One technical communicator, quoted in Ede and Lunsford (1990, p. 66), sums it up this way: "No one here loses a job because of incompetence; they lose jobs if they can't work with others." Following are nine guidelines for establishing and maintaining good relationships while collaborating:

- *Arrive on time and be prepared to work.* Show up on time, have all your papers and other materials ready, and know what your goal is for the meeting. Everyone is late or unprepared once in a while, but conscientious professionals seem to get caught in fewer traffic problems than other people.
- *Meet your deadlines.* Another aspect of professionalism is to have your work ready on time. If you know you will not be able to, give as much warning as possible so that the group leader can plan around the problem and not waste the time of the other members.
- *Listen carefully.* It's a fact of life: most people would rather talk than listen. But in any group you ought to listen far more than you talk. Listen carefully, trying to understand and interpret what the speaker is saying. Listening will be discussed in more detail in the next section.
- *Ask pertinent questions.* To make sure you understand what is being said, ask questions. Don't be afraid that you will appear ignorant; the brightest people constantly try to understand what they hear and to make connections to other ideas. Asking pertinent questions helps not only

you but also the other members of the group: it forces them to examine speakers' assumptions, evidence, and logic.

- *State your views diplomatically.* Experienced professionals earn respect by thinking well, not by bulldozing other people. You want to appear self-confident but not overbearing. Avoid personal remarks and insults, and avoid overstating your position. A modest qualifier such as "I think" or "it seems to me" is an effective signal to your listeners.

OVERBEARING: My plan is a sure thing; there's no way we're not going to kill Allied next quarter.

DIPLOMATIC: I think this plan has a good chance of success: we're playing off our strengths and Allied's weaknesses.

Notice, in the diplomatic version, the speaker's decision to call it "this plan" rather than "my plan."

PERSONAL: John, you're wrong about the export figures for 1997; we did $32 million, not $22 million.

BETTER: John, are you sure those export figures for 1997 are accurate? I thought I read in the February summary that we did $32 million, not $22 million.

The second version of this sentence is better than the first because it does not accuse John of being wrong; it merely questions his facts. Another advantage is that the second version states the source where the correct figures might be found, thus suggesting how the discrepancy can be resolved without making it a personal issue.

- *Don't get emotionally attached to your own ideas.* You cannot look at your own ideas completely objectively, but it is smart to try. When you meet opposition, try to understand why other members of the group don't share your views. Digging in is usually unwise—unless it is a serious matter of principle—because although you may be right and everyone else wrong, it's not likely.
- *Be supportive, not critical.* Be tolerant of other people's views and working methods. Show them respect and trust them. Doing so is right, both ethically and practically. It is ethical to treat people respectfully, as you want them to treat you. And it is practical: if you anger people, they will go out of their way to oppose you.
- *Share the work space.* In most groups, there is a shared work space: a workstation, a chalkboard, or a flip chart. Don't monopolize the shared space; let everyone have an opportunity to control it.
- *Share your resources with other members of the group.* The more resources everyone has, the more productive the group will be. Make sure you make your resources available to other group members—articles,

books, software, and so on; they might be able to use those resources in ways that had not occurred to you.

Improving Listening Skills

Listening is not the same as hearing. *Hearing* consists of the physiological, sensory processes by which auditory sensations are received by the ear and transmitted to the brain. *Listening* is the psychological process of interpreting and understanding the content of the sensory experience.

Regardless of whether we are working in a collaborative group, listening is a critically important skill. According to Hulbert (1989, p. 3), 45 percent of the time we spend in communicating is devoted to listening, yet our listening skills are very poor. Right after hearing a 10-minute presentation, the average person has understood and remembered about half of what the speaker said; after 48 hours, that figure drops to about 25 percent.

The following discussion presents guidelines to improve your listening skills. The first four guidelines relate to the physical environment and the listener's physical presence:

- *Reduce distractions.* If the temperature is uncomfortable and the room is noisy, or if you're hungry or tired, you won't be able to listen effectively.
- *Face the speaker.* You will concentrate better if you face the speaker. Doing so will also help the speaker, for you will seem more alert and interested in listening.
- *Maintain eye contact.* Looking at the speaker's eyes helps you concentrate on the message. In addition, it gives the speaker more confidence and thus can improve the quality of the information.
- *Reduce irrelevant movements.* Pay attention physically. Don't tap a pencil on the table or play with your keys.

Here are three guidelines to follow as you listen to the speaker:

- *Set aside your preconceptions.* Too often we make up our minds before the speaker says a word. We react to the speaker's gender, age, race, personal appearance, or clothing. Or we decide that we already know what the speaker will say because we know that he or she is advocating environmental protection or handgun interests.
- *Listen to the ideas, not the words.* One barrier to effective listening is to focus on the words rather than the meaning. Don't focus on pronunciation or accents; don't worry about the speaker's grammar errors. Instead, concentrate on the points the speaker is making.
- *Use your listening time productively.* Although people speak at about 150 words per minute, they think at about 500 words per minute. Speakers use the phrase "cruising down Route 350" to describe a listener drifting off

because the speaker is talking so much more slowly than the listener is thinking. Use your time well: think about what the speaker is saying. Examine the evidence and the logic. Compare the speaker's message with what you already know about the subject. Think of questions you might ask later.

Finally, follow these three guidelines if the speaker is talking with you alone or in a small group and you are permitted or expected to respond:

- *Don't interrupt.* If someone is speaking, listen.
- *Paraphrase what the speaker has said.* Use phrases such as the following: "So what you are saying is . . . ." Paraphrasing demonstrates to the speaker that you are listening carefully; in addition, it reduces the chances of misunderstanding.
- *Ask open questions.* If it is appropriate for you to ask questions, include open questions as well as closed ones. A *closed question* calls for a brief response, either yes/no or a simple fact. An *open question* encourages the speaker to elaborate.

CLOSED QUESTIONS: "Did we mail the original or fax it?"
"When did we do the last preventive maintenance on the roller?"

OPEN QUESTIONS: "Why do you think we're getting these results?"
"Where do you see the flat-panel market in five years?"

Gender and Collaboration

As this discussion has suggested, effective collaboration involves two related challenges: maintaining the group as an effective, friendly working unit and accomplishing the task. Scholars of gender and collaboration see these two challenges as representing the feminine and masculine perspectives.

Any discussion of gender studies must begin with a qualifier: when we talk about gender, we are generalizing, not talking about particular people; the differences in behavior between two men or between two women are likely to be greater than the difference between men and women.

The differences in how the sexes communicate and work in groups are traced to our traditional family structure. As the primary caregivers, women learn to value nurturing, connection, growth, and cooperation; as the primary breadwinners, men learn to value separateness, competition, debate, and even conflict (Chodorow, 1978).

For decades, scholars have studied the differences in speech between women and men. Women use more qualifiers and tag questions (such as "Don't you think?") (Tannen, 1990). However, some scholars suggest that women might be using these patterns because it is expected of them, and that they use these

INTERVIEW: TECHNICAL WRITER RONDA WILSON

on the value of collaboration in making documents

Collaboration is important, just to ensure technical accuracy. I try to talk to as many people as possible, and sometimes I get different information. Maybe the electrical person told you one thing in the beginning, and the mechanical person told you another, and they clearly contradict, so you have to get to the bottom of it. Sometimes collaboration actually exposes a real problem with the equipment that has to be fixed.

on the challenge of working collaboratively

I think the biggest challenge is probably the writer's lack of technical knowledge. You're approaching people who are talking another language. In general, I do as much research as I possibly can before talking with subject-matter experts. A mechanical engineer is looking at you and saying, "I've got to explain something that's very technical and very complicated, and this person just isn't a technical person." The engineer either wants to give you the short easy answer that doesn't get you what you want, or wants to get too complex.

on the irony of this communication problem

The irony is, sometimes the person who is not an expert is closer in profile to the actual user of the equipment than the engineer is. The person who designed it is just not going to see things from a user's perspective, so writers without a technical background provide a useful point of view.

on ways to solve the communication problem

I've been to several training sessions on teamwork and collaboration, dealing with ways for different disciplines to see the other perspectives around them. But I really think common sense and politeness will take you most of the way. You try to establish a relationship with the different technical people.

on the value of politeness

These are busy people, so I try not to waste their time. I plan out what I'm going to ask them: this is the first question I'm going to ask, then this one, and we're going to go either from specific to general or from general to specific. Or I'm going to go over there with this diagram, or I'm going to give them a rough draft. Sometimes I will give them a list of questions and then set up an interview time. I think they appreciate a little advance notice, because sometimes you're asking them things that maybe aren't right at the tips of their tongues and they have to research a little bit, and it gives them a chance to do so. Also, they can do the research when they want, work that in whenever, take it home and look at it, and then set up an interview a day or two later.

on gender and collaboration

Most of the technical people are men, so there are times when I think that maybe I would do better if I weren't a woman. I guess, in general, gender does get involved, but once you've established a relationship with someone, gender becomes less important.

Ronda Wilson, a technical writer at SCP Global Technologies in Boise, Idaho, has twelve years' experience as a writer and editor for several engineering and high-tech companies.

patterns mainly in groups that include men (McMillan, Clifton, McGrath, & Gale, 1977). Many experts caution against using qualifiers and tag questions; for some listeners, qualifiers and tags suggest subservience and powerlessness.

The study of women and men in collaborative groups is a newer research subject, but as expected, women appear to value consensus and relationships, show more empathy, and demonstrate superior listening skills (Borisoff & Merrill,

(1987); men appear to be more competitive and more likely to assume leadership roles.

Although many questions remain unanswered, it seems clear that women's communication patterns are more focused on maintaining the group, and men's on completing the task. For instance, women talk more about topics unrelated to the task (Duin, Jorn, & DeBower, 1991). But this talk is central to maintaining group coherence, which is necessary for the group to accomplish its task.

Scholars of gender recommend that all professionals strive to achieve an androgynous mix of the skills and aptitudes commonly associated with women and men. According to William Eddy (1983):

> Traditional male traits of task focus, objectivity, confrontation, and control are clearly important in many situations. . . . But to build and lead groups that attain effectiveness and viability by fully utilizing their human resources, you also need some of the traits traditionally thought of as female. . . . It is not surprising, when you think about it, why an androgynous combination of skills is best.

For an excellent review of the literature, see Lay (1994).

Multiculturalism and Collaboration

Most collaborative groups in industry and in the classroom include people from other cultures. The challenge for all people in collaborative groups is to understand the ways that cultural differences can affect the behavior of the group.

Some 70 percent of the world's cultures value the family, the community, and the corporation more highly than the individual (Thiederman, 1991). As Bosley (1993) points out, this fact would suggest that people from these cultures would be excellent participants in collaborative projects. But this inference is not necessarily accurate.

Collaborative groups in the United States are based on the premise that the individual should speak up. But people from other cultures often find it difficult to assert themselves in collaborative groups in the United States. In a study of college students from other cultures in collaborative groups, Bosley (1993, p. 57) lists several behaviors that differ from those of students from the United States. Students from other cultures

- may be unwilling to respond with a definite "no"
- may be reluctant to admit when they are confused or to ask for clarification
- may avoid criticizing others
- may avoid initiating new tasks or performing creatively

Even the most benign gestures of friendship from a U.S. student can cause confusion. If a U.S. student casually asks a Japanese student about her major and the courses she is taking, the Japanese student might find the question too personal—

but find it perfectly appropriate to talk about her family and her religious beliefs (Lustig & Koester, 1993, p. 234).

This brief discussion of multiculturalism is meant only to point out the range and variety of cultural differences you are likely to encounter—or have encountered already. There is no substitute for immersing yourself in the culture you wish to learn about, but that is usually impractical. The best alternative is to try to remain open to encounters with people from other cultures, without jumping to conclusions about what their behaviors might mean.

A good first step is to read a full-length discussion of multiculturalism. The most respected scholar of multiculturalism in the United States is Edward T. Hall, whose major books include the following:

Beyond Culture. Garden City, NY: Anchor, 1976.

The Hidden Dimension. Garden City, NY: Doubleday, 1969.

Understanding Cultural Differences. Yarmouth, ME: Intercultural Press, 1990 (with Mildred Reed Hall).

For a detailed discussion of multiculturalism, see Chapter 5.

CASE

The Reluctant Collaborator

Some of the projects in your technical communication course are to be completed collaboratively. The instructor has assigned you to a group; the other members of the group are Allison and Ken. Allison is a senior in mechanical engineering; Ken is a junior in computer science. You are a senior in accounting. Along with your collaborative assignments, each member of the group is to submit the two forms presented in this chapter: the self-evaluation form and the evaluation form for the other members of the group. In addition, each member is invited to submit any notes, drafts, or other documents to show the instructor the work he or she contributed to the project.

At your first group meeting, you choose a leader: Allison. As a senior in mechanical engineering, she has had a lot of experience working collaboratively and is happy to take on the task of supervising your group's work. The first meeting, to discuss topics for a research proposal, is scheduled for 7 P.M. the next night in Conference Room 3 in the library.

It is 7:15, the night of the meeting, and Ken has not arrived. Neither Allison nor you had heard from Ken before the meeting with word that he would be late. At 7:30, you and Allison decide to phone Ken. There is no answer.

You and Allison are angry that Ken hasn't arrived or sent a friend to tell the group that he will be late or will have to cancel. The two of you are unwilling to start work on the project without Ken there. You decide to cancel the meeting. Allison tells you she will call him from home later that evening.

At about 9:00 P.M. you get a call from Allison. She reached Ken a few minutes before, and he said he had car trouble that night, and he apologized. Allison asks you if the group could meet the next afternoon at 3:00 in the library. Allison tells you that Ken agreed to her request that each group member bring a list of three possible topics to the meeting. You agree and write the time in your appointment book.

The next day, at 3:15, Ken rushes into the room. "Sorry I'm late," he says with a smile. You and Allison give him a cold look. Allison begins the discussion.

"Okay, we each agreed to bring three topics to the meeting today. Ken, what have you got?"

"That was for today? I'm really sorry. I just blew that off."

Allison turns to you, and you read off your three topics. Allison reads hers. The three of you agree to select one of Allison's topics: a feasibility study of software used in creating engineering drawings. You set up a time for the next meeting, as well as an agenda.

As you conclude the meeting, Allison asks you if you could stay a minute to talk with her. Ken leaves.

"I'm not happy with what I see developing with Ken," Allison says to you. "We're going to be slowed down if he doesn't come to the meetings and he doesn't prepare. What do you think we should do?"

In a memo to your instructor, respond to the following questions: How should you respond to Allison? Should you merely hope that Ken starts to participate more responsibly? Is there some way to delegate tasks that will motivate Ken to participate more actively? Should you and Allison go ahead with the project, letting Ken participate when he chooses to? Should Allison talk with him? Should both you and Allison talk with him together? Should you go to your instructor? See Chapter 16 for a discussion of memos.

WRITER'S CHECKLIST

1. Has your group
   - defined the group's task?
   - chosen a group leader?
   - defined tasks for the group members?
   - established working procedures?
   - established a work schedule?
   - created evaluation materials?
   - involved outsiders?

2. To reduce conflict in collaborating, do you
   - arrive on time and prepared to work?
   - meet your deadlines?
   - listen carefully?
   - ask pertinent questions?
   - state your views diplomatically?
   - avoid getting emotionally attached to your own ideas?
   - act supportively, not critically?
   - share the work space?
   - share your resources?

3. To improve your listening skills, do you
   - reduce distractions?
   - face the speaker?
   - maintain eye contact?
   - reduce irrelevant movements?
   - set aside your preconceptions?
   - listen to the ideas, not the words?
   - use your listening time productively?
   - paraphrase what the speaker has said?
   - refrain from interrupting?
   - ask open questions?

EXERCISES

Several of the following exercises call for you to write a memo. See Chapter 16 for a discussion of memos.

1. Interview a professor in your department who has published a cowritten article. Ask this professor to describe the collaborative techniques used in writing the article and to evaluate the strengths and weaknesses of these techniques. Write a memo to your instructor that briefly summarizes the professor's main points and then analyze the information you gathered. In what ways, and to what extent, do the professor's insights confirm the discussion in this chapter about collaboration? See Chapter 7 for a discussion of interviewing.

2. Interview a technical communicator or technical professional in your community who has participated in a collaborative project. Ask this person to describe the collaborative process and to evaluate the strengths and weaknesses of the process. Write a memo to your instructor that briefly summarizes the technical communicator's main points and then analyzes the information you gathered. In what ways, and to what extent, do the technical communicator's insights confirm the discussion in this chapter about collaboration? See Chapter 7 for a discussion of interviewing.

Collaborative Exercises

3. Form small groups. Each member of your group is to attend the same lecture or view the same interview show on television, such as *Newshour with Jim Lehrer* on the Public Broadcasting System or one of the Sunday morning news shows. Take notes. Then, after the lecture or television show, meet with the other members of the group to discuss what you learned. Concentrate on the following questions:
   • What were the central points established in the lecture or interview?
   • Were the points made clearly or did you have trouble understanding them?
   • What techniques might the speaker use to improve his or her ability to convey the information?
   • Do you think your gender or culture affected how you listened to the lecture or interview?
   • What suggestions can you make that might help you improve your listening skills?
   Write a memo to your instructor presenting your findings.

4. Form groups and perform a collaborative project such as the one described in exercise 3. Each of you will keep a log of your experiences, covering such topics as the following:
   • What procedures were used to assign tasks?
   • What procedures were used to establish policies for meetings and submitting work?
   • How helpful were your contributions—and those of the other group members—both in meetings and in submitting work?
   • How effective were you and the other members in establishing a cooperative and efficient environment for working?

   • Did conflicts arise and, if so, how did you and the other group members try to resolve the situation? How effective were these efforts? How might they have been improved?
   Meet as a group and discuss your logs. Then write a memo to your instructor presenting your findings.

Web Exercise

5. Form small groups for this project, which requires a basic understanding of the Web (see Chapter 4). Your college or university wishes to update its Web site to include a section called "For Prospective International Students." Along with members of your team, first determine whether your school already has information of particular interest to prospective international students. If it does, write a memo to your instructor describing and evaluating the information. Is it accurate? Comprehensive? Clear? Useful? What kinds of information should be added to the site to make it more effective?
   If the school's site does not have this information, perform the following tasks:
   • *Plan.* What kind of information should it include? Does some of this information already exist, or does it all have to be created from scratch? For example, can you create a link to information on how to obtain a student visa, or does this information not exist on the Web? Write an outline of the main topics that should be covered.
   • *Draft.* Write the following sections: "Where to Live on or Near Campus," "Social Activities On or Near Campus," and "If English Is Not Your Native Language." What graphics could you include? Are they already available? What other sites should you link to for these three sections?
   In a memo, present your suggestions to your instructor.

REFERENCES

Allen, N., Atkinson, D., Morgan, M., Moore, T., & Snow, C. (1987). What experienced collaborators say about collaborative writing. *Iowa State Journal of Business and Technical Communication, 1*(2), 70–90.

Bolton, R. (1979). *People skills.* New York: Simon & Schuster.

Borisoff, D., & Merrill, L. (1987). Teaching the college course in gender differences as barriers to conflict resolution. In L. B. Nadler, Nadler, M. K., & Todd-Mancillas, W. R. (Eds.), *Advances in gender and communication research* (pp. 351–361). Lanham, MD: University Press of America.

Bosley, D. (1993). Cross-cultural collaboration: Whose culture is it, anyway? *Technical Communication Quarterly, 2*(1), 51–62.

Chodorow, N. (1978). *The reproduction of mothering: Psychoanalysis and the sociology of gender.* Berkeley: University of California Press.

Couture, B., & Rymer, J. (1989). Interactive writing on the job: Definitions and implications of collaboration. In M. Kogen (Ed.), *Writing in the business professions* (pp. 73–93). Urbana, IL: National Council of Teachers of English.

Duin, A. H., Jorn, L. A., & DeBower, M. S. (1991). Collaborative writing—Courseware and telecommunications. In M. M. Lay & W. M. Karis (Eds.), *Collaborative writing in industry: Investigations in theory and practice* (pp. 146–169). Amityville, NY: Baywood.

Eddy, W. (1983). Qtd. in A. G. Sargent (1983), *The androgynous manager,* in J. Stewart (1990). *Bridges not walls: A book about interpersonal communication* (5th ed.) (pp. 274–282). New York: McGraw-Hill.

Ede, L., & Lunsford, A. (1990). *Singular texts/plural authors: Perspectives on collaborative writing.* Carbondale: Southern Illinois University Press.

Faigley, L., & Miller, T. P. (1982). What we learn from writing on the job. *College English, 44,* 557–569.

Hulbert, J. (1989). Barriers to effective listening. *Bulletin of the Association of Business Communication, 52*(2), 3–5.

Killingsworth, M. J., & Jones, B. G. (1989). Division of labor or integrated teams: A crux in the management of technical communication? *Technical Communication, 36*(3), 210–221.

Lay, M. M. (1994). The value of gender studies to professional communication research. *Journal of Business and Technical Communication, 8*(1), 58–90.

Lustig, M. W., & Koester, J. (1993). *Intercultural competence.* New York: HarperCollins.

McMillan, J. R., Clifton, A. K., McGrath, D., & Gale, W. S. (1977). Women's language: Uncertainty or interpersonal sensitivity and emotionality? *Sex Roles, 3,* 545–549.

Tannen, D. (1990). *You just don't understand.* New York: William Morrow.

Thiederman, S. (1991). *Profiting in America's multicultural marketplace.* New York: Macmillan.

Using Electronic Communication Tools

Drafting Tools on the Word Processor 69
  The Outline Feature 69
  Templates and Styles 69
  Automated Hypertext Linking and Web-Conversion Tools 72
Revision Tools on the Word Processor 72
  Spell Checkers 72
  Grammar Checkers 73
  Thesaurus Software 73
Working Collaboratively Using Electronic Tools 74
  Electronic Mail 74
  Groupware 77
  Other Electronic Tools 80
Using Web Pages 81
  Understanding Hypertext 82
  How to Use Web Pages 82
  Ethics, Copyright Law, and the Web 86
Creating Web Pages 88
  Analyze Your Audience and Purpose 90
  Gather or Create Your Text and Other Information 90
  Create a Site Design 90
  Create a Page Design 92
  Convert the Text and Graphics 93
  Revise and Test the Information 94
  Get the Files to an Internet Server 95
  Register the Site with the Search Engines 95
The Future of the Web 96
Case: Creating a Web Site 96
Writer's Checklist 97
Exercises 97
  Collaborative Exercises 98
  Web Exercises 99

Electronic communication tools have become fundamentally important to the working lives of technical professionals and technical communicators. The use of personal computers on intranets (company networks) and the World Wide Web have radically changed the way people communicate in organizations.

For example, engineers subscribe to different discussion lists and newsgroups on the Internet to keep up with new developments in their fields. They use project-management software to plan and monitor their projects, and they use spreadsheets to calculate the financial implications of different business scenarios. They keep in touch with their colleagues and their clients through electronic mail (e-mail) and send data files across the country and around the world using FTP (file transfer protocol).

The extent to which professionals rely on electronic communication tools is best seen when there is a power failure in the building: people wander out into the halls, for there is very little work they can accomplish.

We all need to understand how these tools work and to be able to use them effectively. This chapter covers six aspects of electronic communication tools, as shown in Figure 4.1.

One of your challenges as a professional will be to keep up with the tools used in your field. To do this effectively, you need to determine your own learning

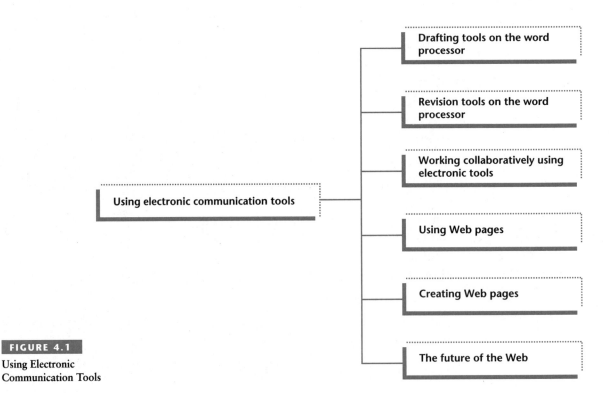

FIGURE 4.1

Using Electronic
Communication Tools

style. Some people learn best in a structured course, such as a short training class. Others prefer detailed how-to books. Still others like to search the Internet or use the online documentation that is part of the tool itself. And some people like to experiment with the tool, learning as they play. Once you have determined how you like to learn, you can pursue the necessary training.

This chapter will not teach you to use a particular tool. For instance, there are no step-by-step instructions on how to use the Alta Vista search engine on the Web. There are too many tools to address, and the software is always changing. Rather, this chapter will provide an overview of some of the major tools, concentrating on the kinds of tasks they make possible and how to use them wisely. The rest of this book also contains discussions of these tools.

Drafting Tools on the Word Processor

Many chapters in this book mention the central role played by word processing in technical communication. Probably you are already familiar with the ways that word-processing software makes your job as a writer easier. You can draft quickly and revise easily. You can move chunks of text around to achieve the most effective organization of your document. And you can copy text from one document to another.

This section discusses how to use three more advanced tools as you draft:

- the outline feature
- templates and styles
- automated hypertext linking and Web-conversion tools

The Outline Feature   Word-processing software lets you look at your document in different ways: as continuous text or as separate pages, for instance. One way to look at your text is through the outline feature. As you plan your text, you might find it helpful to use the outline feature to move a heading up a level or down a level. As you draft your text, you can use the outline feature to shift your perspective on what you have written. For instance, you can view only the first-level headings, or the first- and second-level headings, or all the levels of headings. Being able to move from a close-up view of the document to a bird's-eye view helps you stay on track by making it easy to see the overall structure without having to scroll through the whole document.

Templates and Styles   *Templates* are preformatted designs for different kinds of documents, such as business letters, memos, newsletters, and reports. *Styles* are like small templates; they apply to the design of smaller elements of documents, such as headings and body text.

Templates incorporate all the different design specifications for the document, including typeface, type size, margins, and spacing. Once you select a template, you just type in the information; the document is already designed. Figure 4.2 (Microsoft, 1996) shows a template for a résumé.

Keep in mind, however, that there are three major drawbacks to using templates:

- *Templates encourage you to use a "one size fits all" approach to your writing.* For example, the résumé template in Figure 4.2 might be appropriate for someone with twenty years' experience in business, but it isn't for someone seeking an entry-level professional position. For one thing, the template emphasizes the job-experience category by placing it above the education category. In addition, the template provides little detail in the education section. The point is not that the template is bad or wrong; the point is that it might not be effective for a particular person applying for a particular position. It all depends on the context—your audience and purpose, and the information you want to present. Although it is true that you can modify a template any way you wish, its design and content encourage you to modify your strategy to conform with it. Good writers think through the context, then decide how to write the document.
- *Templates can become boring through overuse.* Many thousands of job applicants are using résumé templates included with the popular word-processing software. The hiring officers at large companies are likely to be unimpressed.
- *Templates sometimes violate guidelines for effective design.* One of the report templates in a popular word processor shows table titles placed beneath the tables, whereas table titles are generally placed above.

The more you rely on templates, the less likely you are to learn how to use the software. If you will be creating more than a couple of memos on the job, for instance, you really ought to invest the time to learn how to create them using the full features of the software so that you can create sophisticated and individual documents when the need arises. Put aside some time to learn the software.

Like templates, styles save time and help ensure consistent formatting throughout a document. They save time in that, as you draft your document, you don't have to stop and add all the formatting each time you want to designate an item as a first-level heading, for example. After you finish drafting, you put your cursor in the text that you want to be a first-level heading and, using a pull-down menu, select that style. The text automatically incorporates all the different specifications of a first-level heading; you don't have to change the typeface, then change the size, then change the spacing, and so forth. Naturally, automating all these specifications helps ensure consistency, for you don't have to change four or five technical specifications every time you want to change the style of an item. And

[Click **here** and type address 1]   [Click **here** and type address 2]

Max Benson

Objective   [Click **here** and type objective]

Experience   1990–1994   Arbor Shoe   Southridge, SC
                    National Sales Manager
- Increased sales from $50 million to $100 million.
- Doubled sales per representative from $5 million to $10 million.
- Suggested new products that increased earnings by 23%.

                    1985–1990   Ferguson and Bardell   Southridge, SC
                    District Sales Manager
- Increased regional sales from $25 million to $350 million.
- Managed 250 sales representatives in 10 Western states.
- Implemented training course for new recruits — speeding profitability.

                    1980–1984   Duffy Vineyards   Southridge, SC
                    Senior Sales Representative
- Expanded sales team from 50 to 100 representatives.
- Tripled division revenues for each sales associate.
- Expanded sales to include mass market accounts.

                    1975–1980   LitWare, Inc.   Southridge, SC
                    Sales Representative
- Expanded territorial sales by 400%.
- Received company's highest sales award four years in a row.
- Developed Excellence in Sales training course.

Education   1971–1975   Southridge State University   Southridge, SC
- B.A., Business Administration and Computer Science.
- Graduated Summa Cum Laude.

Interests   SR Board of Directors, running, gardening, carpentry, computers.

Résumé Tips   Select the text you would like to replace, and type your résumé.

FIGURE 4.2

Résumé Template

if you decide to change the specifications of a style—by adding italics to a heading, for instance—you just change it once; the software automatically changes each instance of the style in the document. Styles are particularly useful as you create collaborative documents, for they make it easier for the different collaborators to achieve a consistent look in the document.

However, just as you have to be cautious when you use templates, you have to think carefully before you accept the default characteristics of a style. For instance, the defaults for different levels of headings in several of the major word-processing programs do not follow accepted guidelines for readability. (See Chapters 13 and 14 for more information on readability.)

Automated Hypertext Linking and Web-Conversion Tools

Word-processing software today is designed to make it easier to print documents in the traditional way or to put them on the Web. Automated hypertext linking and Web-conversion tools are two functions that help you cut down on the time needed to produce documents for the two media.

Automated hypertext linking is a function that recognizes Web addresses (see the discussion later in this chapter) and formats them so that they are active links when you put the document on the Web. This way, you don't have to add the coding to make the link active.

A Web-conversion function is a feature that automatically adds *hypertext codes* (see the discussion later in the chapter) when you save the document. Just as you can save a word-processed document as ASCII text or as a Word or WordPerfect file, now you can save it as a hypertext document, so that when you want to put it on the Web *most* of the work is done—the Web-conversion tools still cannot handle every kind of design or effect you wish to create. You still have to learn hypertext coding if you want to create a highly complex Web page.

Revision Tools on the Word Processor

Just as word processors include tools you can use while you draft, they also include revising tools. These tools, like the drafting tools, should be used carefully. The following section discusses three major tools:

- spell checkers
- grammar checkers
- thesaurus software

Spell Checkers

A *spell checker* alerts you when it sees a word that isn't in its dictionary. That word might be misspelled, or it might be a correctly spelled word that isn't in the dictionary. You can add the word to your dictionary so that the spell checker will recognize it in the future.

Some spell checkers work "on the fly"; that is, as soon as you have typed a word that is not in the dictionary, they highlight it or make a beeping sound. I recommend you turn off this function so that you can concentrate on drafting, then fix the spelling later, as you revise.

Regardless of whether the spell checker works automatically, you are probably aware of the danger of relying on it: a spell checker cannot tell whether you have used the correct word; it can tell you only whether the word you have used is in its dictionary. Therefore, if you have typed "We need too dozen test tubes," the spell checker will not see a problem.

Grammar Checkers   *Grammar checkers* can help you identify and fix potential grammatical and stylistic problems, such as wordiness, subject/verb agreement, and double negatives. Many grammar checkers identify abstract words and suggest more specific ones or point out sexist terms and provide nonsexist alternatives. Figure 4.3 shows a screen from a popular grammar checker.

Sentence:

> Cognitive psychologists have known for some decades (see Samuels and Eisenberg [1981] for a summary) that comprehension is determined by internal and external factors.

Suggestions:

> Consider **decade** instead of **decades.**
> --------------------OR--------------------
> Consider **are** instead of **is.**

FIGURE 4.3

Advice from a Grammar
Checker

As you can see from Figure 4.3, grammar checkers are frequently wrong. There is nothing wrong with *decades* in the original sentence, and both of the suggestions recommended by the software would make the sentence incorrect. Many writing teachers advise that unless you are a capable and experienced writer who knows when to ignore bad advice, you shouldn't use a grammar checker at all.

Thesaurus   A *thesaurus* lists synonyms or related words. An electronic thesaurus program
Software   has the same strengths and weaknesses as a printed one: if you know the word you are looking for, but can't quite think of it, the thesaurus will help you remember it. But the listed terms might not be related to the key term closely enough to function as synonyms. Unless you are aware of the shades of difference, however, you might be tempted to substitute an inappropriate word. For example, the entries for the word *famous* in one electronic thesaurus include *infamous* and

notorious. If you use either of them as a synonym for *famous,* you could embarrass yourself badly.

The revision tools discussed in this section cannot replace a careful reading by you and by other people. These tools don't understand your subject, your audience, and your purpose. They cannot identify unclear explanations, contradictions, inaccurate data, inappropriate tone, and so forth. Use the tools, but don't rely on them. Revise your document yourself, then get help from someone you trust.

Working Collaboratively Using Electronic Tools

As discussed in Chapter 3, technical communication is often produced collaboratively. Electronic communication tools are becoming more and more useful in helping people in the same location and at different sites work together effectively.

These tools are changing rapidly. Regardless of what form these tools take, they all allow you to share documents on an intranet or the Internet.

This section discusses three topics, as shown in Figure 4.4.

Electronic Mail   Electronic mail (e-mail) is a tool for sending textual messages on a computer network. The Congressional Office of Technology Assessment has estimated that by the year 2000, some 140 billion pieces of mail—two-thirds of the nation's total—will be sent by e-mail, not by the Postal Service. E-mail is the most basic electronic tool for working collaboratively because it allows people to send a text file back and forth on a private network or the Internet. Some 50 million people use e-mail (Ellsworth & Ellsworth, 1996).

E-mail offers four chief advantages over the interoffice delivery of paper documents or the U.S. Postal Service.

- *E-mail is fast.* On a local area network (LAN) such as an intranet, delivery usually takes less than a second. On the Internet, e-mail can travel halfway around the world in a few seconds.

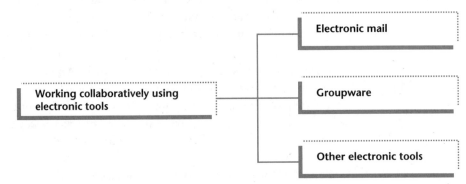

FIGURE 4.4

Working Collaboratively
Using Electronic Tools

- *E-mail is cheap.* Once the network is in place, it doesn't cost anything to send a message to one person or a thousand. You don't pay per copy, as you do with photocopies, because you're not using any paper.

- *E-mail is easy to use.* Once you learn how to use your particular e-mail software, it is easy to send mail either to one person or to a group. You create an address book with the names and addresses of the people or groups you write to, then assign each one a nickname. Then, if you want to send the message to, say, a group of coworkers scattered across the country, you just type in the group's nickname, such as "Salesforce," and the message automatically goes to each person in the group. The message sits in the mailbox until the recipient accesses the message and reads it. It is also simple to respond to an e-mail message or to forward it to a third party—or many third parties.

- *E-mail is digital.* E-mail can be read and erased or printed out, but it can also be stored like any other electronic file. Therefore, it can be used in other documents. For this reason, e-mail is a convenient way for people in different places to collaborate.

Because of these advantages, e-mail has replaced many interoffice memos as well as some phone calls. Keep in mind, however, that there will always be a place for hard-copy interoffice mail, for two reasons:

- In some environments, people do not have access to an intranet or the Internet or do not know how to use it.

- For legal reasons, people sometimes need original hard copies, on letterhead stationery or memo forms, complete with signatures.

Guidelines for Writing E-Mail

As you write e-mail, follow two guidelines:

- *Use appropriate formality.* In some organizations, managers expect e-mail messages to be as formal as paper-based documents; in other organizations, managers expect them or even want them to be quite informal. Learn the expectations in your organization. If the e-mail messages you read sound just like memos or letters, you know that the company considers e-mail as a practical way to communicate relatively formal messages. However, if you see writers using *emoticons,* the most popular of which is the smiley face : -) , at the end of a comment that is meant to be taken ironically or as a joke, or using abbreviations (such as *BTW* for *by the way*), you know that the organization sees e-mail not only as a way to communicate but also as a way to foster group cohesiveness. Before you begin to send e-mail messages, read those written by other group members.

Be sure not to confuse informality with inarticulateness. Even when e-mail is used as an informal medium, unclear writing is never acceptable. Send your message only when it is clear. Keep revising it until it says what you want it to, just as you would with any other kind of communication.

- *Realize that e-mail, like print, is permanent.* Many people do not realize that most networks are archived. That is, all the activity on the network is backed up on some kind of tape or disk system. Therefore, the message that you send to your colleague is probably stored somewhere, even if the recipient has deleted it. For this reason, do not write in an e-mail message anything that you would not write in a letter or memo. A simple test: would you be unhappy if your message inadvertently were sent to the entire company or appeared in tomorrow's newspaper? Businesses exist that know how to search computer archives to retrieve "erased" e-mail messages and other kinds of computer files.

E-mail is being treated just like other forms of written communication in court cases. According to the 1986 Electronic Communications Privacy Act, the organization that established the e-mail network may look at all e-mail traffic on that network without violating the privacy rights of employees using it (Crowe, 1994, p. 31).

Netiquette

To use e-mail effectively, you should know the basics of *netiquette*—etiquette on a network. These seven guidelines will help you communicate effectively and appropriately.

- *Don't waste bandwidth.* Keep the message brief so that it doesn't clog up the network or the recipient's in-basket. When you reply to another e-mail, don't quote long passages from it. Instead, establish the context of the original e-mail by paraphrasing it briefly or by including a short quotation from the original e-mail. (When you excerpt a small portion of an e-mail, add a phrase such as <snip> at the start and the end of the quotation.) When you do quote, delete the routing information from the top, as well as the signature block from the bottom. And make sure that everyone who is to receive a copy of your e-mail really needs to read it.

- *Take some care with your writing.* E-mail is informal, but messages shouldn't be sloppy. Although text-editing functions on many e-mail systems are much more limited than on a word processor, don't embarrass yourself by sending messages that you haven't at least proofread.

- *Don't flame.* To flame is to scorch a reader with scathing criticism, usually in response to something that person has said in a previous message. Flaming is rude. When you are really angry, keep your hands away from the keyboard.

- *Use the subject line.* Readers like to be able to decide whether they want to read the message. The subject line that appears on each incoming message helps them decide.

- *Make your message easy on the eyes.* Use uppercase and lowercase letters as you do in other forms of correspondence, and skip lines between paragraphs. Don't use italics or underlining; they will appear as bizarre characters. Instead, use uppercase letters for emphasis. Keep the line length to under 65 characters so that your lines do not get broken up if your reader has a smaller screen.

- *Don't repost a message without the writer's permission.* When you receive a message, you shouldn't post it to another person without the permission of the writer. Such a posting might be illegal (the courts haven't decided yet), but it is certainly unethical.

- *Don't send a message unless you have something to say.* Resist the temptation to write a message that says, in effect, that you agree with another message. If you can add something new, fine, but don't send a message just to be part of the conversation.

Figure 4.5 on page 78 shows an e-mail that violates some of these guidelines. The writer is a technical professional working for a microchip manufacturer. Figure 4.6 on page 79 is a revised version of this e-mail message.

Groupware   Groupware, a category of software that helps people work together, lets people at the same location or at different locations plan, draft, revise, and track a document. For instance, a complicated project at a company might require the participation of engineers, marketing specialists, accountants, technical communicators, and legal staff. Groupware makes this kind of collaboration much more convenient than it has been in the past.

There is no single definition of groupware, and some commentators include e-mail in their definitions. I distinguish e-mail from groupware, however, in two major ways:

- *E-mail is limited to text.* E-mail can accommodate only ASCII text, that is, only letters and numbers and a limited number of special characters. You cannot use the full range of special characters. However, many e-mail programs let you "attach" formatted files such as word-processing or graphics files.

The writer does not use the subject line.

The writer does not make clear in the first paragraph what his purpose is.

The writer has not double-spaced between paragraphs to make the e-mail easier to read and has used a long line length that will be broken up on some e-mail programs.

The writer has not proofread.

The writer slips into the passive voice (see Chapter 14) when giving instructions to his readers.

The writer uses uppercase letters and an accusatory tone throughout the e-mail. Obviously, he is angry.

To: Supers and Leads
Subject:

   LATELY, WE HAVE BEEN MISSING LASER REPAIR FILES FOR OUR 4MEG WAFERS. AFTER BRIEF INVESTIGATION, I HAVE FOUND THE MAIN REASON FOR THE MISSING DATA.
   OCCASIONALLY, SOME OF YOU HAVE WRONGLY PROBED THE WAFERS UNDER THE CORRELATE STEP AND THE DATA IS THEN COPIED INTO THE NONPROD STEP USING THE QTR PROGRAM. THIS IS REALLY STUPID. WHEN DATE IS COPIED THIS WAY THE REPAIR DATA IS NOT COPIED, IT REMAINS UNDER THE CORRELATE STEP.
   TO AVOID THIS PROBLEM, FIRST PROBE THE WAFERS THE RIGHT WAY. IF A WAFER MUST BE PROBED UNDER A DIFFERENT STEP, THE WAFER IN THE CHANGE FILE MUST BE RENAMED TO THE *.* FORMAT.
   EDITING THE WAFER DATA FILE SHOULD BE USED ONLY AS A LAST RESORT. IF THIS BECOMES A COMMON PROBLEM, WE COULD HAVE MORE PROBLEMS WITH INVALID DATA THAT THERE ARE NOW.
   SUPERS AND LEADS: PLEASE PASS THIS INFORMATION ALONG TO THOSE WHO NEED TO KNOW.

ROGER VANDENHEUVAL

FIGURE 4.5

E-Mail Message That Violates Netiquette Guidelines

- *E-mail is asynchronous.* E-mail doesn't let two people look at the same document at the same time. One person sends it to another person, who can revise it, then send it back. But the two people are not communicating in real time.

Groupware refers to a broad range of software. Some products perform limited, specialized tasks, such as sending electronic "post-it" notes to someone's computer; the software can even be set so that the note breaks through the recipient's screen saver and appears on the screen. Other products are highly versatile and complex. One lets an organization create an electronic database of all kinds of files—documents, spreadsheets, graphics—that any authorized user at any location can access, download, or revise. The software keeps a complete record of all actions that have occurred and instantly implements any changes made by any user.

You may already be familiar with one kind of groupware: collaborative-writing programs such as Lotus Notes™, Daedalus Integrated Writing Environment™, or CommonSpace™. Different programs offer different features, yet many programs permit synchronous discussion; that is, different people can work on the document at the same time by typing at their keyboards, and everyone's screen

The writer uses the subject line effectively.

The first paragraph provides a clear statement of the writer's purpose. Note too that the writer is using uppercase and lowercase letters.

The writer double-spaces between paragraphs to aid in legibility and uses a short line length that will not be broken up by other e-mail programs.

The writer has removed the errors and has diagnosed the problem more diplomatically.

The correct method is stated more clearly.

The writer promises to provide copies of the specification.

The writer concludes politely.

> **To:**   Supers and Leads
> **Subject:**   Missing Laser Repair Files for 4meg Wafers
>
> Supers and Leads:
>
> Lately, we have been missing laser repair files for our 4meg wafers. In this e-mail I want to briefly describe the problem and recommend a method for solving it.
>
> Here is what I think is happening. Some of the wafers have been probed under the correlate step; this method copies the data into the Nonprod step and leaves the repair data uncopied. It remains under the correlate step.
>
> To prevent this problem, please use the probing method outlined in Spec 344-012. If a wafer must be probed using a different method, rename the wafer in the CHANGE file to the *.* format. Edit the wafer data file only as a last resort.
>
> I'm sending along copies of Spec 344-012. Would you please pass this e-mail and the spec to all of your operators.
>
> Thanks. Please get in touch with me if you have any questions.
>
> Roger Vandenheuval

FIGURE 4.6

E-Mail That Adheres to Netiquette Guidelines

shows the dialogue. In most groupware programs, the computer screen is divided into two parts. The top half of the screen displays the ongoing discussion, with each comment identified by speaker; the bottom half of the screen is the area where the writer drafts his or her own comments. When the writer wishes to add these comments to the group discussion, he or she presses a "send" button.

Many groupware functions are now incorporated in office productivity suites. These suites, which combine common business applications such as word processing, presentation graphics, spreadsheets, and databases, are making it easier to work collaboratively. The groupware functions on these suites allow collaboration across local networks or on the World Wide Web. However, there are still compatibility problems: one company's product doesn't necessarily "understand" another's.

Figure 4.7 on page 80 shows two screens from a groupware product called DOORS, manufactured by Quality Systems & Software (1996). The first screen shows the form on which the user makes a change in a company document stored in a database. The second screen shows the change implemented in section 3.1.4

a. Form on which changes are entered

b. The changes are implemented

Screens from a Group-
ware Program

of the document. The description of the DOORS software from the company's
Web site includes the following explanation of this feature:

> DOORS keeps an exact record of all changes to modules, and to objects such as re-
> quirements stored in each. Each object has a traffic-signal colored change bar just to
> the left of its text: green if the text is as it was in the last baseline, yellow if it changed
> but has been saved, and red if it has changed this session since the last save. Clicking
> on the change bar pops up the history window. Previous versions of the requirement
> can be read and if desired copied.

Other Electronic   In a cover story from late 1996, *PC Magazine* (Hill, Ozer, & Mace, 1996, p. 103)
Tools   describes a business scenario that could actually take place with existing communication tools:

> Your PC alerts you to an incoming call; it's a colleague phoning via Internet from Mex-
> ico City. He's been coediting a memo with the CPA in Atlanta and wants you to take a
> look. A couple of mouse clicks later, you're both staring at it. You and your colleague
> continue to talk as you make a few changes—still through an Internet connection.
>
>   You realize that you'll need legal advice on your side to push this through. You
> opt for a videoconference—now through the corporate intranet. A few seconds later,
> your in-house counsel appears on your screen, and you begin to talk things over. After-
> wards, you reconnect with your colleague, finish the memo, and transfer a copy to
> everyone involved.

While few companies had all these communication tools in place in 1997, the
number of organizations using them now is increasing at a rapid rate. Among
the most popular communication tools are the following:

- *FTP (file transfer protocol).* As discussed in Chapter 7, FTP has for some years been a major research tool, for it allows you to download files that are stored on other computers on the Internet. As a collaborative tool, FTP is a way for authors to transfer files back and forth across the Internet.

- *Telnet.* Also called *remote log-in,* Telnet is a tool that lets you connect your computer to another one, using your terminal as if it were connected to the remote computer. As mentioned in Chapter 7, Telnet is a popular research tool. For collaborating authors, Telnet lets one person use the databases and software stored on a colleague's computer. In addition, Telnet lets you log onto your own desktop computer from home or on the road.

- *Whiteboards.* Whiteboard software lets people at different locations draw on the screen, as if they were all in a room with a whiteboard. Anything drawn on one screen is displayed immediately on every person's screen. The image can be printed or saved as a file. Whiteboard software is used often by engineers, who create and revise drawings as they think through their designs.

- *Videoconferencing.* Standard videocams can be connected to computers, enabling people to see each other as they talk. Videoconferencing is a valuable tool for helping people who do not know each other well, for vision and words together make for a much richer communication medium than words alone. Through the visual contact afforded by videoconferencing, people can create a better working relationship that helps make the collaboration productive.

Using Web Pages

Chapter 7 discusses how to use the Internet in research using e-mail, Telnet, FTP, discussion lists, and newsgroups. This section discusses one portion of the Internet: the World Wide Web (or simply, the Web).

Since its inception in 1991, the World Wide Web has grown at a pace that has far outstripped everyone's predictions. The Web is now a gigantic publishing medium, made up of millions of Web "pages" housed on hundreds of thousands of computer servers all around the world. Using a program called a *browser,* people can access multimedia files. Nobody has precise figures about Web usage because it changes too quickly and because there is no Internet headquarters. An informed guess (Ellsworth & Ellsworth, 1996) is that in 1996 some 25 million people were Web users.

Who makes Web pages? Everyone, it seems. Governments, research institutions, publishers, interest groups, all kinds of large and small businesses, and probably a number of students in your technical communication class. In fact, some college courses require that students make their own Web sites. All you need to have a basic, primarily text-based site on the Web is about one megabyte of storage space on an Internet server; many Internet service providers make this space available as part of the contract when you acquire an e-mail account.

Much of the material on the Web is extraordinarily interesting and valuable: you can view photographs from NASA's latest trip to Saturn, visit major art museums, read e-zines (electronic magazines), get up-to-the-minute stock quotes, and witness scientific experiments carried out by means of virtual reality. Many of the world's greatest minds are using the Web to publish their research and creative works.

However, because the Web is not supervised in any consistent way, it contains material that is biased, inaccurate, and sometimes even silly. Is the world a significantly richer place because there is a Web site with audio clips of people popping many different varieties of bubblewrap? The Web also seems to be the favorite medium for spreading rumors. Microsoft has not in fact purchased the Vatican, Snapple does not contribute money to the Ku Klux Klan, and the President is not a cyborg. Spending time on the Web requires sifting through many bad sites to get to the good ones. But that is the price we all pay for an open publishing medium in which millions of people have a voice.

Because the Web is already so important as a means of publishing significant new information, professional people need to understand how to use the Web. In addition, they should understand the basic process of creating a Web site. The following discussion of using and creating Web sites begins with a brief explanation of hypertext, which is the central concept underlying communication on the Web.

Understanding Hypertext

Chapter 20 contains a discussion of hypertext as it is used in instructional training software and information booths. In understanding how hypertext applies in discussing the Web, you need to know only that it refers to linking. When you look at a Web site, you see links: words and phrases that are underlined or that appear in a different color than the rest of the text. Several kinds of elements—words, buttons, or graphics—can serve as links. When you click on a link, you go to another file on your site, another file on another site, or sometimes to a different place in the document you are already in.

Figure 4.8 shows the basic structure of a simple Web site. Almost all sites consist of a home page—the point of entry to the site—as well as other pages. A page refers to a file; this file might fit on one screen or extend to a number of screens.

How to Use Web Pages

The Web or some variant of it will almost certainly be the major medium for publishing technical information in our working lifetimes. To access the Web,

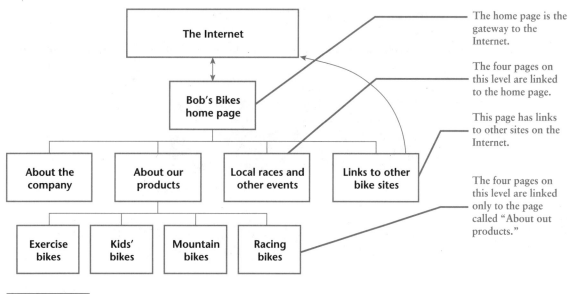

The home page is the gateway to the Internet.

The four pages on this level are linked to the home page.

This page has links to other sites on the Internet.

The four pages on this level are linked only to the page called "About out products."

FIGURE 4.8

Structure of a Basic Web Site

you need a connection to the Internet through your university, a local Internet service provider, or a national company such as America Online or CompuServe. Your university is likely to offer the fastest connection; dialing in to a local Internet service provider or a national company requires a modem, which slows down the transmission rate.

You will also need a *browser*, by which you connect to Web sites. Your university already has the browser installed; if you link to the Web through a service provider, it will supply the browser. At this time, most people use one of two major browsers: Netscape Navigator (http://home.netscape.com) and Microsoft Internet Explorer (http://www.msn.com). Some of the major word-processing programs also function as browsers.

Browsers differ somewhat in how they work, but essentially they all perform the same function. They all help you get to the sites you want to visit, then print files or save them to your own disk. Many browsers also let you use FTP and Telnet, connect to Usenet newsgroups (see Chapter 7), and send e-mail. Figure 4.9 on page 84 shows the home page of the Netscape Navigator site.

There are two basic ways to use a browser to find Web sites:

- *Enter the address of the site you wish to visit.* Every site has a unique address, called a *Uniform Resource Locator* (URL). If you know the URL of the site you want, you type it in the appropriate place in your browser.

The "Netsite" shows you the address—the URL (Uniform Resource Locator)—of the page you are viewing.

Pressing "Net Search" brings you to the search engines.

FIGURE 4.9

The Netscape Navigator Home Page

- *Use a* search engine *to find a site to visit.* Common search engines include Lycos, Alta Vista, and Excite. If you know the name of the organization or person but not the URL, you enter a keyword or phrase. For example, if you want to find the site for the Centers for Disease Control, you enter the key phrase "centers disease control." In this case, you are using the search engine like a book index. If you want to search the Web by subject area, you can choose one of the search engines that classifies sites by subject. For instance, you want to find companies that produce groupware products. You select the category "business," then the subcategory "computers," then the subsubcategory "groupware." When you have clicked down to a sufficiently specific category, you ask the search engine to display the results: a list of links to companies that sell groupware.

Using search engines can be frustrating, for each one works differently. Some search engines pick up keywords from a site's title; others pick up every word on every page in the site. Some search engines provide evaluative comments about the site; others do not. Often you will find that your search yields hundreds or even thousands of "hits," a situation that is hardly any better than a search that yields none. Most search engines include directions on how to narrow the search; read this information carefully, for it can save you a lot of frustration.

Figure 4.10 shows a portion of the home page of the search engine Yahoo!, on which you enter the keywords for your search. Notice the subject listings that can be used for a search.

Figure 4.11 shows the first page of the results of a search for the term "centers disease control."

Here you have selected the search engine Yahoo!.

You type in your keywords here, then press "Search."

Yahoo! lets you select a general subject area before you search.

FIGURE 4.10

The Yahoo! Home Page

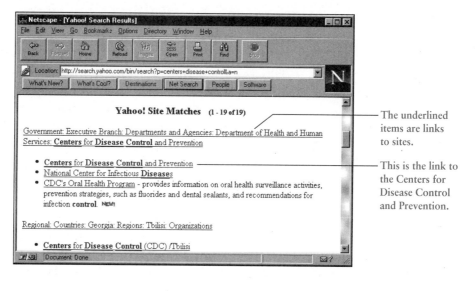

The underlined items are links to sites.

This is the link to the Centers for Disease Control and Prevention.

FIGURE 4.11

First Screen of the Results of a Search for "centers disease control"

When you click one of the sites on the search-results page, you go to the home page for the site, as shown in Figure 4.12 on page 86.

A quick look at Figure 4.12 shows some of the common functions you can perform from most browsers:

- *Navigating.* The back button returns you to the site you looked at most recently. After you move backward, you can use the forward button, which moves you forward one site at a time.

- *Printing.* You can print a copy of the site you are viewing.

The Bookmarks menu.

The Print button.

The File menu.

The Back and Forward buttons.

The Centers for Disease Control and Prevention Home Page

- *Saving.* If you want to save the page to a floppy disk or your hard drive, pull down the File menu and select "Save As." You type an address, such as C:\info.cdc, and the text from the page is saved to your C drive with the file name "info.cdc." You can save the page as plain text or as HTML-coded text. (See the following section on ethics and copyright for a discussion of your ethical and legal obligations; see "Convert the Text and Graphics" later in this chapter for information on HTML.)

- *Adding a bookmark.* If you think you might want to visit this site again, pull down the Bookmark menu, and add a *bookmark*. A bookmark, as the name suggests, lets you get back to the site without having to remember its URL or having to search for it.

Ethics, Copyright Law, and the Web

The Web is unlike other publishing media in that the content is remarkably easy to steal. Users can download digital versions of the text and graphics and manipulate them any way they want. An unethical user can take graphics and republish them without permission from the person who owns the copyright. Or a user can modify text from a Web site and claim it as his or her own. These practices are illegal as well as unethical, for digital material is covered by the same copyright protection that applies to printed material. All text and graphics on the Web are protected, regardless of whether the person who put the material on the site included a copyright symbol. Unless the copyright owner specifically says that you may use the material, you must seek written permission, just as you would for printed material.

However, saying that Web sites are protected just like other kinds of documents is just the beginning of the discussion, not the end. Benedict O'Mahoney

(1997) has written a thoughtful essay on some of the complicated issues involved in interpreting copyright law. Here are just five of the puzzles he addresses:

- *Is viewing a Web page the same as copying it?* If you do not save the page to a disk, probably not, although some courts have held that merely viewing a page entails copying the source code from the server to your own computer's RAM. If you save a page to a disk, you probably are copying it. Is this legal? The answer depends on how you interpret the fair use doctrine (see Chapter 2).

- *Is the design of a Web page protected by copyright law?* Some would say no, for what viewers see on the screen is a function of their hardware and software. In addition, viewers can customize the image. However, the design of a Web page is an original work, created by a person, and thus should be protected, regardless of how viewers might change it after it is transmitted.

- *Are lists of links protected?* Many Web sites contain lists of links to other sites. Is each link protected by copyright? No. Is the whole list of links protected? Probably, if the person showed some originality in creating the list. Courts have held that phone books are not protected, for alphabetized lists of names, addresses, and phone numbers do not show originality. But a list of links to resources for agriculture students, for example, probably is protected, because the author of the list had to do some original thinking in creating categories in which to place the individual links. Therefore, copying some other site's list of links and putting it on your own site is probably illegal (and certainly unethical).

- *May you link to anyone else's Web site? May anyone link to yours?* The Web was initially envisioned as an open environment in which anyone can link to anyone else. In theory, this is a sensible idea. But what if some organization doesn't want you to link to its site, because it doesn't want the extra traffic on its server? Or what if you discover that a hate group has written a complimentary paragraph about you and linked to your site? Is it your responsibility to find out who has linked to you, or should the site that wishes to link to you notify you and get permission to do so? Netiquette (and common courtesy) would suggest that you notify an organization if you wish to link to its site and that you abide by its wishes: don't link if the organization says no. If the organization welcomes the link to its site, acknowledge its courtesy on your own site, and be sure to link to the home page (not to a second- or third-level page) so that viewers will be coming into that site through the front door.

- *Are composite Web pages legal?* A composite Web page is a page made up of links to elements on other organizations' sites. For example, a composite page can consist of links to graphics on one organization's site, sound samples on a second organization's site, and text on a third organization's. Common sense would suggest that this composite Web page is illegal; unless formal permission has been obtained for each of these elements, it is a clear violation of copyright. But the fair-use provision of the copyright law permits copying for some purposes, such as commentary or education, within some limits. Therefore, some composite pages might be legal under some interpretations.

As this brief discussion suggests, questions of ethics and legality are extremely complex and will remain unresolved for many years. Over the next decade the courts will doubtless hear many cases in which copyright law has to be reinterpreted in light of the unique technical, economic, and social implications of electronic media. An excellent source of information, news, and analysis of these debates is the World Wide Web Consortium (http://www.w3.org/pub/WWW).

Creating Web Pages

Learning how to create Web pages is becoming so important in the workplace that you ought to understand the basics of the process. You might find yourself involved in creating a Web site a lot sooner than you think, for all kinds of companies and other organizations use the Web for advertising and customer support.

The following discussion is not a step-by-step tutorial, but rather an explanation of the concept involved in the process. Many excellent tutorials in books and on the Web explain the process in detail and even provide links to downloadable software tools. (*Downloadable* means that the server computer can send the files directly to your own computer.) To find these tutorials, see one of the many "resource sites" for technical communication (such as the site for this book) or search for "Web tutorials." Perhaps the single most important site for everything related to creating Web pages is the WWW Developer's Virtual Library (http://www.stars.com), which contains links to over 1,000 sites.

The eight basic steps in creating a Web site are shown in Figure 4.13. Keep in mind, however, that building and maintaining a site is not a linear process. As

FIGURE 4.13

Steps in Creating a
Web Site

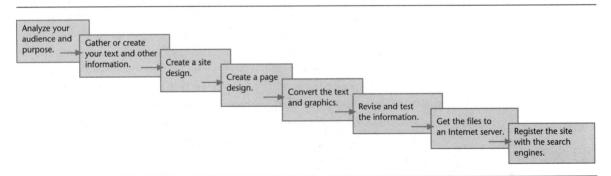

INTERVIEW: WEBMASTER JEFF SAVIN

on his education

I have a bachelor's degree in computer science. In school, I learned a number of programming languages and about computers in general, but nothing about the demands of making a Web site.

on the biggest challenge in making Web pages

More and more Web sites need to be linked to databases for applications such as data-capturing forms that the viewer fills out. A few years ago it was hard to find tools that allowed you to create dynamic pages. Today's tools are a lot more sophisticated and more readily available.

on guidelines for designing a page

We don't limit pages by the number of bytes, and we don't say you have to be able to get from one place to another in one or two clicks. But generally we like to use a lot more smaller pages rather than a few longer pages. We try not to be too graphic-intensive, although we do use many graphics; we use the latest graphics-compression techniques to make the file size smaller. And people always have the option of turning the graphics off if they prefer. In general, we try to make sure that we create sites that can be read by the software that is out there. A lot of people are using old browsers that can't handle new Web page designs and the new HTML code. In some cases we offer two versions of the pages—one for people with current browsers, one for people with older browsers.

on determining a client's needs

Some clients have very definite ideas about what they want their site to look like; they might even have it drawn out on paper. In these cases, I go over their ideas with them, pointing out features that I would or would not recommend. Some clients have no idea at all. I have to interview them to find out what their goal is in having a Web site: are they trying to sell products or services, or do they just want their name out there? I usually find it easier if the clients don't already have the whole site planned out. Then we can work together to come up with a good solution, and they're more receptive to ideas that they hadn't considered.

on keeping up with the field

Web Week has a lot of good news and information. *InfoWorld* is another good source of information. Those are the two main journals for Web developers. Plus there are a number of good journals for users of the Internet.

on looking at other people's sites

In the beginning all I did was look at as many sites as I could, just to see what people were doing. Now I don't normally do that. I'll look at another site only if someone tells me there's something really special about it.

on where the profession is going over the next couple of years

The technology is changing so rapidly, it's hard to know. I try to stay up with the journals and see where I might fit in, but two years is an eternity in Internet time.

on his sources of job satisfaction

One source is seeing a satisfied customer. When a customer looks at the site and is incredibly pleased with it, that gives me a great feeling. The other source of satisfaction is knowing that the site was done well and a lot of people are viewing it and like it.

Jeff Savin is president of WaveMaker, Inc. (http://www.wavemaker.com), a consulting company in Flagstaff, Arizona.

with any kind of writing, you will find yourself going forward in the process, then doubling back and changing things. You will write some text, then design it, then go back and rewrite. In fact, a Web site is never really done, for you will probably be adding, deleting, and revising information as long as it exists. The needs of your audience might change, and the information you present almost certainly will. If you have spent even a little time looking at Web sites, you probably have seen the "under construction" graphic, with the little fellow swinging his sledgehammer.

Analyze Your Audience and Purpose

As with any technical communication task, your first goal is to understand your audience—the people who will be viewing the site—and their purposes for visiting. A site meant to advertise a new movie has a very different function from a site for a domestic-abuse shelter. Who are your viewers? Why would they visit the site? What kinds of information do they seek? Are they looking mainly for links to other sites? Do they need to download information to their own computers? What kinds of computer equipment are they likely to have?

Although some sites provide information and some seek only to entertain, most viewers want some combination of the two. A site for an automobile manufacturer will have hard information (specifications and other technical information about the products) as well as advertising content (video clips, animation, virtual-reality "walks" around the showroom).

Until you understand your viewers—their ages, interests, perceptions, and reasons for visiting your site—you are not yet ready to create the site. See Chapter 5 for more information about analyzing your audience.

Gather or Create Your Text and Other Information

Although you can create a text-only Web site, most people want to include graphics or some other media, either to make the site more interesting visually or to communicate in ways that text alone cannot accomplish. For instance, many universities have sites that include maps, photographs, music, and video clips of events on campus. The first step in creating a site, therefore, is to gather the material you want to put on the site. And remember, any information to be put on the Web has to be in digital format. For instance, you will have to scan a map and save it as a graphic file.

Create a Site Design

Almost all sites consist of a home page—the main file—as well as other pages. A big challenge is to design the site so that people can navigate among the different pages easily and quickly.

Some kinds of information lend themselves to a shallow design, as shown in Figure 4.14. If the home page clearly displays the second-level pages clearly—by displaying each link as an item in a traditional outline format, for example—the viewer will find it easy to navigate the site. Electronic phone books and bibliographies are often presented using a shallow design.

Some kinds of information call for a deeper design, as shown in Figure 4.15. In a design such as this, the viewer has to navigate through one level to get to a deeper level; the viewer cannot link directly from the home page to the third level, for example. Some *Webmasters* design their sites so that a viewer does not have to click more than twice to get from one place on the site to another.

A Shallow Site Design

However, for large, complex sites containing dozens of pages, this guideline is impractical. See NASA's site (www.nasa.gov) for an example of a deep design.

Another popular design alternative uses *frames;* that is, the screen is divided into several sections or windows. One frame displays the site map—the master list of all the pages on the site—and the other frame displays the page the viewer is looking at. The advantage of frames is that the viewer is less likely to get lost because the site map is always visible. However, frames have two disadvantages: some browsers cannot display frames, and frames reduce the size of the screen displaying the information you want to see. Figure 4.16 on page 92 shows a portion of a screen with frames.

As shown in Figure 4.16, the left-hand portion, with the round buttons, is the site map; it stays on the screen at all times to help viewers know where they are in the site. Notice the arrow on the top button: QSS HOME. This arrow points to the right-hand portion of the screen, the section that contains the large QSS logo and the text, which is the home page for the site. If the viewer were to click

A Deeper Site Design

FIGURE 4.16

A Web Page with Frames

on another button, a different page would appear on the right-hand side of the screen.

You should know two other important design principles that affect the whole site:

- *Build in navigational tools.* Be sure your viewers know where they are at all times. Include a site map, which is a structured list of the site pages, much like a multilevel outline or a block diagram (see Chapter 12). Include informative headers and footers (see Chapter 11), and include links back to the home page (if the site consists of more than two levels). Don't let a viewer get to the bottom of a page and not know how to proceed.
- *Include a "what's new" section.* In most cases, you don't want the viewer to visit your site only once. To communicate your message effectively, you want to encourage return visits. Adding a "what's new" section can give viewers a reason to come back.

There are no rules for designing sites: the needs of the audience and the nature of the information you wish to show dictate your design. There are an infinite number of choices, and before you're satisfied you probably will go through a dozen.

Create a Page Design Like a printed document, a Web site must be designed: you want it to be attractive and functional. One complication in creating a Web site is that you cannot completely control the appearance of the site the way you can with a document. When you produce a brochure, you can control every aspect of what the reader sees. On the Web, however, the viewer's browser, computer, and monitor help determine what he or she sees. Therefore, you have to anticipate potential problems and create a design that avoids as many of them as possible. For instance, some viewers have equipment that cannot display graphics; for this reason, you have to design the site so that it communicates its message even if the graphics are missing.

To a large extent, however, the principles of good Web page design are very similar to the principles of good page design for printed documents (see Chapter 11). For instance, you should start with a page grid, use white space liberally, and use typography effectively.

However, some design principles are unique to Web documents. Here are five examples:

- *Accommodate the equipment limitations of your audience.* If your audience consists of people who are using low-end hardware or software and therefore cannot receive graphics or frames, avoid these features or at least provide ways for your audience to view versions of your information that are suited to their hardware.

- *Don't go overboard with the fancy features.* It's fun to put in graphics that blink and spin and fly across the screen, but your primary goal is probably to convey information to the viewer; don't let your message be overshadowed by showy effects. Keep in mind, too, that graphics files are very large compared to text files, and they therefore take a long time to download, especially if your viewers are using modems.

- *Keep pages relatively short.* Although people disagree about how long a page should be, most authorities agree (Horton, 1994; Lynch, 1996) that a page of information should not be longer than two or three pages of text. You don't want your viewer to be scrolling down endlessly. If your text must be longer, use separate, smaller files to which the viewer can link from the home page, or use *jump links* so viewers can jump from the top of the page directly to the section they wish to read.

- *Include contact information and a way to e-mail you.* Include information about how to get in touch with your organization: addresses, phone numbers, and fax numbers. Most browsers make it easy to include a form that enables a viewer to e-mail you directly from the site.

- *Create an informative header and footer.* The top and bottom of your pages are very important in establishing your credibility. Use your organization's logo, and design the header and footer to be informative and dignified. Remember: a lot of information on the Web is not credible. Make sure your viewers know that they are visiting the official site of your organization, and that it was created by professionals. Figure 4.17 (Lynch, 1996) on page 94 shows an informative footer.

Convert the Text and Graphics

For information to be transmitted on the Web, it has to meet certain technical standards. For example, text must be written in HTML (hypertext markup language),

FIGURE 4.17

Informative Footer

a coding system. In HTML, for example, a paragraph of text must be preceded by the following tag: <p>. Graphics must also meet standards; for instance, a graphics file must be in either a ".GIF" or ".JPG" format. HTML appears to be a stable standard for text, but the technical standards for nontextual media are expected to change as the Web evolves.

How do you convert the material for use on the Web? To add HTML tags to text, you can type the tags in by hand or use *conversion programs.* As the name suggests, conversion programs examine your word-processed text and convert it to HTML. The current generation of word-processing software includes conversion software; that is, you can create text in the normal way, then instruct the software to save it as HTML text, ready to be put on the Web. However, many people who use a conversion program are not fully satisfied with it and have to tinker with the HTML code before the file is satisfactory. Table 4.1 shows two short passages of coded text, as well as how the passages would appear on the Web.

For formatting graphics, the situation can be a little trickier, for you sometimes have to open your graphics file in a graphics program, then save it in one of the correct formats. The best, most current information on technical standards for the Web is, of course, on the Web.

Revise and Test the Information

Once you think you have the page the way you want it to appear, test it as you would a traditional document (see Chapter 20) to make sure it accomplishes your purposes. Determine whether the technical aspects of the page work. For instance, does the page load correctly when you enter the URL on your browser? Do all the links work? Does the e-mail form for getting back to you work? And determine whether the site works as a piece of communication. Do readers

TABLE 4.1	HTML Codes

The Coded Text	How the Passage Would Appear on the Web
<P> This is the simplest coding: to begin and end a paragraph.</P>	This is the simplest coding: to begin and end a paragraph.
<P>Here is how you would make a numbered list:</P> <OL> <LI>item 1 <LI>item 2 <LI>item 3 </OL>	Here is how you would make a numbered list: 1. item 1 2. item 2 3. item 3

This symbol is used to begin a paragraph.

This symbol begins an ordered (numbered) list.

This symbol begins a "list" item.

This symbol ends the ordered list.

understand the main point? Do they understand how the different pages that make up the site work together?

In addition to this kind of usability testing, however, you also have to test to see how the site appears with different kinds of computers, monitors, and browsers. You might find, for example, that one browser doesn't understand the tables you have created, displaying instead a mass of meaningless, unformatted numbers. You will have to figure out a different way to display that information (see the Web tutorials for technical solutions to such problems). Or you might find that a browser doesn't display your graphics, so you have to add textual descriptions for viewers using that browser.

Get the Files to an Internet Server

As you create your Web site, you are working on your own computer. After you have created the site and tested it, you have to get the files to an Internet server, a computer that is connected to the Internet. If you are using a local or national Internet service provider, they will show you how to use FTP to transport your files to the server. If the Internet server is in your own organization, you might use FTP or even carry a floppy disk down the hall.

Once you get the site up on the Web, test it again: make sure the site is professional and attractive and that its technical aspects work.

Register the Site with the Search Engines

Once your site is up and running, you will want to publicize it. In part, you can do this in the traditional way: add the URL to all your product information and advertising, as do many companies in their television and magazine ads. However, you also want to make it easy for people to find you through the search engines. Although some search engines have programs that roam the Web and automatically record the addresses of new sites, you should notify the search engines. When you look at the site of a search engine such as Alta Vista, you will see that it has a link to a page where you register your site by listing its keywords, describing its subject matter, and writing its URL. There are even sites that disseminate this kind of information to a number of search engines, so you do not have to repeat the information for each one.

The Future of the Web

It is risky to predict the future of the Web, for computer technology is changing much more rapidly than any other similar communication technology ever changed. (You probably know this already if you own a three-year-old computer that doesn't run the software you want to use.)

But one trend appears to be clear: the Web will become integrated with other electronic media in homes. The giants of the communication industry already understand that the key to their future success is to integrate the Web with other electronic devices in the home, such as television. New, inexpensive Web-capable hardware that connects to your television set is only the beginning. The real breakthrough will be when the Web is delivered to homes through either the television cable or by satellite, for these media allow a much broader band-width than modems do; in other words, Web designers will be able to use as much video as they want, just as television or movie makers do. You will not have to wait interminably while the information trickles through your modem onto your screen.

The main point of this chapter, however, is that although the high-tech tools are truly spectacular, none of them is of any value without an intelligent, thought-

CASE

Creating a Web Site

Create a Web site for your major, student organization, or community group. (You do not actually have to put the site on the Web; you can create it and test it on a personal computer that has a browser.) Follow these steps:

- Analyze your audience and purpose. Who would view the site, and what would you hope to achieve with the site?
- Gather or create your text and other information. If you do not have a sufficient amount of information available in digital format, you can still proceed with building the site.
- Design the site. What sort of design is appropriate for the information you wish to communicate and for the needs of your audience?
- Convert the text and graphics. What sorts of tools are most useful for the conversion process?
- Revise and test the information. Other students in the class might help you with testing the site.

Keep a log of the activities each of you carried out. Be prepared to describe your experiences in an oral presentation to the class. Which tasks were easy to accomplish? Which were difficult? What surprised you about the process?

ful person who knows what they can and cannot do. To use the tools effectively, you have to be sensitive to the needs of your audience and aware of your purpose, for communication is really about people, not about tools.

Computer Tools

1. Have you considered using the outline feature during the planning and drafting stages?

2. Have you used templates and styles carefully?

3. If you used a spell checker, did you also proofread the document for spelling errors?

4. If you used a grammar checker or thesaurus, have you thought about the advice offered by the software, and have you also revised the document without the software?

E-mail

1. Is the tone appropriate?

2. Have you written the message carefully and revised it?

3. Have you avoided flaming?

4. Have you used the subject line?

5. Have you used uppercase and lowercase letters?

6. Have you skipped lines between paragraphs?

7. Have you kept the line length to under 65 characters?

8. Have you checked with the writer before reposting his or her message?

Creating Web Pages

1. Have you analyzed your audience and purpose before planning your Web site?

2. Have you gathered your text and graphics, making sure the graphics are in digital form?

3. Have you created a site design that will make it easy for your viewers to get the information they need?

4. Have you created a page design that is clear and professional?

5. Have you converted your text to HTML and your graphics to an appropriate format?

6. Have you revised and tested the information?

7. Have you gotten the files to an Internet server?

8. Have you registered the site with search engines?

EXERCISES

Several of the following exercises call for you to write a memo. For a discussion of memos, see Chapter 16.

1. The writer of the following e-mail, Bill, is the primary technical supervisor on the production line in a microchip manufacturing facility. He is responding to Larry, a supervisor who reports to him. Larry has sent him the following note: "Bill, I can't seem to find a spec that describes coat tracks. Some of the new hires don't know what they are. What should I tell them?" Bills' reply:

> Larry—
>
> Coat tracks are the machines used in the first step of the photo process. The wafers coem to coat to have a layer of a photosensitive resist applyed. This requires several operations to be done by the same machine. First the wafer is coated with a layer of primer or hmds. This ensures that the resist will adhere to the wager. The wafer is then caoted with resist. There are 5 different types of resist in use, each has its won characteristics, and all are used on different levels and part types. I dont really have time to go into all the details now. The photo resist must be applyed ina uniform layer as the unfioromity of the resist can effect several other steps to include exsposure on the stepper to the etch rate on a lam. To insure the proper unfiromity and resist volumes the tracks are inspected at 24 hour intervals, all the functions are checked and partical monitors are ran to ensure proper operation and cleanlyness. After the wafer is coated it is soft baked, this rids the wafer of solvents in the primer and resist and also hardens the resist. The wafers are then ready to go to the next step at the p&e or the stepper.

If, in fact, there is no specification that Bill can give to Larry, what would be the best way to communicate the information to him? What impression will his e-mail make on Larry?

2. Evaluate the following e-mail posted to a discussion group. How would you revise it to make it more effective?

> To: TECHWR-L List
> Subject: Binders v. Updates
>
> On July 13, Barbara Ackley wrote:
>
> >We always prefer using three-ring binders to pub-
> >lishing updated manuals. Admittedly, not every
> >one is going to keep their binder up to date, but
> >the cost differential is just too great for us to
> >overlook.
>
> I have to agree with Barbara.
>
> Regards,
> John Altow, SysPerfect Corporation

3. Louise and Karen work for the same manufacturing company. Louise, a senior engineer, is chairing a committee to investigate ways to improve the hiring process at the company. Karen, a technical editor, also serves on the committee. The excerpts quoted in Louise's e-mail are from an e-mail written by Karen to all members of the committee in response to Louise's request for all members of the committee to describe their approach to evaluating job-application materials. How would you revise Louise's e-mail to make it more effective?

> To: Karen
> From: Louise
>
> Sometimes I just have to wonder what you're thinking, Karen.
>
> >Of course, it's not possible to expect perfect
> >resumes. But I have to screen them, and
> >last year I had to read over 200. I'm not looking for
> >perfection, but as soon as I spot an error I make
> >a mental note of it and, when I hit a second and
> >then a third error I can't really concentrate on the
> > writer's credentials.
>
> Listen, Karen, you might be a sharp editor, but the rest of us have a different responsibility: to make the products and move them out as soon as possible. We don't have the luxury of gazing at the writing to

see if we can find errors. I suggest you concentrate on what you were hired to do, without imposing your "standards" on the rest of us.

> >From my point of view, an error can include a
> >misused tradmark.

Misusing a "tradmark," Karen? Is that Error Number 1?

Louise Wilkins

Collaborative Exercises

4. Form small groups and describe and evaluate your college's or university's Web site. A different member of the group might carry out one of the following tasks:
   - E-mail the site's Webmaster, asking questions about the process of creating the site. How involved with the content and design of the site was the Webmaster? What is the Webmaster's role in maintaining the site?
   - Analyze the kinds of information the site contains and determine whether the site is intended primarily for faculty, students, alumni, or prospective students.
   - Determine the overlap between information on the site and information in printed documents published by the school. In those cases in which the media overlap, is the information on the site merely a duplication of the printed information, or has it been revised to take advantage of the unique capabilities of the Web?

   In a memo to your instructor, communicate the conclusions from your research and recommend how the site might be improved.

5. Form groups according to major and analyze the Web site for a major professional organization in your field. Each member of the group might investigate one of the following questions:
   - Who is responsible for creating and maintaining the site?
   - What are the major kinds of information provided on the site?
   - What is the major function of the organization? Whom does it serve? What services does it provide? What activities does it sponsor?

- What is the role of students in the organization? Are there special activities or opportunities for students? Can you think of ways to increase the role of students in the organization?

Present your findings to the appropriate officer at the organization and to your instructor.

Web Exercises

6. Some of the e-mail sent by students to other students on your college's network has been inappropriate: sloppy and inarticulate, and sometimes even rude and offensive. The college is looking for a set of netiquette guidelines that it can endorse to try to reduce the problems. Using a search engine, find two different descriptions of netiquette. In a memo to your instructor, compare and contrast the two descriptions. Explain which one seems more relevant to your uses of e-mail.

7. Using a search engine, locate two different manufacturers of groupware software. In a brief memo to your instructor, describe how each company defines groupware and briefly describe some of the major products they offer, giving your recommendation for the groupware most appropriate for your class.

8. Find a Web site for a company that makes consumer products, such as a soft-drink manufacturer, a movie studio, or an auto maker. In a brief memo to your instructor, describe some of the most interesting and creative features of the site. What is the ratio of information to entertainment in the site? Overall, how effective is the site?

REFERENCES

Crowe, E. P. (1994). *The electronic traveller: Exploring alternative online systems.* New York: Windcrest/McGraw-Hill.

Ellsworth, J. H., & Ellsworth, M. V. (1996). *The new Internet business book.* New York: Wiley.

Hill, J., Ozer, J., & Mace, T. (1996). Connecting on the Net. *PC Magazine* (October 8), 101–203.

Horton, W. K. (1994). *Designing and writing online documentation* (2nd ed.). New York: Wiley.

Lynch, P. (1996). *Web style manual.* <http://info.med.yale.edu/caim/StyleManual_Top.html> (1996, Dec. 24).

Microsoft Corporation. (1996). Professional résumé template. Microsoft Word for Windows 95, Version 7.0a. Redmond, WA: Microsoft Corporation.

O'Mahoney, B. (1997). Bleeding edge: Internet issues. *Copyright page.* <www.benedict.com> (1997, Apr. 19).

Quality Systems & Software. (1996). DOORS. <http://www.qssinc.com/tour6.html> (1996, Sept. 19).

Quality Systems & Software. (1997). *Home page.* <http://www.qssinc.com/index.shtml> (1997, Mar. 6).

Analyzing Your Audience

Understanding the Writing Situation 101

Primary and Secondary Audiences 101

Basic Categories of Readers 102

   The Expert *103*
   The Technician *104*
   The Manager *104*
   The General Reader *108*

Individual Characteristics of Readers 110

   Education *110*
   Professional Experience *111*
   Job Responsibilities *112*
   Personal Characteristics *112*
   Personal Preferences *112*
   Attitudes toward You and Your Subject *113*
   Cultural Characteristics *114*
   Document Use *114*
   Physical Environment *115*

The Audience Profile Sheet 115

Accommodating the Multiple Audience 118

Understanding Multiculturalism 118

   Categories of Cultural Patterns *119*
   Writing to Multicultural Audiences *124*

Case: Analyzing an Audience *125*

Writer's Checklist 126

Exercises 127

   Collaborative Exercise *127*
   Web Exercise *128*

The content and form of every technical document you write are determined by the situation that calls for that document: your audience and your purpose. Understanding the writing situation helps you devise a strategy to meet your readers' needs—and your own.

This chapter covers how to analyze your audience. The discussion begins with how to understand the basic categories of readers—experts, technicians, managers, and general readers. Then the chapter explains how to analyze the individual characteristics of your readers. Finally, the chapter discusses how to address multicultural audiences.

Understanding the Writing Situation

The concepts of audience and purpose are hardly unique to technical communication; most everyday communication is the product of the same two-part environment. When a classified advertisement describes a job for prospective applicants, the writing situation of the advertiser is clearly identifiable:

Audience: prospective applicants

Purpose: to describe the job opening so that qualified persons will apply

Once you have defined the two basic elements of your writing situation, you must analyze each before deciding on the document's content and form.

Although you might assume that purpose would be your primary consideration, it's better to start by analyzing your audience; examining its characteristics often improves your understanding of your purpose. The separation of audience and purpose is to some extent artificial—you cannot think about one without thinking about the other—but it is nonetheless useful. The more thoroughly you can define first your audience and then your purpose, the more exactly you will be able to tailor your document to the situation.

Purpose is discussed in detail in Chapter 6.

Primary and Secondary Audiences

Identifying and analyzing the audience can be difficult. You have to put yourself in the position of people you might not know, reading something you haven't yet started to write. Most writers want to concentrate first on the content: what they have to say. Feeling most comfortable with content, they want to write something first and shape it later.

Resist this temptation. If you have written mostly for teachers, you probably haven't had to think much about the audience. In most cases, you have some idea of what your teachers want to read, and you have received explicit guidelines and expectations for the assignments. And you realize that most teachers usually know more about the subject than their students do.

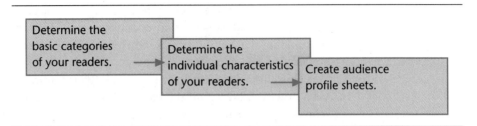

FIGURE 5.1

Analyzing Your Audience

In business and industry, however, you will often have to write to different audiences, some of whom you will know little about and some of whom will know little about your technical field.

In addition, these different audiences may well have very different purposes in reading what you have written. Readers are often classified into two categories:

- A *primary audience* that consists of people who have a direct role in responding to your document. They might be people who use your information in doing their jobs. They might evaluate and revise your document, or they might act on your recommendations. An executive who decides whether to authorize building a new production facility is a primary reader. So is the treasurer who has to determine whether the organization can pay for it.

- A *secondary audience* that consists of people who need to know what is being planned, such as salespeople who want to know where a new facility will be located, what products it will produce, and when it will be operational. A secondary audience does not have a direct role in responding to your document.

As shown in Figure 5.1, the first step in analyzing your audience is to make general assumptions about them by classifying them into one of several basic categories. The second and third steps, to determine the individual characteristics of each reader you can identify and to create audience profile sheets, are discussed in the next sections.

Basic Categories of Readers

Although everyone is unique, a useful first step is to try to classify each reader on the basis of his or her knowledge of the subject you are writing about. In general, every reader could be classified into one of four categories, as shown in Figure 5.2.

Of course, these categories are generalizations. For example, everyone is a general reader in almost every field outside of his or her own specialty. On the job, however, most people fit into one or perhaps two categories. Ellen DeSalvo, for example, a Ph.D. in materials engineering, would be an expert in

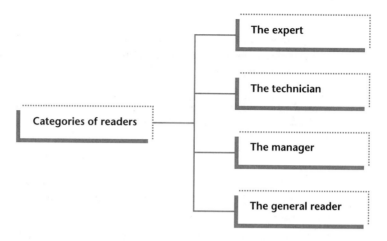

FIGURE 5.2
Basic Categories of
Readers

that particular field, and perhaps she is also the manager of the materials group at her company. Off the job, she might be an expert on Native American folk art, having studied it seriously for many years as a hobby.

For this reason, classifying readers into categories is simply a first step in helping you understand how to tailor a particular document to them. It is not a useful way to understand them as people. That part of audience analysis is discussed later in this chapter.

The Expert The expert is a highly trained individual with an extensive theoretical and practical understanding of the field. Often the expert carries out basic or applied research and communicates those research findings. Here are some examples:

- a *physician* who is trying to understand how the AIDS virus works and who delivers papers at professional conferences and writes research articles for scholarly journals
- an *engineer* who is trying to devise a simpler and less expensive test for structural flaws in composite materials
- a *forester* who is trying to plan a strategy for dealing with the threat of forest fires

In short, almost everyone with a postgraduate degree—and many people with an undergraduate degree in a technical field—is an expert in one area. However, not everyone with a degree is an expert, and many experts have no formal advanced training.

Because experts share a curiosity about their subject and a more or less detailed understanding of the theory in their field, they usually have no trouble understanding technical vocabulary and formulas. Therefore, when you write to them, you can get right to the details of the technical subject, without spending time sketching the fundamentals.

In addition, most experts are comfortable with long sentences, if the sentences are well constructed and no longer than necessary. Like all readers, experts appreciate graphics, but they can understand more sophisticated diagrams and graphs than most readers can.

Figure 5.3 is a passage from an article on how artificial intelligence (AI) can be used to produce engineering labor standards (Yazici, Benjamin, & McGlaughlin, 1994, p. 303). This passage illustrates the needs and interests of the expert reader.

The Technician   The technician has practical, hands-on skills. The technician takes the expert's ideas and turns them into real products and procedures. The technician fabricates, operates, maintains, and repairs mechanisms of all sorts, and sometimes teaches other people how to operate them. An engineer having a problem with an industrial laser will talk over the situation with a technician. After they agree on a possible cause of the problem and a way to try to fix it, the technician will go to work.

Like experts, technicians are very interested in their subject, but they know less about the theory. They work with their heads and their hands. Technicians have a wide variety of educational backgrounds; some have a high school education, while others have attended trade schools or earned an associate's degree or even a bachelor's degree. When you write to technicians, keep in mind that they do not need complex theoretical discussions. They want to finish a task safely, effectively, and quickly. Therefore, they need schematic diagrams, parts lists, and step-by-step instructions. Most technicians prefer short or medium-length sentences and common vocabulary, especially in documents such as step-by-step instructions.

Figure 5.4 on page 106, an excerpt from the user's guide for a personal computer (Micron, 1996, p. 2-6), illustrates the needs and interests of the technician.

The Manager   The manager is harder to define than the technical person, for the word *manager* describes what a person does more than what a person knows. A manager makes sure an organization operates smoothly and efficiently. For instance, the manager of the procurement department at a manufacturing plant sees that raw materials are purchased and delivered on time so that production will not be interrupted. The manager of the sales department of that same organization sees that salespeople are out in the field, creating interest in the products and following up leads. In other words, managers coordinate and supervise the day-to-day activities of the organization.

Upper-level managers, known as executives, address longer-range concerns. They foresee problems years ahead by considering questions such as the following:

- Is current technology at the company becoming obsolete?
- What are the newest technologies?

The discussion assumes a strong background in the relationship between artificial intelligence and the manufacturing process. The general reader would be unable to follow this discussion because of its highly technical vocabulary ("automated knowledge acquisition tools," for instance).

The discussion refers frequently to the published literature on the subject; expert readers would be familiar with some—or all—of the literature cited here. (The bracketed numbers 1, 2, and 25 are citations.)

The discussion uses fairly long sentences, not because the writers want to show off but because the material is complicated.

The discussion refers to a fairly complex graphic.

Further implementation of AI-based tools can be found in manufacturing control. For instance, in a study done for the Department of Defense, an automated knowledge acquisition tool was designed to evaluate the hardness of weapon systems. The tool, called KNACK, interacts with the design engineers and program managers and automatically generates the knowledge base code to evaluate the performance of the weapon system. KNACK evaluates whether the weapon system is "hard" in relation to the effect of a specific nuclear weapon being investigated [2]. Examples of the knowledge acquisition tools to assist the task analysis can be found in [1] and [25].

These studies showed the importance of AI techniques for production planning and control. The transfer of knowledge can be facilitated by the use of knowledge-based systems that can result in improved production. The AI-based approach can enhance the knowledge transfer within the organization and provide better planning and control of products and resources.

Automated generation of shop-floor job descriptions and relevant labor time estimates can be a valuable alternative to the traditional methods of developing engineering labor standards. Fig. 1 presents a conceptual framework for incorporating AI-based techniques in this domain. Data from the shop floor on parts, machines and processes are processed to establish the method, to generate the necessary job elements, to compute times associated with the job elements, and to estimate standard job times for the application and provide a complete job detail sheet. An AI-based module composed of automated knowledge acquisition and codification can assist in better integrating the functions of methods engineers and computer programmers, and in reducing delays and conflicts between engineers and programmers.

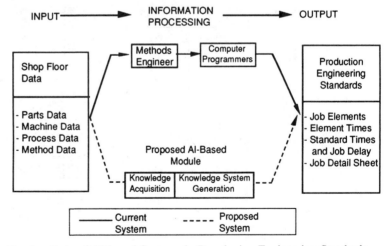

Fig. 1.   Role of AI-based Systems in Developing Engineering Standards.

FIGURE 5.3

Writing Addressed to an Expert Audience

Writing addressed to technicians often includes very little text. This excerpt about jumper settings on a personal computer is a diagram and a table.

Micron Millennia User's Guide

Jumper Settings

Configuration

The jumpers on the Micron Millennia are preset at the factory and in most cases do not need to be changed. However, if certain functions need to be changed, the jumpers may need to be reconfigured. Figure 2-4 shows an example of a jumper.

Figure 2-4. Example of a Jumper and Jumper Block

Refer to the following tables to set the jumpers.

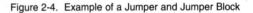

CPU Freq. (MHz)	Host Bus Freq. (MHz)	CPU Clock Multiplier	Jumpers					
			W1	W2	W4	W5	W6	W13
200	66	3	Open	Short	Open	Short	Open	Short
166	66	2.5	Open	Short	Open	Short	Short	Short
150	60	2.5	Short	Open	Short	Short	Short	Short
133	66	2	Open	Short	Open	Open	Short	Short
120	60	2	Short	Open	Short	Open	Short	Short
100	66	1.5	Open	Short	Open	Open	Open	Short
90	60	1.5	Short	Open	Short	Open	Open	Short
75	50	1.5	Short	Short	Short	Open	Open	Open

Figure 2-5. CPU Speed Selection

2-6

Micron Electronics, Inc.

FIGURE 5.4

Writing Addressed to Technicians

- How expensive are they?
- How much would they disrupt operations if they were adopted?
- What other plans would have to be postponed or dropped altogether?
- When would the conversion start to pay for itself?
- What has been the experience of other companies that have adopted the new technologies?

Executives are concerned with these and dozens of other broad questions that go beyond the day-to-day managerial concerns.

Because management is a popular college major, many managers today have studied general business, psychology, and sociology. Often, however, managers start out in a technical area. An experienced chemical engineer, for instance, might manage the engineering division of a consulting company. Although he has a solid background in chemical engineering, he earned his managerial position because of his broad knowledge of engineering and his ability to work well with colleagues. In his daily routine he might have little opportunity to use his specialized engineering skills.

Although generalizing about the average manager's background is difficult, identifying the manager's needs is easier. Managers want to know the bottom line. They have to get a job done on schedule; they don't have time to study a theory the way an expert does. Rather, managers must juggle constraints—financial, personnel, time, and informational—and make logical and reasonable decisions quickly. And they have to communicate with their own supervisors.

When you write to a manager, try to determine his or her technical background; then choose an appropriate vocabulary and sentence length. Regardless of the individual's background, however, focus on the practical information the manager will need. For example, if you are a research-and-development engineer who is describing a new product line to the sales manager, you might begin with some theoretical background so that the sales representatives can communicate effectively with potential clients. For the most part, however, you should concentrate on the product's capabilities and its advantages over the competition.

If you know that your reader will take your information and use it in a document that will be addressed to higher-level managers or executives, make your reader's job easier. Include an executive summary (see Chapter 10) and use frequent headings (Chapter 13) to help your reader see the major points you are making. Ask your reader if there is an organizational pattern, a format, or a strategy for writing the document that will help him or her in using your document as source material.

Figure 5.5 on page 108, an executive summary (D'Ottavi, 1993) from a report entitled "Manufacturing Bidirectional Valves: A Feasibility Study," illustrates the interests and needs of the manager.

This executive summary provides little technical information about the technology—how the valves work—and concentrates on what the manager cares about: the demand for the product, the costs of researching and developing it, and the market value for it.

The writer uses a simple vocabulary; the only term that the general reader might not understand is "break-even analysis," which the intended reader would surely understand.

The writer concludes with a clear recommendation.

Executive Summary

Advances in biomechanical engineering have created a demand for small, quick-acting, bidirectional hydraulic valves. Presently, no valves exist commercially that can satisfy bidirectional flow requirements. Experts speculate that federal and private funding for biomechanical projects will exceed $2 million by 1999. Other specialty valves similar to bidirectional valves have proven to be highly patentable, resulting in fifteen years or more of exclusive rights to manufacture and sell. Comparable valves can cost between $100 and $300.

This report describes an investigation that examined the demand, the production costs, and the market value for bidirectional valves.

Presently, the demand for these valves is low, due to the fairly limited number of applications. In addition, their future use, although expected to be high, is uncertain at this time.

The initial investment to research and develop bidirectional valves is estimated to be $42,500. The cost to manufacture the valves is estimated at $81 each.

The estimated market value is $200 each. According to a break-even analysis, revenue from 360 valves would equal the initial investment costs plus the costs per valve. These estimates are very conservative, since the actual production costs would be significantly lower if a high volume of valves were manufactured.

Because bidirectional valves are versatile (they can also be used and sold as unidirectional valves), investment in the research and development of bidirectional valves would be a relatively low risk. In addition, the initial investment costs are no higher than those for a unidirectional valve. Considering that technological development of devices that use bidirectional valves could increase dramatically, a valve prototype available for customer testing and evaluation could prove highly profitable.

I therefore recommend that we research and develop bidirectional valves. However, because of the uncertain future demand for them, I recommend delaying their manufacture.

FIGURE 5.5

Writing Addressed to a Manager

The General Reader

Often you will have to address the general reader, sometimes called the layperson. A nuclear scientist reading about economics is a general reader, as is a homemaker reading about new drugs used to treat arthritis.

The layperson reads out of curiosity or self-interest. The typical article in the magazine supplement of the Sunday paper—for example, on attempts to increase the populations of endangered species in zoos—will attract the general reader's attention if it seems interesting and well written. The general reader may also seek specific information that will bring direct benefits: someone interested in buying a house might read articles on new methods of alternative financing.

The passage focuses on the human aspects of the technology, not on the technical aspects. It concentrates on the girl, Dawn Price, and the researcher, Richard Foulds.

The passage uses an informal tone, including idiomatic phrases and contractions ("What's more" in paragraph 2).

The passage requires no previous knowledge about the subject.

The passage gradually introduces technical vocabulary, such as "luminance valleys" and "edge detection." A technical term—*bits*—is defined in a parenthetical definition.

FIGURE 5.6

Writing Addressed to the General Reader

Dawn Price is telling a joke about three men on a train—a Cuban, a Russian, and an American. The joke relies on timing, body language, and audience rapport.

The difference here is that Dawn, who is 12, is deaf, as are the three schoolmates she is entertaining. What's more, her schoolmates cannot see her directly. They are watching a personal computer screen showing an outlined image of Dawn performing sign language.

Dawn and her friends are communicating through a sign language telephone that is being developed at the A. I. duPont Institute, a children's hospital in Wilmington, Delaware.

The system is necessary because sign language requires fluid motion, so much so that deaf people have difficulty understanding the jerky pictures that are transmitted by even the most expensive video phones.

"We look not for the quality of the picture, but for the quality of the movement," said Richard A. Foulds, director of the University of Delaware's Applied Science and Engineering Laboratories, who heads the development of the project.

The sign language telephone operates with computers, not telephones. It is a custom-designed circuit board that fits into an IBM-compatible PC. Users need a video camera or camcorder, which can be plugged into the back of the system. As the camera takes in images of a person performing sign language, the circuit board strips out unimportant information.

The only elements retained are the "luminance valleys," which correspond to the edges of a person's fingers, clothing, and facial features.

The result is a black-and-white outlined drawing of the person. Mr. Foulds said transmitting only the essential elements of an image—so-called "edge detection"—reduces the number of bits, or pieces of information, in the original image 3,500 times.

In writing for a general audience, use a simple vocabulary and relatively short sentences when you are discussing subjects that might be confusing. Translate jargon into standard English idiom. Use analogies and examples to clarify your discussion. Discuss the human angle—how the situation affects people. Sketch in any special background—historical or ethical, for example—so that your reader can follow your discussion easily. Concentrate on the implications for the general reader. For example, in discussing a new substance that removes graffiti from buildings, focus on its effectiveness and cost, not on its chemical composition.

Figure 5.6 (Schwartz, 1994), the beginning of a newspaper article addressed to the general reader, illustrates the interests and needs of the general reader.

Table 5.1 on page 110 summarizes the previous discussion of how to write for the basic categories of readers.

TABLE 5.1	Basic Categories of Readers	
Audience	**Reasons for Reading**	**Guidelines for Writing**
EXPERT	To gain an understanding of the theory and its implications	Include theory, technical vocabulary, citations, and sophisticated graphics.
TECHNICIAN	To gain a hands-on understanding of how something works or how to carry out a task	Include graphics. Use common words, short sentences, and short paragraphs. Avoid excessive theory.
MANAGER	To learn the bottom-line facts to aid in making decisions	Focus on managerial implications, not technical details. Use short sentences and simple vocabulary. Put details in appendices.
GENERAL READER	To satisfy curiosity and for self-interest	Use short sentences and paragraphs, human appeal, and an informal tone.

Individual Characteristics of Readers

Classifying your readers into categories is a good first step, but it is a very imprecise process. Sometimes it's difficult to be more precise because you don't know who the readers will be. You might be writing a report for your supervisor, who you know is likely to distribute it, but you don't know to whom. Or you might be addressing an audience of several hundred or even several thousand people who share some characteristics but are not a unified group.

Still, it is a good idea to try to find out as much information as you can about the individual characteristics of your readers. Figure 5.7 shows nine aspects to consider for each person who you know will be getting a copy of your document.

Education Think not only about the degree the person has earned, such as a bachelor of science in nutrition; think also about what kinds of courses, and concentrations of courses, the person has taken. For example, a metallurgical engineer who earned a B.S. in 1977 has a much different technical background from the person who earned the same degree in 1997, because the field has changed so much in recent years.

Don't forget, too, formal and informal course work on the job; a reader who has not studied business in college might have studied it in on-site workshops or training programs.

Knowing your readers' educational backgrounds will help you determine how much supporting material to provide, what level of vocabulary to use, what kind of sentence structure and length to use, what type and number of graphics to include, and whether to provide such formal elements as a glossary or an executive summary.

FIGURE 5.7

Individual Characteristics
of Readers

How do you discover your readers' educational backgrounds? It isn't easy. You cannot ask people to send you their current résumés. But you can try to learn as much as possible in conversation with your colleagues. The longer you stay at your organization, and the more you talk with people, the more you will learn.

Professional Experience Although formal education in school and on the job is significant, a person's professional experience is equally important. As a person's professional career progresses, of course, work experience becomes even more critical than formal education. A reader who has been out of school for a decade or more may have had a variety of professional experiences. A nurse might have represented her hospital on a community committee to encourage citizens to give blood, might have worked with the hospital administration to choose vehicles for the emergency medical staff, and might have contributed to the planning for the hospital's new delivery room or hospice. In short, her range of experience might have provided several areas of competence or even expertise.

If one of your readers has experience in a subject related to the one you are writing about, you might want to adjust the content and style of your document accordingly. In addition, you should ask that person about tactics to try or to avoid, both in carrying out the project and in writing about it.

Job Responsibilities   Learn as much you can about your readers' job responsibilities. If you are writing a document for internal distribution, take a copy of your company's organization chart and circle the boxes that represent your readers. This process will give you a clear idea of your readers' different technical and managerial areas of responsibility—useful information as you plan the content, organization, and style of the document. Write in one sentence the major job responsibility for your primary reader. Then think about how your document will help that person accomplish it. For example, if you are writing a feasibility study of several means of cooling the air for a new office building, you know that your reader—an upper-level manager—will have to worry about electricity or other utility costs over the lifetime of the cooling system. Therefore, in your report you need to make clear how you are estimating utility costs in the future.

Personal Characteristics   Do not overlook your readers' personal characteristics. Do the ages and genders of your readers tell you anything important about how they will read and interpret your document? Can you draw inferences about your readers' knowledge of and attitudes toward your topic? Do your readers have any other personal characteristics you should consider, such as vision impairment, that would affect the way you write and design your document?

Personal Preferences   In addition to personal characteristics, people have individual tastes and biases: some rational and reasonable, some not. One person hates to see the first-person pronoun *I*. Another finds the word *interface* distracting when the writer isn't discussing computers. A third reader appreciates a descriptive abstract on the title page of documents; a fourth relies on executive summaries. One good way to find out a person's preferences is to read that person's own documents. With careful study, you will find that everyone's writing is as distinct as a fingerprint.

Common sense dictates that you accommodate as many of your readers' preferences as you can. Sometimes you can't, of course, either because the preferences contradict one another or because the special demands of the subject won't permit you to accommodate them. But try to avoid alienating or distracting your readers.

Attitudes toward You and Your Subject

Try to learn as much as you can about your readers' attitudes toward you and your subject.

- *Attitudes toward you.* If things are going your way, most people will like you because you are hardworking, intelligent, and cooperative. However, some people will dislike you because you are hardworking, intelligent, and cooperative. Why is this? I don't know.

    If a reader's animosity toward you is irrational or unrelated to the current project, you can do little other than try to repair the damage. You can try to earn that person's respect and trust by meeting him or her on some neutral ground, perhaps by discussing other, less volatile projects or some shared interest: gardening, skiing, science-fiction novels.

- *Attitudes toward your subject.* If possible, discuss the subject thoroughly with your primary readers to determine whether they are positive, neutral, or negative. Figure 5.8 explains basic strategies for responding to a particular type of reader attitude.

FIGURE 5.8

Responding to Your Readers' Attitudes toward Your Subject

If . . .	Try this . . .
Your reader is neutral or positively inclined toward your subject	• Write the document so that it responds to the needs of your readers; make sure the vocabulary, level of detail, organization, and style are appropriate for them.
Your readers are hostile to the subject or your approach to it	• Try to find out what the objections are and then answer them effectively. Explain clearly why the objections are either not valid or not as important as the benefits. • Organize the document so that the actual recommendation follows the explanation of the benefits. This strategy encourages the hostile reader to understand your argument rather than reject it out of hand. • As you make your case, avoid describing the subject as a dispute. Seek areas of agreement and concede points. Avoid trying to persuade your readers; people don't like to be persuaded, because it threatens their ego. Instead, suggest that there are new facts that need to be expressed. People are less reluctant to change their minds when they realize that there are additional facts to be considered.
One of your readers was instrumental in creating the policy or procedure that you are arguing is ineffective	• Be diplomatic when you discuss the shortcomings of the present system, but be especially careful if there is a chance of offending one of your readers. People can be very defensive about their ideas, and sometimes they are most defensive about their worst ideas. You don't want someone to dig in and do everything to oppose you just because you were tactless. When you address such a reader, don't write, "The present system for logging customer orders is completely ineffective." Instead, write, "While the present system has worked well for many years, new developments in electronic processing of orders might enable us to improve the speed and reduce the errors substantially."

Cultural Characteristics   What you know about your readers' cultural characteristics can help you appeal to their interests and avoid confusing or offending them. As discussed later in this chapter, cultural characteristics can affect virtually every aspect of a reader's comprehension of a document and perception of the writer. If you are writing to a multicultural audience, you'll want to modify your vocabulary, sentence length and structure, and use of graphics.

Document Use   What will readers do with the document? Will they

- file it?
- skim it?
- read only a portion of it?
- study it carefully?
- modify it and submit it to another reader?
- attempt to implement recommendations?

FIGURE 5.9

Questions to Ask about the Individual Characteristics of Readers

Characteristic	Questions to Ask about Your Readers
Education	• What do the readers know about the subject from their formal or informal education?
Professional experience	• What do the readers know about the subject from their experience on the job?
Job responsibilities	• What are your readers' major areas of technical and managerial responsibility?
Personal characteristics	• What are the ages and genders of your readers? • Do they have any disabilities that affect how they read?
Personal preferences	• Do your readers have preferences or biases about writing that you should consider?
Attitudes toward you and your subject	• Do your readers have any attitudes toward you that you should consider? • Do your readers have any attitudes toward your subject or your approach to it that you should consider?
Cultural characteristics	• Do your readers have cultural characteristics that you should consider as you create text or graphics?
Document use	• What do you know about how your readers will be using your document that will help you decide how to write or deliver it?
Physical environment	• What do you know about the physical environment in which your readers will use the document that will help you design or produce it?

- use it to perform a test or carry out a procedure?
- use it as a source document for another document?

Just as you have a purpose in writing the document, each reader will have a purpose in reading it. If only 1 of 15 readers needs detailed information, you must provide it, but you don't want to make the other 14 people wade through it. Determine some way to make the details available, but clearly set them off from the rest of the document; an appendix might do the job. If you know that your reader wants to use your status report as raw material for a report to a higher-level reader, try to write your report so that it requires little rewriting. You might use your reader's own writing style and put the file on the network so that your work can be merged with the new document without retyping.

Physical Environment Technical documents are often formatted in a special way or constructed of special materials to improve their effectiveness in particular physical environments. For example, documents used in poorly lit places are printed with a type larger than normal. Some documents will be used on ships or on aircraft or in garages, where they might be exposed to wind, salt water, and grease. Special waterproof bindings, oil-resistant or laminated paper, coded colors, and unusual-sized paper are just some of the special formats and materials that you might have to consider. A user's manual for a computer system should be bound so that it can lie flat on a table for a long period. See Chapter 11 for a more detailed discussion of how to design a document for use in different environments.

Figure 5.9 summarizes the discussion of the individual characteristics of readers.

The Audience Profile Sheet

To help you analyze your audience, you might create an audience profile sheet. You could then fill out the sheet for each primary reader and each secondary reader (but only if there are a few).

To understand how to use the audience profile sheet, assume you work in the drafting department of an architectural engineering firm. You know that the company's computer-assisted design (CAD) equipment is out of date and that recent CAD technology would make it easier and faster for the draftspeople to do their work. You want to persuade your company to authorize the purchase of a CAD workstation that costs about $4,000.

Your primary reader is Harry Becker, manager of the Drafting/Design Department. Your secondary reader is Tina Buterbaugh, manager of the Finance Department.

Figures 5.10 on page 116 and 5.11 on page 117 show two audience profile sheets: one for Harry Becker and one for Tina Buterbaugh.

Audience Profile Sheet

1. Reader's name and job title.

   Name. <u>Harry Becker</u>   Job Title. <u>Manager, Drafting/Design Dept.</u>

2. Kind of reader.   Primary <u>x</u>   Secondary _____

3. Reader's educational background.
   Formal education. <u>BS, Architectural Engineering, Northwestern, 1985</u>

   Training Courses and Workshops. <u>CAD/CAM Short Course 1988; Motivating Your Employees</u>
   <u>Seminar, 1989; Writing on the Job Short Course, 1992.</u>

4. Reader's professional background (previous positions or work experience).
   <u>Worked for two years in a small architecture firm. Started here 12 years ago as a draftsperson.</u>
   <u>Worked his way up to Assistant Manager, then Manager. Instrumental in the Wilson project,</u>
   <u>particularly in coordinating personnel and equipment.</u>

5. Reader's chief responsibilities on the job. <u>Supervises a staff of 12 draftspersons. Approves or</u>
   <u>denies all requests for capital expenditures over $1,000 coming from his dept. Works with</u>
   <u>employees to help them make the best case for the purchase. After approving or denying</u>
   <u>the request, forwards it to Tina Buterbaugh, Manager, Finance Dept., who maintains all capital</u>
   <u>expenditure records.</u>

6. Reader's personal characteristics. <u>N/A</u>

7. Reader's likes. <u>Straightforward, simple documents, lots of evidence, clear structure.</u>

   Reader's dislikes. <u>Long, complicated documents full of technical terms.</u>

8. Reader's attitude toward you and the subject of the document.
   Positive <u>x</u>   Neutral _____   Negative _____

   Why? In what ways? <u>Strongly in favor of the request, but he knows that his department has</u>
   <u>authorized a lot of computer expenditures, and is skeptical because of an unwise purchase last</u>
   <u>year. His attitude is, if someone wants to buy it, let them make a good case for it.</u>

9. Reader's cultural characteristics. <u>Nothing unusual.</u>

10. How the reader will use the document.
    Skim it. _____   Study it. <u>x</u>
    Read a portion of it. _____   Which portion? _____
    Modify it and submit it to another reader. <u>x</u>
    Attempt to implement recommendations. _____
    Use it to perform a task or carry out a procedure. _____
    Use it to create another document. _____
    Other? _____ Explain. _____

11. Reader's physical environment. <u>Normal corporate environment.</u>

FIGURE 5.10
Audience Profile Sheet

Audience Profile Sheet

1. Reader's name and job title.

   Name. <u>Tina Buterbaugh</u>   Job Title. <u>Manager, Finance Dept.</u>

2. Kind of reader.   Primary _____   Secondary <u> x </u>

3. Reader's educational background.

   Formal education. <u>B.S., Finance, USC, 1982; MBA, Wharton, 1987.</u>

   Training Courses and Workshops. <u>Don't know, but she is on the training and development</u>
   <u>committee.</u>

4. Reader's professional background (previous positions or work experience).
   <u>Not sure, but I know she worked for 6 years as Director of Finance for a large retailer.</u>

5. Reader's chief responsibilities on the job. <u>In general, in charge of long-range financial planning. In</u>
   <u>this case, she records all capital expenditures and, at the end of the year, will decide the</u>
   <u>department's capital budget for next year.</u>

6. Reader's personal characteristics. <u>NA</u>

7. Reader's likes. <u>Same as Harry Becker.</u>

   Reader's dislikes. <u>Long, complicated documents full of technical terms.</u>

8. Reader's attitude toward you and the subject of the document.
   Positive _____   Neutral <u> x </u>   Negative _____

   Why? In what ways? <u>She has no feelings about proposals this small ($4,000), and she knows Harry</u>
   <u>Becker is good at forcing the writer to make a good case, but the company has spent a lot lately on</u>
   <u>computers. She expects Harry to ask for a bigger capital budget next year.</u>

9. Reader's cultural characteristics. <u>I know she lived in Paris for five years and has traveled extensively.</u>

10. How the reader will use the document.
    Skim it. <u> x </u>   Study it. _____
    Read a portion of it. _____   Which portion? _____
    Modify it and submit it to another reader. _____
    Attempt to implement recommendations. _____
    Use it to perform a task or carry out a procedure. _____
    Use it to create another document. _____
    Other? _____ Explain. <u>She will probably spend no more than 5 minutes on it.</u>

11. Reader's physical environment. <u>Normal.</u>

FIGURE 5.11

Audience Profile Sheet

117

Accommodating the Multiple Audience

You will be writing to a multiple audience far more often than writers did even a couple of decades ago, for two reasons:

- *The knowledge explosion.* Two decades ago, there was little difference between writing to a single reader and writing to multiple readers. Most readers, whether technical people or managers, shared a basic understanding of the organization's product or service. In fact, most managers had risen from the technical staff. Today, however, the knowledge explosion and the increasing complexity of business operations have created a fairly wide gap between what the technical staff knows and what the managers know.

- *New technology for communication.* The photocopy machine has made it simple and inexpensive to send copies of most documents to dozens of people. Electronic mail makes it even simpler and less expensive.

If you think your document will have a number of readers, consider making it *modular.* That is, break it up into different components addressed to different kinds of readers. A modular report, for example, might contain an executive summary for the managers who don't have the time, knowledge, or desire to read the whole report. It might also contain a full technical discussion for expert readers, an implementation schedule for technicians, and a financial plan in an appendix for budget officers.

Strategies for accommodating the multiple audience are discussed in Chapter 10, as well as in Chapters 15–20, which treat different kinds of technical documents.

Understanding Multiculturalism

Our culture and our workforce are becoming increasingly diversified culturally and linguistically, and our businesses are relying more on exports. Technical communicators and technical professionals need to communicate effectively with various groups:

- nonnative speakers of English in the United States
- nonnative speakers of English outside the United States
- speakers of other languages who read texts translated from English into their own languages

Thrush (1993) tells the story of a U.S. elevator company installing Korean-made escalators in Mexico. Even though the U.S. company had the Korean manufacturer translate the manuals into English, the style and organization of the documentation were so different from what the Mexican engineers and technicians expected that they couldn't understand it.

Consider these facts about our dependence on global trade (Lustig & Koester, 1993, pp. 5–6):

- One-third of U.S. corporate profits are earned in international trade.
- The 23 largest U.S. banks do almost half of their business overseas.
- Four of five new jobs in the United States are created to produce goods and services for foreign trade.
- Current U.S. investments overseas are worth more than $300 billion.

And consider these facts about the demographics of the United States (Lustig & Koester, 1993, pp. 9–10):

- In California in the 1980s, the Asian population increased by 1.5 million; the Latino population, by 3.1 million. Today, 43 percent of that state's population is nonwhite; by the year 2000, it is expected to be more than 50 percent nonwhite.
- Children in U.S. schools speak more than 150 languages; the figure in Los Angeles alone is 108.
- By the year 2000, white males are expected to be a minority in the U.S. workforce.

Effective communication requires an understanding of the different categories of cultural patterns.

Categories of Cultural Patterns

First, what are cultural patterns? *Cultural patterns* are the "shared beliefs, values, and norms that are stable over time and that lead to roughly similar behaviors across similar situations" (Lustig & Koester, 1993, p. 98). Unless you have lived in a different culture, you are probably unaware of how powerful and pervasive cultural patterns are. In fact, more than half of U.S. businesspeople on long overseas assignments return home early because of their inability to adapt to the foreign culture (Ferraro, 1990).

As shown in Figure 5.12 on page 120, this section discusses five basic categories of cultural patterns, focusing on information useful to technical communicators and technical professionals.

Values and Beliefs

Cultures differ significantly in gender roles and attitudes toward time:

- *Gender roles.* In most Western cultures, women play a much greater role in the workplace than they do in many Middle Eastern and Asian

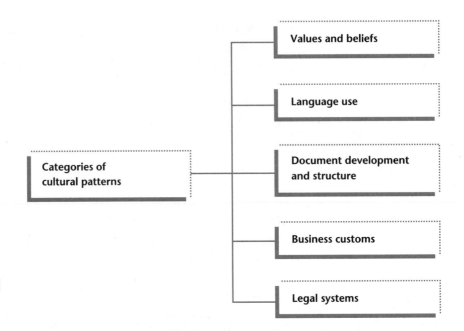

FIGURE 5.12

Categories of Cultural Patterns

cultures. For this reason, women appear more frequently in marketing and advertising materials in Western countries than in others.

- *Attitudes toward time.* Western cultures often view time as a linear concept; once the time has passed, it is lost forever. The focus is on the future and the past. Many other cultures see time as a cycle, with no beginning and no end; the focus is on the present, and a meeting that extends two hours beyond its scheduled length is no problem (Limaye & Victor, 1991). The Hopi language, for instance, does not have tenses, and the Hopi culture has no concept of past, present, and future. For the Hopi, time merely "gets later" (Littlejohn, 1992, p. 209). If you refer to the fast pace of business when addressing readers in India, they might not understand what you mean.

Language Use

Every culture is unique in its use of language and the mechanics of language. Some of the most important differences among cultures are in the following three areas:

- *Vocabulary and sentence length.* In the United States, readers expect simple vocabulary and short sentences. In most European countries, readers expect more complicated vocabulary and long sentences.
- *Numbers.* The United States is the only major country that has not adopted the metric system. In addition, Americans use periods to separate whole numbers from decimals, and commas to separate the thousands from the hundreds. Much of the rest of the world reverses this usage.

United States: 3,425.6

Europe: 3.425,6

- *Dates.* We write out and abbreviate dates differently from most other cultures:

United States: March 2, 1998 3/2/98

Europe: 2 March 1998 2/3/98

Japan: 2 March 1998 98/2/3

The European style—2 March 1998—is gaining popularity in the United States.

Document Development and Structure

Many aspects of document development and structure are culturally determined. You should consider the following eight factors before you start to write for a multicultural audience:

- *Individual and group orientation.* The typical U.S. professional is highly individualistic in outlook; the average Asian professional is highly group oriented. Therefore, U.S. writers use *I* often and refer to themselves more than to the organization they represent; Asian writers do not express a personal viewpoint but refer to their organization frequently.
- *The nature of evidence.* In most Western countries, evidence is empirical and testable. In many other cultures, such as the Japanese, tradition, authority, and group consensus are more important than empirical evidence.
- *Organization.* People in the United States tend to structure documents deductively: from the general to the particular. In much of the rest of the world, the organization is inductive: from the particulars to the general.
- *Directness and indirectness.* Readers in the United States expect explicit, direct documents with the structure clearly spelled out. Many European and Asian cultures value indirectness and digression.
- *Graphics.* Graphics that are perfectly clear in one culture can be confusing or meaningless in another. The garbage-can icon on the Macintosh computer has often been cited as a graphic that does not translate well, for garbage cans are shaped differently and made of different materials around the world.
- *Document design.* Different cultures have different attitudes toward space that affect their use of white space in their documents. Fisk (1992) discusses differences in page design in the United States, England, and Japan: the English use a denser, more crowded design than Americans, and the Japanese design documents with a focal point in the center of the page.

INTERVIEW: TRANSLATOR GEORGE P. RIMALOWER

on translating by machine and by humans

In our company, translations are done exclusively by highly trained professional translators who are native speakers of the target language. We strongly believe that sophistication and perfection cannot be obtained through machine translation. Even though the world of machine translations is improving rapidly, a machine will never be able to grasp cultural issues and nuances. An alternative solution to bridging the bilingual and bicultural gap would be a machine translation accompanied by human editing. However, this process is very labor intensive and, in most instances, it would be easier and faster for the translator to translate a document from scratch.

on why translating is more than just translating individual words

A year or two ago, McDonald's printed take-out bags decorated with flags from around the world. On the bag was additional text translated into the different languages of the flags depicted. Had McDonald's taken the time to consult with a cultural expert, they would have avoided committing the major *faux pas* of printing bags that contained a Saudi flag with holy Koran scripture. This was extremely offensive to the Saudis. To print the Saudi flag, which contains holy scripture, on a disposable bag that would wind up in a garbage heap was sacrilegious. Another example that comes to mind is the Chevy Nova. When it was introduced to the Latin-American market, no one would buy it. No one at General Motors realized that "No va" means "doesn't go" in Spanish.

on internationalization

In the development process of a project, consideration should be given to the possibility of going global and, therefore, internationalizing the product. This is a fancy way of saying that at the development stage the product has to be designed for easy adaptation to international market requirements. If you haven't internationalized the product, then the localization to a specific geographic market can pose a problem. For example, one area of localization is the conversion of measurements (metric system, British and U.S. systems), colors, dates (day first followed by the month and vice versa, depending on the country), calendars (Gregorian used in Israel; Imperial used in Japan, etc.), currencies, temperatures (Celsius versus Fahrenheit), translation of icons, etc., to a specific market. Localization is an expensive process, and those expenses can be significantly reduced if the product has been internationalized from its inception. Considering all these issues at the beginning makes the globalization process much easier.

- *Format.* In business letters, many cultures use distinctive patterns for the salutation, the opening and closing paragraphs, and the complimentary close. In Japan, for example, business letters usually begin with a reference to the season, such as "The snow has now disappeared, and the buds are starting to appear on the branches."
- *Amount of detail.* Cultures are classified as high context and low context. In a *high-context culture,* such as the Japanese, the meaning is communicated more implicitly than explicitly. The reader is left to infer the details through the context and tone of the document; therefore, documents tend to be relatively short. In a *low-context culture,* such as that of the United States, the message is stated explicitly and fully supported by details; therefore, documents tend to be longer.

INTERVIEW *(continued)*

on domestic localization for products sold in the United States

If you look at U.S. demographics, we are really a melting pot. With so many non-English-speaking consumers, domestic localization can be a major selling point. We are living in a very competitive environment. If you do not speak English and need to purchase a certain product, you would most likely choose the product that has packaging instructions written in your language. By applying simple localization rules, the various ethnic groups can be easily reached.

on the aspects of culture that are vulnerable to miscommunication

Issues such as the use of icons in written text can be considered offensive or misleading to many cultures. For example, in Australia, Greece, or Turkey, the familiar "thumbs up," both physical and printed, is considered very offensive. Similarly, the American gesture of "OK" means "zero" or "worthless" in France, while in Japan it means you want your change in coins. Pitfalls such as these must be carefully avoided to get the right message across.

Idiomatic expressions are not always translatable. One idiom can mean something quite different in another language; therefore, cultural equivalents

must be used. Other aspects to consider are the level of formality, and foods that are acceptable or unacceptable in other cultures. Using this information in your favor gives you an edge over your competitors.

on the process of translating

The very first thing to consider before translating is the target audience and localization issues that need to be addressed. We then select a team of subject-specific translators/editors. The linguists chosen are not only proficient in the language pair combination, but are native speakers of the target language. They are degreed linguists who frequently hold dual degrees in the subjects they translate. After a document is translated, the material goes to an editor to ensure content and continuity and to guarantee that sensitive cultural issues are appropriately addressed. The final step is an in-depth quality control review by our project managers. This includes proofreading, formatting, and incorporation of any special requests the client may have.

George P. Rimalower is CEO of ISI Translation Services in Valley Village, California.

Business Customs

Business customs vary greatly from culture to culture. Differences include the following:

- *Business protocol.* This category includes greetings, business dress, and customs regarding gifts.
- *Formality and informality.* The level of formality used in addressing a person varies greatly from culture to culture.
- *Nonverbal communication.* Innocent hand gestures in one culture are highly offensive in another. The space that people maintain between one another also differs significantly from culture to culture. In many European cultures, the distance between two people talking is much smaller

than in the United States. Seating arrangements in offices and at meetings also differ from one culture to another.

- *Punctuality.* Some cultures require absolute punctuality in business meetings. Other cultures expect you to be 10 or 20 minutes late.
- *Bribery.* Bribery is still required for doing business in many cultures. This reality has serious implications—involving ethics and U.S. law—that your organization has to consider (Martin & Chaney, 1992). For a good introduction to the topic, see Velasquez (1992).

Legal Systems

When you distribute documents in foreign countries, you must adhere to their laws regarding safety, health, and the environment. In addition, you must be aware of local laws regarding timing of delivery, form and terms of payment, the role of consultants, censorship, and authority to make binding decisions. Recently, a group of American businesspeople trying to invest in a Russian company were surprised to learn that they could not meet in their hotel because they had not obtained the permit required for meetings of more than 10 people.

Writing to Multicultural Audiences

The preceding discussion of the major areas of difference among cultures cannot answer questions about how to write for any particular multicultural audience; all it can do is point out the areas you need to investigate. The most valuable piece of advice is to do your homework on the culture of your readers; read everything you can about the society. In addition, as you plan the document, get help from someone native to the culture of your readers. This person can help you prevent blunders that might confuse or offend your readers.

In general, the following eight guidelines will help you improve the success of your writing:

- *Limit your vocabulary.* Make it a goal that every word you use will have only one meaning. See Chapter 14 for a discussion of Simplified English, a group of languages that rely on a limited vocabulary.
- *Define abbreviations and acronyms in a glossary.* Don't count on your readers to know what a GFI—ground fault interrupter—is, because the abbreviation is based on English vocabulary and word order. See Chapter 10 for a discussion of glossaries.
- *Avoid jargon unless you know your readers are familiar with it.* For instance, your readers might not know what a *graphical user interface* is.
- *Avoid idioms.* Don't write anything immediately after watching a game on television. You'll end up telling your Japanese readers that your company plans to put on a full-court press, and they will have no idea what you're talking about. One person new to American culture got in the carpool lane on the expressway because he wanted to get his car washed.

- *Keep sentences short.* There is no magic number, but try to stay within a range of 20 to 25 words.

- *Use the active voice whenever possible.* The active voice—in which the actor is the grammatical subject of the sentence—is easier for nonnative speakers of English. For more information on voice and other aspects of sentence construction, see Chapter 14.

- *Be sensitive to the culture when creating graphics.* Keep in mind the culture's values, particularly regarding gender roles. See Chapters 3 and 12.

- *Have the document reviewed by someone from the culture.* Even if you had help planning the document, have it reviewed before you send it out or get it printed.

CASE

Analyzing an Audience

You work in the Customer Service department at Greenlawn, a company that applies chemical fertilizers, herbicides, and pesticides to lawns. Recently Greenlawn received a letter from Gwen Smith, a customer with a five-year-old child and an infant. Ms. Smith is very concerned about the safety of the chemical you spray on her lawn; recently she read an article in the newspaper about a man who died suddenly after playing a round of golf on a course that had been treated by a similar company. The article stated that although commercial lawn treatments are generally safe, every year two or three people die because of an unusual vulnerability to the chemicals.

Your supervisor, Helen Lewis, has drafted the following letter to Ms. Smith and has asked you to review it. "Do you think this responds to Ms. Smith's concerns?" Helen asks you. Write a memo to Helen Lewis evaluating how effectively the letter responds to the concerns of its reader. See Chapter 16 for a discussion of memos.

GREENLAWN   April 5, 19—

Dear Ms. Smith,

Thank you for inquiring about the safety of the Greenlawn program. The materials purchased and used by professional landscape companies are effective, non-persistent products that have been extensively researched by the Environmental Protection Agency. Scientific tests have shown that dilute tank-mix solutions sprayed on customers' lawns are rated "practically nontoxic," which means that they have a toxicity rating equal to or lower than such common household products

(Continued)

(Case continued)

as cooking oils, modeling clays, and some baby creams. Greenlawn applications present little health risk to children and pets. A child would have to ingest almost 10 cupsful of treated lawn clippings to equal the toxicity of one baby aspirin. The child's stomach would rupture before the lawn clippings proved dangerous!

Research published in the *American Journal of Veterinary Research* in February 1984 demonstrated that a dog could not consume enough grass treated at the normal rate of application to ingest the amount of spray material required to produce toxic symptoms. The dog's stomach simply is not large enough.

A check at your local hardware or garden store will show that numerous lawn, ornamental, and tree-care pesticides are available for purchase by homeowners either as a concentrate or combined with fertilizers as part of a weed-and-feed mix. Label information shows that these products contain generally the same pesticides as those programmed for use by professional lawn-care companies, but contain higher concentrations of these pesticides than found in the dilute tank-mix solutions applied to lawns and shrubs. By using a professional service, homeowners can eliminate the need to store pesticide concentrates and avoid the problems of improper overapplication and illegal disposal of leftover products in sewers or household trash containers.

On the basis of these facts, I am sure that you will be pleased to know that the Greenlawn program is a safe and effective way to protect your valued home landscape. I have also enclosed some additional safety information. I encourage you to contact me directly should you have any questions.

Sincerely,

Helen Lewis
Branch Manager

WRITER'S CHECKLIST

Following is a checklist for analyzing your audience. Remember that your document might be read by one person, several people, a large group, or several groups with various needs.

1. Have you filled out a reader's profile sheet for your primary and secondary audiences that considers
   ❑ education?
   ❑ professional experience?
   ❑ personal preferences?
   ❑ attitudes toward you and your subject?
   ❑ cultural characteristics?
   ❑ use of the document?
   ❑ physical environment?

2. In planning to write to a multicultural audience, have you considered differences in
   ❑ values and beliefs?
   ❑ use of language and mechanics?

❑ document development and structure?

❑ business customs?

❑ laws?

3. In writing for a multicultural audience, did you

❑ limit your vocabulary?

❑ define abbreviations and acronyms in a glossary?

❑ avoid jargon unless you know that your readers are familiar with it?

❑ avoid idioms?

❑ keep sentences short?

❑ use the active voice whenever possible?

❑ use sensitivity in creating graphics?

❑ have the document reviewed by someone from the culture?

EXERCISES

Several of the following exercises call for you to write a memo. See Chapter 16 for a discussion of memos.

1. Choose a 200-word passage from a technical article addressed to an expert audience. Rewrite the passage so that it is clear and interesting to the general reader. Submit the original passage along with your memo.

2. Audience is the primary consideration in many types of nontechnical writing. Choose a one- or two-page magazine advertisement for an economy car, such as a Subaru, and one for a luxury car, such as a Mercedes. In a memo to your instructor, contrast the audiences for the two ads by age, sex, economic means, hobbies, interests, and leisure activities. In contrasting the two audiences, consider the explicit information in the ads—the writing—as well as the implicit information—hidden persuaders such as background scenery, color, lighting, angles, and the situation portrayed by any people photographed. Submit the ad along with your memo. (Submit color photocopies or the original ads from the magazines.)

3. Fill out an audience profile sheet about yourself, as if you were the reader of your own writing. Then fill out an audience profile sheet about your instructor. What are the major differences between the two profiles?

4. Interview a student from another culture. If you do not know someone, consult the international-student office, the English Department, or the English as a Second Language Office. Choose one of the topics discussed in this chapter, such as gender roles, notions of time, or document standards, and write a memo to your instructor comparing and contrasting your culture's view with that of the other student. (For a discussion of comparison and contrast, see Chapter 8.)

5. The following passage is for an advertisement from a translation service. Revise the passage to make it more appropriate for a multicultural audience.

> If your technical documents have to meet the needs of a global market but you find that most translation houses can't handle the huge volume, accommodate the various languages you require, or make your deadlines, where do you turn?
>
> Well, your search is over. Translations, Inc. provides comprehensive translations in addition to full-service documentation publishing.
>
> We utilize ultrasophisticated translation programs that can translate a page in a blink of an eye. Then our crack linguists comb each document to give it that personalized touch.
>
> No job too large! No schedule too tight! Give us a call today!

Collaborative Exercise

6. Form small groups for this project on multiculturalism. The student-exchange program at your college wishes to write information booklets about each of the countries in which students from your school study. Research one country and write a 2,000-word guide that explains the culture for students preparing to study there. Begin by studying the Central Intelligence Agency's fact book about the country (http://www.odci.gov/cia/publications) and any materials that the student-exchange program can provide. One group member might concentrate on the country itself: the land, the culture, and the language. Another student might concentrate on the city in which the college or university is located. A third group member might focus on the college or university itself: its policies, its student makeup, its opportunities, the expectations for students in that country, and so forth. Do not overlook Web sites in researching the country, the city, and the school.

Web Exercise

7. Form small groups and study two sites on the Web that advertise competing products: two manufacturers of cars or television shows or music producers. Have each person in the group work alone, comparing and contrasting the two sites according to three bases:

- the kind of information they provide: hard, technical information or more emotional advertising
- the use of multimedia such as animation, sound, or video
- the amount of interactivity they invite, that is, the extent to which you can participate in activities while you visit the site

After each person has separately studied the sites and taken notes about the three points, come together as a group. Each person will share his or her findings, then the group will discuss the differences among the impressions of its members. What aspects of the sites seemed to cause the most difference in how the group members reacted to it? Which aspects seemed to elicit the most consistent reactions? In a brief memo to your instructor, describe and analyze the way the two sites were perceived by the different members of the group.

REFERENCES

D'Ottavi, J. (1993). *Manufacturing bidirectional valves: A feasibility study.* Unpublished manuscript, Drexel University.

Ferraro, G. P. (1990). *The cultural dimensions of international business.* Englewood Cliffs, NJ: Prentice Hall.

Fisk, M. J. (1992). People, proxemics, and possibilities for technical writing. *IEEE Transactions on Professional Communication, 35*(3), 176–182.

Limaye, M. R., & Victor, D. A. (1991). Cross-cultural business communication research: State of the art and hypotheses for the 1990s. *Journal of Business Communication, 28*(3), 277–299.

Littlejohn, S. (1992). *Theories of human communication* (4th ed.). Belmont, CA: Wadsworth.

Lustig, M. W., & Koester, J. (1993). *Intercultural competence.* New York: HarperCollins.

Martin, J. S., & Chaney, L. H. (1992). Determination of content for a collegiate course in intercultural business communication by three Delphi panels. *Journal of Business Communication, 29*(3), 267–283.

Micron Electronics, Inc. (1996). *Micron Millennia: PCI-Bus system user's manual.* Boise, ID: Micron Electronics, Inc.

Schwartz, E. I. (1994, August 21). A work in progress: Sign language telephones. *New York Times,* p. 3: 7.

Thrush, E. A. (1993). Bridging the gaps: Technical communication in an international and multicultural society. *Technical Communication Quarterly, 2*(3), 271–283.

Velasquez, M. (1992). *Business ethics: Concepts and cases* (3rd ed.). Englewood Cliffs, NJ: Prentice Hall.

Yazici, H., Benjamin, C., & McGlaughlin, J. (1994). AI-based generation of production engineering labor standards. *IEEE Transactions on Engineering Management, 41*(3), 302–309.

Determining Your Purpose and Strategy

Determining Your Purpose 130

Determining Your Strategy 131

Ethical Constraints *131*
Legal Constraints *131*
Informational Constraints *132*
Personnel Constraints *132*
Financial Constraints *133*
Time Constraints *133*
Format and Tone Constraints *134*

Planning Your Tactics 134

Proposing a Strategy 136

Proposing a Small Project *137*
Proposing a Large Project *137*

Case: Making a Q-and-A Sheet *138*

Writer's Checklist 140

Exercises 140

Collaborative Exercise *140*
Web Exercise *140*

Chapter 5 introduced two important elements of writing: the audience and the purpose of a document. This chapter begins with a discussion of how to determine your purpose, and follows with guidelines for planning a strategy and proposing it to your managers.

Determining Your Purpose

Once you have identified and analyzed your audience, examine your general purpose in writing. Ask yourself this simple question: "What do I want this document to accomplish?" When your readers have finished reading what you have written, what do you want them to *know* or *believe?* What do you want them to *do?* Think of your writing not as an attempt to say something about the subject but as a way to help others understand it or act on it. Think of your writing as an attempt to persuade your readers, to convince them to hold a particular belief or take a particular action.

To define your purpose clearly, think in terms of verbs. Isolate a single verb that represents your purpose, and keep it in mind throughout the writing process. (Of course, in some cases a technical document has several purposes, and therefore you might want to choose several verbs.) Following are examples of verbs that indicate typical purposes in technical documents. The list has been divided into two categories: verbs used to communicate information to your readers and verbs used to convince them to accept a particular point of view.

Communicating Verbs	*Convincing Verbs*
to explain	to assess
to inform	to request
to illustrate	to propose
to review	to recommend
to outline	to forecast
to authorize	to evaluate
to describe	
to define	
to summarize	

This classification is not absolute. For example, *to review* could in some cases be a *convincing verb* rather than a *communicating verb*: one writer's review of a complicated situation might be very different from another's review of the same situation. Following are a few examples of how you can use these verbs in clarifying the purpose of your document. (The verbs are italicized.)

- This report *describes* the research project that is intended to determine the effectiveness of the new waste-treatment filter.

- This report *reviews* the progress in the first six months of the heat-dissipation study.
- This letter *authorizes* the purchase of six new PCs for the Jenkintown facility.
- This memo *recommends* that we create the Web site as soon as possible.

As you devise your purpose statement, remember that your real purpose might differ from your expressed purpose. For instance, if your real purpose is to persuade your reader to lease a new computer system rather than purchase it, you might phrase the purpose this way: *to explain the advantages of leasing over purchasing.* As mentioned in Chapter 5, many readers don't want to be *persuaded* but are willing to learn new facts or ideas.

Determining Your Strategy

Once you have analyzed your audience and determined your purpose, the next step is to put the two together and create a strategy for writing the document. *Strategy* refers to your game plan: how you are going to create a document that accomplishes your goals. Specifically, planning a strategy involves determining the constraints you must work within.

As a student writer, you work within a number of limits or constraints: the amount of information you can gather for a paper, the length and format, the due date, and so forth. In business, industry, and government, you will be working within similar constraints. Figure 6.1 on page 132 shows seven sets of constraints.

Ethical Constraints　As discussed in Chapter 2, we all work within a web of rights and responsibilities. Our greatest responsibility is to our own sense of ethical behavior. Our ethical standards are challenged when we are asked to lie or mislead in documents.

In most cases, you have some options when you feel you are being asked to act unethically. Some organizations and professional communities have a published code of conduct. In addition, many large companies have ombudspersons, people whose job is to help employees resolve ethical and other conflicts within the organization. Don't feel that you are obligated to do whatever your supervisor tells you to. If you think that you are being asked to act unethically, consider the issue carefully and then take advantage of the resources available to you.

Legal Constraints　You are obligated to abide by all applicable laws concerning labor practices, environmental issues, fair trade, consumer rights, and so forth. If you think that the action recommended by your supervisor has potential legal implications for you or your company, meet with your organization's legal counsel and, if necessary, attorneys outside the organization. Chapter 2 discusses legal issues at greater length.

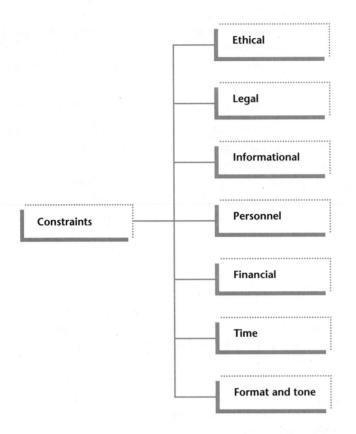

Constraints

Informational Constraints You might face constraints on the kind or amount of information you can use in the document. Most often, the information you need is not available. For example, you might want to recommend that your organization buy a piece of equipment, but you can't find objective evidence that it will do the job. You have advertising brochures and testimonials from satisfied users, but apparently nobody has performed the kinds of controlled experiments or tests that would convince the most skeptical reader.

What do you do? You tell the truth. You state exactly what the situation is, weighing the available evidence and carefully noting what is missing. Your most important credential on the job is credibility; you will lose it if you unintentionally suggest that your evidence is better than it really is. In the same way, you don't want your readers to think that you don't realize your information is incomplete; they will doubt your technical knowledge.

Personnel Constraints As discussed in Chapter 3, much technical communication is written collaboratively by teams of writers, editors, production specialists, and subject-matter experts. One constraint you will face in working on large projects is lack of access to as many collaborators as you need. Every organization hires as few people as it can, of course. For this reason, you must have a good idea of the kind of personnel you will need, and the sooner you can communicate your needs to management, the more likely you are to get what you need.

Many organizations hire technical communicators and other personnel project by project, not as regular employees. In such cases, present a clear and persuasive proposal to hire the personnel you need. However, you will probably have to make do with fewer people than you want, and everyone will have to work harder and perform some tasks for which they are ill suited.

Financial Constraints   Related to personnel constraints are financial constraints, for if you had unlimited funds, you could hire all the personnel you need. But financial constraints can affect other categories of resources, too. For instance, a printing expert can tell you, almost to the penny, what the extra costs will be to move up a grade in paper, or to add color to a document, or to use a comb binding instead of saddle binding. Producing large, complex documents—especially in large quantities—is an exercise in precise estimating. If your organization has a technical-publications department, consult it for advice. Otherwise, work closely with print shops and service bureaus.

Time Constraints   People, especially writers, never think they have enough time to do their jobs. In fact, many managers don't allot enough time for their employees to do a good job putting together all the necessary memos, letters, reports, and manuals. Most working professionals resign themselves to taking work home in the evenings or on the weekends. Some people come in early to do their writing before things get hectic.

The first piece of information you need, of course, is the deadline for the document. Sometimes a document will have a number of deadlines—for intermediate reviews, progress and status reports, and so forth. Once you know when a particular document (or part of it) is due, work backward to devise a schedule. For example, if a major report is due on a Friday, you know that you must have it essentially done by the day before, because you want to let it sit overnight before you proofread it for the last time. If it has to be done by Thursday, you might want to give yourself at least four days for drafting and revising. That means you want to finish your outlining by the previous Friday.

Keep in mind that few people can accurately estimate the time a particular activity will take. As you probably know intuitively, tasks almost always take longer than estimated. People call in sick, information is lost or delayed in the mail—whatever can go wrong will go wrong, as Murphy's Law states. When you collaborate, the number of potential problems increases dramatically, because when one person is delayed, other people may lack the necessary information to proceed, causing a logjam. In addition, friction between people who are working hard can slow down a project substantially.

Creating a schedule is discussed in Chapter 3.

Format and Tone Constraints

You will be expected to work within one more set of constraints:

- *Format.* Format constraints are limitations placed on the size, shape, or style of the document, either implicitly or explicitly, by the organization for which you work or the organization to which you are writing. For example, many organizations have guidelines on how to present different kinds of information—all tables and figures are presented at the end of the report, or the names of the people receiving copies of your memo are listed in alphabetical order, or by hierarchy. You can best determine the organizational constraints by finding out whether a company style guide exists. If not, the company might use an externally prepared style guide, such as the *Chicago Manual of Style* or the *U.S. Government Printing Office Style Manual.* If your organization has no official style guide, check similar documents on file to see what other writers have done, and, of course, talk with more experienced coworkers.

  If you are writing to another company, learn what you can about its preferences. If you are responding to a request for a proposal, determine whether the company specifies length, organization, and style. Many government agencies, for example, require that potential suppliers submit their proposals in two separate parts: the technical proposal and the budget.

- *Tone.* You already know that you write one way to an organizational superior and another way to a peer or a subordinate. When addressing superiors, use a formal, polite tone. When addressing peers or subordinates, use a less formal tone but be equally polite. Politeness is common courtesy, and it is common sense: you should not bully people. If you obnoxiously tell the janitor to change the bulb in the ceiling fixture above your desk, you might wait a few weeks for it, or the next day you might find that you have a new bulb and a large footprint in the middle of your desk blotter.

Planning Your Tactics

You have analyzed your audience, determined your purpose, and learned under what constraints you will be working. Now it is time to put all these factors together and to think about what the document will eventually look like and how to get from here to there.

For illustration, let's go back to the writing situation described in Chapter 5: you want to persuade your two readers to authorize the purchase of some CAD equipment. Your primary reader, Harry Becker, is the head of your department. Your secondary reader is Tina Buterbaugh, the head of finance. Your audience analysis suggests that only Harry Becker knows much about CAD equipment. However, because Tina Buterbaugh has been at the company for a number of years, she probably understands the basics of the technology: what the equipment is used for, why it makes draftspeople much more efficient, and so on.

In this case, your purpose in writing is clear: to convince Harry Becker that your proposal is reasonable, so that he will authorize the purchase of the equipment. But you might want to phrase that purpose differently when it appears in the document: *to explain the advantages of acquiring new CAD equipment*, rather than *to persuade you to authorize me to purchase new CAD equipment*.

Think about the constraints you will be working under:

- *Ethical.* You can't think of any ethical constraints. The project is a straightforward proposal to upgrade equipment.

- *Legal.* Again, nothing.

- *Informational.* As far as you know, obtaining the necessary information should be no problem because the CAD system you are interested in has been on the market for some months and has been reviewed in several professional journals. In addition, you know draftspeople at other companies who have purchased it, will talk with you about their experience, and will let you try out the equipment.

- *Personnel.* The project is relatively simple and should not require any personnel. You can research, write, and produce the document by yourself. However, you realize that it is a good idea to talk with other draftspeople during the planning stage and hear their thoughts about your draft. In addition, you decide to show the final draft to Karen, one of the technical writers, who often critiques and proofreads documents for the draftspeople.

- *Financial.* There are no financial constraints for your proposal. Ordinarily, a recommendation to spend $4,000 would be routine. However, in light of an unwise recent project that wasted $50,000, you know that Harry has become careful about authorizing capital expenditures, even small ones; he knows that next year's budget will depend to a great extent on the success of this year's purchases.

- *Time.* The only time constraint is the familiar problem: finding the time to write the document. There is no external deadline, because this is a self-generated project; however, you have not received authorization to spend company time on the proposal. Therefore, you will probably have to work on it for a few evenings at home.

- *Format and tone.* Although your company doesn't have a style manual that describes internal proposals, the format has remained fixed over the years, and there are plenty of models to consult.

What does your analysis of audience and your determination of purpose and constraints tell you about how to write the document? Because your readers will be particularly careful about large capital expenditures, you must clearly show that the type of equipment you want is necessary and that you have recommended the most effective and efficient model. In addition, your proposal will have to answer a number of questions that will be going through your readers' minds:

- What system is being used now?
- What is wrong with that system, or how would the new equipment improve our operations?
- On what basis should we evaluate the different kinds of available equipment?
- What is the best piece of equipment for the job?
- How much will it cost to purchase (or lease), maintain, and operate the equipment?
- Is the cost worth it? At what point will the equipment pay for itself?
- What benefits and problems have other purchasers of the equipment experienced?
- How long would it take to have the equipment in place and working?
- How would we go about getting it?

Your readers' preferences about writing style indicate that you should be straightforward, direct, and objective. Avoid technical vocabulary and unnecessary technical details, but make clear the practical advantages of the new equipment. In addition, it would be a good idea to include an executive summary for Tina Buterbaugh's convenience.

Proposing a Strategy

After you know what you want to do, consider gaining management's approval before proceeding.

You needn't seek approval for every kind of project, of course. Sometimes you simply do your research and write your document. But the larger and more complex the project and the document, the more sense it makes to be sure that you are on the right track before you invest too much time and effort. For the sake of simplicity, projects can be classified into two categories: small and large.

Small and *large* mean different things in different contexts. At a three-person software company, a $5,000 project might be large; at Microsoft, a $5,000,000 project would be relatively small. Despite these enormous differences of scale, these categories can help show the range of possibilities in proposing a strategy.

Proposing a Small Project

You are planning the CAD equipment project. You have a good understanding of your audience, purpose, and strategy, and a general outline is taking shape in your mind. But before you actually start to write an outline or gather all the information you will need, it's a good idea to spend another 10 or 15 minutes making sure your primary reader agrees with your thinking. You don't want to waste days or even weeks working on a document that won't fulfill its purpose.

Submitting to your primary reader a statement of your understanding of the audience, purpose, and strategy is a means of establishing an informal contract. Although readers can change their minds—in response to new developments or after more thought—provisional approval is better than no approval at all. Then, if the project is subsequently revised or canceled, at least you can prove that you have been using your time well.

Writing such a statement has an additional benefit: it helps redefine the situation. What was earlier an idea you were thinking about is now a project, and this redefinition makes it easier to receive authorization to use company resources.

Some writers are reluctant to submit such a statement to the primary reader, either because they don't want to be a nuisance or because they're worried that it might reveal a serious misunderstanding of the writing situation. There may be some people out there who don't want to be bothered, but I haven't met any. Why should they object? It doesn't take more than a minute to read your brief statement, and if there *has* been a misunderstanding, it is far easier to remedy at this early stage. Having the document come out right the first time is in everyone's interest.

Your statement can serve a different purpose as well: if you want your reader's views on which of two strategies to pursue, you can describe each one and ask your reader to state a preference.

What should this statement look like? It doesn't matter. You can send an e-mail or write a memo, as long as you clearly and briefly state what you are trying to do. Figure 6.2 on page 138 is an example of the statement you might submit to your boss about the CAD equipment.

In composing this statement, the writer began with her audience profile sheets for the two principal readers (see Chapter 5, Figures 5.10 and 5.11). She describes a logical, rational plan for proposing the equipment purchase.

Once you have received your primary reader's approval, you can feel confident in starting to gather and interpret information and then writing your document. These tasks will be discussed in the next six chapters.

Proposing a Large Project

Most organizations require that large, complex projects be planned more carefully than small, less complex ones. Within your organization, there might be written guidelines on what kinds of projects require what kinds of planning documents, or the guidelines might be unwritten.

Harry:

Tell me if you think this is a good approach for the proposal on CAD equipment.
Outright purchase of the complete system will cost more than $1,000, so you
would have to approve it and send it on for Tina's approval. (I'll provide leasing
costs as well.) I want to show that our CAD hardware and software are badly out
of date and need to be replaced. I'll be thorough in recommending new equipment,
with independent evaluations in the literature, as well as product demonstrations.
The proposal should specify what the current equipment is costing us and show
how much we can save by buying the recommended system.

I'll call you later today to get your reaction before I begin researching what's
available.

Renu

FIGURE 6.2

**Proposal for a Small
Project**

Many kinds of planning documents exist. Complex projects usually call for
a *proposal*, a formal statement that answers the following questions:

- What do you plan to do?
- Who will participate in the project?
- How do you plan to do the project?
- Why do you plan to do it that way?
- When do you plan to do the project?
- Where do you plan to do the project?
- What are your credentials to do the project?

See Chapter 15 for a discussion of proposals.

CASE

Making a Q-and-A Sheet

You are an assistant to Gilbert F. Casellas, Chairman of the U.S. Equal Employment Opportunity
Commission (EEOC). The EEOC has published a number of fact sheets, such as the one on sexual
harassment printed here. Mr. Casellas tells you that he has received a number of requests for a
question-and-answer sheet discussing sexual harassment that can be posted in workplaces. "Would
you mind taking care of this for me?" he asks you. "Make sure it is easy to understand, even for
nonnative speakers. And be sure you use examples so it doesn't sound like legalese." Write a one-
page question-and-answer sheet that can be distributed to companies and other organizations that
request it. (For more information on the EEOC, visit its site at http://www.eeoc.gov.)

FACTS ABOUT SEXUAL HARASSMENT

Sexual harassment is a form of sex discrimination that violates Title VII of the Civil Rights Act of 1964.

Unwelcome sexual advances, requests for sexual favors, and other verbal or physical conduct of a sexual nature constitute sexual harassment when submission to or rejection of this conduct explicitly or implicitly affects an individual's employment, unreasonably interferes with an individual's work performance or creates an intimidating, hostile, or offensive work environment.

Sexual harassment can occur in a variety of circumstances, including but not limited to the following:

- The victim as well as the harasser may be a woman or a man. The victim does not have to be of the opposite sex.
- The harasser can be the victim's supervisor, an agent of the employer, a supervisor in another area, a coworker, or a nonemployee.
- The victim does not have to be the person harassed but could be anyone affected by the offensive conduct.
- Unlawful sexual harassment may occur without economic injury to or discharge of the victim.
- The harasser's conduct must be unwelcome.

It is helpful for the victim to directly inform the harasser that the conduct is unwelcome and must stop. The victim should use any employer complaint mechanism or grievance system available.

When investigating allegations of sexual harassment, EEOC looks at the whole record: the circumstances, such as the nature of the sexual advances, and the context in which the alleged incidents occurred. A determination on the allegations is made from the facts on a case-by-case basis.

Prevention is the best tool to eliminate sexual harassment in the workplace. Employers are encouraged to take steps necessary to prevent sexual harassment from occurring. They should clearly communicate to employees that sexual harassment will not be tolerated. They can do so by establishing an effective complaint or grievance process and taking immediate and appropriate action when an employee complains.

WRITER'S CHECKLIST

1. What is your purpose in writing? What is the document intended to accomplish?

2. Is your purpose consistent with your audience's needs?

3. How does your understanding of audience and purpose determine your strategy: the scope, structure, organization, tone, and vocabulary of the document?

4. Will you have to accommodate
   ❏ ethical constraints?
   ❏ legal constraints?
   ❏ informational constraints?
   ❏ personnel constraints?
   ❏ financial constraints?
   ❏ time constraints?
   ❏ format or tone constraints?

5. Have you proposed your project to your primary readers to see if they approve of it?

EXERCISES

Several of the following exercises call for you to write a memo. For a discussion of memos, see Chapter 16.

1. Locate a fund-raising letter from a charitable organization, such as the American Cancer Society, the March of Dimes, or the Special Olympics. Write a memo to your instructor describing the different tactics used in the letter to persuade you to give money to the cause. Which tactics work well, and which do not? Do not limit your analysis to the argument made by the words themselves; consider also the design of the letter, the appearance of the type, and the type of paper used. If the letter includes any graphics or the envelope included any materials other than the letter itself, analyze them as well.

Collaborative Exercise

2. Form small groups for this exercise on audience. Choose two articles on the same subject, one from a general-audience periodical, such as *Time* or *Newsweek*, and one from a technical journal, such as *Science, Journal of Visual Languages and Computing, Management Science*, or *Journal of Microcomputer Applications*. Have each member of the group work alone in studying each article, focusing on the following questions:
   a. What is the background of the article's audience likely to be? Does the article require that the reader have specialized knowledge?
   b. What is the author's purpose? In other words, what is the article intended to accomplish?
   c. How do the differences in audience and purpose affect the following elements in the article?
      • scope and organization
      • sentence length and structure
      • vocabulary
      • number and type of graphics
      • references within the articles and at the end
      • headings and white space

Meet and discuss your analyses of the two articles. Do each of you agree more in your analyses of the general-audience article or of the technical article? What are the main areas of difference in your analyses? What do you think accounts for the differences? Write a memo to your instructor that not only compares and contrasts the two articles from the point of view of the authors' assessment of the writing situation but also describes and analyzes the differences among each group members' analyses. (For more information on comparison and contrast, see Chapter 8.) Submit the two articles (or photocopies of them) along with your memo.

Web Exercise

3. Form small groups to study the Web site of a large organization such as the American Red Cross (www.redcross.org), the White House (www.whitehouse.gov), or Microsoft Corporation (www.miscrosoft.com). Have each person in the group study the site separately, considering the following questions about the site:
   • What are the various purposes of the site? (Briefly describe the major pages the site contains.)
   • How do the writers and designers of the site try to accomplish these purposes? What kinds of graphics or animation do they use? What techniques do

they use in the textual portions of the site (for example, comment on word choice, level of formality, paragraph development, and so forth)?

- How effective is the site in achieving these purposes? What techniques are ineffective, and how might the site be improved?

Come together as a group and share your perceptions about the site. Draft a memo to your instructor presenting your findings, paying particular attention to those aspects of the site that elicited different perceptions of the site's purpose.

REFERENCE

U.S. Equal Employment Opportunity Commission. (1997). Facts about sexual harassment. <http://www.eeoc.gov/facts/fs-sex.html> (1997, Apr. 8).

CHAPTER

7

Researching Your Subject

The Role of Research in the Problem-Solving Process 143

Characteristics of Good Information 144

Choosing a Topic 145

Generating Ideas about the Topic 146
　Asking Journalistic Questions *147*
　Brainstorming *148*
　Freewriting *148*
　Talking *149*
　Sketching *150*

Conducting Secondary Research 152
　Understanding the Research Media *153*
　Using Basic Research Tools *156*
　Researching Government Information *160*
　Understanding the Information *161*
　Evaluating the Information *169*
　Evaluating Internet Sources *170*

Conducting Primary Research 171
　Inspections *171*
　Experiments *171*
　Field Research *172*
　Interviews *173*
　Letters of Inquiry *176*
　Questionnaires *176*

An Example of How to Plan a Research Strategy 180
　Deciding on a Research Sequence *182*
　Carrying Out Secondary Research *182*
　Planning Primary Research *184*

Case: Creating a Research Guidebook for Your Major *185*

Writer's Checklist 186

Exercises 186
　Collaborative Exercise *187*
　Web Exercise *187*

Chapters 5 and 6 discussed planning a writing project: analyzing the document's audience and determining its purpose, then planning a writing strategy that accommodates the audience's needs and your time and other constraints. The next step in most writing projects is to find and evaluate the technical information that will go into the document.

This chapter covers the techniques of *primary research* and *secondary research*. Primary research involves creating technical information yourself. Secondary research is the process of collecting information that other people have discovered or created. The most common ways to perform secondary research are to read books and journals and search the Web, but researchers also talk with colleagues, consult databases, and attend conferences.

This chapter presents secondary research first because rarely would you conduct primary research before secondary research. To design the experiments or the field research that goes into primary research, you should have a thorough understanding of the information that already exists about the subject.

This chapter explains how to generate ideas and how to perform secondary research: understanding the different research media, using basic research tools, researching government information, and understanding and evaluating the information you have gathered. The chapter then discusses techniques of primary research, including how to conduct inspections, experiments, field research, and interviews; how to write letters of inquiry; and how to administer questionnaires. The chapter concludes with a description of the process a group of students might use in planning a research strategy for an investigative report.

The Role of Research in the Problem-Solving Process

Sometimes students think they write research reports so they can show their teachers that they know how to write research reports. In the working world, you do research as part of the larger task of solving problems. Nobody has ever said, "I have an easy week coming up; I think I'll write a report."

In the working world, you don't have to look for problems and opportunities; they present themselves all the time. For instance, you are on a business trip, during which you use your laptop to make a slide presentation for a potential client. Presentations are on your mind, and during a free hour you visit an exhibition of office products and see a new device for projecting slide presentations from a laptop. You bring back some product literature. After talking with your colleagues about the new product, you conclude that it might be worthwhile to investigate the need for this device at your company. You will want to answer questions such as these:

- How many presentations do we make from laptops?
- What are the advantages and disadvantages of the way we currently handle slide projection?

- What are our technical and financial needs for a projection device?
- What are the available products?
- What are the advantages and disadvantages of each product?
- Which product would be best for us?

Each of these questions calls for research. Here is how you might research several of them:

- *How many presentations do we make from laptops?* You contact a few colleagues within the company to find out how many business trips your company's employees go on, how many times they take a laptop, and how often they make presentations from the laptop.

- *What are the advantages and disadvantages of the way we currently handle projection?* You ask an assistant to review some recent trip reports to see whether there is any mention of how your colleagues make presentations from a laptop. If that technique doesn't yield useful information, you prepare a survey or contact three or four colleagues who are frequent travelers and who represent different groups within your company: females and males, more experienced and less experienced, more computer literate and less computer literate.

- *What are the available products?* You could visit dealers, but they will want to sell you what they carry, not necessarily what is best for you. You could review product literature, but, again, the information might be biased. So you review the literature in professional journals, focusing on articles appearing in the most respected ones. You look for articles written by unbiased sources who have no financial stake in the outcome. You look for articles that include some sort of benchmark testing—the same task is performed on each competing product, and the evaluators measure the results and evaluate the output.

You do all kinds of secondary and primary research: conduct interviews, send out surveys, use online bibliographies to find journal articles, post questions to an Internet newsgroup on the subject, and search the Web. With more information than you can possibly use, you start to write—to try to figure out what all the information means. In the process, you realize that you still have some questions, that some of the information is incomplete, some contradictory, and some just unclear. You go back and do more research, then more writing.

Characteristics of Good Information

There is no shortage of information; the challenge is to find good information. Look for information that is

- *accurate.* If you estimate that people in your company make 500 trips a year but the actual figure is 50, you might end up buying equipment that you should be renting.

- *unbiased.* Obviously, you want sources that have no financial interest in your decision. But you also want to watch out for other kinds of bias; some sources might favor one kind of computer operating system, a bias that prevents them from offering a helpful analysis.

- *comprehensive.* You want to hear from different kinds of people—in terms of gender, cultural characteristics, and age—and people representing all views of the topic. The last person you interview might be the first one to point out something you never thought of: if the projection device you decide to buy weighs more than eight pounds, he would never take it on an airplane, although he would take it on a trip by car.

- *current.* If your information about laptop presentation devices is from 1995, it might as well be from 1955.

- *clear.* You want information that is easy to understand; otherwise, you'll waste time figuring it out, and you might misinterpret it.

Later, this chapter discusses how to evaluate the information according to these five criteria.

Choosing a Topic

Few professionals get to choose their topics, whereas students often do. This freedom is a mixed blessing. An instructor's request that you "come up with an appropriate topic" is for many people frustratingly vague; they would rather be told what to write about—how computers have changed the process of manufacturing, for instance—even though the topic might not interest them. If you don't spend weeks agonizing over the decision, however, the freedom to choose your own topic and approach is a real advantage: you're more likely to want to read and write about a topic that interests you and, therefore, more likely to do a better job.

Try to avoid topics such as the legal drinking age, the draft, and abortion, which don't relate to most writing situations on the job. Ask yourself what you are really interested in: perhaps something you are studying at school, some aspect of your part-time job, something you do during your free time. Browse through three or four recent issues of a news magazine or professional journal, and you will find dozens of articles that suggest interesting, practical topics involving technical information.

Table 7.1 on page 146 shows the titles of several articles from the February 17, 1997, *Newsweek* that suggest possible topics.

TABLE 7.1	Possible Topics for a Research Paper
Article Title	**Possible Subject for a Report**
"Too Close for Comfort"	• Problems integrating the civilian and military air-traffic control system • Skill level and training of Air National Guard pilots
"Hey, Can You Spare a 'Euro'?"	• Problems creating a common European currency
"The Holocaust in the Dock"	• The secrecy of the Swiss banking industry • Recovering lost assets of Holocaust victims
"Trouble in the Ranks"	• Racism in the military • Sexual harassment in the military
"Look Who's Talking"	• Sexual harassment in online chat rooms • First Amendment rights on the Internet

Many of these subjects are too broad for a brief research report of 10 to 15 pages, but they are good starting points. Think about your local community. When you read about the online chat rooms, think about the computer labs on your campus. How is your school attempting to keep up with hardware and software advances? What is the school's policy on permissible language or topics in online chat forums? How does your school allocate computer resources? If you are permitted to create a Web site and its popularity starts to slow down the network, what is your school's policy? Good topics are everywhere.

Generating Ideas about the Topic

Once you have a topic, it is time to start generating ideas about it. In this discussion, the task of generating ideas is explained before the task of doing research, but in fact you will be generating ideas throughout the writing process. Although you will generate ideas before you do the research, the research you read will force you to reevaluate your ideas and create new ones. Even after you are well under way drafting the document, you will revise and add ideas.

By *ideas* I don't mean formal, page-length explanations that are ready to be placed into the document, although if you have these in mind after choosing a topic, all the better. By *ideas* I mean subtopics of the overall topic you are writing about. For instance, if you plan to write about a new technology that mixes water with oil to increase the efficiency of power plants and reduce pollution, one of your ideas probably will be the theory behind the new technology, because almost certainly your readers will need to know it. However, you might also think of other ideas—such as the history of this technology, its cost, its advantages and disadvantages—that might not make their way into the document.

At this stage, many ideas might be little more than phrases or questions; logic tells you that you ought to know about them, and you remember reading about them, but you are not prepared to write about them yet. Generating ideas is a way to start mapping out what to include in the document, where it will go, and what additional information you will need.

How do people generate ideas? However they can. Many people don't even know what process they use. For some, the best ideas come when they aren't consciously thinking about their topic. A billboard, a television commercial, a conversation with a neighbor—anything can trigger an idea. The brain works on a problem when we are driving or taking a shower or sleeping. Some creative people keep a pad and pen on their nightstand because sometimes they wake up in the middle of the night with a great idea.

Even though we can't completely understand or control the process of generating ideas, we can help it along. Figure 7.1 shows five techniques that have proven useful for many people. Note that these techniques are used throughout the writing process. Good writers rethink their documents at every stage.

Asking Journalistic Questions   The journalistic questions are the *who, what, when, where, why,* and *how* of the topic. If, for instance, you think you might be interested in writing about digital audiotape (DAT), ask the following questions:

- Who invented DAT? Who uses it?
- What is DAT?

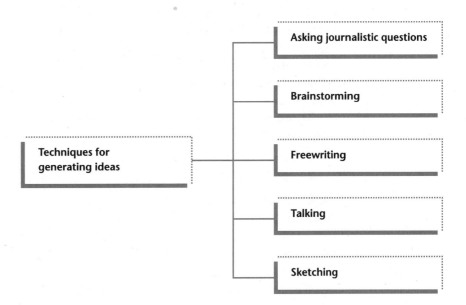

FIGURE 7.1

Techniques for
Generating Ideas

- When was DAT invented? When was it licensed? Will it become popular?
- Where was it invented? Where is it used now? Where will it be used in the future?
- Why is DAT used? What advantages does it offer over analog audiotape or compact discs? Why is it not widely used now?
- How does DAT work? What is its principle of operation?

Answering these basic questions will help you understand how much you know about the topic and what kind of information you need to acquire.

Brainstorming   When you brainstorm you spend 10 or 15 minutes listing ideas about your subject. List them as quickly as you can, using short phrases, not sentences. For instance, if you are describing global warming, the first five items in your brainstorming list might look like this:

- definition of global warming
- role of chlorofluorocarbons
- political issues
- effect on smokestack industries
- international agreements to limit global warming

As the word *brainstorming* suggests, you are not trying to impose order on your thinking. You will probably skip from one idea to another. Some ideas will be subsets of larger ideas. Don't worry about it; you will straighten out the logic later.

Brainstorming has one purpose: to free your mind so that you can think of as many ideas as possible that relate to your subject. When you have done your initial research and start to construct an outline, you will probably find that some of the items you listed do not in fact belong in the document. Just toss them out. The advantage of brainstorming is that for many people it is the most effective and efficient way to catalog what might be important to the document. Ironically, using a more structured way of generating ideas would cause you to miss more of them.

Brainstorming is actually a way of classifying everything you know into two categories:

- information that probably belongs in the document
- information that does not belong in the document

Once you have decided what material probably belongs, you can start to write the outline.

Freewriting   You might have used freewriting in a previous writing class. Freewriting consists of writing without plans or restrictions. If your topic is the technology for mixing oil and water in power plants, you write whatever comes into your mind about

Insurance rates have gone up over 10% a year for last 5 years. Other problem: the loss of employees to preventable health problems. Check out health-promotion programs. How much do they cost? Are they administered by us or by a subcontractor? How about reduced-rate memberships at health clubs? Will we have further liability if someone gets hurt in the programs? How much would we save—in premiums and reduced health problems? Is there an organization that rates these programs? First we have to find out how many employees would be interested. Maybe the insurance company could help us with some of this information.

FIGURE 7.2

Freewriting Text

it. You don't prepare an outline, stop to think about sentence construction, or consult a reference book. You just make the pen or the cursor move.

The freewriting text seldom becomes a part of the final document. Your explanations might be only half-correct; they almost certainly will be expressed unclearly. And half of what you write might be merely phrases or questions that you think you might need to answer. But something valuable might be happening: the mere act of trying to make sense helps you determine what you do and do not understand. And one phrase or sentence might spark an important idea.

Figure 7.2 is a brief freewriting text created in about five minutes by a student in a technical-communication course. She was working for a company considering a health-promotion program.

This freewriting text is a mess: a series of incomplete thoughts, and more questions than answers. Yet the writer did start to work out some of the problems her company would face (keeping the cost down, motivating employees, avoiding liability for injuries during the program) and some possible solutions to research (asking the insurance company for assistance and finding out if there is an organization that oversees the different programs). By starting to think through the problems her company would face, the writer is of course starting to think about her own problem: what kind of information her readers will need. Not bad for five minutes' work.

Talking Talking with someone about your topic is an excellent way to find out what you already know about it and to generate new ideas. Talking with another student is a good choice, because you will be able to return the favor. Have the person ask questions of you as you speak. If you start with your main idea—"Our company is thinking about instituting some sort of health-promotion program for its employees"—the person could ask why. You respond, "Because our health insurance premiums are too high and because too many of our employees suffer from preventable health problems." The person might then ask, "How high are the premiums? What is the rate of increase?" or "What percentage of your employees get sick? What is an acceptable rate? Is the rate changing?" A careful listener doesn't have to be a detective, just someone trying to understand what you are saying and what it means.

As the person asks questions, you will quickly get a sense of how clearly you understand your topic, what additional information you will need, and what questions your readers will want answered. You will also find yourself making new connections among ideas; the mere act of forcing yourself to put your ideas into sentences for someone else helps you synthesize the information.

Sketching   The four techniques for generating ideas that have been discussed so far—asking journalistic questions, brainstorming, freewriting, and talking with someone—are all nonhierarchical. That is, when you use them, you do not worry about the

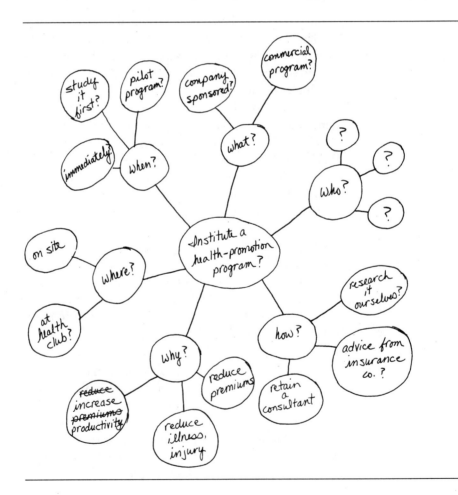

FIGURE 7.3

Clustering

relative importance of ideas; minor ideas are mixed in with major ideas. However, some people prefer a more structured way to generate ideas.

Sketching ideas—making a visual representation rather than a verbal one—can help you start to structure the document. Sketching has an additional benefit: it can help you see what information is weak or missing. Two kinds of sketches are used often in the professional world:

- *Clustering.* Write the main idea or a main question in the middle of the page. Then draw a circle around it. Write second-level ideas around the main idea. Then add third-level ideas around the second-level ideas. Figure 7.3 shows a cluster for the report on health-promotion programs. The writer has started with the journalistic questions. Notice that the writer doesn't yet know how to answer the question "Who?" She has to go back to her sources to find out who actually administers these kinds of programs.

- *Branching.* There is only one major difference between clustering and branching: the movement from larger to smaller ideas is from the center to the perimeter in the cluster format and from top to bottom in the branching format. Figure 7.4 shows branching.

As you think more about your topic, you will extend and revise your sketch. You might want to rethink the sketch if, for example, one of the branches contains

FIGURE 7.4

Branching

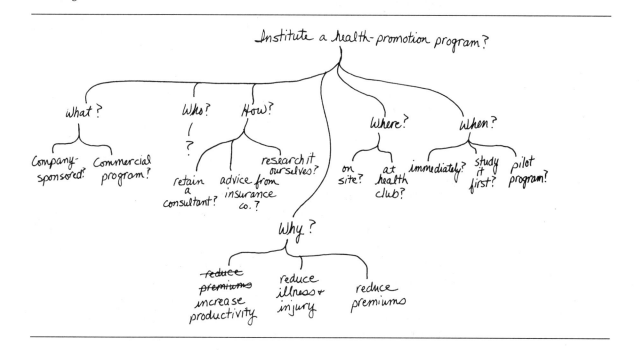

many more subbranches than the others. A short branch might signal that you need more research; or it might just be a smaller idea that is fine as it is. A long branch could signal that you went into much more detail on that one idea than you did on the other ideas; or it could merely be a more complicated idea. Counting branches and clusters provides no sure answers.

If as a student you realize that your topic would require much research and time—weeks to wait for an interlibrary loan, for example—consider changing to a topic that you know more about. Even though many writers like to have a precise topic in mind before they begin to search the literature, sometimes it is best to start with a general topic and see whether information is readily available. For example, you might want to write about computers but be unsure about what aspect to focus on. The basic research techniques described here will enable you to discover, quickly and easily, what aspects of computers will make the most promising topics. The advantage of being flexible about your topic is obvious: in 15 minutes you can discover dozens of topics from which to choose.

Conducting Secondary Research

For students, the best place to begin a search for additional information is a college or university library. Most college libraries have substantial reference collections and receive the major professional journals. Large universities have more comprehensive collections. Many large universities have specialized libraries that complement selected graduate programs, such as those in zoology or architecture. In addition, large cities often have special scientific or business libraries that are open to the public.

For working professionals, the best place to begin a search might be the organization's information center. An information center is the organization's library, a resource that collects different kinds of information that are critical to the organization's operations. The mission of the information center is to help employees get the information they need. At some information centers, the staff does all the searching for the employee. For instance, you can e-mail a question such as "How many cell phones were sold in the United States last year?" and the staff will find an answer and get back to you. At other information centers, the staff provides guidance as you carry out the search.

At both public and private libraries, the most important information sources are the reference librarians. They are invariably willing to suggest new ways to search for what you need—specialized directories, bibliographies, or collections that you didn't know existed. They can also assist you in using online databases. Perhaps most important, they will tell you if the library doesn't have the information you need and suggest other libraries to try. Reference librarians can save you much time, effort, and frustration. Don't be afraid to ask them questions when you run into problems.

Understanding the Research Media

We are in the midst of a massive and rapid transformation in the way we create, publish, and use information. I am referring, of course, to the transition from printed information to searchable digital information. For several decades, a great deal of technical information has in fact been produced digitally: for example, publishers have transformed typescript into digital code in producing a book. But the book was then printed and distributed in much the same way that books have been printed and distributed for centuries: if you want to read a book, you go to the library or buy a copy. Today, however, most technical information is being distributed through one or several digital media. Figure 7.5 shows the five major media used to publish information today. Often, information is published simultaneously in several different media. For instance, the federal government is likely to publish census information in printed reports and books, on CDs, and on Web sites.

In general, digital information is easier and more convenient to use than printed information for three reasons:

- Digital information can be searched, sorted, and saved to a disk, so you do not have to make photocopies or take notes. You do not introduce spelling errors into your document as you take notes, then type the information.

- Digital information is more current than printed information because it is easier to produce.

- Digital information is cumulative. That is, for indexes and other information that is published periodically, new information is integrated with

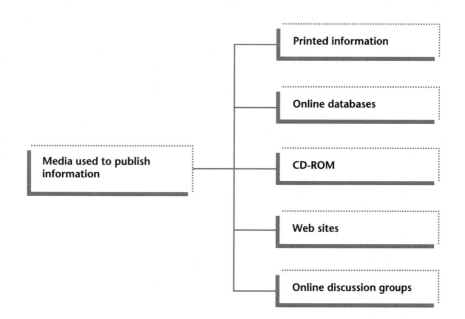

FIGURE 7.5

Media Used to Publish Information

the old; therefore, you do not have to search through back issues of the periodical because there are no back issues.

This is where the reference librarian comes in. No book can describe accurately what technical information exists at any given moment, or in what media it is available. You need to ask the reference librarian to help you find information. The transformation from print to online is well under way, and it will continue into the twenty-first century. Some existing media will disappear, some will change, and some new ones will appear. The way you use information tools will also continue to change from month to month.

Printed Information

Books, journals, reports, and other print media will continue to be produced because in some important ways printed information is still convenient to use: it is portable, and you can write on it. Company newsletters will continue to appear on paper for some time, even though they are also being published on company intranets. For documents that do not need periodic updates, print remains a useful and popular medium. And to find it you will continue to use online catalogs, as you do now.

Online Databases

Most college and university libraries—and even many public libraries—have facilities for online information searching. Libraries today lease access to different database services, with which they communicate by computer. The largest database service is DIALOG Information Services, which offers electronic access to articles, conference proceedings, news, and statistics from millions of documents in scientific, technical, and medical literature, as well as business, trade, and academic studies. DIALOG also provides access to more than 100 full-text newspapers and thousands of magazines and journals.

For more information on database services, see the *Directory of Online Databases* (New York/Cuadra/Elsevier), a quarterly publication that describes thousands of databases. The journal *Online* is also devoted to the subject. For more information on how to use databases, see Kathleen L. Nolan's *Gale Directory of Databases* (Detroit: Gale, 1997, 2 vols.). Or search the Web, using the "reference" category of a search engine.

CD-ROM

CD-ROM is likely to become the dominant form of information storage in all research libraries; it has already become a standard technology in business and industry, where its relatively low cost and small size make it the best storage medium.

Currently, CDs are used primarily to store research tools such as indexes, abstract services, and reference texts. However, full-text disks—holding the full texts of journal articles and books—are fast coming on the market. *Fulltext Sources Online* (Needham Heights, MA: BiblioData), a directory published twice a year, lists thousands of journals, newspapers, newsletters, and television and radio transcripts available in full text online.

Searching a bibliography on CD calls for the use of keywords and the Boolean operators, words such as *and* and *or*. Using Boolean operators, you can broaden or narrow your search to find an appropriate number of items. For instance, if you search for "nursing and technical communication," you will find information that includes both of these terms. If you use *or* instead of *and*, you will find information that includes either of the two terms. Different CD sources use Boolean operators differently, so be sure to read the instructions.

Because the same reference source often is available both online and on CD, online directories often include the CD sources. For more information on how to use CD-ROM, see *Gale Directory of Databases*.

Web Sites

No medium for delivering information has become more prominent more quickly than the World Wide Web, the point-and-click multimedia portion of the Internet. As discussed in Chapter 4, the Web contains millions of files on hundreds of thousands of servers around the world. Private companies, organizations, governments, and individuals all have Web sites.

To use the Web to find information, you either enter the address of the site you want to visit or use a search engine to find the narrowest category of information, such as "computers, operating systems, UNIX," that can help you find sites on your topic. When you find information you like, you can print it from the screen or digitally save the information to your computer.

Many Web sites also include files that you can download through FTP. In some cases, FTP requires authorization; that is, you have to have a password that lets you into the other computer. In other cases, however, FTP is open to anyone. This is called anonymous FTP.

As you probably know, information found on the Web varies greatly in quality. See the discussion of ways to evaluate Internet information later in this chapter.

Online Discussion Groups

There are two major forums for online discussions: *Usenet newsgroups* and *listserv discussion groups*.

Usenet newsgroups, sometimes called bulletin boards, publish e-mail messages sent by members of the group. Newsgroups give participants an opportunity to discuss issues, ask questions, and get answers. Usenet consists of thousands of

INTERVIEW: SCIENCE WRITER JOSEPH CHEW

on starting to use the Internet
About 1988, while I was working here at Lawrence Berkeley Laboratory, it became obvious that I should use it, because that's how the international accelerator physics community talks to each other. It's a very geographically scattered bunch of people.

on why he uses the Internet
Partly for amusement and partly to fulfill our public information mission, to correct some of the more outrageous things people say about science and to point them toward various sources of information.

on the Internet as a teacher
I've probably learned a lot more on the Internet by making mistakes than by searching for information, because a lot of people who know something have to be provoked in some way into telling you.

on the excitement of the Internet
I think the most exciting thing about it is the presence of live experts. An FAQ [frequently asked question] file is really nothing more than an encyclopedia, and using it is nothing more than looking things up, but actually participating in the dynamics of the information-building processes, that's

useful and tremendously exciting. What you're really talking about is a professional conference without the cheap wine.

on the role of the Internet
What's the purpose of the Internet? Is it a peer-to-peer forum for experts trying to do work, or is it essentially a teaching forum? As more people get onto it, it gravitates to the latter, just by sheer force of numbers. It doesn't belong to the UNIX wizards who originated it, it doesn't even belong to the scientists who have been using it for 10 years. It belongs by sheer force to AOL [America Online] and Prodigy and .edu [educators].

on the future of the Internet
I really think that the general direction of the future is the World Wide Web and the graphical interfaces in particular, because you can just present so much more information so much more easily with text and graphics together.

Joseph Chew is the scientific coordinator of the Accelerator and Fusion Research Division at Lawrence Berkeley Laboratory in Berkeley, California.

newsgroups, organized according to seven basic categories, including computer science, science, recreation, and business. In a Usenet newsgroup, mail is not sent to individual computers but stored on databases, which you then access. In addition, some newsgroups offer computer files for you to copy. These can include graphics as well as text.

Listserv discussion groups are like newsgroups in that they publish e-mail messages sent by people who subscribe to the service. The basic difference is that listservs send e-mail messages to every person who subscribes. The mail comes to you; you don't go to it, as you do with a newsgroup.

The two forms of discussion vary greatly in quality. Discussion groups on topics such as astronomy tend to contain high-level discussions by experts from universities, research institutes, and the government. Discussion groups on topics such as UFOs or *Star Trek* tend not to.

Using Basic Research Tools There is a tremendous amount of information in the different media for you to find. The trick is to learn how to find what you want. This section discusses the five basic research tools shown in Figure 7.6.

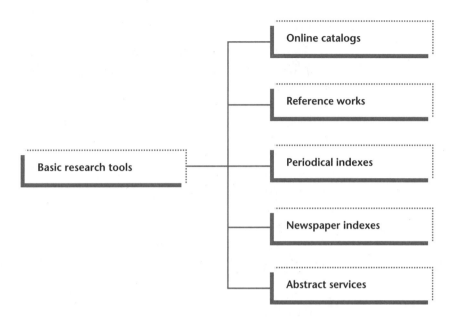

FIGURE 7.6

Basic Research Tools

Online Catalogs

In most libraries, the card catalog has been replaced by an online catalog of almost all the library's holdings: its books, microforms, films, compact discs and phonograph records, tapes, and other materials. Most online catalogs work essentially the same way, even though their screens and commands differ. Study the instructions for the system you will be using, and read the on-screen commands.

An online catalog does what the card catalog did—and much more. You can search for an item if you know its subject, title, or author, or any key term from one of these three. For instance, if you know that the title of the book you want contains the word *quantitative,* you can use an online catalog to locate the title *The Visual Display of Quantitative Information.* And the display indicates not only the standard bibliographic information about the book—author, publisher, call number, and so forth—but also whether the book is on the shelves or checked out.

Using an online catalog, you can modify the search to include or exclude different parameters. For instance, you can limit the search by a number of parameters:

- medium (such as books, reports, or videotapes)
- language (such as English)
- publication date (such as 1995 or later)
- publisher (such as St. Martin's Press)

The online catalog lists and describes the holdings at one particular library or a group of libraries. Using the Internet, you can search the catalogs of hundreds of major libraries around the world. See the discussion of Telnet (remote log-in) in Chapter 4.

Reference Works

Some books listed in the catalog have call numbers preceded by the abbreviation "Ref." These books are part of the reference collection, a separate grouping of books that normally may not be checked out of the library. Many reference works in the library also exist on the Web.

The reference collection in print and online includes the general dictionaries and encyclopedias, biographic dictionaries (*International Who's Who*), almanacs (*Facts on File*), atlases (*Rand McNally Commercial Atlas and Marketing Guide*), and dozens of other general research tools. In addition, the reference collection contains subject encyclopedias (*Encyclopedia of Banking and Finance*), dictionaries (*Psychiatric Dictionary*), and handbooks (*Biology Data Book*). These specialized reference books are especially useful when you begin a writing project; they provide an overview of the subject and often list the major works in the field.

How do you know if there is a dictionary of the terms used in a given field, such as nutrition? Searching the Web is probably the fastest and most accurate way to find out. Check, too, the following reference books that list the many others available. Among these guides-to-the-guides are the following:

- Harris, S. (1994). *The New York Public Library book of how and where to look it up*. New York: Macmillan.
- Mullay, M., & Schlicke, P. (Eds.). (1993). *Walford's guide to reference material* (6th ed.). 3 vols. London: Library Association.
- Sheehy, E. P. (Ed.). (1986). *Guide to reference books* (10th ed.). Chicago: American Library Association.

The most comprehensive guide-to-the-guides is Sheehy's *Guide to Reference Books* (also see its recent supplements), a valuable resource that lists bibliographies, indexes, abstract services, dictionaries, directories, handbooks, encyclopedias, and many other sources. However, Sheehy is updated infrequently. And obviously, these guides do not provide current information about reference works on the Web. For that information, go to the Web itself.

To find information on the Web, use a search engine and go to its "reference" section, where you will find numerous sites, such as the following, that contain links to excellent collections of reference works online:

- Where the Wild Things Are (http://www.sau.edu/CWIS/Internet/Wild/index.htm)
- CyberStacks(sm) (http://www.public.iastate.edu/~CYBERSTACKS/)
- The Internet Public Library (http://ipl.sils.umich.edu/ref/)

Another strategy is to use a search engine to find a category closely related to your topic. For instance, if you are writing about helicopters, you could search the categories in Yahoo! by clicking on "science," then "aviation and aeronautics," then "organizations," then "American Helicopter Society." There you will see a set of links to other sites about helicopters. Many of these sites list only online information, but others include valuable bibliographies of printed information as well.

Periodical Indexes

Periodicals are the best source of information for most research projects because they offer recent, authoritative discussions of limited subjects. The biggest challenge in using periodicals is identifying and locating the dozens of relevant articles that are published each month. Although only half a dozen major journals may concentrate on your field, a useful article might appear in one of a hundred other publications. A periodical index, which is simply a listing of articles classified according to title, subject, and author, can help you determine which journals you want to locate.

There are periodical indexes in all fields. The following brief list gives you a sense of the diversity of titles:

- *Applied Science & Technology Index*
- *Business Periodicals Index*
- *Readers' Guide to Periodical Literature*
- *Engineering Index*

Once you have created a bibliography of articles you want to study, you have to find them. Check your library's online catalog or *serials holding catalog,* the printout that lists all the journals your library receives. If your library does not have an article you want, you can use one of two techniques for securing it:

- *Interlibrary loan.* Your library uses an online directory to learn which nearby library has the article. That library makes a photocopy of the article and sends it to your library. The advantage of an interlibrary loan is that it is a free (or very inexpensive) service to you. The disadvantage is that it can take two weeks or longer for the article to arrive at your library.
- *Document-delivery services.* If you are in a hurry, you can log on to a document-delivery service, such as UnCover (http://www.carl.org/uncover/unchome.html), which searches a database of 17,000 periodicals. If the service has the article, it faxes it to you, often within an hour, for a charge of about ten dollars.

Newspaper Indexes

Many major newspapers around the world are indexed by subject. The three most important indexed U.S. newspapers are the following:

- *The New York Times*
- *The Christian Science Monitor*
- *The Wall Street Journal*

Volume and citation number   Author name   Article title   Chronology

ABSTRACT:

26:5577   1841-67

Charland, Jean-Pierre. LE RESEAU D'ENSEIGNEMENT PUBLIC BAS-CANADIEN, 1841-1867: UNE INSTITUTION DE L'ETAT LIBERAL [The public school system in Lower Canada, 1841-67: an institution of the liberal state]. *Revue d'Histoire de l'Amerique Française* [Canada] 1987 40(4): 505-535. Liberal principles such as equality under the law underlay the creation and organization of Lower Canada's public school system. The increasing centralization of authority, in the person of the superintendent and Conseil de l'Instruction Publique, seemingly contradictory to liberal tenets, was necessary to resist the efforts of Catholic clergy to take control of education. After 1867, the clergy gained greater control of the schools. Based on contemporary government documents and other sources; 71 notes.   M. Schumacher

Title translation — Periodical title — Year, volume, issue and page numbers

Country of publication

Abstract

Documentation — Abstracter

FIGURE 7.7

Guide to the Use of an Abstract Service

The first two are highly reputable general newspapers; the third is the authoritative source on business, finance, and the economy. Numerous newspapers are now available on the Web, including the above. Many of them are free and many can be searched electronically.

Abstract Services

Abstract services are like indexes but also provide abstracts: brief summaries of an article's important results and conclusions. (See Chapter 10 for a discussion of abstracts.) In most cases, reading the abstract will enable you to decide whether to search out the full article. The title of an article, alone, is often a misleading indicator of its contents.

Some abstract services, such as *Chemical Abstracts,* cover a broad field, but many are specialized rather than general. *Adverse Reaction Titles,* for instance, covers research on the subject of adverse reactions to drugs. Figure 7.7, an excerpt from a guide included in *America: History and Life,* explains how to use that journal. Most abstract journals are organized in this way.

As is the case with periodical indexes, abstract services are going online.

Researching Government Information

The U.S. government is the world's biggest publisher. In researching any field of science, engineering, or business, you are likely to find that a government agency or department has produced a relevant brochure, report, or book.

Government publications usually are not listed in the indexes and abstract journals. The *Monthly Catalog of United States Government Publications,* available on paper, on CD, and on the Web (http://www.access.gpo.gov/su_docs/dpos/adpos400.html), provides extensive access to these materials. The digital versions of the *Monthly Catalog* are fully searchable.

Printed government publications are usually cataloged and shelved separately from the others. They are classified according to the Superintendent of Documents system, not the Library of Congress system. See the reference librarian or the government documents specialist for information about finding government publications in your library.

You can also access many government sites and databases on the Internet. Two Web sites are particularly useful:

- Fedworld Information Network (http://www.fedworld.gov). This searchable site contains more than 15,000 files on numerous fields in business, health, safety, and the environment. The site also links to the U.S. Government Information Locator Service, a database of federal government information.
- GPO Access (http://www.access.gpo.gov). This searchable site offers links to hundreds of sites for all three branches of the federal government, including dozens of departments and agencies.

Patricia Cruse and Sherry DeDecker have created an excellent site, "How to Effectively Locate Federal Government Information on the Web" (http://gort.ucsd.edu/pcruse/universe/intro.html) that clearly explains how to research government information.

For additional information on government publications, consult the following three printed guides:

Bailey, W. G. (1993). *Guide to popular U.S. government publications* (3rd ed.). Englewood, CO: Libraries Unlimited.

Morehead, J., & Fetzer, M. K. (1992). *Introduction to United States government information sources* (4th ed.). Englewood, CO: Libraries Unlimited.

Schwarzkopf, L. C. *Government reference books.* Englewood, CO: Libraries Unlimited. (A biennial publication.)

Keep in mind, however, that books cannot keep up with the rapid transition from paper to screen. Check the two Web sites listed above, and consult the reference librarian or government document librarian.

Understanding the Information

Once you have gathered your books, articles, and electronic information, you need to assimilate them. When you have a good understanding of what you have, you can go on to the next stage—evaluating the information.

To understand the information you have gathered, you will want to skim it and take notes.

Inexperienced writers sometimes make the mistake of trying to read every potential source. All too often, they get only halfway through one of their several books when they realize they must start writing immediately to submit their reports on time. Knowing how to skim is invaluable.

To skim a book, read the following parts:

- *the preface and introduction:* to get a basic idea of the writer's approach and methods
- *the acknowledgments section:* to learn about assistance the author received from other experts in the field, or about the author's use of important primary research or other resources
- *the table of contents:* to get an idea of the scope and organization of the book
- *the notes at the ends of chapters or the end of the book:* to understand the nature and extent of the author's research
- *the index:* to get a sense of the coverage of the information you need
- *a few paragraphs from different portions of the text:* to gauge the quality and relevance of the information

To skim an article, focus on the following elements:

- *the abstract:* to get an overview of the article's content
- *the introduction:* to get a broad understanding of the article's purpose, main ideas, and organization
- *the notes and references:* to understand the nature and extent of the author's research
- *the headings and several of the paragraphs:* to get a sense of the article's organization and the quality and relevance of the information

Skimming will not always tell you whether a book or article is going to be useful. A careful reading of a work that looks useful might prove disappointing. However, skimming can tell you if the work is *not* going to be useful: because it doesn't cover your subject, for example, or because its treatment is too superficial or too advanced. Eliminating the sources you don't need will give you more time to spend on the ones you do.

Note-taking is often the first step in writing the document. Your notes will provide the vital link between what your sources have said and what you are going to say. Because you will refer to them over and over, it is smart to take notes logically and systematically.

The best way to take notes is electronically. If you can download files from the Internet, download bibliographic references from a CD-ROM database, and take notes on a laptop computer, you will save a lot of time and prevent many errors.

But if you do not have access to these electronic tools, buy two packs of note cards: one large (4" × 6" or 5" × 8"), the other small (3" × 5"). There is nothing special about ready-made note cards; you can make your own out of scrap

Information to Record for a Book	**Information to Record for an Article**
• Author	• Author
• Title	• Title of the article
• Publisher	• Periodical
• Place of publication	• Volume
• Year of publication	• Number
• Call number	• Date
	• Pages on which the article appears
	• Call number

FIGURE 7.8

Bibliographic
Information to Record

paper. The major advantage of using cards is that they are easy to rearrange later, when you want to outline the report.

On the small cards, record the bibliographic information for each source from which you take notes. Figure 7.8 shows the information to record for books and articles. If you print information from a Web site, in most cases the bibliographic information will be printed automatically for you at the top of the page.

On the large cards, write your notes. To simplify matters, write on one side of a card only and limit each card to a narrow subject or discrete concept so that you can easily reorder the information to suit the needs of your document.

Most note-taking involves three kinds of activities: paraphrasing, quoting, and summarizing.

Paraphrasing

A paraphrase is a restatement, in your own words, of someone else's words. *In your own words* is crucial: if you simply copy someone else's words—even a mere two or three in a row—you must use quotation marks.

What kind of material should you paraphrase? Any information that you think might be useful in writing the report: background data, descriptions of mechanisms or processes, test results, and so forth.

To paraphrase accurately, follow these four steps:

- *Study the original until you understand it thoroughly.*
- *Rewrite the relevant portions of the original.* If you find it easiest to rewrite in complete sentences, fine. If you want to use fragments or merely list information, make sure you haven't compressed the material so much that you'll have trouble understanding it later. Remember, you might not be looking at the card again for a few weeks.
- *Title the card so that you'll be able to identify its subject at a glance.* The title should include the general subject and the author's attitude or approach to it, such as "Criticism of open-sea pollution-control devices."

- *Include the author's last name, a short title of the article or book, and the page number of the original.* You will need this information later to cite your source.

Figure 7.9 shows two paraphrased note cards based on the following discussion (adapted from Lovgren, 1994, pp. 87–88). The author is describing his business: creating metaphors to be used in graphical user interfaces on computers. The student has paraphrased each paragraph on a separate note card.

> How do you find good metaphors? To be usable, a metaphor must be in the user's sphere of knowledge. The lack of understanding of this requirement frequently causes problems in interface design. Many of the interface metaphors I have seen are metaphors within the knowledge of the developer, but not within the knowledge of the user.
>
> You have to talk to the users, not just to other developers. In my consultancy, we use a technique called contextual interviews to gather this information. I'll explain it by walking you through a session with a hypothetical potential user I'll call Sally.
>
> The first thing we do is phone Sally and schedule a two-hour meeting at her place of employment. We always try to meet the person where he or she works. When we interview users outside their work environment, they give us an overly organized picture of their work, apparently forgetting that real work tends to be less structured and is often interrupted. Also, we have found that two hours is just about the right time.

FIGURE 7.9

Paraphrased Notes

Longer sessions tend to distort results because the session is exhausting for both interviewer and interviewee.

Before beginning the interview, we ask Sally if she has any comments, questions, or issues of a general nature. Our intent is to put her at her ease. As part of the preinterview, we inform Sally that the interview will be audiotaped.

We then ask Sally what she would be doing if we weren't here and ask her to start the task. While she is doing it, we ask her to describe what she is doing, why, and for whom. We also ask what kinds of difficulties she might run into that we don't see.

As follow-up, we prepare a complete report of the interview, including an annotated transcript, sketches, and copies of any forms, calendars, and other supplemental material. In annotating the transcript, I mark the nouns Sally used, the actions she mentioned, and any relationships among the nouns.

After repeating this process for several other potential users, we generally get a sense of the metaphors that might be appropriate. For example, if several people in addition to Sally refer to a "red book," and the system is to manage information in this book, then the red book becomes a likely metaphor. It works because when anybody in the organization mentions this term, everyone knows what they are talking about.

Notice that a heading provides a focus for each card. The student has omitted the information he does not need, but he has recorded the necessary bibliographic information so that he can document his source easily or return to it. There is no one way to paraphrase: you have to decide what to paraphrase—and how to do it—on the basis of your analysis of the audience and the purpose of your report.

Quoting

Sometimes you will want to quote a source, either to preserve the author's particularly well-expressed or emphatic phrasing or to lend authority to your discussion. In general, do not quote passages of more than two or three sentences, or your report will look like a mere compilation. Your job is to integrate an author's words and ideas into your own work, not merely to introduce a series of quotations.

The simplest form of quotation is an author's exact statement:

As Jones states, "Solar energy won't make much of a difference in this century."

To add an explanatory word or phrase to a quotation, use brackets:

As Nelson states, "It [the oil glut] will disappear before we understand it."

Use ellipses (three spaced dots) to show that you are omitting part of an author's statement:

ORIGINAL STATEMENT:	"The generator, which we purchased in May, has turned out to be one of our wisest investments."
ELLIPTICAL QUOTATION:	"The generator . . . has turned out to be one of our wisest investments."

For more details on the mechanics of quoting, see the entries under "Quotation Marks," "Ellipses," and "Brackets" in Appendix A. For a discussion of different styles of documentation, see Chapter 10.

Summarizing

Summarizing, a more comprehensive form of taking notes, is the process of rewriting a passage to make it shorter while still retaining its essential message.

The two main reasons to summarize are:

- *To learn a body of information.* When you skim a source and find it useful, you might want to summarize it so that you can integrate it with other information in creating your document.
- *To create one or more of the summaries that will go into your document.* Most long technical documents contain several kinds of summaries:

  - a letter of transmittal that provides an overview of the document
  - an abstract, a brief technical summary
  - an executive summary, a brief nontechnical summary directed to the manager
  - a summary at the end that draws together a complicated discussion

These summaries are discussed in Chapters 8 and 10.

The present discussion covers the process of summarizing printed information that you uncover in your research and might use in your document. Although the technique described here will prove useful as you prepare the abstract and the different kinds of summaries in your own writing, the focus in this section is on extracting the essence of a passage by summarizing it.

The process of summarizing information consists of the following seven steps:

1. *Read the passage carefully several times.* Before you proceed, you must thoroughly understand what the writer is saying.

2. *Underline the key ideas.* Most writers put their main ideas in a few obvious places: titles, headings, topic sentences, transitional paragraphs, concluding paragraphs. The bodies of the paragraphs usually explain and exemplify the main points, and therefore are not the most likely places to find the general ideas of the passage.

3. *Combine the key ideas.* Take a short break and then study what you have underlined. Paraphrase the underlined ideas. Don't worry about your grammar, punctuation, or style; if you will be submitting the summary to another reader, you can revise later. At this point, you just want to see if you can reproduce the essence of the original.

4. *Check your draft against the original for accuracy and emphasis.* Reread the original to make sure your summary is accurate and reflects the writer's emphasis. In checking for accuracy you need to pay attention to a wide variety of matters, from getting individual statistics and names correct to ensuring that your version of a complicated concept faithfully represents the original. Checking for proper emphasis means getting the proportions right; if the original devotes 20 percent of its space to a particular point, your draft should not devote 5 percent or 50 percent of its space to that point.

5. *If necessary, edit your summary for style.* If your purpose in summarizing is to learn the material yourself, you don't need to edit for style. However, if you will be submitting the summary to another reader, now is the time to revise it for grammar, punctuation, usage, style, and spelling. See Chapters 13 and 14 for more information on the revision process.

6. *Record the bibliographic information carefully.* Even though a summary might contain all your own words, you still must cite it, because the main ideas are someone else's. If you don't already have the bibliographic information in an electronic form, put it on note cards.

7. *If necessary, add keywords.* If you are creating an abstract (see Chapter 10), or if for some other reason readers might want to retrieve your summary by using keywords, list the keywords now.

Figure 7.10 (based on McComb, 1991, pp. 19–21) is a narrative history of television technology addressed to the general reader. Figure 7.11 on page 169 is a

---

A Brief History of Television

Although it seems as if television has been around for a long time, it's a relatively new science, younger than rocketry, internal medicine, and nuclear physics. In fact, some of the people that helped develop the first commercial TV sets and erect the first TV broadcast antennas are still living today.

The Early Years

The first electronic transmission of a picture was believed to be made by a Scotsman, John Logie Baird, in the cold month of February 1924. His subject was a Maltese Cross, transmitted through the air by the magic of television (also called "Televisor" or "Radiovision" in those days) the entire distance of ten feet.

To say that Baird's contraption was crude is an understatement. His Televisor was made from a cardboard scanning disk, some darning needles, a few discarded electric motors, piano wire, glue, and other assorted odds and ends. The picture reproduced by the original Baird Televisor was extremely difficult to see—a shadow, at best.

Until about 1928, other amateur radiovision enthusiasts toyed around with Baird's basic design, whiling long hours in the basement transmitting Maltese Crosses, model airplanes, flags, and anything else that would stay still long enough under the intense light required to produce an image. (As an interesting aside, Baird's lighting for his 1924 Maltese Cross transmission required 2,000 volts of power, produced by a roomful of batteries. So much heat was generated by the lighting equipment that Baird eventually burned his laboratory down.)

FIGURE 7.10

Original Passage

(Continued)

(*Figure 7.10 continued*)

Baird's electromechanical approach to television led the way to future developments of transmitting and receiving pictures. The nature of the Baird Televisor, however, limited the clarity and stability of images. Most of the sets made and sold in those days required the viewer to peer through a glass lens to watch the screen, which was seldom over seven by ten inches in size. What's more, the majority of screens had an annoying orange glow that often marred reception and irritated the eyes.

Modern Television Technology

In the early 1930s, Vladimir Zworykin developed a device known as the iconoscope camera. About the same time, Philo T. Farnsworth was putting the finishing touches on the image dissector tube, a gizmo that proved to be the forerunner to the modern cathode ray tube or CRT—the everyday picture tube. These two devices paved the way to the TV sets we know and cherish today.

The first commercially available modern-day cathode ray tube televisions were available in about 1936. Tens of thousands of these sets were sold throughout the United States and Great Britain, even though there were no regular television broadcasts until 1939, when RCA started what was to become the first American television network, NBC. Incidentally, the first true network transmission was in early 1940, between NBC's sister stations WNBT in New York City (now WNBC-TV) and WRGB in Schenectady.

Postwar Growth

World War II greatly hampered the development of television, and during 1941–1945, no television sets were commercially produced (engineers were too busy perfecting radar, which, interestingly enough, contributed significantly to the development of conventional TV). But after the war, the television industry boomed. Television sets were selling like hotcakes, even though they cost an average of $650 (based on average wage earnings, that's equivalent to about $4,000 today).

Progress took a giant step in 1948 and 1949 when the four American networks, NBC, CBS, ABC, and Dumont, introduced quality, "class-act" programming, which at the time included Kraft Television Theatre, Howdy Doody, and The Texaco Star Theatre with Milton Berle. These famous stars of the stage and radio made people want to own a television set.

Color and Beyond

Since the late 1940s, television technology has continued to improve and mature. Color came on December 17, 1953, when the FCC approved RCA's all-electronic system, thus ending a bitter, four-year bout between CBS and RCA over color transmission standards. Television images beamed via space satellite caught the public's fancy in July of 1962 when Telstar 1 relayed images of AT&T chairman Frederick R. Kappell from the U.S. to Great Britain. Pay-TV came and went several times in the 1950s, 1960s and 1970s; modern-day professional commercial videotape machines were demonstrated in 1956 by Ampex; and home video recorders appeared on retail shelves by early 1976.

The first electronic transmission of a picture was produced by Baird in 1924. The primitive equipment produced only a shadow. Although Baird's design was modified by others in the 1920s, the viewer had to look through a glass lens at a small screen that gave off an orange glow.

Zworykin's iconoscopic camera and Farnsworth's image dissector tube—similar to the modern CRT—led in 1936 to the development of modern TV. Regular broadcasts began in 1939, on the first network, NBC. Research stopped during WWII, but after that, sales grew, even though sets cost approximately $650, the equivalent of $4,000 today.

Color broadcasts began in 1953; satellite broadcasting began in 1962; and home VCRs were introduced in 1976.

Key terms: television, history of television, NBC, color television, satellite broadcasting, video cassette recorders, Baird, Zworykin, Farnsworth.

FIGURE 7.11

Summary of the Original Passage

summary that includes the key terms. This summary is 10 percent of the length of the original.

Evaluating the Information

Earlier, this chapter made the point that good information is accurate, unbiased, comprehensive, current, and clear. Consider the following factors in evaluating information.

- *Authorship*. Do you recognize the author of the information? Can you learn about the author's credentials and current position from a biographical note? If this information is not included in the document itself, can you find it in a who's who or by searching for other books by the author or other journal articles?

- *Publisher*. What is the publisher's reputation? A book should be published by a reputable trade, academic, or scholarly publisher. A journal should be sponsored by a professional association or university. Read the list of editorial board members; they should be well-known names in the field. Trade publications—magazines about a particular industry or group—often promote the interests of that industry or group. Don't automatically assume the accuracy of information in trade publications for loggers or environmentalists. If you have doubts about the authority of a book or journal, ask the reference librarian or a professor in the appropriate field.

- *Knowledge of the literature*. Does the author appear to be current with the major literature in the field? Is there a bibliography? Are there notes throughout the document?

- *Accuracy and verifiability of the information.* Does the information appear to be accurate and verifiable? Does the author clearly describe the methods and theories used in creating the information? Are these methods and theories appropriate to the topic?
- *Timeliness.* Does the document rely on recent data? Was the document published recently?

Evaluating Internet Sources

The most difficult kind of material to evaluate is information from the Internet. The Internet's greatest strengths—that it is open to anyone who is connected, and that one person's voice is heard as loudly as that of anyone else—pose a great challenge when it comes to evaluating sources of information. In most cases, information appears on the Internet without passing through the formal review procedure that is characteristic of books and professional journals. Therefore, you have to be particularly careful.

For the most part, the five bases for evaluating research information discussed in the previous section apply to Internet sources as well. The following discussion is based on Kirk's (1996) suggestions.

- *Authorship.* If you do not recognize the author of the information, did you find the site by linking from another reputable site? Does the site contain links to other reputable sites? Does the site contain biographical information—the author's current position and credentials? Can you use a search engine to find other references to the author's credentials, or other documents by the author?
- *Publishing body.* The publishing body is the server on which the document is stored. Can you determine the publishing body's identity from headers or footers? Is the publishing body reputable in the field? If the document resides in a personal account on an Internet service provider, an account that is not associated with a professional organization, perhaps the author is writing outside his or her field of expertise. Keep in mind that many Internet sites exist largely for public relations or overt advertising. For instance, the home page for the White House links to a page about the president and vice president called "Their accomplishments, their families, and how to send them electronic mail." Obviously, you're not going to find a hard-hitting critique of the administration on its own site. The same holds true for corporations and other organizations.
- *Knowledge of the literature.* Analyze the Internet source as you would any other source. Often, references to other sources will take the form of links.
- *Accuracy and verifiability of the information.* Analyze the Internet source as you would any other source.
- *Timeliness.* Was the document created recently? Was it updated recently?

Another excellent discussion is Hope N. Tillman's "Evaluating Quality on the Net" (1997).

Conducting Primary Research

Although the library and the Web offer a wealth of secondary sources, often you will need to conduct primary research in addition to consulting the work of others. Inspections, experiments, and field research are three common primary-research techniques. To inspect something is, of course, to look at it, either with or without the use of specialized tools. To experiment is to test a theory about why something happened, or whether something would happen. For instance, an engineer could devise an experiment to test whether cars with antilock brakes handle better in the rain than do cars with conventional brakes. To conduct field research is to study people (or other animals) working in their natural environment.

Inspections
Regardless of the field you work in, you are likely to read many sentences that begin "An inspection was conducted to determine . . . ." A civil engineer can often determine what caused the cracking in a foundation by inspecting the site. An accountant can learn a lot about the financial health of an organization by inspecting the company's financial records.

These people are looking at a site, an object, or a document (for example, the financial records) and applying their knowledge and professional judgment to what they see. Sometimes the inspection techniques are more complicated. A civil engineer inspecting foundation cracking might want to test hunches by taking a soil sample back to the lab and testing it. An accountant checking the books might need to perform some computerized analyses on the information.

When you carry out an inspection, be sure to take good notes. Try to answer the appropriate journalistic questions—*who, what, when, where, why,* and *how*—as you inspect, or as soon after as possible. Where appropriate, photograph or sketch the site or print out the output from computer-assisted inspections. You will probably need the data later for your document.

Experiments
Learning to conduct the many kinds of experiments used in a particular field can take months or even years. Therefore, this discussion can serve only as a brief introduction.

In many cases, conducting an experiment involves four phases, as shown in Figure 7.12.

Conducting Experiments

- *Establishing a hypothesis.* A hypothesis is an informed guess about the relationship between two factors. For instance, to determine the relationship between gasoline octane and miles per gallon, you can set up an experiment with the following hypothesis: "A car will get better mileage with 89 octane gas than with 87 octane gas."

- *Testing the hypothesis.* In most cases, you need an experimental group and a control group—two groups identical except for the condition you are studying: in this case, the gasoline. The control group would be a car running on 87 octane. The experimental group would be an identical car running on 89 octane. The experiment would consist of driving the two cars over an identical course at the same speed—preferably in some sort of controlled environment such as a laboratory—over a given distance, such as 1,000 miles. At the end of the 1,000 miles you would measure how much gasoline each car used and calculate the miles per gallon. The results would either support or refute your original hypothesis.

  An experiment as neat as this one is often difficult to set up. For instance, you probably wouldn't be able to get two identical cars and run them for 1,000 miles in a lab. So you do the best you can: you run your own car with two or three tanks of 87 octane and calculate your miles per gallon. Then you run it on the same amount of 89 octane and do the same calculations. Of course, you try to make sure that you are subjecting the car to the same driving conditions to control other variables that could affect the results. The important point is that you make the conditions for both groups as similar as possible.

- *Analyzing the data.* You must try to understand whether your data show merely a correlation—one factor changing along with another—or a causal relationship. For instance, nonsmokers tend to live seven years longer than smokers. The tobacco industry claims this is only a statistical coincidence, but most people see the relationship as causal: smoking causes people to die sooner. Another example: we know that sports cars are involved in more fatal accidents than sedans, but we don't know whether the car has much to do with that fact, or whether driving habits are the important factor. Analyzing the data is a challenge because it is tempting to see what you want to see.

- *Reporting the data.* When researchers explain their findings in a report, they try to explain as clearly as they can what they did, why they did it, what they saw, what it means, and what ought to be done next. For more information on writing reports, see Chapters 16–17.

Field Research Whereas an experiment is quantitative (it yields statistical data that can be measured), field research is usually qualitative; it yields data that cannot be measured or cannot be measured as precisely as experimental data. Often in field research you seek to understand the quality of an experience. For instance, you might want to understand how a new seating arrangement would affect group dynamics in a classroom. You could design a study in which you observed and recorded the classes and perhaps interviewed the students and the instructor about their reactions. Then you could do the same in a traditional classroom and compare the results.

Some kinds of studies have both quantitative and qualitative elements. In the case of the classroom seating arrangements, you could create some quantitative measures, such as the number of times students talked with each other or the length of the interchanges between them. In addition, you could distribute questionnaires to elicit the opinions of the students and the instructor. If you used these quantitative measures on a sufficiently large number of classrooms, you could gather valid quantitative information.

When you are doing quantitative or qualitative studies on the behavior of animals—from rats to monkeys to people—try to minimize two common problems:

- *The effect of the experiment on the behavior you are studying.* When a television camera crew arrives on the scene to cover a protest demonstration, the protest becomes more animated. Similarly, if you are studying the effects of the classroom seating arrangement, you have to try to minimize the effects of your presence. For instance, make sure that the camera is placed unobtrusively, and that it is set up before the students arrive, so they don't see the process. Still, any time you bring in a camera, you can never be sure that what you witness is typical. Even an outsider who sits quietly taking notes can disrupt typical behavior.

- *Bias in the recording and analysis of the data.* This problem can occur because researchers want to confirm their hypotheses. For instance, in an experiment to determine whether word processors help students write better, the researcher is likely to see improvement where other people don't. For this reason, the experiment should be designed so that it is *double blind.* That is, the students doing the writing to be studied shouldn't know what the experiment is about, so they won't change their behavior to support or negate the hypothesis. And when the data are being analyzed, they should be disguised so that researchers don't know whether they are examining the results from the control group or the experimental group. If the former wrote in ink and the latter used word processors, the control papers should be typed on a word processor, so that all the papers look identical.

Conducting an experiment or field research is relatively simple; the hard part is designing your study so that it accurately measures what you want to measure.

Interviews   Interviews are extremely useful when you need information on subjects too new to have been discussed in the professional literature or inappropriate for widespread publication (such as local political questions). Most students are in-experienced at interviewing and hence are reluctant to do it. Interviewing, like any other communication skill, requires practice. The following discussion explains how to make interviewing less intimidating and more productive.

Determine these three pieces of information in choosing a respondent, a person to interview:

1. *What you want to find out.* Only then can you begin to search for a person who can provide the necessary information.

2. *Who could provide the information.* The ideal respondent is an expert willing to talk. Many times your topic will dictate whom you should ask. If you are writing about research at your university, for instance, the logical choice would be a researcher involved in the project. Sometimes, however, you might be interested in a topic about which a number of people could speak knowledgeably, such as the reliability of a particular kind of office equipment or the reasons behind the growth in the number of adult students. Use directories, such as local industrial guides, to locate the names and addresses of potential respondents.

3. *Whether the person is willing to be interviewed.* On the phone or in a letter, state what you want to ask about; the person might not be able to help you but might be willing to refer you to someone who can. And be sure to explain why you have decided to ask the respondent; a well-chosen compliment works better than admitting that the person you really wanted to interview is out of town. Don't forget to mention what you plan to do with the information, such as write a report or give a talk. Then, if the person is willing to be interviewed, set up an appointment at his or her convenience.

Preparing for the Interview

The interview is likely to progress smoothly if you make the following preparations:

1. *Do your homework.* Never give the impression that you are conducting the interview to avoid doing other kinds of research. If you ask questions that are already answered in the professional literature, the respondent might become annoyed and uncooperative.

2. *Prepare good questions.* Good questions are clear, focused, and open.

   – Be clear. The respondent should be able to understand what you are asking.

UNCLEAR:	Why do you sell Trane products?
CLEAR:	What are the characteristics of Trane products that led you to include them in your product line?

   The unclear question can be answered in a number of unhelpful ways: "Because they're too expensive to give away" or "Because I'm a Trane dealer." The clear version of the question forces the respondent to think about specific reasons for selling Trane products as opposed to (or in addition to) other brands.

– Be focused. The question must be sufficiently narrow to be answered briefly. If you want additional information, you can ask a follow-up question.

UNFOCUSED: What is the future of the computer industry?

FOCUSED: What will the American chip industry look like in 10 years?

– Be open. Your purpose is to get the respondent to talk. Don't ask a lot of questions that have yes or no answers.

CLOSED: Do you think the federal government should create industrial partnerships?

OPEN: What are the advantages and disadvantages of the federal government's creating industrial partnerships?

Conducting the Interview

Remember the following four guidelines as you begin the interview:

- *Arrive on time.*
- *Thank the respondent for taking the time to talk with you.*
- *Repeat the subject and purpose of the interview and what you plan to do with the information.*
- *If you wish to tape the interview, ask permission.* Taping makes some people uncomfortable.

As you conduct the interview, keep in mind the following four guidelines:

- *Take notes.* Write down important concepts, facts, and numbers, but don't take such copious notes that you are still writing when the respondent finishes an answer. It is important to maintain eye contact with the respondent.
- *Start with prepared questions.* Because you are likely to be somewhat nervous at the start, you might forget important questions. Therefore, it is wise to begin with prepared questions.
- *Be prepared to ask follow-up questions.* Listen carefully to the respondent's answer. Be ready to ask a follow-up question or request a clarification. Have your other prepared questions ready, but be willing to deviate from them. In a good interview, the respondent probably will lead you in directions you had not anticipated.
- *Be prepared to get the interview back on track.* Gently return to the point if the respondent begins straying unproductively, but don't interrupt rudely or show annoyance.

Follow these three guidelines as you conclude the interview:

- *Thank the respondent.* After all your questions have been answered (or you have run out of time), thank the respondent again.
- *Ask for a follow-up interview.* If a second meeting would be useful—and you think the person would be willing to talk with you further—ask for the second meeting now.
- *Ask for permission to quote the respondent.* If you might want to quote the respondent by name, ask permission now.

Following Up after the Interview

Do two things after the interview:

- *Write up the important information.* Do it while the interview is fresh in your mind. (This step is unnecessary, of course, if you have recorded the interview.) If you will be printing a transcript of an interview, make the transcript now.
- *Send a brief thank-you note.* Within a day or two of the interview, show the respondent that you appreciate the courtesy extended to you and that you value what you have learned. In the letter, confirm any previous offers you made, such as to send a copy of the report.

Figure 7.13 is a portion of a transcript of an interview with an attorney specializing in information technology. The interviewer is a student who is writing about legal aspects of software ownership. Notice how the student prompts the attorney to expand her answers. Also notice how the student responds to the attorney's answers, making the interview more of a discussion.

Letters of Inquiry   A letter of inquiry is often a useful alternative to a personal interview. If you are lucky, the respondent will provide detailed and helpful answers. However, the person might not understand what information you are seeking or might not want to take the trouble to help you. Also, you can't ask follow-up questions in a letter, as you can in an interview. Although the strategy of the inquiry letter is essentially that of a personal interview—persuading the reader to cooperate and phrasing the questions carefully—inquiry letters in general are less successful, because the reader has not already agreed to provide information and therefore sometimes does not respond. In addition, an inquiry letter doesn't give you the opportunity to follow up, as an interview does.

For a full discussion of inquiry letters, see Chapter 19.

Questionnaires   Questionnaires enable you to solicit information from a large group of people. Although they provide a useful and practical alternative to interviewing dozens

Q. Why is copyright ownership important in marketing software?

A. If you own the copyright, you can license and market the product and keep other people from doing so. It could be a matter of millions of dollars if the software is popular.

Q. Shouldn't the programmer automatically own the copyright?

A. If the programmer wrote the program on personal time, he or she should and does own the copyright.

Q. So "personal time" is the critical concept?

A. That's right. We're talking about the "work-for-hire" doctrine of copyright law. If I am working for you, anything I make under the terms of my employment is owned by you.

Q. What is the complication, then? If I make the software on my machine at home, I own it; if I'm working for someone, my employer owns it.

A. Well, the devil is in the details. Often the terms of employment are casual, or there is no written job description or contract for the particular piece of software.

Q. Can you give me an example of that?

A. Sure. There was a 1992 case, *Aymes* v. *Bonelli*. Bonelli owned a swimming pool and hired Aymes to write software to handle recordkeeping on the pool. This was not part of Bonelli's regular business; he just wanted a piece of software written. The terms of the employment were casual. Bonelli paid no health benefits, Aymes worked irregular hours, usually unsupervised— Bonelli wasn't a programmer. When the case was heard, the court ruled that even though Bonelli was paying Aymes, Aymes owned the copyright because of the lack of involvement and participation by Bonelli. The court found that the degree of skill required by Aymes to do the job was so great that, in effect, he was creating the software by himself, even though he was receiving compensation for it.

Q. How can such disagreements be prevented? By working out the details ahead of time?

A. Exactly. The employer should have the employee sign a statement that the project is being carried out as work-for-hire, and should register the copy-right with the U.S. Copyright Office in Washington. Conversely, employees should try to have the employer sign a statement that the project is not work-for-hire, and should try to register the copyright themselves.

Q. And if agreement can't be reached ahead of time?

A. Then stop right there. Don't do any work.

Excerpt from an
Interview

of people spread over a large geographical area, questionnaires rarely yield completely satisfactory results for three reasons:

- *Some of the questions will misfire.* No matter how careful you are, some questions are not going to work: the respondents will misinterpret them or supply useless answers.
- *You won't obtain as many responses as you want.* The response rate will almost never exceed 50 percent; in most cases, it will be closer to 10 percent or 20 percent.
- *You cannot be sure the respondents are representative.* In general, people who feel strongly about an issue are much more likely to respond than are those who do not. For this reason, be careful in drawing conclusions based on a small number of responses to a questionnaire.

When you send a questionnaire, you are asking the recipient to do a favor. If the questionnaire requires only two or three minutes to complete, of course you are more likely to receive a response than if it requires an hour. Your goal, then, should be to construct questions that will elicit the information you need as simply and efficiently as possible.

Creating Effective Questions

Effective questions are unbiased and clearly phrased:

- *Use unbiased language.* Don't ask "Should we protect ourselves from unfair foreign competition?" Instead, ask "Are you in favor of imposing tariffs on men's clothing?"
- *Be specific.* If you ask "Do you favor improving the safety of automobiles?" only an eccentric would answer no. You are more likely to get a useful response if you ask "Do you favor requiring automobile manufacturers to equip new cars with side-impact air bags, which would raise the price $300 per car?"

As you prepare the questions, keep in mind that there are several formats from which to choose:

- *Multiple choice*

  Would you consider joining a company-sponsored sports team?
  Yes_____   No_____

  How do you get to work? (Check as many as apply.)
  _____my own car   _____train
  _____car/van pool   _____walk
  _____bus   _____other (please specify) _____

- *Likert scale.* Likert-scale questions consist of a statement to which the respondent registers the degree of agreement or disagreement.

The flextime program has been a success in its first year.

strongly disagree _____ _____ _____ _____ _____ _____ strongly agree

Although opinions differ, most statisticians recommend using an even number of possible responses (six, in this case); with an odd number, too many respondents choose the middle response, which does not provide useful data.

- *Semantic differentials.* Semantic differential questions call for the respondent to register a response along a continuum between a pair of opposing adjectives. Usually, semantic differential questions are used to gauge a person's subjective response to a task, an experience, or an object.

  simple   ____ ____ ____ ____ ____ ____   difficult
  attractive   ____ ____ ____ ____ ____ ____   ugly
  interesting ____ ____ ____ ____ ____ ____   boring
  neat   ____ ____ ____ ____ ____ ____   sloppy

- *Ranking.* Ranking questions call for the reader to indicate a priority among a number of alternatives.

  Please rank the following work schedules in order of preference. Put a 1 next to the schedule you would most like to have, a 2 next to your second choice, and so on.
  8:00–4:30 _____   9:00–5:30 _____
  8:30–5:00 _____   flexible   _____

- *Short answer.* Short-answer questions call for the reader to respond briefly using phrases or sentences.

  What do you feel are the major advantages of the new parts-requisitioning policy?
  1. _____
  2. _____
  3. _____

- *Short essay.* Short-essay questions call for a longer response.

  The new parts-requisitioning policy has been in effect for a year. How well do you think it is working?

  _____
  _____
  _____
  _____
  _____

The primary advantage of an essay question is that sometimes you will discover information you never would have found using closed-end questions. But keep in mind that you will receive fewer responses if you ask for essay answers; moreover, essays, unlike other kinds of answers, cannot be quantified precisely. A simple statement with a specific number in it—"Seventy-five percent of the respondents own two or more PCs"— helps you make a clear and convincing case. However thoughtful and persuasive an essay answer may be, it is subject to the interpretations of different readers.

After you have created the questions, write a letter or memo to accompany the questionnaire. A letter to someone outside your organization is basically an inquiry letter (sometimes with the questions themselves on a separate sheet); therefore, it must clearly indicate who you are, why you are writing, what you plan to do with the information, and when you will need it. (See Chapter 19 for a discussion of inquiry letters.) For people within your organization, a memo accompanying a questionnaire should provide the same information.

Testing the Questionnaire

Before you send *any* questionnaire, show it and the accompanying letter or memo to a few people who can help you see problems in them. After you revise the materials, test them on people whose backgrounds are similar to those of your real respondents. Revise the materials a second time, and, if possible, test them again. Remember, after you send the questionnaire, you cannot revise it and send it to the same people.

Sending the Questionnaire

After drafting the questions and testing them, administer the questionnaire. Determining the recipient can be simple or difficult. If you want to know what the residents of a particular street think about a proposed construction project, your job is easy. But if you want to know what mechanical-engineering students in colleges across the country think about their curricula, you will need background in sampling techniques to isolate a representative sample.

Be sure to include a self-addressed, stamped envelope with questionnaires sent to people outside your organization.

Figure 7.14 shows a sample questionnaire.

An Example of How to Plan a Research Strategy

This chapter has covered a lot of territory in describing how to carry out research. This section provides an example of how you might go about planning research for a collaborative report to be submitted at the end of the course. First, a description of the context follows.

Working with two other students in the class, you have tentatively decided on a topic: crime on your campus. An assault took place on campus a few weeks ago, and everyone is concerned about it. You're not exactly sure what aspect of the topic to focus on, so you decide to do some preliminary research to see whether you can narrow the focus to make it manageable for a brief report.

September 6, 19XX

To: All employees
From: William Bonoff, Vice-President of Operations
Subject: Evaluation of the Lunches Unlimited food service

As you may know, every two years we evaluate the quality and cost of the food service that caters our lunchroom. We would like you to help by sharing your opinions about the food service. Your anonymous responses will help us in our evaluation. Please drop the completed questionnaires in the marked boxes near the main entrance to the lunchroom. Thank you for your cooperation.

1. Approximately how many days per week do you eat lunch in the lunchroom?
   0 _____   1 _____   2 _____   3 _____   4 _____   5 _____

2. At approximately what time do you eat in the lunchroom?
   11:30–12:30 _____   12:00–1:00 _____   12:30–1:30 _____   varies _____

3. It is always easy to find a clean table.
   Strongly Disagree _____ _____ _____ _____ _____ _____ Strongly Agree

4. The Lunches Unlimited personnel are polite and helpful.
   Strongly Disagree _____ _____ _____ _____ _____ _____ Strongly Agree

5. Please comment on the quality of the different kinds of food you have eaten in the lunchroom.
   a. Hot meals (daily specials)
      excellent _____   good _____   satisfactory _____   poor _____
   b. Hot dogs and hamburgers
      excellent _____   good _____   satisfactory _____   poor _____
   c. Sandwiches
      excellent _____   good _____   satisfactory _____   poor _____
   d. Salads
      excellent _____   good _____   satisfactory _____   poor _____
   e. Desserts
      excellent _____   good _____   satisfactory _____   poor _____

6. What *foods* would you like to see served that are not served now?
   _____

7. What *beverages* would you like to see served that are not served now?
   _____

8. Please comment on the prices of the foods and beverages served.
   a. Hot meals (daily specials)   d. Desserts
      too high _____ fair _____ a bargain _____   too high _____ fair _____ a bargain _____
   b. Hot dogs and hamburgers   e. Beverages
      too high _____ fair _____ a bargain _____   too high _____ fair _____ a bargain _____
   c. Sandwiches
      too high _____ fair _____ a bargain _____

9. Would you be willing to spend more money for a better-quality lunch, if you thought the price was reasonable?   yes, often _____   sometimes _____   not likely _____

10. Please provide whatever comments you think will help us evaluate the catering service.
    _____

FIGURE 7.14

Questionnaire

181

Your instructor wants you to create an organizational context for your study; that is, for the purposes of the assignment you are to create an organization in which you have a role. When you write the proposal and the completion report, you will assume that role. Rather than being a student in the technical-communication class, you will be pretending to be associated with this organization, and the instructor will also assume whatever role you assign him or her. This organization can be a real or imagined company or government organization or campus group.

Deciding on a Research Sequence

Many topics can be approached through both primary and secondary research, and that is the case with campus crime. Which should you do first? After discussing this question with your group, you decide to do secondary research first—find out what others have written and said about campus crime before you start interviewing people and distributing questionnaires. Why? Because you and your group have not decided what aspect of the topic you wish to study. The secondary information will help you understand the topic better. If you went straight to primary research techniques, your results might be uninformed and unfocused because you would not know exactly what aspects of campus crime to concentrate on.

Carrying Out Secondary Research

After discussing the topic with the reference librarian, you decide to focus on three keywords: *universities, colleges,* and *crime.* But a preliminary search through the CD-ROM version of *Readers' Guide to Periodical Literature* under "universities and crime" and "colleges and crime" yields few citations. After thinking about it for a while, one member of your group suggests trying "campus crime." This phrase works.

Searching the Web

You start on the Web. Using Yahoo! and the phrase "campus crime," you find one category and 23 sites. The category is "Society and Culture: Crime: Law Enforcement: Campus Police Departments," which leads you to more than 100 campus sites. Among the sites is the University of Oklahoma Campus Safety Patrol, which is described on the search-results page as "100+ public safety Web pages online including crime prevention, personal safety, fire safety, first aid and health, and more." You visit the site, which indeed has hundreds of links to organizations specializing in crime prevention, campus safety, and other law-enforcement pages, including county and state police organizations, the FBI, the Department of Justice, and the Bureau of Alcohol, Tobacco and Firearms. You click back to the Yahoo! search-results page to look at some of the 23 site matches. You see several private organizations, including the following three:

- Security on Campus, Inc., "dedicated to providing college and university communities with crime safety information"

- Campus Outreach Services, an "organization that provides presenters on date-rape and experts on model sexual assault policy and procedure prevention for high schools and colleges"
- Aegis Protection Group, "consulting, seminars, publications, and training videos on campus crime prevention, campus parking management, campus fire safety, campus occupational safety, and police training"

You bookmark the search-results page to study later.

Searching Databases at the Library

The reference librarian suggests four CD-ROM databases to begin your search: PAIS (Public Affairs Information Service, Inc.), ERIC (Educational Resources Information Center), Criminal Justice Abstracts, and GPO (Government Printing Office). Using "campus crime" as your keywords, you locate well over 100 books, articles, and reports. Here are just a few:

O'Kane, J. B., Fisher, R. M., & Green, L. (1994). Mapping campus crime. *Security Journal* v. 5, n. 3, pp. 172–180. This article discusses "the utility of collecting call-for-service data from university and municipal police departments. . . . The use of computer mapping technology allowed for a distributive analysis of crime clustering, and facilitated an examination of the spatial and situational patterns of specific types of campus crime. Site observations of campus crime 'hot spots' helped generate potential intervention strategies intended to manipulate the environmental context in which certain types of campus crime occur. The result was the construction of a powerful crime prevention information base."

Mullendore, P. (1991). The Federal Campus Crime Reporting Act: Challenges and opportunities for campus and local law enforcement. *Journal of California Law Enforcement* v. 25, n. 4, pp. 110–116. This article "discusses student safety and college campus security, with particular attention to U.S. federal and California legislation."

Lederman, D. (1995, Feb. 3). Colleges report rise in violent crime: Crime data from 796 colleges and universities. *Chronicle of Higher Education,* v. 41, n. 21, pp. A31–A42. "Data on 1993 campus crime from 796 higher education institutions with enrollments over 5,000, required by federal law, reflects a rise in robberies, aggravated assaults, arrests for drug and weapons violations though property crimes and murders decreased. Interpretation of data, seen as providing an incomplete picture, is debated."

Kirkland, C. J, & Siegel, D. G. (1994). Campus security: A first look at promising practices. Office of Educational Research and Improvement, Washington, D.C.: Programs for the Improvement of Practice. "This booklet is designed to assist colleges and universities as they develop policies, procedures, and programs to increase safety, respond appropriately to victims of crime, and fully comply with federal requirements established by the Crime Awareness and Campus Security Act of 1990. Part 1 presents 11 topics for helping define comprehensive, exemplary campus security programs. . . . Part 2 presents some examples of practices, shared by campus officials across the country, that may be helpful to other colleges and universities, categorized according to Part 1's topic areas. . . ."

You decide to discuss the different topic areas with the other group members. When you have narrowed the topic, you will mark which articles you want to study.

Searching the Library's Online Catalog

Using the subject listing for "campus crime," you discover one book that might be relevant:

Fossey, R., & Smith, M. C. (1996). *An administrator's guide for responding to campus crime: From prevention to liability.* San Francisco: Jossey-Bass.

You check it out of the library.

Your group decides to spend a week reading about a dozen articles and the book and deciding on a narrower topic. You decide not to carry out any primary research at this point; when you have a better focus, you can plan the primary research.

Planning Primary Research   After studying the secondary research, you decide to focus on the Crime Awareness and Campus Security Act of 1990. Before you began this project, none of you were aware of this law: why it was enacted, what it requires, how it is applied. You wonder how widely known the law is on campus, and whether a better awareness of crime would reduce the incidence of crime. You decide to carry out three primary research tasks:

- Go to the campus safety office and get a copy of the Safety and Security Annual Report. Study it according to the principles of effective document design.
- Distribute a brief questionnaire to students in the technical communication class to determine their awareness of the report's existence and their knowledge of crime on campus.
- Interview the Director of Campus Safety to learn how he or she applies the law, collects the statistics, assembles the report, and distributes it. Does this office offer other information or services to help students prevent crime on campus?

Your group settles on a context for the project: you will be a group reporting to the Student Senate. Your task is to investigate how your school applies the law and publicizes information about crimes on campus. Your purpose will be to report the findings of your investigation and, if appropriate, propose recommendations on how to improve the quality of the school's efforts to publicize crime information.

As this brief discussion has made clear, planning and carrying out research requires a clear understanding of your topic. But the catch is that a clear understanding of your topic requires that you plan and carry out research. In fact, the process of researching a topic is cyclical: you learn more about your topic, which means you go out and do more research, which makes you see your topic differently. You will probably find that you do research throughout the entire process, constantly refining your topic as you proceed. By the time you finish your report at the end of the semester, you will understand why experienced writers know that they are not done doing their research until they submit the document.

Once you have performed your basic primary and secondary research, you can turn your attention to shaping the document. Chapters 8–10 discuss techniques of drafting the text. Chapters 11 and 12 discuss document design, layout, and graphics. Keep in mind, however, that because you will probably do more research, it is smart to keep all your research in order. Don't toss out any information you have gathered, even if you think you will never need it again.

CASE

Creating a Research Guidebook for Your Major

Information sources for students in your major are changing so rapidly that new majors in your department need a current guidebook describing how to carry out research. The chair of your department has asked your group to assemble a guidebook that can be distributed at the library and the department office. Different members of your group might concentrate on the following tasks:

- interviewing professors for their suggestions about the best sources of information for undergraduate majors
- using the printed information in the library: dictionaries, directories, handbooks, specialized encyclopedias, indexes, and so forth
- using the online resources in the library: online databases and CD-ROMs
- using the Web

Compile your findings into a guidebook. Be sure to submit the electronic file to your instructor; the manual will be out of date by next semester.

WRITER'S CHECKLIST

Have you

1. focused your topic?

2. used techniques to generate ideas?

3. devised a strategy for conducting secondary and primary research?

4. consulted the appropriate reference books, including periodical indexes, newspaper indexes, abstract services, government publications, guides to business and industry, online databases, CD-ROM, and the Web?

5. studied the information by skimming, taking notes, and summarizing?

6. in evaluating information, been especially careful about
   ❑ the author's credentials?
   ❑ the publishing body?
   ❑ the author's knowledge of literature in the field?
   ❑ the accuracy and verifiability of the information?
   ❑ the timeliness of the information?

7. if appropriate,
   ❑ conducted inspections?
   ❑ conducted experiments?
   ❑ performed field research?
   ❑ conducted interviews?
   ❑ sent letters of inquiry?
   ❑ administered questionnaires?

EXERCISES

Several of the following exercises call for you to write memos. See Chapter 16 for a discussion of memos.

1. Check with your instructor about whether this assignment should be done individually or in a small group. Choose a topic on which to write a report for this course. (See Chapters 15–17 for information on reports.) Make sure the topic is sufficiently focused so that you will be able to cover it in some detail. Subjects that can be focused into topics include the following:
   - *A campus problem, such as a shortage of parking space.* What are the symptoms and causes of the problem? What effects has the problem had on life at the college and its surrounding area? Has the college or a community group attempted to investigate the problem? What have been the findings? Conduct your own analysis of the problem, draw conclusions, and recommend what should be done next.
   - *The evolving role of the Internet as an information source.* Which aspects of the Internet will prove most useful for researchers? Which resources already in place are most likely to change over the next few years? What new services are likely to evolve? What possible problems—technical, political, economic, social, ethical—must be faced if the potential of the Internet is to be realized?
   - *Alternative fuels and energy sources for automobiles.* What alternatives are being developed now? What are their strengths and weaknesses? What factors encourage their development? What are the social or political barriers to their widespread introduction? Which is most likely to succeed in the marketplace?
   - *Job prospects in a particular field.* What data exist on job prospects, both locally and nationally? What are the trends? Does one aspect of the field appear to have better prospects than others? What steps can a student take to ensure the best preparation for the job market? How well does the curriculum meet the needs of industry?
   - *The evolving role of communication skills for the technical professional.* Determine the different skills and backgrounds of technical professionals in your field. What kinds of documents do they make? What kinds of equipment do they use? Where did they learn their skills? How well equipped is your college to prepare tomorrow's technical professionals to communicate in the workplace? What steps could be taken to improve their preparation?
   a. Using the Web and one of the guides-to-the-guides, plan a strategy for researching your topic.
      - Which guides, handbooks, dictionaries, and encyclopedias contain the background information you should read first?
      - Which basic reference books discuss your topic?
      - Which major indexes and abstract journals cover your topic?

- Which of these basic tools are available online in your library? In which media (for instance, online bibliographies, CD-ROM, the Web) are they available?

b. Write down the call numbers or Internet addresses of the three indexes and abstract journals most relevant to your topic.

c. Make a preliminary bibliography of two books, five articles, and five Web sites that relate to your topic.

d. Find one of the works listed in your bibliography and write a brief assessment of its value.

e. Paraphrase any three paragraphs, each on a separate note card, from the first two pages of an article listed in your bibliography. Also note at least two quotations, each on a separate card: one should be a complete sentence, and one an excerpt from a sentence. Include a photocopy of the original when you submit the assignment, with the sentences you have quoted highlighted or underlined.

f. Write a 500-word summary of an article in your bibliography. Include a photocopy of the article when you submit the assignment.

g. Using a local industrial guide, list five people who might have firsthand knowledge of your topic. Arrange for and carry out an interview with one of the five.

2. Revise the following questions from interviews to make them more effective. In a brief paragraph for each, explain why you have revised them as you have.

a. What is the role of communication in your daily job?

b. Do you think it is better to relocate your warehouse or go to just-in-time manufacturing?

c. Isn't it true that it's almost impossible to train an engineer to write well?

d. Where are your company's headquarters?

e. Is there anything else you think I should know?

3. Revise the following questions from questionnaires to make them more effective. In a brief paragraph for each, explain why you have revised them as you have.

a. Does your company provide tuition reimbursement for its employees? Yes_____ No_____

b. What do you see as the future of bioengineering?

c. How satisfied are you with the computer support you receive?

d. How many employees work at your company?
5–10_____   10–15_____   15 or more_____

e. What kinds of documents do you write most often?
memos_____   letters_____   reports_____

Collaborative Exercise

4. Form small groups for this project about search engines. As a group, choose a topic in the news, such as an advance in computer technology, and create a list of keywords to research the topic. Then have each member of the group choose a different search engine and keep a log of what kinds of information the search engine uncovers. As a group, write a memo to your instructor explaining your findings. Does one search engine work better than the other for this topic? Can you draw any inferences about the strengths and weaknesses of the different search engines you investigated?

Web Exercise

5. Using a search engine such as Yahoo!, InfoSeek, or MetaCrawler, answer the following questions:

a. What are the three largest or most important professional organizations in your field? For example, if you are a mechanical engineering major, an important professional organization is the American Society of Mechanical Engineers.

b. What are the three most important journals read by people in your field?

c. What are the three most important electronic discussion groups or news services read by people in your field?

d. What are the date and location of a professional meeting this year for people in your field?

e. Name and describe, in one paragraph each, three major issues being discussed by practitioners or academics in your field. For instance, nurses might be discussing the effect of managed care on the quality of medical care delivered to patients.

Present your findings in a memo to your instructor. At the end of your response to each of the five questions or statements, provide the URL (the uniform resource locator, or address) of the site, or the address of the discussion group or news service from which you derived the information.

REFERENCES

Kirk, E. (1996). Evaluating information found on the Internet. *Milton's Web.* <http://milton.mse.jhu.edu:8001/research/education/net.html> (1997, Apr. 20).

Lovgren, J. (1994). How to choose good metaphors. *IEEE Software, 11*(3), 86–88.

McComb, G. (1991). *Troubleshooting and repairing VCRs* (2nd ed.). Blue Ridge Summit, PA: TAB/McGraw-Hill.

Tillman, H. N. (1997). Evaluating quality on the net. <http://www.tiac.net/users/hope/findqual.html> (1997, Apr. 20).

CHAPTER

8

Developing the Argument

A Brief Review of Outlining 190
  Grouping Similar Items 190
  Ordering the Items in the Groups 191
  Organizing the Groups 191
  Avoiding Common Logical Problems 191
  Choosing an Outline Format 192

Guidelines for Drafting 193

How to Develop an Argument 195
  Understanding Your Audience's Goals 196
  Understanding the Nature of Persuasive Evidence 196
  Presenting Yourself Effectively 197
  Understanding Political Realities 198

Basic Patterns of Organizing Arguments 198
  Chronological 199
  Spatial 200
  General to Specific 203
  More Important to Less Important 206
  Comparison and Contrast 207
  Classification and Partition 211
  Problem-Methods-Solution 217
  Cause and Effect 221

Introducing and Concluding the Body 224
  Introducing the Body 224
  Concluding the Body 227

Case: Introducing a Document 229

Writer's Checklist 230

Exercises 231
     Collaborative Exercise 233
     Web Exercise 234

At this point in the writing process, you know to whom you are writing, and why, and you have performed most of the primary and secondary research. Although you will likely revise your analysis of your audience and purpose, and you will probably continue to gather information throughout the entire writing process, now it is time to start developing your argument.

This chapter begins with a brief review of outlining and guidelines for drafting the body, focusing on how to use the computer effectively as a drafting tool. The chapter then turns to a discussion of argument, explaining how to craft a convincing case for your point of view. Next, the chapter discusses basic patterns for organizing the argument. The chapter ends with a discussion of ways to introduce and conclude the body.

A Brief Review of Outlining

Chapter 7 offered strategies for generating ideas and gathering more information, concentrating on using the Web and the library; performing experiments, field research, and inspections; conducting interviews; and sending questionnaires. After gathering a mass of information, most writers create an outline, a process that is fundamental to organizing an effective argument.

Creating an outline involves five main tasks, as shown in Figure 8.1.

Grouping Similar Items   If you used branching or clustering to generate ideas, you have already grouped similar items. But if you used brainstorming, you have not. Start by looking at the first item on your brainstorming list and determining what major category it belongs to. For instance, it could be an item that belongs in the background section. Scan the rest of the list, noting other items that belong in the background section. Link them. Then go back to the top of the list and determine the logical category of the next item. Link it to other items that belong in the same category. Repeat this process until all the items are either grouped into logical categories or discarded.

FIGURE 8.1

Steps in Outlining

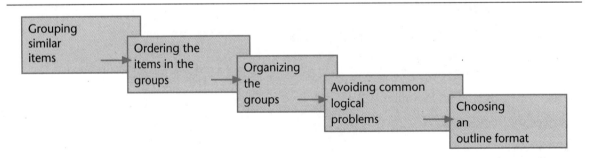

Even though the outline is incomplete, some writers like to start drafting the document at this point. They want to see what they have to say, and the only sure way to do that is to start to write. Once they have started drafting, they feel better able to sequence items. They juggle paragraphs instead of juggling outline items. Other writers feel more comfortable working from a more refined outline. For them, the next step is to sequence the items in each group.

Ordering the Items in the Groups

In some cases, sequencing the items in each group is a simple matter—one item might precede another in time, or perhaps one item is more important than another and therefore should come first. Likewise, one item in the group might represent the problem, another the methods for solving it, and the next the solution. Later, this chapter offers a detailed discussion of eight basic patterns of organizing an argument.

Organizing the Groups

To organize the groups, follow the same logical processes you used to organize the items within each group. Often your readers will expect a particular structure. For instance, the body of a report begins with an introduction, followed by the detailed discussion and the major findings. If your material does not lend itself to a traditional pattern of organization, you'll need to create one that your readers will find clear and easy to follow.

Avoiding Common Logical Problems

As you create and refine your outline, avoid two common logical problems:

- faulty coordination
- faulty subordination

Faulty coordination occurs when a writer equates items that are not of equal value or not of the same level of generality. Figure 8.2 shows a portion of an outline with faulty coordination and a corrected version.

Faulty subordination occurs when an item is made a subunit of a unit to which it does not belong, as shown in Figure 8.3 on page 192.

Faulty Coordination	**Proper Coordination**

Common household tools
1. screwdrivers
2. drills
3. claw hammers
4. ball-peen hammers
	Common household tools
	1. screwdrivers
	2. drills
	3. hammers
	    3.1 claw hammer
	    3.2 ball-peen hammer

FIGURE 8.2

Faulty Coordination and Proper Coordination

Faulty Subordination	Proper Subordination
Power sources for lawnmowers 1. manual 2. gasoline 3. electric 4. riding mowers	Power sources for lawnmowers 1. manual 2. gasoline 3. electric

FIGURE 8.3

Faulty Subordination and Proper Subordination

In Figure 8.3, "riding mowers" is out of place because it is not a power source of a lawnmower. Whether it belongs in the outline at all is another question, but it certainly doesn't belong here.

A second kind of faulty subordination occurs when only one subunit is listed. The solution is to incorporate the single subunit into the unit: in Figure 8.4, item 1.1 is deleted and restated as item 1.

Single Subunit	Proper Subordination

Types of sound-reproduction systems
1. records
   1.1 phonograph records
2. tapes
   2.1 cassette
   2.2 open reel
   2.3 DAT
3. compact discs
	Types of sound-reproduction systems
	1. phonograph records
	2. tapes
	   2.1 cassette
	   2.2 open reel
	   2.3 DAT
	3. compact discs

FIGURE 8.4

Single Subunit and Proper Subordination

Choosing an Outline Format

If you are creating an outline purely for your own use, you don't have to worry about its format, as long as you can understand it and work effectively from it. However, you should use one of the standard outline formats if your outline will be read by someone else, or if your document will contain a table of contents, which is basically an outline of the document. (See Chapter 10 for a discussion of tables of contents.)

Two kinds of outlines are common:

- topic outlines
- sentence outlines

In a topic outline, the entries are phrases, such as "increased worker productivity." In a sentence outline, the entries are sentences, such as "A major advantage of health-promotion programs is that they increase worker productivity." Relatively few writers use sentence outlines, because they know they will have to revise the sentences later. The main advantage of the sentence outline is that it keeps you honest: if you don't know what to say about an item, you will realize your uncertainty as soon as you try to come up with a sentence.

If you will be submitting your outline to someone, you will have to make one more decision: whether to use a traditional alphanumeric system or the decimal system. The two systems are shown in Figure 8.5.

An advantage of the decimal system is that it is simple to use and to understand. You don't have to remember how to represent the different heading levels. In the same way, your readers can easily see what level they are reading.

Use whichever system helps you stay organized—or is required by your organization.

Traditional Outline	Decimal Outline
I.	1.0
A.	1.1
1.	1.1.1
2.	1.1.2
B.	1.2
1.	1.2.1
a.	1.2.2
b.	2.0
(1)	
(2)	2.1 etc.
(a)	
(b)	
2.	
C. etc.	

FIGURE 8.5

Traditional and Decimal Outlines

Guidelines for Drafting

People use different methods for drafting. Some start with a detailed outline; others work with only a very brief, general one. Still others work from sketches of graphics. Whatever technique you use, the following eight guidelines are suggestions to help you efficiently create a draft that you can revise into a final document.

- *Get comfortable.* Drafting is hard work, so you might as well get comfortable. Figure out the best conditions for you. Obviously, a good chair set at the right height for the keyboard is important, and so is good light that doesn't reflect off the screen. But everything else about your environment is a matter of personal preference. Some people work best in quiet; others concentrate better with music.

- *Draft right on the outline.* One advantage of drafting on a computer is that you can draft on the outline. By using the outline, you are less likely to lose sight of your overall plan. Before drafting, make a copy of your outline. Then draft on one copy of it. With this technique, you can easily refer to a clean copy of the outline. Many writers like to print out a paper copy of their outline to place next to their keyboard.

- *Start with the easiest topics.* There is no rule that says you have to start writing at the beginning of the document. Many writers like to start with a section from the middle, usually a technical section that they are comfortable writing about.

- *Draft quickly.* Many students make the mistake of trying to draft slowly so that their first draft will be good enough to serve as a final draft. It probably won't be. Expressing ideas on paper is so complicated that you shouldn't expect to say what you mean—and say it well—the first time. Don't even try. Just try to make your fingers keep up with your brain. Turn the phrases from your outline into paragraphs.

- *Don't stop to get more information or to revise.* Set a timer and draft without stopping. Many writers prefer one hour; some, two. When you hit an item in your outline that you don't understand or that requires more research, just go to the next item. Don't worry about clumsy sentence structure or odd spelling. Your goal is to create a big, rough draft.

- *Try invisible writing.* Turn the contrast knob to darken the screen. This technique encourages you to look at your hardcopy outline or the keyboard. You don't stop typing so often because you are less tempted to revise what you have just written. If you get nervous because you don't see the words on the screen, turn the contrast knob enough so that you can see the text faintly; in this way you will see that words are appearing on the screen but you won't be tempted to stop and revise.

- *Use abbreviations.* To speed up the drafting, use the *search-and-replace* function, which lets you find any phrase, word, or series of characters and replace it with any other writing. For example, you could type *w/* instead of the word *with* throughout the draft. Or if you will need to use the word *potentiometer* a number of times in your document, you can type **po* each time. Then, during the revising stage, you can instruct the computer to change every *w/* to *with* or every **po* to *potentiometer.* The search-and-replace function also can help you reduce the number of misspellings because you have to spell the word correctly only once. One word of caution: if you don't use an asterisk or some other unusual character to mark your abbreviation, you might accidentally turn every *po* into *potentiometer.* The word *potential* will become *potentiometertential.* After you make this mistake once in a long document, you will understand why it is called *search-and-destroy.*

- *When your time is up, stop in the middle of a section.* When it is time to stop drafting, many writers find it helpful to stop in the middle of a paragraph, or even in the middle of a sentence. When you start to draft again, you will find it easy to conclude the idea you were working on. This technique will help you avoid writer's block—the mental paralysis that can set in when you stare at a blank page—to which you might be more vulnerable if you had to start a new major section.

For a discussion of how to use word processing software effectively, see Chapter 4.

How to Develop an Argument

In technical communication, as in other forms of nonfiction writing, you use different kinds of evidence to develop your argument as you draft and revise a document. An *argument* is simply your arrangement of facts and judgments about some aspect of the world. When you create an argument, you are asking your readers to accept your set of facts and judgments and, sometimes, to act according to your recommendations. An argument can be as short as a paragraph or as long as a multivolume report; it can take many forms, and it can discuss almost any kind of issue. Here are some examples:

FROM A DESCRIPTION OF A SITE:
Features A, B, and C characterize the site.

FROM A STUDY OF WHY A COMPETITOR IS OUTSELLING YOUR COMPANY:
Its dominance can be attributed to the following four major factors: A, B, C, and D.

FROM A FEASIBILITY STUDY CONSIDERING FOUR COURSES OF ACTION:
Alternative A is better than alternatives B, C, and D.

FROM A SET OF INSTRUCTIONS FOR PERFORMING A TASK:
The best and safest way to perform the task is to complete task A, then task B, and so on.

To develop an effective argument, you must

- understand your audience's goals
- understand the nature of persuasive evidence
- present yourself effectively
- understand political realities

Understanding Your Audience's Goals

Chapter 5 discussed how to use audience profile sheets to help you think about how individual readers will receive your message and use the document you are preparing. In addition to seeking information about particular readers, however, you should think about your audience's broader goals. Certainly, most people want their company to prosper. Yet most people are also concerned about their own welfare and interests within the company. If your arguments recognize people's needs for *security* and *recognition,* they are more likely to be effective.

People's need for security leads them to prefer safe actions to controversial ones, especially if the controversial ones have a significant chance of jeopardizing their status or power within the organization or threaten their livelihood. Therefore, people oppose an action if they perceive that it might hurt them personally, even if it seems certain to help the organization. Another aspect of security is workload; most people resist an argument that calls for them to do more work.

People's need for recognition leads them to defend their own actions, even if they would oppose the same actions by someone else. For this reason, you can expect opposition from the person or people associated with any action or decision your argument criticizes. It is wise, therefore, to present your argument so that you are not openly criticizing anyone. Rather, present it as a response to the company's present and future needs. Look forward, not back, and be diplomatic.

Understanding the Nature of Persuasive Evidence

Every argument calls for a different kind of evidence, but people most often react favorably to four kinds of evidence:

- *"Commonsense" arguments.* In this phrase, the word *commonsense* has a special meaning: "most people would think that . . ." For example, the following sentence presents a commonsense argument that binge drinking is harmful:

  Binge drinking cannot be good for you; it makes you sick while you are doing it and for quite a few hours afterward.

  A commonsense argument, therefore, appeals to a person's understanding and experience of the way the world works. It says, "I don't have hard evidence to support my conclusion, but it stands to reason that . . ." In this case, the argument is that behavior that the body rejects is probably not healthy. If your audience's sense of commonsense matches yours, your argument is likely to be persuasive.

- *Numerical data.* Numerical data—statistics—are generally more persuasive than commonsense arguments.

  Compared with students who do not binge-drink, binge drinkers are three times as likely to have unprotected sex; female students who binge-drink are twice as likely as their non–binge-drinking peers to be the victims of sexual abuse and rape.

- *Illustrations.* An example or illustration makes the abstract more concrete and therefore more vivid and more memorable.

  John M. began to binge-drink when he was a junior in high school. Three or four nights a week he and his friends would get together and each drink 6 to 12 beers. Because they were underage and wanted to hide their activities from their parents, they did their drinking away from home—in alleys, in parking lots, at the school athletic fields. One night, as they were walking back from the athletic fields . . .

  Illustrations are often used in conjunction with numerical data. The illustration provides a memorable incident, and the numerical data show that the illustration is part of a pattern, not a fluke or some sort of coincidence. For example, the writer using the illustration of a teenage boy hit by a car when he is binge-drinking needs some numerical data to prove that a significant number of binge drinkers suffer similar fates; otherwise, the illustration, although powerful, doesn't support the argument.

- *Expert testimony.* A message from an expert is more persuasive than the same message from someone without credentials. A well-researched article on binge-drinking written by a respected public-health official and published in a reputable medical journal is likely to be persuasive. When you make arguments, you will often cite expert testimony from published sources, Web sites, or interviews you have conducted. When you use information you found on the Internet, be particularly careful, as explained in Chapter 7.

Presenting Yourself Effectively

Just as testimony from an expert is more persuasive than the same testimony from a nonexpert, an argument from a person the audience likes is more persuasive than the same argument from a person the audience dislikes. As you create your argument, try to project the following characteristics:

- *Cooperativeness.* As discussed in Chapter 3, you will work with groups much of the time in your professional career. Show a cooperative spirit by focusing on the goals and values of the organization, not on your own interests.

- *Moderation.* Be moderate in your judgments. The problem you are describing will not likely spell doom for your organization, and the solution you propose will not solve all the company's problems.

- *Fair-mindedness.* Acknowledge opposite points of view, even as you offer counterarguments. Don't let your readers think that you dismiss everyone else's opinions.

- *Modesty.* If you fail to acknowledge that you don't know everything, someone else will be sure to volunteer that insight.

The following paragraph shows how a writer can demonstrate the qualities of cooperativeness, moderation, fair-mindedness, and modesty:

In the first three sentences, the writer acknowledges the problems with his recommendation.

The use of "I think" adds an attractive modesty: the recommendation might be unwise.

The recommendation itself is moderate; the writer does not assert that the plan will save the world.

In the last two sentences, the writer shows a spirit of cooperativeness by moving the focus from his analysis of the alternatives to the company's goals.

This plan is certainly not perfect. For one thing, it calls for a greater up-front investment than we had anticipated. And the return-on-investment through the first three quarters is likely to fall short of our initial goals. However, I think this plan is the best of the three alternatives for the following reasons. . . . Therefore, I recommend that we begin planning immediately to implement the plan. I am confident that this plan will enable us to enter the flat-screen market successfully, building on our fine reputation for high-quality advanced electronics.

Understanding Political Realities

Choose your battles. There are some goals that you can reasonably hope to achieve, and others that you can't; don't use all your energy and sacrifice all your credibility on a losing cause.

The same advice applies in the workplace. If you know that a particular course of action is clearly in the best interests of the company, but you also know that upper management disagrees or that the company can't afford to do it, don't be a martyr to the cause. Instead, try to figure out what aspects of the course of action might be achieved through some other means, or scale back the idea so that it is affordable.

Two big exceptions to this rule are: matters of ethics or safety. As discussed in Chapter 2, under certain circumstances compromise is unacceptable.

Basic Patterns of Organizing Arguments

To create a successful argument, you need to understand your audience's goals and the nature of persuasive evidence, present yourself effectively, and understand the political realities. But how do you actually organize the information to make your argument? Every argument is different and calls for a unique organizational pattern. Long, complex arguments often contain a number of organizational patterns. For instance, one of your arguments might be a comparison and contrast of two pieces of equipment, but within that argument you might organize the features of the two pieces of equipment from more important to less important.

It is useful to think about basic organizational patterns for two reasons:

- *The basic patterns can serve as templates for you to modify.* There is no reason to reinvent the wheel. If a standard pattern will work well with the information you want to convey, start with it, then modify it as necessary.

- *Your readers will understand the basic patterns.* Because they are accustomed to the patterns, they will be able to concentrate better on your message.

This section discusses the eight organizational patterns shown in Figure 8.6.

Chronological The chronological pattern—the time pattern—commonly describes events. Here are some examples of documents that use chronology as an organizing principle in some sections:

- In an *accident report,* you describe the events in the order they occurred.
- In a *trip report,* you describe what you did on the trip.
- In the background section of a *report,* you describe the events that led to the present situation.

Organizational Patterns

- In a *reference manual,* you explain how to carry out a task by describing the steps in order.
- In the methods section of a *journal article,* you describe the actions you took in carrying out the experiment.

When writing a chronologically organized argument, follow three guidelines:

- *Provide signposts.* If the passage is more than a few hundred words long, use headings. Use words such as *step, phase, stage,* and *part,* and consider numbering them. Add substantive phrases to increase clarity:

  Phase One: Determining Our Objectives
  Step 3: Installing the Lateral Supports

  You can also use signposts at the paragraph and sentence levels. Transitional words such as *then, next, first,* and *finally,* used in topic sentences and in support sentences, help your reader follow your discussion. For more information on transitional words and phrases, see Chapter 13.
- *Consider using graphics to complement the text.* As Chapter 12 suggests, graphics can clarify and emphasize chronological passages. Flowcharts are common in writing for all kinds of readers, from the most expert to the general reader.
- *Analyze events where appropriate.* Chronology is perhaps the easiest pattern to use because the narrative provides its own momentum; you just tell your story. Remember, however, that chronology is limited in that it tells the reader what happened but doesn't explain why or how it happened, or what it means. For instance, the largest section of an accident report is usually devoted to the chronological discussion, but the report is of little value if it lacks sections that explain what caused the accident, who bears liability for the accident, and how such accidents can be prevented.

Figure 8.7 (based on Bell, 1993, pp. 32–35) shows a chronologically organized passage.

Spatial   The spatial pattern is used commonly to describe objects and physical sites. Here are some examples of documents that might use spatial organization in some sections:

- In an *accident report,* you describe the physical scene of the accident.
- In a *feasibility study* about building a facility, you describe the property on which it would be built.
- In a *proposal* to design a new microchip, you describe the new chip.

INTRODUCTION

The writer introduces the topic.

The following discussion explains the process by which Panasonic builds semicustom bicycles. What is a semicustom bicycle, and how does it differ from a custom bicycle? A custom bicycle is one whose frame is built to the buyer's individual measurements (torso length, leg length, and arm length) and style of riding (racing, touring, or off road). The frame is then fitted with the buyer's choice of components. A custom bicycle costs $2,500 or more.

Panasonic's semicustom bicycles are halfway between a custom bike and a mass-produced one. For a road bike, say, customers choose one of 10 styles of frame from 15 sizes of chrome-molybdenum or chromoly steel (from 46 cm to 60 cm in 1-cm increments) or 13 sizes of aluminum or carbon fiber. They specify one of three head angles for the front fork, six variations in the stem length for the handlebars (to fine tune for torso length), four in the handlebar width (to adjust for shoulder width), and two in gearing. Finally, customers choose among 191 color schemes of up to three colors each, and decide which of two places to put a decal bearing the owner's name in five styles of type. Figure 1 shows that Panasonic offers more than 11 million variations on its semicustom bikes.

The writer uses a graphic to show the number and kind of variations from which the customer can choose.

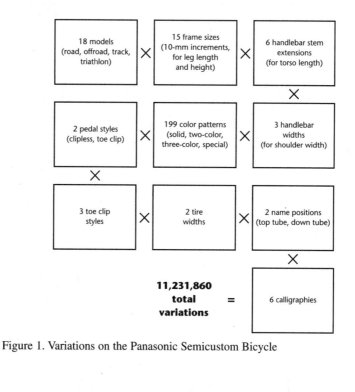

Figure 1. Variations on the Panasonic Semicustom Bicycle

(Continued)

FIGURE 8.7

Argument Organized Chronologically

(Figure 8.7 continued)

Assembling a semicustom bike entails the six steps shown in Figure 2:

Figure 2. Steps in Assembling a Semicustom Bike

The steps of the process are presented in sequence.

Taking the Customer's Order

The customer places an order at one of 1,500 bike shops in Japan. The customer's body measurements are taken and entered, along with his or her preferences, on a detailed order form. That form is then faxed directly to the factory, where operators enter the information into a computer.

Fabricating the Frame

The computer generates a CAD/CAM mechanical drawing of the bicycle and assigns a bar code number to each component. A veteran craftsman is assigned to oversee the fabrication of the bicycle. The three main tubes are automatically cut to length with a machine tool; the computer calculates the exact tube lengths and miter angles and sets the cutters at the exact position.

Inspecting the Frame

Although the details of fabricating the frames differ with the material chosen by the customer, each frame undergoes an automatic three-dimensional measuring-system check. All dimensions and data are processed by a computer to yield a good/no-good decision on a video display.

Painting the Frame

If the frame passes inspection, robots prime and paint the frame, according to the customer's specifications, with six thin coats of paint. For two- and three-color jobs, the craftsman assists the robots. Counting the curing time for the paint, the process of fabricating and painting the frame takes five days.

Assembling the Components

The craftsperson then assembles and attaches the components—the gears, brakes, pedals, and so on. The different components are stored at parts stations

(Figure 8.7 continued)

on shelves. Lights at each station are wired into the computer system. As each order is processed, the proper station lights up, enabling the craftsperson to find the correct components quickly and accurately. The craftsperson then inspects the completed bicycle.

Shipping the Bicycle

The completed bicycle is partially disassembled for compactness, then shipped to the shop where the order was taken. The shop owner assembles the bike and notifies the customer, who also receives a printout of the CAD mechanical drawing of the bike's specifications and a personal thank-you letter from the company.

Conclusion

Panasonic produces 50 to 60 semicustom bikes per day. The production time is 8 to 10 working days, about twice that of a mass-produced bicycle. At a price of about $1,300, the semicustom bike is about four times the price of a mass-produced bike, but only half the price of a custom bike.

- In the results section of a *journal article* about the effectiveness of an anti-toxin, you describe the tissue sample on which you tested the antitoxin.

When you write a spatially organized argument, follow three guidelines:

- *Provide signposts.* Just as with chronologically organized passages, help your readers follow the argument by using words and phrases indicating location (*to the left, above, in the center*) in headings, topic sentences, and support sentences.

- *Consider using graphics to complement the text.* Diagrams, drawings, photographs, and maps add a visual dimension to the text.

- *Analyze events where appropriate.* As is the case with chronology, a spatial organization doesn't explain itself; you have to do the analysis. A diagram of a floor plan cannot explain why the floor plan is effective or ineffective.

Figure 8.8 on page 204 is an excerpt from a process description in which the argument is organized spatially.

General to Specific The general-to-specific pattern is fundamental to technical communication. Based on the idea that readers need a general understanding of a subject before they

DESCRIPTION OF A TROPICAL-FISH AQUARIUM

Tropical fish require an artificial environment that simulates their natural habitat. Although aquariums vary in size, shape, and material, they all contain certain features. This paper describes a suitable microcosm for most tropical fish (usually bright-colored, fresh- and warm-water fish, originally from the tropics). The three critical aspects of the environment are the following:

- aquascape (gravel, plants, and decorative items)
- water
- light

These aspects of the environment will be described from the bottom of the tank to the lid. This order reflects the process you would follow in setting up the tank: you would establish the aquascaping, add the water, and set up the light.

The writer states and justifies his pattern of organization.

Aquascape

Aquascaping is the process of beautifying and enhancing an aquarium in order to simulate a natural habitat. Three items are important in aquascaping a freshwater aquarium: gravel, plants, and decorative objects. The plants and decorative objects promote breeding by providing a hiding place for fish eggs. If the eggs are not hidden and protected, the parents and other fish in the aquarium will eat them. Plants and decorative items also provide the fish with security. Less active fish need quiet, hidden areas; agile, good swimmers need open, turbulent areas.

Gravel

Gravel is not only decoration; it is also important for anchoring plants to the aquarium floor. Gravel 2 to 3 millimeters in diameter is best because it is small enough to anchor plants, yet not large enough to trap uneaten food particles. The gravel should make a bed about 2 inches thick. A light-colored gravel is recommended because gravel color will be an indicator of overfeeding. Excess food will decay and turn black very rapidly.

Plants

Plants, whether live or artificial, are not essential for the well-being of fish, but they . . .

Decorative Items

Decorative items should be thoroughly cleaned . . .

Water

Water conditioning, pH, and temperature are important to the survival of the fish. Many people believe that fish acquire their oxygen from separating the hydrogen and oxygen that bond to create the water, but they don't. They use the

FIGURE 8.8

Argument Organized Spatially

(Figure 8.8 continued)

oxygen that is dissolved in the water from the atmosphere. Distilled water (water without dissolved solids and oxygen) is unsuitable for aquarium use.

Conditioning

If the aquarium water is not properly conditioned, the fish . . .

pH

Water pH is the measure of alkalinity and acidity . . .

Temperature

Most tropical fish thrive between 75° and 80° F. . . .

Light

If the proper amount of light is not provided for the fish and the plants, both will die. Fish do see colors that help them to identify food, as well as fish of the opposite gender. Fish are diurnal (awake during the day), and most species require 12 hours of light. However, direct sunlight is harmful. The ideal area for an aquarium using sunlight would be a northern exposure. Artificial light is just as good as sunlight. Artificial light can be turned on when the room is dark. This artificial light illuminates the tank and provides beauty. Too much light allows green algae to grow.

Conclusion

Tropical fish are, in general, more fragile than nontropical fish, and therefore they need greater care. An aquarium consisting of aquascaping to simulate the natural environment, properly conditioned water of the right temperature and alkalinity, and adequate light will enable the tropical fish to prosper.

can understand and remember the details, the general-to-specific pattern is used in many kinds of technical documents:

- In a *process description,* you describe the overall process before you describe each step in detail. (In the detailed descriptions of each step, you will probably use other organizational patterns.)
- In a *report,* you provide an executive summary—an overview for managers—before the body of the report.
- In a set of *instructions,* you provide general information about the tools and materials needed and about safety measures before providing the step-by-step instructions.

- In a *memo,* you describe the background before you provide the details of your message.

When you write an argument organized from general to specific, follow these two guidelines:

- *Provide signposts.* In the introduction, explain that you are using the general-to-specific pattern. If appropriate, incorporate the words *general* and *specific* or other appropriate terms in the major headings or at the start of the text for each item you are describing.
- *Consider using graphics to complement the text.* Diagrams, drawings, photographs, and maps often add a visual dimension to the text.

Figure 8.9 is an example of an argument organized according to the general-to-specific pattern. The document is a description of types of optical telescopes used in astronomy.

More Important to Less Important

The more-important-to-less-important pattern is useful when you want to direct your readers' attention to the most critical information first. This organizational pattern recognizes that readers of technical communication often want the bottom line—the most critical information—on top.

The writer provides general information about the history, size, and operating principle of telescopes used in astronomy.

INTRODUCTION

The first telescope was invented in 1608 by the Dutch optician Hans Lippershey, who mounted two lenses in a tube. One year later, Galileo refined this design and began to study space. He soon discovered, for instance, that Jupiter has several moons. In 1668, Isaac Newton invented a telescope that used a mirror. His design is the basis of modern astronomical optical telescopes.

Optical telescopes vary tremendously in size and in power. A small telescope has a 1" diameter lens; a large one, 20" or even 30". But both instruments work essentially the same way: they use a lens to collect and focus light waves. In simple astronomical optical telescopes, photographic film is used to record the image created by the lens. In more sophisticated ones, computers are used to enhance the process of amplifying, interpreting, and recording the image.

Three basic types of optical telescopes are used in astronomy: refracting telescopes, reflecting telescopes, and refracting-reflecting telescopes.

The Refracting Telescope
The refracting telescope . . .

The writer begins his detailed discussion of the types of optical telescopes.

FIGURE 8.9

Argument Organized from General to Specific

Here are examples of documents that use the more-important-to-less-important pattern in some sections:

- In an *accident report,* you describe the three most important factors that led to the accident.
- In a *feasibility study* about building a facility, you describe the major reasons that the site is appropriate.
- In a *proposal* to design a new microchip, you describe the major applications for the new chip.
- In the conclusion section of a *journal article* about the effectiveness of an antitoxin, you describe the major conclusions of the experiment.

When writing an argument organized from more important to less important points, follow three guidelines:

- *Provide signposts.* Tell your readers how you are organizing the passage. For instance, in the introductory text of a proposal to design a new microchip, write something like the following: "These three applications for the new chip are arranged from most important to least important."

    Be straightforward. If you have two very important points and three less important points, present them that way: group the two big points together and label them, as in "Major Reasons to Retain Our Current Management Structure." Then present the less important factors as "Other Reasons to Retain Our Current Management Structure." Being straightforward is effective for two reasons: the material is easier to follow because it is clearly organized, and your credibility is increased.

- *Make clear in your discussion why one point is more important than another.* It is not enough to say that you will be arranging the items from more important to less important; you have to explain why the more important point is in fact more important.

- *Consider using graphics to complement the text.* Diagrams and lists are often used to suggest levels of importance.

Figure 8.10 on page 208, from a memo written by an executive at a company that sells equipment to manufacture semiconductors, shows the more-important-to-less-important organizational structure.

Comparison and Contrast

Typically, the comparison-and-contrast organizational pattern is used to evaluate two or more options. This pattern lies at the heart of the feasibility study (see Chapter 17 for more information). Here are examples of documents that use the comparison-and-contrast pattern as an organizing principle in some sections:

- In a *feasibility study* about building a facility, you compare and contrast two sites.

A THREE-POINT PROGRAM TO IMPROVE SERVICE

What we have learned from the recent conference of semiconductor purchasers and from our own focus groups is that customers expect and demand better service than the industry currently provides. By this I don't mean returning phone calls, although that sort of common courtesy remains important. I mean something much more ambitious and difficult to attain: helping our customers do their jobs by anticipating and addressing their total needs. For this reason, I have formed a Customer Satisfaction Panel, chaired by Maureen Bedrich, whose job will be to develop policies that will enable us to improve the quality of the service we offer our customers.

I have asked the panel to consider three major areas:

- improving the ease of use of our equipment
- improving preventive and corrective maintenance
- improving our compatibility with other vendors' products

Improving the Ease of Use of Our Equipment

The most important area to improve is user friendliness. When we deliver a new product, we have to sit down with the customers and explain how to integrate the new product into their manufacturing processes. This process is time consuming and costly not only for us but also for them. For this reason, we must explore the option of automating this integration process. Automated process control, which is already common in such industries as machine-vision, would allow our customers to integrate our semiconductors much more efficiently. In addition, it would help them determine, through process control, whether our equipment is functioning according to specification.

Improving Preventive and Corrective Maintenance

The second most important area for study is improving preventive and corrective maintenance. Our customers will no longer tolerate down times approaching 10 percent; they will accept no more than 2 percent to 3 percent. In the semiconductor equipment field, preventive maintenance is critical because gases used in vapor-deposition systems periodically have to be removed from the inside of the equipment. Customers want to be able to plan for these stoppages to reduce costs. Currently, we have no means of helping them plan.

We also have to assist our customers with unscheduled maintenance. Basically, this means modularizing our equipment so that the faulty module can be pulled out of the equipment and replaced on the spot. This redesign will take many months, but I'm sure it will become a major selling point.

Improving Our Compatibility with Other Vendors' Products

Finally, we have to accept the fact that no one in our industry is likely to control the market, and so we have to make our products more compatible with other manufacturers'. Making our products more compatible means that we must be willing to put our people on-site to see what the customers' setup is and help them determine how to modify our product to fit in efficiently. We can no longer offer a "take-it-or-leave-it" product.

I hope you will extend every effort to work constructively with Maureen and her committee over the coming months to ensure that we improve the overall service we offer our customers.

The writer suggests his organizational pattern in the topic sentence.

The writer again indicates his organizational pattern.

FIGURE 8.10

Argument Organized from More Important to Less Important

- In a *memo,* you compare and contrast the three finalists for a position in your department.
- In a *proposal* to design a new microchip, you compare and contrast two different strategies for designing the chip.
- In an internal *report* describing a legal challenge that your company faces, you compare and contrast several options for responding.

Writing a comparison-and-contrast argument requires two basic steps:

- *Establishing criteria for the comparison and contrast.* To compare and contrast two or more items, you must determine the criteria: the standards or needs against which you will study the items. For instance, if you need to choose an elective course to take next semester, your only criterion might be the time that it is offered: it must be offered at 10 o'clock on Mondays, Wednesdays, and Fridays. For this criterion, MWF 10 is its *required characteristic,* sometimes called a *minimum specification.* However, you probably have other criteria: you would like the course to be interesting, to look good on your transcript, and so on. That the course be interesting is an example of a *desired characteristic,* sometimes called an *evaluative criterion.* If necessary, you might take a course that holds little interest for you—provided it meets your required characteristic.

   Almost always, you will need to consider several criteria in carrying out a comparison and contrast. For a recommendation report on which kind of computer your company should buy, for example, you would probably consider a number of criteria, some of which have required characteristics and some of which have desired characteristics. For instance, a required characteristic might be that you be able to attach the computer to your company's network. If a particular computer cannot be attached, you will not consider it. But most of the criteria might call for desired characteristics, such as ease of operation, reliability, and accessibility of maintenance and service personnel. You will evaluate each option—each computer for sale—first by eliminating those that fail to meet the required characteristics and then by comparing and contrasting the remaining options according to the desired characteristics.

- *Organizing the discussion.* Two basic patterns are used for organizing the discussion: *whole by whole* and *part by part.* Figure 8.11 on page 210 shows the difference between these two patterns. In this figure, two printers—Model 5L and Model 6L—are being compared and contrasted according to three criteria: price, resolution, and print speed.

The whole-by-whole structure is effective if you want to focus on each item as a complete unit rather than on individual aspects of different items. For example, suppose you are writing a feasibility report on purchasing an expensive piece

Whole by Whole	Part by Part
Model 5L • price • resolution • print speed Model 6L • price • resolution • print speed	Price • Model 5L • Model 6L Resolution • Model 5L • Model 6L Print Speed • Model 5L • Model 6L

FIGURE 8.11

Whole-by-Whole and
Part-by-Part Patterns

of laboratory equipment. You have narrowed your choice to three models, each of which meets your required characteristics. Each model has its advantages and disadvantages. Because your organization's decision on which model to buy will depend on an overall assessment rather than on any single aspect of the three models, you choose the whole-by-whole structure; it gives your readers a good overview of the different models.

In contrast, the part-by-part structure focuses on the individual aspects of the different items. Detailed comparisons and contrasts are more effective when they use the part-by-part structure. For example, suppose you are comparing and contrasting the three pieces of laboratory equipment. The one factor that distinguishes the three models is reliability: one model has a much better reliability record than the other two. The part-by-part structure lets you create a section on reliability to highlight this aspect. Comparing and contrasting the three models in one place in the document rather than in three places emphasizes your point. You sacrifice a coherent overview of the different items for a clear, forceful comparison and contrast of an aspect of the three models.

Of course, you can have it both ways; if you want to use a part-by-part pattern to emphasize particular aspects, you can begin the discussion with a general description of the different items. The general description provides the overview of a whole-by-whole discussion, whereas the part-by-part discussion emphasizes particular aspects of the options you are comparing and contrasting.

Once you have chosen the overall pattern—whole-by-whole or part-by-part—you must decide how to organize the second-level items. That is, in a whole-by-whole passage, you have to sequence the "aspects"; in a part-by-part passage, you have to sequence the "options." For most documents, a more-important-to-less-important pattern will work well because readers want to get to the bottom line as soon as possible.

For some documents, however, other patterns might work better. People who write for readers outside their own company often reverse the more-important-to-less-important pattern because they want to make sure their audience reads the whole discussion. This pattern is also popular with writers who are delivering bad news. If, for instance, you want to explain why you are recommending that your organization not go ahead with a popular plan, the reverse sequence enables you to show your readers the problems with the popular plan before you present the plan you recommend. Otherwise, you might surprise the readers, and they might start to formulate objections in their minds before you have had a chance to explain your position.

Figure 8.12 on page 212 (Murphy, 1994), an excerpt from a report, shows the part-by-part form of the comparison-and-contrast organizational structure.

Classification and Partition

Classification is the process of assigning items to categories. For instance, all the students at a university could be classified by sex (males and females), age (18 years old, 19, and so forth), major (nursing, forestry), and any number of other characteristics. You can also create categories within categories. For instance, within the category of students majoring in business at your college or university, you can create subcategories: male business majors and female business majors.

Here are examples of documents that use classification as an organizing principle:

- In a *feasibility study* about building a facility, you classify sites into two categories: domestic and foreign.
- In a *journal article* about ways to treat a medical condition, you classify the treatments as surgical and nonsurgical.
- In a description of a major in a *college catalog,* you classify courses as required or elective.

Partition is the process of breaking a unit into its components. A stereo system could be partitioned into the following components: cassette deck, CD player, tuner, amplifier, and speakers. Each component is separate, but together they form a stereo system. Each component can of course be partitioned further.

Partition is used in descriptions of objects, mechanisms, and processes, as explained in Chapter 9. Here are examples of documents that use partition in some sections:

- In an *equipment catalog,* you use partition to describe one of your products.
- In a *proposal,* you use partition to describe an instrument you propose developing.
- In a *brochure* describing how to use a product, you use partition to describe its features.

The writer introduces her topic by defining the two items she will compare and contrast.

The writer presents her main recommendation and a cross-reference to the sections where she discusses it in more detail.

The writer explains the criteria by which she will compare and contrast the two kinds of debit cards.

The writer is using the part-by-part pattern. For each criterion, she compares and contrasts the two cards.

ONLINE AND OFFLINE DEBIT CARDS: FIRST-YEAR COSTS OF IMPLEMENTATION AT JOHN'S MARKET

The debit card, a plastic card that transfers funds directly from a consumer's bank account to a retailer's bank account, is becoming very popular in our industry. At your request, I have investigated the two basic forms of debit cards: online and offline.

Debit-card transactions are either online (instantaneous like automatic-teller-machine transactions) or offline (in which the transaction takes about two days, like a check transaction). This paper focuses on the costs and benefits of implementing online or offline debit card programs throughout John's Market's 35 stores.

I recommend that John's Market implement the online debit-card system as soon as possible. See the recommendations section for more details.

Costs to the Company

Costs of the debit-card program are evaluated on the basis of four criteria:

- new equipment
- fees
- bill payment and delinquency
- handling the transaction

New Equipment

The online debit card requires the purchase of a point-of-sale (POS) terminal for each cash register. These POS terminals, configured like an ATM keypad, cost about $400 each. The 35 stores have an average of 10 cash registers each, meaning that 350 POS terminals would need to be purchased. The start-up cost for equipment would total about $140,000 if the online program were adopted throughout John's chain.

No new equipment purchases would be necessary to implement the offline debit card because it operates on John's existing credit-card system. The debit card is run through the credit-card scanners already connected to each register. The transaction is stored and processed at a later time.

Fees

A one-time $5,000 hookup fee to access telecommunications and the ATM network would be charged for each store that implements the online program. This communication fee would total $175,000 if the online program were implemented at all 35 stores. Banks would thereafter assume the responsibility for monthly transaction fees and pass that cost to their members in the form of monthly ATM transaction fees.

The offline program requires John's Market to pay a 5 percent service charge on each purchase to the banks that guarantee the debit cards. This is much like a bank/

(Figure 8.12 continued)

The writer explains the procedures for bill payment and delinquency for the two kinds of cards.

The writer explains the procedures for handling the transaction with the two kinds of cards.

merchant relationship with credit-card transactions. Customer response to offline debit cards has been high in test areas. An average of 20 percent of customers adopted the offline payment system when offered. This means John's could expect to pay about $750,000 a year in service charges on its $75 million gross sales.

Bill Payment and Delinquency

John's Market would receive instant payment. . . .

Handling

The costs associated with handling . . .

Conclusions

The online debit program would require initial investments of $140,000 for equipment and $175,000 for hookup fees. This card could pay for part of its own setup costs in staff reductions in the accounting department. A $525,000 savings could be expected in staff reductions. The card would not only improve cash flow in the stores but also yield an estimated savings of $210,000 in the first year.

The offline debit card program would not require setup costs, but banks would assess service charges for guaranteeing their cards. Service charges could be expected to total $750,000 the first year. Cash would flow one day faster with the online program, and staffing costs may increase with increased use of the offline program. The offline debit card would cost an estimated $750,000 in the first year.

Cost in Year One

Type of Card	Equipment	Fees	Payment/ Delinquency	Handling	Total Cost
Online Card	+140,000	+175,000	N/C	-525,000	-210,000
Offline Card	N/C	+750,000	N/C	N/C	+750,000

N/C = No Change

When writing an argument organized by classification or partition, follow these six guidelines:

- *Choose a basis of classification or partition that is consistent with your audience and purpose.* Never lose sight of your audience—backgrounds, skills, and reasons for reading—and your purpose. If you are writing a brochure for hikers in a particular state park, you will probably classify the local snakes according to whether they are poisonous or nonpoisonous, not according to size, color, or any other basis.

- *Use only one basis of classification or partition at a time.* If you are classifying graphics programs according to their technology—paint programs and draw programs—do not include another basis of classification, such as cost.

- *Avoid overlap.* In classifying, make sure that no single item could logically be placed in more than one category; in partitioning, make sure that no listed component includes another listed component. Overlapping generally occurs when you change the basis of classification or the level at which you are partitioning a unit. In the following classification of bicycles, for instance, the writer introduces a new basis of classification that results in overlapping categories:

  – mountain bikes
  – racing bikes
  – touring bikes
  – ten-speed bikes

  The first three items share a basis of classification: the type of bicycle. The fourth item has a different basis of classification: number of speeds. Adding the fourth item is illogical because a particular ten-speed bike could be a mountain bike, a touring bike, or a racing bike.

- *Be inclusive.* When you classify or partition, be sure to include all the categories. For example, a classification of music according to type would be incomplete if it included popular and classical music but not jazz. A partition of an automobile by major systems would be incomplete if it included the electrical, fuel, and drive systems but not the cooling system.

  If your purpose or audience requires that you omit a category, tell your readers what you are doing. If, for instance, you are writing in a classification passage about recent sales statistics of General Motors, Ford, and Chrysler, don't use "American car manufacturers" as a classifying tag for the three companies. Although all three belong to the larger class "American car manufacturers," there are other American car manufacturers. Rather, classify General Motors, Ford, and Chrysler as "the big three" or refer to them by name.

- *Arrange the categories in a logical sequence.* After establishing your categories and subcategories of classification or the components in a partition, arrange them according to some reasonable plan: chronology (first to last), spatial development (top to bottom), importance (most to least), and so on.

- *Consider using graphics to complement the text.* Block diagrams are commonly used in classification arguments; drawings and diagrams are often used in partitions arguments.

In Figure 8.13, a classification of nondestructive testing techniques, the writer uses classification and subclassification effectively in introducing nondestructive testing to a technical audience.

The writer classifies nondestructive testing into four categories. Notice that she justifies the sequence she has chosen.

TYPES OF NONDESTRUCTIVE TESTING

Nondestructive testing of structures permits early detection of stresses that can cause fatigue and ultimately structural damage. The least sensitive tests isolate macrocracks. More sensitive tests identify microcracks. The most sensitive tests identify slight stresses. All sensitivities of testing are useful because some structures can tolerate large amounts of stress—or even cracks—before their structural integrity is threatened.

Currently there are four techniques for nondestructive testing, as shown in Figure 1. These techniques are presented from least sensitive to most sensitive.

Figure 1. Types of Nondestructive Testing

Body-Wave Reflection

In this technique, a transducer sends an ultrasonic pulse through the test material. When the pulse strikes a crack, part of the pulse's energy is reflected back to the transducer. Body-wave reflection cannot isolate stresses: the pulse is sensitive only to relatively large cracks.

Surface-Wave Reflection

The transducer generates an ultrasonic pulse that travels along the surface of the test material. Cracks reflect a portion of the pulse's energy back to the transducer. Like body-wave reflection, surface-wave reflection picks up only macrocracks.

(Continued)

(Figure 8.13 continued)

Because cracks often begin on interior surfaces of materials, surface-wave reflection is a poor predictor of serious failures.

Ultrasonic Attenuation

The transducer generates an ultrasonic pulse either through or along the surface of the test material. When the pulse strikes cracks or the slight plastic deformations associated with stress, part of the pulse's energy is scattered. Thus, the amount of the pulse's energy decreases. Ultrasonic attenuation is a highly sensitive method of nondestructive acoustic testing.

There are two methods of ultrasonic attenuation. One technique reflects the pulse back to the transducer. The other uses a second transducer to receive the pulses sent through or along the surface of the material.

The writer introduces a second level of classification.

Acoustic Emission

When a test specimen is subjected to a great amount of stress, it begins to emit waves; some are in the ultrasonic range. A transducer attached to the surface of the test specimen records these waves. Current technologies make it possible to interpret these waves accurately for impending fatigue and cracks.

Notice that the writer could have used another basis for classification: sensitivity. The four techniques range from very sensitive to less sensitive.

Figure 8.14 (Larson, 1990, p. 674) is an example of partition.

THE EAR

The two main responsibilities of the ear are hearing and balance. The ear has three parts: outer ear, middle ear, and inner ear. The outer ear, the part that we see, is composed of the pinna or auricle (the folds of skin and cartilage that we usually refer to as the ear) and the outer ear canal, which delivers sound to the middle ear. Within the outer ear canal are wax-producing glands and hairs that protect the middle ear.

The writer partitions the ear into three components and arranges them spatially.

The function of the middle ear is to deliver sound to the inner ear, where it is processed into a signal that our brain recognizes. The middle ear is a small cavity with the eardrum (tympanic membrane) on one side and the entrance to the inner

FIGURE 8.14

Argument Organized by Partition

(Figure 8.14 continued)

The writer partitions the middle ear into three components.

ear on the other. Within the middle ear are three small bones known as the hammer (malleus), anvil (incus), and stirrup (stapes) because of their shapes. These bones act like a system of angular levers to conduct sound vibrations into the inner ear. The hammer is attached to the lining of the eardrum, the anvil is attached to the hammer, and the stirrup links the anvil to the oval window, the opening to the inner ear.

As the drawing makes clear, the writer is partitioning and subpartitioning the ear.

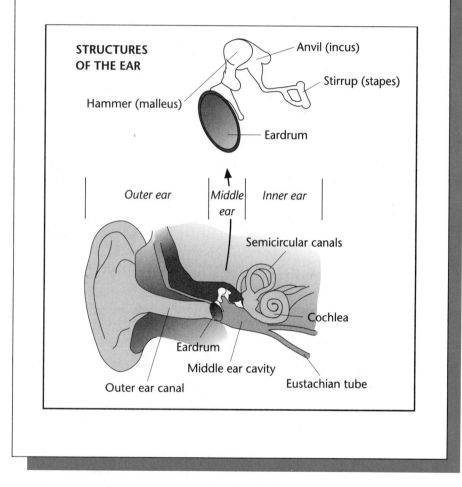

For more examples of partition, see Chapter 9, which covers descriptions of objects, mechanisms, and processes.

Problem-Methods-Solution

The problem-methods-solution pattern is an excellent way to organize an argument about most kinds of problem-solving projects. It is easy to write—and easy to read—because it reflects the logic used in carrying out a project.

The three components of this pattern are simple to identify:

- *Problem.* The problem is a description of what was not working (or not working as effectively as it should), or what opportunity exists for improving current processes.
- *Method.* The method is the procedure you performed either to confirm your analysis of the problem or to solve it, or to exploit the opportunity that exists.
- *Solution.* The solution is a statement of whether your analysis of the problem was correct, or of what you discovered or devised to solve the problem or capitalize on the opportunity.

Here are some examples of documents that use problem-methods-solution as an organizing principle:

- In a *proposal,* you describe a problem in your business, how you plan to carry out your research, and how your end product (an item or a report) can be used to solve the problem.
- In a *completion report* about a project to improve a manufacturing process, you describe the problem that motivated the project, the methods you used to carry out the project, and the findings: the results, conclusions, and recommendations.
- In a *journal article* about ways to treat a medical condition, you describe the problem (the disadvantages of the current techniques), your methods, and your results and conclusions.

The problem-methods-solution pattern is discussed in more detail in Chapters 15–17.

When you write a problem-methods-solution argument, follow these four guidelines:

- *In describing the problem, be clear and specific.* Don't say that "our energy expenditures are getting out of hand." Instead, say that "the energy usage has increased 7 percent in the last year" and that "the utility costs have risen 11 percent." Then calculate the total increase in energy costs.
- *In describing your methods, make sure your readers understand not only exactly what you did but also why you did it that way.* Most technical problems can be approached using several methods; therefore, you might have to justify your choices. Why, for example, did you use a *t*-test in calculating the statistics in an experiment? If you can't defend your choice, you lose credibility.
- *In describing the solution, don't overstate.* Be careful about claiming that "this project will increase our market share from 7 percent to 10 percent within 12 months." The world is too complicated for that sort of self-assurance. Instead, be cautious: "This project promises to increase our

market share significantly: an increase from 7 percent to 10 percent or even 11 percent is quite possible." This way, your document won't come back to haunt you if things don't turn out as well as you had anticipated.

- *Choose a logical sequence.* The most common sequence, naturally, is to start with the problem and conclude with the solution. Sometimes, however, you might want to present the problem first and then go directly to the solution, leaving the methods for last. This sequence deemphasizes the methods, an appropriate strategy if your readers either already know them well or have little need to understand them. When readers want to focus on the solution, you can begin with the solution and then discuss the problem and methods.

  Different sequences work equally well so long as you provide some kind of preliminary summary to give readers an overview and provide headings or some other design elements (see Chapter 11) to help readers find the information they want.

The problem-methods-solution argument in Figure 8.15 is an excerpt from a newsletter article written by the president of a company that manufactures personal computers.

THE PROBLEM

Earlier this year, we were proud to offer the industry's largest array of add-on multimedia products for both our own computers and those of other manufacturers. Our offerings in cards, CD-ROM drives, speakers, and other peripherals were unrivaled in both quantity and quality. And the response was terrific: in our first three months we sold more than 12,000 multimedia kits and 58,000 other peripheral units.

But growing pains soon became apparent: we logged more than 9,000 multimedia-related customer-support calls in that same period. What was the cause of this unprecedented customer-support problem? After considerable analysis of our customer-support data, we concluded that two factors were at work:

- Add-on multimedia kits, even those meant for our own computers, were not necessarily compatible with the hardware or the software our customers were using. We heard too many horror stories about how the kits were installed properly, but when the customer tried to reboot, the operating system was gone.
- Some 70 percent of the customers were novices, as opposed to less than 40 percent for our other product lines, and our documentation was simply inadequate to the task.

The writer describes the problem in detail and analyzes its causes.

The writer speculates on the causes of the problem.

FIGURE 8.15

Argument Organized by the Problem-Methods-Solution Pattern

(Continued)

(Figure 8.15 continued)

Meeting the Challenge

We recognized that being a pioneer in the industry had its costs: we were the first to encounter the problems that are now pervasive in the industry and well publicized in the literature. And because we were first, we took our lumps from the trade journals for the resulting problems with customer satisfaction.

We instituted a four-point plan to meet the challenge:

The writer describes the methods—the steps the company took to solve the problem.

- We instituted a new quality-control program. Now every product is treated just the way a customer treats it. It is taken out of the box, plugged in, and turned on. We make sure that the printer setup is accurate and that the hardware and the bundled software are compatible. At our weekly audit meetings, we review that week's quality-control data; each team leader is now empowered to stop production to investigate a recurrent or unexplained problem.
- We expanded our use of novices in our preproduction focus groups and in the quality-control program. We are concentrating on learning how the novice uses our products; in our expansion into the family market we expect to find that an increasing percentage of our customers are first-time computer owners.
- We instituted a Process-Improvement Team, a group of 12 veteran employees committed to improving customer support and customer satisfaction. Among the first innovations of the Process-Improvement Team was the creation of more than 200 documents to assist users with common problems encountered when installing our kits and using common software. These documents are on our Web site and can be faxed to customers at no charge when they call a special toll-free number.
- We instituted a Quality Team of 15 employees charged with the responsibility of seeking Manufacturing's ideas about quality and efficiency standards.

The Results

The writer describes the solution—the data on the reduction in the problem.

These measures have been in place for only two months, and it is too early to declare total victory, but the preliminary data are encouraging. Customer-support calls on our multimedia kits are down more than 15 percent the last two months. As reported by Customer Support, the incidence of catastrophic problems—such as destruction of the operating system—is down more than 30 percent. The increased use of novices in design and in focus groups has led to three interface improvements that were noted in a *PC Week* article earlier this month. The work of the Quality Team has resulted in a 7 percent decrease in rejection rates of our multimedia kits.

In short, I think we are on the right track. But quality improvement is a frame of mind and a commitment, not a goal that can ever be reached. I pledge to you that we shall continue to strive to make RST the best place to buy PCs and PC-related products.

Cause and Effect   Technical communication often involves cause-and-effect arguments. Sometimes you will reason forward: cause to effect. If we raise the price of a particular product we manufacture (cause), what will happen to our sales (effect)? The government has a regulation that we may not use a particular chemical in our production process (cause); what will we have to do to keep the production process running smoothly (effect)? Sometimes you will reason backward: from effect to cause. Productivity went down by 6 percent in the last quarter (effect); what factors led to this decrease (causes)? The federal government has decided that used-car dealers are not required to tell potential customers about the cars' defects (effect); why did the federal government reach this decision (causes)?

Cause-and-effect reasoning, therefore, provides a way to answer the following two questions:

- What will be the effect(s) of X?
- What caused X?

Here are examples of documents that use cause and effect as an organizing principle:

- In an *environmental impact statement,* you argue that the proposed construction would have three important effects on the ecosystem.
- In the recommendation section of a *report,* you argue that your recommended solution would improve operations in two major ways.
- In the introduction to a *journal article,* you argue that your topic is worthy of study because it helps readers understand an important but overlooked aspect of the subject you are addressing.

The following discussion of creating causal links is based on the theory of philosopher Stephen Toulmin (1980).

The Three Elements of a Causal Argument

A causal argument has three elements, as shown in Figure 8.16 on page 222.

The *claim* is the idea that you wish to prove or make compelling. Here are examples of claims:

- Our company should institute flextime.
- Our company should not institute flextime.
- The search for alternative fuels has been hindered by myopic government policy.
- By the end of next year, we will have the new rolling machine in full production.
- The key to success in our industry is commitment to the customer.

The claim is the conclusion you want your readers to accept—and, if appropriate, act on—after they have read your discussion.

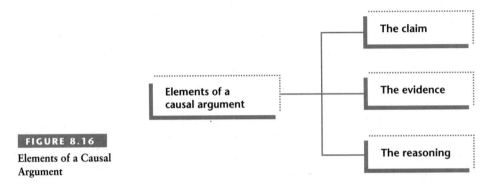

FIGURE 8.16

Elements of a Causal
Argument

The *evidence* is the information you want your audience to consider as they read your discussion. For a discussion of the merits of flextime, for example, the evidence might include the following:

- The turnover rate of our female employees is double that of our male employees.
- Replacing a staff-level employee costs us about one-half the employee's annual salary; replacing a professional-level employee, a whole year's salary.
- At exit interviews, more than 40 percent of our female employees under the age of 38 state that they quit because they cannot work and also be home for their school-age children.
- Other companies have found that flextime significantly decreases the turnover rate among female employees under the age of 38.
- Other companies have found that flextime has additional benefits and introduces no significant problems.

The *reasoning* is the way you derived the assertion from the facts. In the discussion of flextime, the reasoning involves three links:

- Flextime appears to have reduced the problem of high turnover among younger female employees at other companies.
- Our company is similar to other companies.
- If flextime has proven helpful at other companies, it is likely to prove helpful at our company.

To convince readers by your argument, you have to achieve four goals.

- *Readers must understand your evidence and your reasoning.* You have to communicate your evidence clearly, using whatever verbal and visual resources you have. And you have to express your reasoning clearly as well.
- *Readers must accept that your evidence is accurate and relevant to the subject.* If you are not a well-known authority writing to a general reader, you have to cite your sources or use other appropriate techniques to

establish your credibility. And you have to show readers that these are the appropriate evidence for this discussion.

- *Readers must believe that you have not left out any relevant evidence.* This is a challenge, because research standards and practices vary from field to field. In general, however, common sense is the best guide. For instance, if one of your readers believes, from talking to his neighbor, that flextime causes new problems—decreased car-pooling opportunities and increased utility bills, for example—but you haven't addressed these potential problems, that reader might doubt your claim. You have to meet the skeptical or hostile reader's possible objections to your case by qualifying your claim or conceding its drawbacks.

- *Readers must accept that your reasoning is logical.* How do you do that? If there were an easy answer, the world would be much more peaceful than it is. But again, use common sense. If your argument depends on the idea that your company is like the other companies that have found flextime to be successful, you have to decide whether to make that point explicitly. If all the examples you can cite are from the same industry as yours, from companies of about the same size, with employees of approximately the same background, it might be necessary only to list some of those companies. But if the companies you can cite are not apparently similar to yours, you might want to explain why you think that a plan that worked well for them is likely to work well for you, despite the differences.

Logical Fallacies

As you create causal arguments, be careful to avoid three basic logical errors:

- *Inadequate sampling.* Inadequate sampling is the fallacy of drawing false conclusions on the basis of an insufficient sample of cases. It would be illogical to draw the following conclusion:

    The new Gull is an unreliable car. Two of my friends own Gulls, and both have had reliability problems.

    You would have to study a much larger sample and compare your findings with those for other cars in the Gull's class before reaching any valid conclusions.

    Sometimes, however, you have only incomplete data. In these cases, qualify your conclusions:

    The Martin Company's Collision Avoidance System has performed according to specification in extensive laboratory tests. However, the system cannot be considered effective until it has performed satisfactorily in the field. At this time, Martin's system can only be considered very promising.

- *Post-hoc reasoning.* The complete phrase—*post hoc, ergo propter hoc*—means "After this, therefore because of this." The fact that A precedes B does not mean that A caused B. The following statement is illogical:

  There must be something wrong with the new circuit breaker in the office. Ever since we had it installed, the air conditioners haven't worked right.

  The air conditioners' malfunctioning *might* be caused by a problem in the circuit breaker; but malfunctioning might be caused by inadequate wiring or by problems in the air conditioners themselves.

- *Oversimplifying.* The logical fallacy of oversimplifying occurs when a writer omits important information in establishing a causal link.

  The way to solve the balance-of-trade problem is to improve the quality of the products we produce.

  Although improving quality is important, international trade balances are determined by many factors, including tariffs and currency rates, and therefore cannot be explained by simple cause-and-effect reasoning.

Figure 8.17, from a discussion of why mammals survived but dinosaurs did not (Jastrow, 1977, pp. 104–105), illustrates a very effective cause-and-effect argument.

Introducing and Concluding the Body

Drafting the arguments in the body of a document involves using and modifying the basic organizational patterns described in this chapter. Two more elements of the body—the introduction and conclusion—are also fundamental to the success of your argument.

Introducing the Body

An introduction has one main goal: to help readers understand your argument by explaining *what* information you are going to present, *how* you are going to present it, and *why* you choose to present it that way. If you have clearly communicated this information, your readers will be more willing to read the document, better able to understand it, and better able to remember it.

Your document can have one introduction (at the beginning) or several introductions (one at the beginning and one at the start of each major section).

Every document calls for a different kind of introduction. A brochure on backyard pool safety obviously needs little introduction: a brief review of the statistics on injuries and deaths. But the introduction of a scholarly article, which needs to conform to the practices in a particular academic field, might require many elements. To draft the introduction to the body of your document, use your common sense and study similar documents to understand the conventions in your field and the expectations of your readers.

FIGURE 8.17

Argument Organized by the Cause-and-Effect Pattern

The writer asserts here that one reason ancestral mammals were more intelligent than the dinosaurs is that their environment demanded great intelligence. The facts are that the mammals were outnumbered by their predators and had to search for food at night, when especially sharp sensory perception is required. The reasoning, which remains implicit, is the general theory of natural selection: that only those mammals with great intelligence survived and reproduced, and that this fact improved the intelligence of these mammals.

Why were the ancestral mammals brainier than the dinosaurs? Probably because they were the underdogs during the rule of the reptiles, and the pressures under which they lived put a high value on intelligence in the struggle for survival. These little animals must have lived in a state of constant anxiety—keeping out of sight during the day, searching for food at night under difficult conditions, and always outnumbered by their enemies. They were Lilliputians in the land of Brobdingnag. Small and physically vulnerable, they had to live by their wits.

The nocturnal habits of the early mammal may have contributed to its relatively large brain size in another way. The ruling reptiles, active during the day, depended mainly on a keen sense of sight; but the mammal, who moved about in the dark much of the time, must have depended as much on the sense of smell, and on hearing as well. Probably the noses and ears of the early mammals were very sensitive, as they are in modern mammals such as the dog. Dogs live in a world of smells and sounds, and, accordingly, a dog's brain has large brain centers devoted solely to the interpretation of these signals. In the early mammals, the parts of the brain concerned with the interpretation of strange smells and sounds also must have been quite large in comparison with their size in the brain of the dinosaur.

Intelligence is a more complex trait than muscular strength, or speed, or other purely physical qualities. How does a trait as subtle as this evolve in a group of animals? Probably the increased intelligence of the early mammals evolved in the same way as their coats of fur and other bodily changes. In each generation, the mammals slightly more intelligent than the rest were more likely to survive, while those less intelligent were likely to become victims of the rapacious dinosaurs. From generation to generation, these circumstances increased the number of the more intelligent and decreased the number of the less intelligent, so that the average intelligence of the entire population steadily improved, and their brains grew in size. Again the changes were imperceptible from one generation to the next, but over the course of many millions of generations the pressures of a hostile world created an alert and relatively large-brained animal.

Figure 8.18 on page 226 shows the seven critical questions you should consider answering for your readers in an introduction.

- *What are the key terms that will be used in the argument?* Define all key terms that you will discuss in your document.

  Phantom risk refers to alleged but uncertain risks associated with a scientific or technological practice. For instance, weak magnetic fields represent a phantom risk in that the scientific community is uncertain whether they represent any health risk at all, or, if they do, what the threshold for such risks might be.

  If the reader needs to know a key term to understand the introduction, the definition should be presented early in the introduction. See Chapter 9 for more information on definitions.

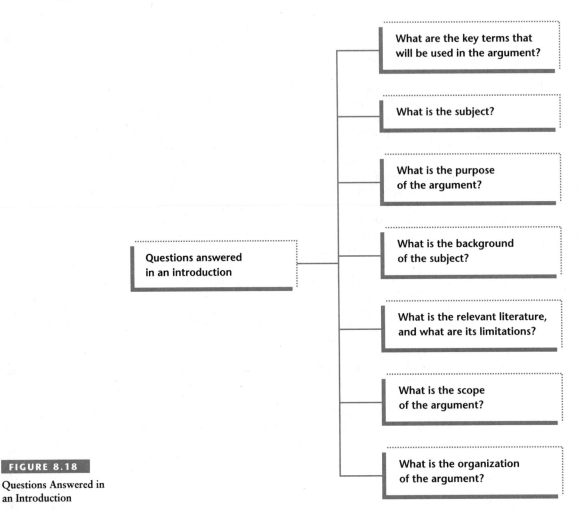

FIGURE 8.18

Questions Answered in
an Introduction

- *What is the subject?* Answer this question directly, even though your
  readers might already know.

  This report describes the relationship between the courts and the scientific community on the subject of phantom risk. We live in an age of tremendous technological
  advances, and some new products and practices might pose health risks. The conflict explored in this report is that whereas the courts require certainty—they need
  to know whether a plaintiff was or was not injured by a technology—the scientific
  community often cannot supply a definitive answer.

- *What is the purpose of the argument?* Explain what you hope to achieve
  in the discussion.

  This report has two main purposes: to summarize the conflict between the legal and
  scientific communities and to propose three principles to guide the future debate on
  the legal ramifications of phantom risk.

- *What is the background of the subject?* Explain the background: the information that readers need to understand the discussion. Many of your readers might not be up-to-date on the subject.

   It is inevitable that some people will be injured by technological advances, and one of the foundations of our legal system is that victims be empowered to seek redress of their injuries through the legal system. However, science does not offer the kind of certainty that the legal system requires. Are breast implants dangerous? Cautious scientists would have to admit that they don't know, but different constituencies answer yes and no with self-assurance and conviction. When courts render false-positive judgments—by finding that harmless products and services have caused injuries—they put companies out of business and deprive people of their products and services. When courts render false negatives—by finding that harmful products and services have not injured anyone—they endanger the public. Currently, there is no way to reconcile the different cultures and practices of the legal and scientific communities.

- *What is the relevant literature, and what are its limitations?* For books, journal articles, proposals, and reports, show your readers that you have done the necessary research on your topic.

   A number of researchers have studied the topic of phantom risk, but most of the studies have examined individual cases from either the legal or scientific perspective. For instance, the asbestos issue has been discussed from the scientific perspective by Wilkins (1995), Thomas (1996), and Rivera (1996), and the legal perspective by Halloran (1997), Bradford (1996), and De Moss (1995). The edited collection by Foster et al. (1996) is the first to attempt a detailed look at the conflict; however, it offers piecemeal opinions about the different phantom-risk issues but does not provide a theoretical framework for reconciling the two cultures.

- *What is the scope of the argument?* Describe the scope of your discussion: what you are including or excluding.

   This report focuses on four phantom-risk issues: electrical and magnetic field risks, spermicide cancer risks, asbestos risks, and secondhand smoke risks. It reviews the scientific evidence and the legal opinions for each issue. Where appropriate, it discusses the social policy resulting from the legal actions.

- *What is the organization of the argument?* Explain how you are organizing your argument so readers can concentrate on the information without worrying about what will come next.

   In the sections that follow, I treat each phantom-risk issue separately, as a case study. After a background discussion, I fill in the scientific consensus, followed by the legal precedents and, where appropriate, public policy. I conclude the article with a recommendation that we discuss three principles for reconciling the legal and scientific approaches to phantom-risk issues.

Concluding the Body

In discussions of writing, the word *conclusion* has two meanings. One meaning refers to the inferences drawn from technical data. If, for instance, you are doing a feasibility report on whether to switch from paper-based to online documentation, you might conclude that the change would be unwise at this time. This kind of conclusion is discussed in Chapter 17.

Conclusion also refers to the final part of a document or a section of a document. The following discussion concerns this second sense of the word.

Although some kinds of documents, such as parts catalogs, do not generally have a conclusion, most do. When drafting a conclusion, be sure you answer the appropriate questions from the four shown in Figure 8.19.

- *What are the main ideas communicated in the argument?* Readers can forget material, especially material from the beginning of a long document. Therefore, it is a good idea to summarize important ideas in a paragraph. The following examples are from a report on reuse-adoption programs: companywide programs to increase the reuse of systems and software.

  Our analysis yielded two main conclusions. First, reuse adoption is good engineering and good business; it can reduce expenses and increase productivity and quality. Second, reuse adoption is a complex procedure that requires a substantial amount of planning and supervision. If the program is carried out casually or thoughtlessly, it can backfire and cause more problems than it solves.

- *What should be done next?* Offer recommendations on the course of future work.

  I recommend that we convene an *ad hoc* committee to study the feasibility of converting to a reuse-adoption plan. Specifically, we should carry out a phase I analysis to determine the answers to the following four questions:

  1. Is demand in our market sufficient for the kinds of products that are appropriate for reuse adoption?
  2. Do we have sufficient assets—code, designs, templates, and so on—to justify reuse adoption?

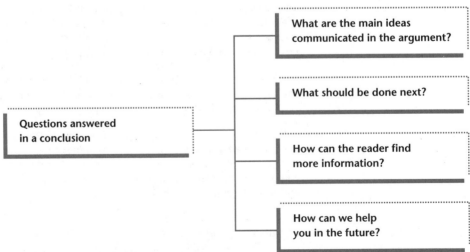

Questions answered in a conclusion

- What are the main ideas communicated in the argument?
- What should be done next?
- How can the reader find more information?
- How can we help you in the future?

FIGURE 8.19

Questions Answered in a Conclusion

3. Are technological developments in our industry sufficiently predictable to justify reuse adoption?
4. Are industry standards and government standards sufficiently predictable to justify reuse adoption?

- *How can the reader find more information?* Sometimes this is a sales message; sometimes it is not.

   I have asked Corporate Information to get the six articles listed in the references section. These articles will be gathered and routed next week. Please read them and be prepared to offer any questions or concerns at our meeting on August 15, at which time we will discuss the recommendations.

- *How can we help you in the future?* Offer to provide services if you can.

   I think reuse adoption offers the promise to improve our business in a number of ways. Please feel free to get in touch with me, either before or after you read the articles, to discuss this initiative.

CASE

Introducing a Document

You are an intern at the U.S. Department of Labor. Your supervisor, Allen Young, has asked you to study the document on methylene chloride that is excerpted below. The full document, which can be found at http://www.osha.gov/oshpubs/osha3144.pdf, is intended for a managerial and general audience. "It's got a disclaimer statement at the front of the document, and a background statement," Allen tells you, "but I think it needs an introduction. Would you mind taking a look at it and drafting an introduction?" Write an introduction for this document.

BACKGROUND

When established under the authority of the *Occupational Safety and Health Act of 1970*, the Occupational Safety and Health Administration (OSHA) had 2 years to adopt existing federal standards or national consensus standards[1] so it would have standards in place to enforce. OSHA chose to adopt existing federal standards issued under the *Walsh-Healey Public Contracts Act*, which were derived from threshold limit values of the American Conference of Governmental Industrial Hygienists and consensus standards from standards-developing organizations such as the American National Standards Institute (ANSI).

(Continued)

(Case continued)

For methylene chloride, OSHA adopted an ANSI standard under Subpart Z of *Title 29 Code of Federal Regulations* (CFR), Part 1910.1000 to ensure that employee exposure did not exceed 500 parts per million parts of air (500 ppm) as an 8-hour time-weighted average (TWA)—i.e., the average exposure during an 8-hour period.

Since 1971, however, industrial experience, new developments in technology, and emerging scientific data clearly indicate that this limit did not adequately protect worker health. The agency realized the need to better control worker exposure to methylene chloride due to its harmful health effects.

Methylene chloride, also called dichloromethane, is a volatile, colorless liquid with a chloroformlike odor. Inhalation and skin exposure are the predominant means of exposure to methylene chloride. Inhaling the vapor causes mental confusion, lightheadedness, nausea, vomiting, and headache. With acute, or short-term exposure, methylene chloride acts as an anesthetic; continued exposure may cause staggering, unconsciousness, and even death. High concentrations of the vapors may cause irritation of the eyes and respiratory tract and aggravate the symptoms of angina. Skin contact with liquid methylene chloride causes irritation and burns. Splashing methylene chloride into the eyes causes irritation. Studies on laboratory animals indicate that long-term (chronic) exposure causes cancer.

Methylene chloride is used in various industrial processes in many different industries: paint stripping, pharmaceutical manufacturing, paint remover manufacturing, metal cleaning and degreasing, adhesives manufacturing and use, polyurethane foam production, film base manufacturing, polycarbonate resin production, and distribution and formulation of solvents.

The agency adopted the methylene chloride final rule on January 10, 1997, as published in the *Federal Register*. The rule became effective on April 10, 1997.

[1]Consensus standards are developed by private, standards-developing organizations and are discussed and substantially agreed upon through consensus by industry, labor, and other representatives.

WRITER'S CHECKLIST

The following four questions relate to planning the argument:

1. Have you analyzed your audience's goals?

2. Is your evidence persuasive?

3. Have you presented yourself effectively?

4. Have you analyzed the political realities?

The following checklists cover the eight organizational patterns discussed in this chapter, as well as introductions and conclusions.

Chronological and Spatial

Have you
- [ ] provided signposts, such as headings or transitional words or phrases?
- [ ] considered using graphics to complement the text?
- [ ] analyzed events where appropriate?

General to Specific

Have you

- ❑ provided signposts, such as headings or transitional words or phrases?
- ❑ considered using graphics to complement the text?

More Important to Less Important

Have you

- ❑ explained clearly that you are using this organizational pattern?
- ❑ made clear in your discussion why the first point is the most important, the second is the second most important, and so forth?
- ❑ considered using graphics to complement the text?

Comparison and Contrast

Have you

- ❑ included the necessary criteria?
- ❑ chosen a structure—whole by whole or part by part—that is appropriate for your audience and purpose?
- ❑ chosen appropriate organizational patterns for your second-level items?

Classification and Partition

Have you

- ❑ chosen a basis consistent with the audience and purpose of the document?
- ❑ used only one basis at a time?
- ❑ avoided overlap?
- ❑ included all the appropriate categories?
- ❑ arranged the categories in a logical sequence?

Problem-Methods-Solution

Have you

- ❑ described the problem clearly and specifically?
- ❑ if appropriate, justified your methods?
- ❑ avoided overstating your solution?
- ❑ sequenced the discussion in a way that is consistent with the audience and purpose of the document?

Cause and Effect

Have you

- ❑ clearly expressed your claim, evidence, and reasoning?
- ❑ sequenced your claim, evidence, and reasoning appropriately for your audience and purpose?
- ❑ avoided inadequate sampling?
- ❑ avoided *post-hoc* reasoning?
- ❑ avoided oversimplifying?

The following questions cover introductions and conclusions.

Introductions

Have you

- ❑ explained the subject of the document?
- ❑ explained the purpose of the discussion?
- ❑ explained the background of the subject?
- ❑ reviewed the relevant literature, explaining its limitations?
- ❑ explained the scope of the discussion?
- ❑ explained the organization of the discussion?
- ❑ defined all key terms that will be used in the discussion?

Conclusions

Have you

- ❑ summarized the main ideas in the document?
- ❑ recommended what should be done next?
- ❑ explained how the reader can find more information?
- ❑ described how you can help?

EXERCISES

Several of the following exercises call for you to write a memo. See Chapter 16 for a discussion of memos.

The first nine exercises call for you to write an argument using a different organizational pattern. For each of these nine exercises, write an argument of 500 to 1,000 words on one of the topics specified or on a topic of your choice. For each exercise, attach a statement indicating your audience and purpose, as well as the type of document in which it might be included. For example, "This argument is addressed to first-semester college students. Its purpose is to help them understand the policies they are to follow in using the university's mainframe. The argument would be part of the *Student Handbook*."

1. Using the chronological organizational pattern:
   a. how to register for courses at your college or university
   b. how to learn to use a software package
   c. how to buy the right car for your needs
   d. how to prepare for a job interview
   e. how to determine job prospects in your field

2. Using the spatial organizational pattern:
   a. your bicycle
   b. your car's dashboard
   c. the room in which you are sitting
   d. the space shuttle
   e. the remote-control device from a television or stereo set

3. Using the general-to-specific organization:
   a. advances in manufacturing technology
   b. alternatives to incarceration for nonviolent criminals
   c. pay scales for general practitioners and medical specialists
   d. cooperative education and internships for college students
   e. energy efficiency in computers and printers

4. Using the more-important-to-less-important organizational pattern:
   a. the reasons you chose your college or major
   b. the effects of acid rain on a particular area
   c. the reasons you should (or should not) be required to study a foreign language (or learn a computer programming language)
   d. the three most important changes you would like to see at your school
   e. the reasons that recycling should (or should not) be mandatory

5. Using a comparison-and-contrast organizational pattern:
   a. lecture classes and discussion classes
   b. the tutorials that come with two different software packages
   c. manual transmission and automatic transmission automobiles
   d. black-and-white and color photography
   e. two different word-processing programs

6. Using the problem-methods-solution pattern:
   a. how you solved a recent problem related to your education
   b. how you went about deciding on a recent major purchase, such as a car, a personal computer, or a bicycle
   c. how you would propose reducing the time required to register for classes or to change your schedule
   d. how you would propose increasing the ties between your college or university and local business and industry

7. Using classification:
   a. foreign cars
   b. smoke alarms
   c. college courses
   d. personal computers
   e. cameras

8. Using partition:
   a. a student organization on your campus
   b. an audio cassette tape
   c. a portable radio
   d. a bicycle
   e. a guitar

9. Using causal reasoning—either forward or backward:
   a. women serving in combat in the military
   b. the price of gasoline
   c. the emphasis on achieving high grades in college
   d. the prospects for employment in your field

10. In one paragraph for each, describe the logical flaws in the following items:
    a. The election couldn't have been fair—I don't know anyone who voted for the winner.
    b. Increased restrictions on smoking in public are responsible for the decrease in smoking.
    c. Since the introduction of cola drinks at the start of this century, cancer has become the second greatest killer in the United States. Cola drinks should be outlawed.
    d. The PC industry is enjoying record profits; just look at the profits of Compaq and Micron.
    e. The value of the Americans with Disabilities Act has to be questioned; since its passage, the number of discrimination suits based on disabilities has actually increased.
    f. The phenomenal growth of the Internet can be explained by the fiber-optic revolution of the 1980s.

11. Write a memo to your instructor evaluating the effectiveness of the introduction (page 233) from a student report about how audits are conducted. What works well in the introduction? How might the introduction be improved?

12. The following conclusion is from the student report from which the introduction in exercise 11 is excerpted. Write a memo to your instructor evaluating the conclusion's effectiveness. What aspects of it are effective? How would you improve it?

INTRODUCTION

An audit is a formal examination and verification of financial accounts. According to the tenth edition of *Montgomery's Auditing* textbook, historians believe that audits were conducted as early as 4000 B.C. in Babylonia. Governments were concerned with establishing controls, including audits, to reduce errors and fraud in the tax collection system. Auditing has been seen in history ever since. The Bible (generally believed to cover the period of time between 1800 B.C. and A.D. 95) refers to internal controls and surprise audits. The earliest records in English-speaking countries are from England and Scotland. They have accounting records and references to audits dating back to A.D. 1130.

Auditing was fairly slow to progress in the United States. In the late 1800s the U.S. railroad companies employed auditors, and in 1887 the American Institute of Accountants, now the American Institute of Certified Public Accountants (AICPA), was established. In 1935 audits became more common for two reasons:

- The Securities Act of 1933 and the Securities Exchange Act of 1934, which required listed companies to file audited financial statements, were enacted
- The AICPA collaborated with the New York Stock Exchange to improve reporting standards

Today all major corporations conduct audits every year. Audits are used to analyze the financial condition of a company. Audits are usually conducted by certified public accountants who work for independent accounting firms such as Arthur Anderson. The purpose of performing an audit is to identify any problematic areas, such as embezzlement. An audit consists of six steps:

1. understanding the client's business
2. planning the audit
3. evaluating internal accounting controls
4. performing compliance tests
5. performing substantive tests
6. preparing the financial statement and issuing an opinion

All major corporations and most minor corporations require an audit to be conducted every year. Audits are the primary tool used to ensure that financial irregularities such as embezzlement, conflict of interest, and inaccurate pricing are kept to a minimum.

Collaborative Exercise

13. Form small groups for this project about argument. Go to the background section from a government regulation (U.S. Department of Labor, 1997) found in

the case on page 229. Write a memo to your instructor describing and analyzing the arguments in this passage. Which techniques discussed in this chapter do the authors of this excerpt use? How effectively do they use them? What would you do differently? Present your findings in a memo to your instructor. To see the full document, go to http://www.osha.gov/oshpubs/osha3144.pdf.

Web Exercise

14. Form small groups for this project about argument. Many professional organizations, interest groups, and private corporations publish reports on the Internet. For example, many manufacturers of computer equipment include reports about their industry or their products on their Web sites. Find a report on the Web, then write a memo to your instructor in which you describe and analyze its pattern of development. What pattern—or combinations of patterns—does it use? How effective is the development? How might it be improved? When you submit the memo, be sure to include the report itself.

REFERENCES

Bell, T. E. (1993). Bicycles on a personalized basis. *IEEE Spectrum, 30*(9), 32–35.

Bryzek, J., Petersen, K., & McCulley, W. (1994). Micromachines on the march. *IEEE Spectrum, 31*(5), 20–31.

Jastrow, R. (1977). *Until the sun dies.* New York: Norton.

Larson, D. E. (1990). *Mayo Clinic family health book.* New York: William Morrow.

Murphy, D. (1994). *Online and offline debit cards: First-year costs of implementation at John's Market.* Unpublished manuscript, Boise State University.

Toulmin, S. (1980). *The uses of argument.* London: Cambridge University Press.

U.S. Department of Labor. (1997). Methylene Chloride. OSHA 3144. <http://www.osha.gov/oshpubs/osha3144.pdf> (1997, Apr. 10).

Drafting Definitions and Descriptions

Understanding the Role of Definitions 236

Analyzing the Writing Situation for Definitions 237

Determining the Kind of Definition to Write 238
Drafting Parenthetical Definitions *238*
Drafting Sentence Definitions *239*
Drafting Extended Definitions *240*

Deciding Where to Place the Definition 246

Drafting Descriptions 248

Understanding the Role of Descriptions 249

Analyzing the Writing Situation for Descriptions 249

Guidelines for Writing Descriptions 250
Clearly Indicate the Nature and Scope of the Description *250*
Introduce the Description Clearly *251*
Provide Appropriate Detail *251*
Conclude the Description *253*

Analyzing Some Descriptions 255

Case: Annotating a Patent Application *261*

Writer's Checklist 262

Exercises 262
Collaborative Exercise *267*
Web Exercise *267*

This chapter discusses two related techniques fundamental to technical communication: definitions and descriptions. Both use words and, in most cases, graphics, to help readers understand concepts. Definitions generally help readers understand what you mean by a word or phrase. For instance, you might provide a definition of *consumer price index* in a discussion of how inflation is calculated by the government. Or you might provide a definition of *cone penetration test* in a discussion of how geologists analyze soil strata. Descriptions generally provide a fuller picture of an object, mechanism, or process, often going into detail about the components of the object or the stages of the process. For example, you might write a description of a *dark object* in a discussion of discoveries in astronomy, explaining theories of the dimensions and materials of a dark object. Or you might write a description of PET scans in a discussion for clinical nurses, explaining how the process was devised, how it works, and when it is used. As you have probably guessed, definitions are often included in descriptions. The description of how PET scans are done might begin with a definition of a PET scan.

This chapter will discuss definitions first, then explain descriptions.

Understanding the Role of Definitions

The world of business and industry depends on clear definitions. Without written definitions, the working world would be chaotic. For example, suppose you learn at a job interview that the potential employer pays tuition and expenses for employees' job-related education. That's good news, of course, if you are planning to continue your education. But until you study the employee-benefits manual, you won't know with any certainty just what the company will pay for.

Who, for instance, is an *employee*? You would think an employee would be anyone who works for and is paid by the company. But you might find that for the purposes of the manual, an employee is someone who has worked for the company full time (40 hours per week) for at least six uninterrupted months.

Other terms also would have to be defined in the description of tuition benefits. What, for example, is *tuition*? Does the company's definition include incidental laboratory or student fees? What is *job-related education*? Does a course about time management qualify under the company's definition? What, in fact, constitutes *education*? All these terms and many others must be defined for the employees to understand their rights and responsibilities.

Definitions play a major role, therefore, in communicating policies and standards "for the record." When a company wants to purchase air-conditioning equipment, it might require that the supplier provide equipment certified by a professional organization of air-conditioner manufacturers. The organization's definitions of acceptable standards of safety, reliability, and efficiency will provide some assurance that the equipment is high quality.

Definitions, of course, have many uses outside legal or contractual contexts. Two such uses occur frequently.

- *Definitions clarify a description of a new development or a new technology in a technical field.* A zoologist discovering a new animal species, for instance, names and defines it. A scientist devising a new laboratory procedure defines and describes it in an article in a technical journal.

- *Definitions help specialists communicate with less knowledgeable readers.* A manual that explains how to tune up a car includes definitions of parts and tools. A researcher at a manufacturing company uses definitions in describing a new product to the sales staff.

Definitions, then, are crucial in many kinds of technical communication, from brief letters and memos to technical reports, manuals, and journal articles. All readers, from the general reader to the expert, need effective definitions to carry out their jobs.

Definitions, like every other technique in technical communication, require thought and planning. Before you can write a definition to include in a document, you should carry out three steps, as shown in Figure 9.1.

Analyzing the Writing Situation for Definitions

The first step in writing effective definitions is to analyze the writing situation: the audience and purpose of your document.

- *Audience.* Unless you know whom you are addressing and how much that audience knows about the subject, you cannot know which terms to define or the kind of definition to write. Physicists wouldn't need a definition of *entropy*, but lawyers might. Builders know what a Molly bolt is, but some insurance agents don't. If you are aware of your audience's background and knowledge, you can easily devise effective informal definitions. For example, if you are describing a compact-disc player to a group of readers who understand automobiles, you can use a familiar analogy to define the function of the pause button: "The PAUSE button is the clutch pedal of the compact-disc player."

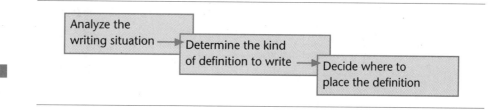

FIGURE 9.1

Steps in Preparing to Write a Definition

- *Purpose.* If you want to give your readers only a basic understanding of a concept—say, time-sharing vacation resorts—a brief, informal definition is usually sufficient. However, if you want your readers to understand an object, process, or concept thoroughly and be able to carry out tasks associated with it, a more formal and elaborate definition is required. For example, a definition of a "Class 2 Alert" written for operators at a nuclear power plant must be comprehensive, specific, and precise.

Determining the Kind of Definition to Write

As the preceding analysis has suggested, definitions can be short or long, informal or formal. Figure 9.2 shows the three basic types of definitions.

Drafting Parenthetical Definitions

A *parenthetical definition* is a brief clarification placed unobtrusively in a sentence. Sometimes a parenthetical definition is only a word or phrase:

> The crane is located on the starboard (right) side of the ship.
>
> Summit Books announced its intention to create a new colophon (emblem or trademark).
>
> United Engineering is seeking to purchase the equity stock (common stock) of Minnesota Textiles.

A parenthetical definition can also take the form of a longer explanatory phrase or clause:

> Motorboating is permitted in the Jamesport Estuary, the portion of the bay that meets the mouth of the Jamesport River.
>
> Before the metal is plated, it is immersed in the pickle, an acid bath that removes scales and oxides from the surface.

Parenthetical definitions are not, of course, comprehensive. They serve mainly as quick and convenient ways of introducing terms. Especially when you are using a parenthetical definition in writing addressed to general readers, make

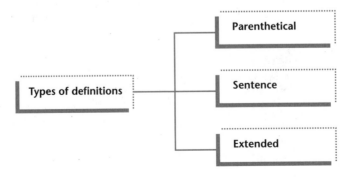

FIGURE 9.2

Types of Definitions

sure that the definition itself is clear. You have gained nothing if readers don't understand your clarification:

> Next, check for blight on the epicotyl, the stem portion above the cotyledons.

If your readers are botanists, this parenthetical definition will be clear (although it might be unnecessary because if they know the meaning of *cotyledons*, they are likely to know *epicotyl*). However, if you are addressing general readers, this definition will merely frustrate them.

Drafting Sentence Definitions

A sentence definition—a one-sentence clarification—is more formal than a parenthetical definition. Usually, a sentence definition follows a standard pattern: the item to be defined is placed in a category of similar items and then distinguished from them:

Item	=	Category	+	Distinguishing Characteristics
A flip flop	is	a circuit		containing active elements that can assume either one of two stable states at any given time.
An electrophorus	is	a laboratory instrument		used to generate static electricity.
Hypnoanalysis	is	a psychoanalytical technique		in which hypnosis is used to elicit information from a patient's unconscious mind.
A Bunsen burner	is	a small laboratory heating device		consisting of a vertical metal tube connected to a gas source.
An electron microscope	is	a microscope		that uses electrons rather than visible light to produce magnified images.

In many cases a sentence definition also includes a graphic. For example, the definitions of electrophorus, Bunsen burner, and electron microscope would probably be accompanied by photographs or drawings.

Sentence definitions are useful when your readers require a more formal or more informative clarification than parenthetical definitions can provide. Writers often use sentence definitions to establish a working definition for a particular document: "In this report, the term *electron microscope* is used to refer to any microscope that uses electrons rather than visible light to produce magnified images." Such definitions are sometimes called *stipulative definitions* because the writer is stating how the term will be used in the document.

In writing sentence definitions, follow four guidelines:

- *Be specific in writing the category and the distinguishing characteristics.* Remember, you are trying to distinguish the item from all other similar items. If you write "A Bunsen burner is a burner that consists of a vertical metal tube connected to a gas source," the imprecise category—"a burner"—defeats the purpose of your definition: many types of large-scale burners use vertical metal tubes connected to gas sources. If you write "Hypnoanalysis is a psychoanalytical technique used to elicit information from a patient's unconscious mind," the imprecise distinguishing characteristics—"used to elicit . . ."—ruin the definition: many psychoanalytical techniques are used to elicit information from a patient's unconscious mind. If your definition describes more than one item, you have to sharpen either the category or the distinguishing characteristics, or both.

- *Don't describe a specific item when you mean to define a general class of items.* If you wish to define *catamaran*, don't describe a particular catamaran. The catamaran you see on the beach in front of you might be made by Hobie and have a white hull and blue sails, but those characteristics are not essential to catamarans.

- *Avoid writing circular definitions, that is, definitions that merely repeat the key words or the distinguishing characteristics of the item being defined in the category.* In "A required course is a course that is required," what does "required" mean? Required of whom, by whom? The word is never defined. Similarly, "A balloon mortgage is a mortgage that balloons" is useless.

   However, you can use some kinds of words in the item as well as in the category and distinguishing characteristics. In the definition of electron microscope given earlier, for example, the word *microscope* is repeated. Here, *microscope* is not the difficult part of the item; readers know what a microscope is. The purpose of defining *electron microscope* is to clarify the *electron* part of the term.

- *Be sure the category contains a noun or a noun phrase rather than a phrase beginning with* when, what, *or* where.

   | INCORRECT: | A brazier is what is used to . . . |
   | CORRECT: | A brazier is a metal pan used to . . . |

   | INCORRECT: | An electron microscope is when a microscope . . . |
   | CORRECT: | An electron microscope is a microscope that . . . |

   | INCORRECT: | Hypnoanalysis is where hypnosis is used . . . |
   | CORRECT: | Hypnoanalysis is a psychoanalytical technique in which . . . |

Drafting Extended Definitions

An *extended definition* is a long, detailed clarification—usually one or more paragraphs—of an object, process, or idea. Often an extended definition begins with a sentence definition, which is then elaborated. For instance, the sentence definition

"An electrophorus is a laboratory instrument used to generate static electricity" tells you the basic function of the device, but it leaves many questions unanswered: How does an electrophorus work? What kinds of procedures can it help people carry out? What are its strengths and limitations? An extended definition would answer these and other questions.

An extended definition is useful, naturally, when you want to give your readers a reasonably thorough understanding of an item. And the more complicated or abstract the item being defined, the greater the need for an extended definition.

There is no one way to "extend" a definition. Your analysis of the audience and purpose of the communication will help you decide which method to use. In fact, an extended definition sometimes employs several of the nine techniques shown in Figure 9.3.

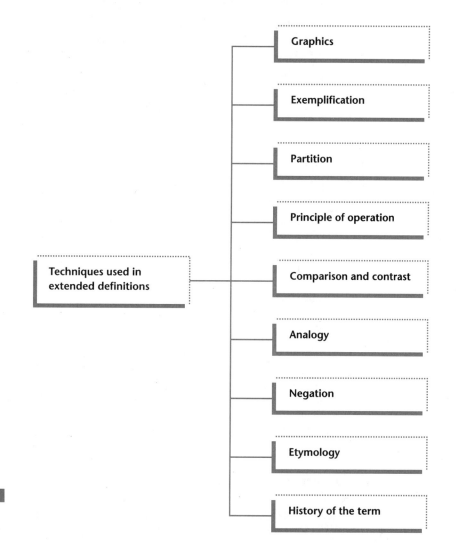

FIGURE 9.3

Techniques Used in Extended Definitions

Graphics

Perhaps the most common way to create an extended definition in technical communication is to create some sort of graphic, such as a photograph, diagram, schematic, or flowchart, and then explain the graphic.

Graphics are particularly useful in defining physical objects. The simplest way to define a two-person saw, for instance, would be to draw a rough diagram. However, illustrations are also useful in defining concepts and ideas. A definition of the meteorological concept of temperature inversion, for instance, might include a diagram. So might definitions of lasers, compound fractures, and buying stocks on margin.

The following excerpt from an extended definition of *parallelogram* shows an effective combination of words and illustrations.

> A parallelogram is a four-sided plane whose opposite sides are parallel with each other and equal in length. In a parallelogram, the opposite angles are the same, as shown in the sketch below.

Exemplification

Exemplification—using an example (or examples) to clarify an object or idea—is particularly useful in making an abstract term easy to understand. The following paragraph is an extended definition of the psychological defense mechanism called *conversion* (Wilson, 1964, p. 84).

> A third mechanism of psychological defense, "conversion," is found in hysteria. Here the conflict is converted into the symptom of a physical illness. In a case of conversion made famous by Freud, a young woman went out for a long walk with her brother-in-law, with whom she had fallen in love. Later, on learning that her sister lay gravely ill, she hurried to her bedside. She arrived too late and her sister was dead. The young woman's grief was accompanied by sharp pain in her legs. The pain kept recurring without any apparent physical cause. Freud's explanation was that she felt guilty because she desired the husband for herself, and unconsciously converted her repressed feelings into an imaginary physical ailment. The pain struck her in the legs because she unconsciously connected her feelings for the husband with the walk they had taken together. The ailment symbolically represented both the unconscious wish and a penance for the feelings of guilt which it engendered.

This extended definition is effective because the writer has chosen a clear and interesting (and therefore memorable) example of the idea he is describing. No other examples are necessary in this case. If conversion were a more difficult concept to describe, an additional example might be useful.

Partition

Partition is the process of dividing a thing or idea into smaller parts so that the reader can understand it easily. (See Chapter 8 for a detailed discussion of partition.) The following example (U.S. Congress, 1995a, p. 133) from a study of technology in nursing is an example of the use of partition.

> A knowledge-based system designed for clinical use, sometimes called a *clinical decision support system* (CDSS), usually involves three components:
>
> 1. *Data on the patient* being diagnosed or treated are either entered into the system manually, captured automatically from diagnostic or monitoring equipment, or drawn from an electronic patient record.
>
> 2. A *knowledge base* contains rules and decision algorithms that incorporate knowledge and judgment about the health problem at hand and alternative tests and treatments for it, mainly in the form of "if-then" statements, such as "if the patient's potassium is less than 3.0 mEq/dl and the patient is on digoxin, then warn the clinician to consider potassium supplementation."
>
> 3. An *inference engine* combines information from both the patient data and the knowledge base to perform specified tasks, outlined in appendix C.

Principle of Operation

The principle of operation is an effective way to develop an extended definition, especially that of an object or process. The following extended definition of a parabolic dish solar-energy system (U.S. Congress, 1995b, pp. 173, 174) is based on the mechanism's principle of operation.

> Parabolic dish systems use a large dish or set of mirrors on a single frame with two-axis tracking to reflect sunlight onto a receiver mounted at the focus. Most commonly, a free piston stirling engine is mounted on the receiver, but hot fluids can also be piped to a central turbine as in the parabolic trough and central receiver systems. Current research is focusing on lowering the cost of the mirror systems through the use of stretched membranes and to improve the reliability and performance of the stirling engine.[4] Stirling engine lifetimes of 50,000 hours (about 10 years) with little or no maintenance are needed and are being developed.[5] In comparison, the typical automobile engine must have minor maintenance every 250 hours or so, and a major overhaul perhaps every 2,500 hours.[6] Parabolic dishes can achieve the highest temperatures (800°C or 1,500°F) and thus the highest efficiencies of concentrating solar thermal systems. Parabolic dish systems currently hold the efficiency record of 31 percent (gross) and 29 percent (net) for converting sunlight into electricity.[7]

---

4. Charles W. Lopez and Kenneth W. Stone. "Design and Performance of the Southern California Edison Stirling Dish," *Solar Engineering,* vol. 2, 1992, pp. 945–952; and Graham T. Reader and Charles Hooper, *Stirling Engines* (New York, NY: E. & F. N. Spon, 1983).

5. Pascal De Laquil III et al., "Solar Thermal Electric Technology," *Renewable Energy: Sources for Fuels and Electricity,* Thomas B. Johansson et al. (Eds.) (Washington, DC: Island Press, 1993).

6. For an automobile, an oil change every 7,500 miles corresponds to 250 hours of operation, assuming an average operating speed of 30 mph. Similarly, a major overhaul every 75,000 miles corresponds to 2,500 hours of operation.

7. William B. Stine, *Progress in Parabolic Dish Technology,* SERI/SP-220-3237 (Golden, CO: Solar Energy Research Institute, June 1989).

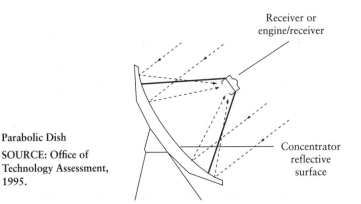

Receiver or
engine/receiver

Parabolic Dish
SOURCE: Office of
Technology Assessment,
1995.

Concentrator
reflective
surface

Comparison and Contrast

Using comparison and contrast, the writer discusses similarities or differences between the item being defined and an item with which the readers are more familiar. The following definition of *digital communication links* is based on a comparison to analog communication links.

> The best way to understand digital communication links is to compare them to analog communication links. In a traditional analog environment such as a telephone system, a person speaks into the phone mouthpiece; the phone transforms the sounds into an analog signal, which is sent along wires to the earpiece in the recipient's phone, which then converts the signals back to sound. In an analog environment, only one signal can be carried on one wire at one time. By contrast, in a digital environment, the caller's signal is digitized—turned into 0's and 1's—almost immediately and *multiplexed* with other signals; that is, the different signals are carried as packets of information on a single set of wires, then separated at their destinations. In addition, in a digital environment, signals do not need to be transmitted along wires; they can be sent as pulses of light on fiberoptic cables or as microwaves between satellites or land stations. Compared to analog signals, digital signals are much less susceptible to noise and require much less amplification along the way.

Analogy

An *analogy* is a specialized kind of comparison. In a traditional comparison, the writer compares the item being defined to a similar item. An electron microscope is compared to a common microscope, or a bit brace is compared to a power drill. In an analogy, however, the item being defined is compared to an item that is in some ways completely different but that shares some essential characteristic.

For instance, the central processing unit of a computer is often compared to a brain. Obviously, these two items are very different—except that the function of the central processing unit in the computer is similar to that of the brain in the body. People often compare computer software to phonograph records, cassette tapes, or compact discs. Again, the differences are many, but the similarity is essential: the phonograph record contains the code that enables the stereo system to fulfill its function, just as the software contains the code that enables the computer to do its job.

The following example, from an extended definition of *computer literacy,* shows how an analogy can clarify an unfamiliar concept.

> Computer literacy is the ability to use computers effectively. If you can operate a personal computer to do word processing or create a database, you are computer literate. If you can operate a digital watch or program a VCR or use an automated teller machine at a bank, you also can be said to be computer literate. To use an analogy, computer literacy is like automotive literacy: if you know how to operate an automobile safely to get from one place to another, you possess automotive literacy. Just as you don't have to understand the principle of the internal-combustion engine to drive a car, you don't have to understand the concepts of RAM and ROM to use a computer.

Negation

A special kind of contrast is sometimes called *negation* or *negative statement.* Negation is the technique of clarifying a term by distinguishing it from a different term with which the reader might have confused it. The following example shows the use of negation.

> An ambulatory patient is not a patient who must be moved by ambulance. On the contrary, an ambulatory patient is one who can walk without assistance from another person.

Negation is rarely the only technique used in an extended definition. In fact, negation is used most often in a sentence or two at the start of a definition. After you state what the item is not, you still have to state what it is.

Etymology

Etymology—the derivation of a word—is often a useful and interesting way to develop a definition.

> The word *mortgage* was originally a compound of *mort* (dead) and *gage* (pledge). The meaning of the word has not changed substantially since its origin in Old French. A mortgage is still a pledge that is "dead" upon either the payment of the loan or the forfeiture of the collateral and payment from the proceeds of its sale.

Etymology is a popular way to begin a definition of acronyms, which are abbreviations pronounced as words.

> SCUBA stands for self-contained underwater breathing apparatus.

> COBOL, which is the Common Business-Oriented Language, was originally invented to . . .

Etymology, like negation, is rarely used alone in technical communication, but it is an effective way to introduce an extended definition.

History of the Term

A common way to define a term is to explain its history. Often the extended definition explains the original use of the term and then describes how the meaning changed in response to historical events or technological advances. The following (Roblee & McKechnie, 1981, pp. 15–16) is an example of the use of the history of a term:

> In general, the common-law definition of *arson* was traditionally the willful burning of the house of another, including all outhouses or outbuildings adjoining thereto. The emphasis was on another's habitation, and his life and safety at the place where he or she resided. Then, many legal issues began to arise. Was a school a dwelling? a jail? a church? The common-law courts began to view the crime of arson as being against the habitation or possessions of another.
>
> Gradually, laws were enacted to plug the loopholes of the common-law definition of *arson*. The first laws brought all buildings or structures into the scope of arson, provided they had human occupancy of any kind on a regular basis. Later, the occupancy requirements were dropped. Today, *arson* is a term applied to the willful and intentional burning of all types of structures, vehicles, forests, fields, and so on.

Figure 9.4 (based on Krol, 1992, p. 159) is an example of an extended definition addressed to a general audience.

Deciding Where to Place the Definition

In many cases, you do not need to decide where to place a definition. In writing your first draft, for instance, you may realize that most of your readers will not be familiar with a term you want to use. If you can easily provide a parenthetical definition that will satisfy your readers' needs, simply do so.

In assessing the writing situation before beginning the draft, however, you will often conclude that one or more terms will have to be introduced, and that your readers will need more detailed and comprehensive clarifications, perhaps sentence definitions and extended definitions. In these cases, you should plan, at least tentatively, where you are going to place the definitions.

Definitions can be placed in four locations:

- *In the text.* The text itself can accommodate any of the three kinds of definitions. Because they are brief and unobtrusive, parenthetical definitions are almost always included in the text; even if a reader already knows what the term means, the slight interruption will not be annoying. Likewise, sentence definitions are often placed within the text. If you want all your readers to see your definition or you suspect that many of them need the clarification, the text is the appropriate location. Keep in mind, however, that unnecessary sentence definitions can bother your readers. Extended definitions are rarely placed within the text, because of

Notice that the term being defined does not have to be introduced in the first sentence.

Nor does the term have to be defined in a traditional sentence definition.

In this paragraph, the writer provides a basic definition of the term by analyzing it.

The writer introduces a comparison.

In describing his strategy for explaining ftp, the writer offers another definition—of anonymous FTP.

Often, you will find information on the Internet that you don't want to examine on a remote system: you want to have a copy for yourself. You've found, for example, the text of a recent Supreme Court opinion, and you want to include pieces of it in a brief you are writing. Or you found a recipe that looks good, and you want to print a copy to take to the kitchen. Or you found some free software that just might solve all your problems, and you want to try it. In each case, you need to move a copy of the file to your local system so you can manipulate it there.

The tool for doing this is *ftp,* named after the application protocol it uses: the "File Transfer Protocol (FTP)." As the name implies, the protocol's job is to move files from one computer to another. It doesn't matter where the two computers are located, how they are connected, or even whether or not they use the same operating system. Provided that both computers can "talk" the FTP protocol and have access to the Internet, you can use the *ftp* command to transfer files. Some of the nuances of its use do change with each operating system, but the basic command structure is the same from machine to machine.

Like *telnet, ftp* has spawned a broad range of databases and services. You can, indeed, find anything from legal opinions to recipes to free software (and many others) in any number of publicly available online databases, or archives, that can be accessed through *ftp.* For a sampling of the archives that you can access with *ftp,* look at the *Resource Catalog* in this book. If you're a serious researcher, you will find *ftp* invaluable; it is the common "language" for sharing data.

Unfortunately, *ftp* is a complex program because there are many different ways to manipulate files and file structures. Different ways of storing files (such as binary or ASCII, compressed or uncompressed) introduce some complications, and may require some additional thought to get things right. First, we will look at how to transfer files between two computers on which you have an account (a log-in name and, if needed, a password). Next, we'll discuss *anonymous ftp,* which is a special service that lets you access public databases without obtaining an account. Most public archives provide anonymous *ftp* access, which means that you can get gigabytes of information for free—without even requiring that you have a log-in name. Finally, we'll discuss some common cases (accessing VMS, VM, DOS, or Macintosh systems), each of which requires some special handling.

Extended Definition

their length. The obvious exception, of course, is the extended definition of a term that is central to the discussion; a discussion of recent changes in workers' compensation insurance will likely begin with an extended definition of that kind of insurance.

- *In footnotes.* Footnotes are a logical place for an occasional sentence definition or extended definition. The reader who needs the definition can find it easily at the bottom of the page; the reader who doesn't need it will ignore it. However, footnotes can make the page look choppy. If you will need more than one footnote for every two or three pages, consider creating a glossary.

- *In a glossary.* A glossary—an alphabetized list of definitions—can accommodate sentence definitions and extended definitions of fewer than three or four paragraphs. It is a convenient collection of definitions that otherwise might clutter the document. A glossary can be placed at the start of a document (for example, after the executive summary in a report) or at the end, preceding the appendices. For more information on glossaries, see Chapter 10.

- *In an appendix.* An appendix is an appropriate place to put an extended definition that is one page or longer. A definition of this length would be cumbersome in a glossary or in a footnote and, unless it explains a crucial term, would be too distracting in the text. Because the definition is an appendix in the document, it will be listed in the table of contents. In addition, you can refer to it—with the appropriate page number—in the text.

Drafting Descriptions

Technical communication not only calls for shared understanding of terms, it is also filled with descriptions—verbal and visual representations of objects, mechanisms, and processes.

- *Objects.* What is an object? The range is enormous: from physical sites such as volcanoes and other kinds of natural phenomena to synthetic artifacts such as hammers. A tomato plant is an object, as is an automobile tire or a book.

- *Mechanisms.* A mechanism is a synthetic object consisting of a number of moving, identifiable parts that work together as a system. A compact-disc player is a mechanism, as is a voltmeter, a lawnmower, a submarine, or a steel mill.

- *Processes.* A process is an activity that takes place over time: the earth was formed, steel is made, animals evolve, plants perform photosynthesis. *Descriptions of processes* differ from *instructions*: process descriptions explain how something happens; instructions tell how to do something. The readers of a process description want to understand the process, *but they don't want to perform the process.* A set of instructions, however, is a step-by-step guide intended to enable the readers to perform the process. A process description answers the question "How is wine made?" A set of instructions answers the question "How do I go about making wine?" Instructions are discussed in Chapter 20.

Understanding the Role of Descriptions

Descriptions of objects, mechanisms, and processes appear in virtually every kind of technical communication. Here are a few examples:

- A company studying the feasibility of renovating an old factory or building a new one creates a report that includes a description of the old factory. The old factory is a complex mechanism. The company's managers can make a rational decision only if they understand it thoroughly.

- A writer who wants to persuade his readers to authorize the purchase of some equipment includes a mechanism description in the proposal to buy the equipment.

- An engineer trying to describe to the research-and-development department the features she would like incorporated in a piece of equipment they are designing includes a mechanism description in her report.

- An engineer trying to describe to the sales staff how a product works, so that they can advertise it effectively, includes a mechanism description in the form of a product specification.

- A company manufacturing a consumer product includes a mechanism description on its Web site to attract buyers.

- A developer who wants to build a housing project includes in his environmental impact statement descriptions of the geographical area and of the process he will use in developing that area.

- An accountant who audits one of her company's branches describes the audit process in her audit report.

Notice that a description rarely appears as a separate document. Almost always, it is part of a larger document. For example, a maintenance manual for a boiler system might begin with a mechanism description to help the reader understand how the system operates.

Analyzing the Writing Situation for Descriptions

Before you begin to write a description, consider carefully how the audience and purpose of the document will affect the way you write it.

- *Audience.* What does the audience already know about the general subject? If, for example, you are to describe an electron microscope, you first have to know whether your readers understand what a microscope is. If you want to describe how industrial robots will affect car manufacturing, you first have to know whether your readers understand the current process. In addition, you need to know whether they understand robotics.

    The audience will determine not only your use of technical vocabulary but also your sentence and paragraph length. Still another audience-related factor is the use of graphics. Less knowledgeable readers need simple graphics; they might have trouble understanding sophisticated schematics or decision charts.

- *Purpose.* What are you trying to accomplish in the description? If you want your readers to understand how a personal computer works, you will write a general description, one that applies to several varieties of computers. If, however, you want your readers to understand a particular computer, you will write a particular description. A general description of personal computers might classify them by size, then go on to describe palmtops, laptops, and desktops. A particular description of a Millennia 200 will describe only that model.

An understanding of your purpose will determine every aspect of the description, including length, amount of detail, and number and type of graphics.

Notice in the following introduction to a description of stress how the writer addresses his readers:

> All police officers are subjected to stress. For a police officer to perform his or her duties effectively—and for the police unit to perform effectively—stress must be recognized and managed.
>
> No officer should be forced to recognize and deal with the symptoms of stress in isolation. It is the responsibility of all police officers, but especially those in managerial positions, to recognize and identify the signs of stress and take an active role in reducing the stress. Otherwise, the stress can become debilitating for both the officer and the unit.
>
> The following description of stress . . .

Guidelines for Writing Descriptions

Because descriptions are written to different audiences for different purposes, they take many shapes and forms. There is no single structure or organization for descriptions. However, the following four guidelines will help you in most situations as you write descriptions:

- Clearly indicate the nature and scope of the description.
- Introduce the description clearly.
- Provide appropriate detail.
- Conclude the description.

Clearly Indicate the Nature and Scope of the Description An obvious point: if the description is to be a separate document, give it a title. If the description is to be part of a longer document, give it a section heading. (See Chapter 13 for detailed discussions of titles and headings.)

In either case, clearly state the subject and indicate whether the description is general or particular. For instance, the title of a general description of an object might be "Description of a Minivan"; of a particular description, "Description of the 1998 Mazda MPV." The title of a general description of a process might be, "Description of the Process of Designing a New Production Car"; the title of a particular description might be, "Description of the Process of Designing the General Motors Saturn."

Introduce the Description Clearly

Give your readers the information they will need to understand the detailed information that will come later in the description.

Introductions to descriptions usually are general: you want your readers to get a broad understanding of the thing you are to describe in greater detail later. Accordingly, at the start of a description you might provide a graphic that helps your readers get the big picture. For example, the description of a process might include a flowchart that summarizes the steps that are to be described in the body of the description; the description of an object such as a bicycle might include a photograph or drawing that shows the major components that will be described in the body.

Table 9.1 on page 252 shows some of the basic kinds of questions you might want to answer in introducing object and mechanism descriptions and process descriptions. Of course, in some cases the answers to one or more of the questions would be obvious and therefore should be omitted.

Figure 9.5 on page 253 shows the introductory graphic accompanying a description of a VCR, as well as a graphic for one of the parts discussed in the detailed description.

Provide Appropriate Detail

The body of a description—the part-by-part or step-by-step section—is essentially like the introduction in that it treats each major part or step as a separate item.

That is, in describing an object or mechanism you define each part and then, if applicable, you describe its function, operating principle, and appearance. The discussion of the appearance should include shape, dimension, material, and physical details such as texture and color (if essential). For some descriptions, other qualities, such as weight or hardness, might also be appropriate. If a part has important subparts, describe them the same way.

In describing a process, you treat each major step as if it were a separate process. Of course, do not repeat your answer to the question about who or what performs the action unless a new agent performs a particular step. But do answer the other principal questions—what the step is, what its function is, and when, where, and how it occurs. In addition, if the step has important sub-steps that the reader must know to understand the process, explain them clearly.

A description therefore resembles a map with a series of detailed insets. A description of a computer system would include a keyboard as one of its parts. The description of the keyboard, in turn, would include the numeric keypad as one of its parts. And the description of the numeric keypad would include the arrow keys as one of its parts. This process of ever-increasing specificity continues as

TABLE 9.1	Questions to Answer in Introducing a Description

For Mechanism and Object Descriptions

- *What is the item?* You might provide a sentence definition: "The Aleutian Islands are a chain of islands curving southwest from the tip of Alaska toward the tip of the Kormandorski Islands of the former Soviet Union."

- *What is the function of the item?* State clearly what the item does: "Electron microscopes magnify objects that are smaller than the wavelengths of visible light." Often, the function of a mechanism is implied in its sentence definition: "A hydrometer is a sealed, graduated tube, weighted at one end, that sinks in a fluid to a depth that indicates the specific gravity of the fluid."

- *What does the item look like?* Include a photograph or drawing if possible (see Chapter 12). If not, use an analogy or a comparison to a familiar item: "The cassette that encloses the tape is a plastic shell, about the size of a deck of cards." Mention the material, texture, color, and the like, if relevant. Sometimes an object is best pictured with both graphics and words.

- *How does the item work?* In a few sentences, define the operating principle of the item: "A tire pressure gauge is a calibrated rod fitted snugly within an open-ended metal cylinder. When the cylinder end is placed on the tire nozzle, the pressure of the air escaping from the tire into the cylinder pushes against the rod. The greater the pressure, the farther the rod is pushed." Sometimes objects do not "work"; they merely exist. For instance, a totem pole has no operating principle.

- *What are the principal parts of the item?* Few objects or mechanisms described in technical communication are so simple that you can discuss all the parts. And, in fact, rarely will you need to describe them all. Some parts will be too complicated or too unimportant to be mentioned; others will already be understood by your readers. A description of a bicycle would not mention the dozens of nuts and bolts that hold the mechanism together; it would focus on the chain, pedals, wheels, and frame.

For Process Descriptions

- *What is the process?* You might write a sentence definition: "Debugging is the process of identifying and eliminating errors within a computer program." Then elaborate, if necessary.

- *What is the function of the process?* For example, write, "The central purpose of performing a census is to obtain up-to-date population figures, which legislators and government agencies use to revise legislative districts and determine revenue-sharing." Make sure the function is clear to your readers; if you are unsure whether they already know the function, state it.

- *Where and when does the process take place?* State clearly the location and time or occasion for the process: "Each year the stream is stocked with hatchery fish in the first week of March." Omit these facts only if you are certain your readers already know them.

- *Who or what performs the process?* Most processes are performed by people, by natural forces, by machinery, or by some combination of the three. In most cases, you do not need to state explicitly that, for example, the young trout are released into the stream by a person; the context makes that clear. But if there is any doubt, include the information explicitly; confusion at this early stage of the process description can ruin the document.

- *How does the process work?* In a few sentences, define the principle or theory of operation of the process: "The four-treatment lawn-spray plan is based on the theory that the most effective way to promote a healthy lawn is to apply different treatments at crucial times during the growing season. The first two treatments—in spring and early summer—consist of quick-acting organic fertilizers to promote rapid growth, and chemicals to control weeds and insects. The late summer treatment . . ."

- *What are the principal steps of the process?* You might name the principal steps of the process, in one or several sentences or in a list. Name the steps in the order in which you will later describe them. The principal steps in changing an automobile tire, for instance, include jacking up the car, replacing the old tire with the new one, and lowering the car back to the ground. Changing a tire also includes secondary steps, such as placing blocks against the tires to prevent the car from moving once it is jacked up, and assembling the jack. Explain or refer to these secondary steps at the appropriate points later in the description.

FIGURE 9.5

Introductory Graphic and
Detailed Graphic

required by the complexity of the item and the needs of the readers. The same principle applies in describing processes: a step might have substeps.

Table 9.2 on page 254 presents guidelines for providing appropriate detail in descriptions.

Conclude the Description

Descriptions generally do not require elaborate conclusions. A brief conclusion is necessary, however, if only to summarize the description and prevent readers from placing excessive emphasis on the part or step discussed last in the body of the description.

A common technique for concluding descriptions of mechanisms and of some objects is to describe briefly how the parts function together. In a description of a telephone, for example, the conclusion might include a paragraph such as the following:

> When the phone is taken off the hook, a current flows through the carbon granules. The intensity of the speaker's voice causes a greater or lesser movement of the diaphragm and thus a greater or lesser intensity in the current flowing through the carbon granules. The phone receiving the call converts the electrical waves back to sound waves by means of an electromagnet and a diaphragm. The varying intensity of the current transmitted by the phone line alters the strength of the current in the electromagnet, which in turn changes the position of the diaphragm. The movement of the diaphragm reproduces the speaker's sound waves.

Process descriptions usually do not require long conclusions. If the description itself is brief—less than a few pages—a short paragraph summarizing the

TABLE 9.2	Providing Appropriate Detail in Descriptions

For Mechanism and Object Descriptions	**For Process Descriptions**

• *Choose an appropriate organizing principle.* Two organizational principles are common:
  – Functional: the way the item works or is used. In a radio, for instance, the sound begins at the receiver, travels into the amplifier, and then flows out through the speakers.
  – Spatial: based on the physical structure of the item: from top to bottom, east to west, outside to inside, and so forth.
  Most descriptions can be organized in various ways. For instance, the description of a house could be organized functionally (the different electrical and mechanical systems) or spatially (top to bottom, inside to outside, east to west, and so on). And a complex description can use a combination of patterns at different levels in the description.
• *Use graphics.* Try to create a graphic for each major part. Use photographs to show external surfaces, drawings to emphasize particular items on the surface, and cutaways and exploded diagrams to show details beneath the surface. Other kinds of graphics, such as graphs and charts, are often useful supplements.
	• *Structure the step-by-step description chronologically.* Discuss the initial step first and then discuss each succeeding step in the order in which it occurs in the process. If the process is a closed system and hence has no first step—such as the cycle of evaporation and condensation—explain to your readers that the process is cyclical, then simply begin with any principal step.
	
	• *Explain causal relationships among steps.* Although the structure of the step-by-step description should be chronological, don't present the steps as if they have nothing to do with one another. In many cases, one step causes another. In the operation of a four-stroke gasoline engine, for instance, each step sets up the conditions under which the next step can occur. In the compression cycle, the piston travels upward in the cylinder, compressing the mixture of fuel and air. In the power cycle, a spark ignites this compressed mixture.
	
	• *Use the present tense.* Discuss steps in the present tense, unless, of course, you are writing about a process that occurred in the historical past. For example, a description of how the earth was formed is written in the past tense: "The molten material condensed . . ." However, a description of how steel is made is written in the present tense: "The molten material is then poured into . . ."
	
	• *Use graphics.* Whenever possible, use graphics to clarify each point. Additional flowcharts are useful, but often other kinds of graphics, such as photographs, drawings, and graphs, are helpful as well. In the description of a four-stroke gasoline engine, you could use diagrams to illustrate the position of the valves and the activity occurring during each step. For example, you could show the explosion during the ignition step with arrows to indicate that the explosion is pushing the piston down within the cylinder.

principal steps is all you need. Following is the conclusion from a description of the four-stroke gasoline engine in operation:

> In the intake stroke, the piston moves down, drawing the air-fuel mixture into the cylinder from the carburetor. As the piston moves up, it compresses this mixture in the compression stroke, creating the conditions necessary for combustion. In the power stroke, a spark from the spark plug ignites the mixture, which burns rapidly, forcing the piston down. In the exhaust stroke, the piston moves up, expelling the burned gases.

For longer descriptions, a discussion of the implications of the process might be appropriate. For instance, a description of the Big Bang—one theory of how the universe began—might conclude with a discussion of how the theory is supported and challenged by recent astronomical discoveries and theories.

Often, a writer concludes a description with some sort of sales message that summarizes the benefits of the product or process and motivates the reader to take further action, such as phoning or writing for more information or to set up a visit from a salesperson.

Analyzing Some Descriptions

A look at some descriptions will give you an idea of how different writers adapt basic approaches to meet the needs of their audiences and purpose.

Figure 9.6 on page 256 (U.S. Congress, 1995b) shows the extent to which a process description can be based on graphics. The topic is drivetrain efficiencies for internal combustion engine vehicles and electric vehicles. The audience for this description is the general reader.

Notice in Figure 9.6 how effectively graphics show the relative efficiencies of an internal combustion engine vehicle (the top row) and an electric vehicle (the bottom row). At each step in the process of turning raw materials into energy to power the vehicles, the efficiency is noted in percentage. Finally, at the right side, the overall efficiencies of the two technologies are compared. The graphics clarify the process and make it interesting.

Figure 9.7 on pages 257–259 (U.S. Environmental, 1991, pp. 45–47) is an excerpt from a discussion of the process of diagnosing IAQ (indoor air quality) problems. The audience consists of building owners and facility managers. After the excerpt reprinted here, the authors provide detailed instructions for each of the steps, starting with the initial walkthrough.

Figure 9.8 on page 260 is an excerpt from a description (Praxis, 1997) of a piece of atmospheric-testing hardware used by the Naval Research Laboratory in its defense Meteorological Satellite Program. The excerpt here, taken from the Web site, consists of the main page for the description.

FIGURE 4-3: Drivetrain Efficiencies

Internal combustion engine vehicles (ICEV) 12.2%

Oil well 96% → Refinery 90% → Transportation and distribution 97% → Engine 14.4% → Transmission 85% → Road load → Wheels

10.3% Overall efficiency

Electric vehicle (EV) 52%

Coal mine 97% → Utility 36% → Transmission and distribution 92% → Charger 90% → Battery 75% → Controller and motor 90% → Transmission 85% → Road load → Wheels

16.6% Overall efficiency

NOTE: An electric drivetrain can be three to four times as efficient as a mechanical ICE drivetrain (e.g., 52 percent for electric vehicles (EVs) versus 12 percent for ICEVs). This efficiency differential drops substantially when the overall fuel chain efficiency for ICEVs and EVs is taken into consideration (16.6 percent for coal-powered EVs versus 10.3 percent for gasoline-powered ICEVs). The fuel chain efficiency for EVs could be much higher if new power generation technologies are deployed. Advanced coal plants might achieve efficiencies close to 50 percent, while efficiencies of 60 percent are possible for advanced natural gas plants. With an advanced natural gas plant the overall fuel chain efficiency for EVs could rise to 27 percent.

SOURCE: John Brogan and S. Venkateswaran, "Diverse Choices for Hybrid and Electric Motor Vehicles," in *Proceedings of the International Conference on Urban EVs* (Stockholm, Sweden: Organization for Economic Cooperation and Development, May 1992).

FIGURE 9.6

A Process Description Based on a Graphic

Diagnosing IAQ Problems

6

This introduction begins by explaining the purpose of performing a diagnostic building investigation. The introduction explains the strategy for conducting the investigation and states who should perform it.

Notice the use of the flowchart to summarize the process.

The goal of the diagnostic building investigation is to identify and solve the indoor air quality complaint in a way that prevents it from recurring and that does not create other problems. This section describes a method for discovering the cause of the complaint and presents a "toolbox" of diagnostic activities to assist you in collecting information.

Just as a carpenter uses only the tools that are needed for any given job, an IAQ investigator should use only the investigative techniques that are needed. Many indoor air quality complaints can be resolved without using all of the diagnostic tools described in this chapter. For example, it may be easy to identify the source of cooking odors that are annoying nearby office workers and solve the problem by controlling pressure relationships (e.g., installing exhaust fans) in the food preparation area. Similarly, most mechanical or carpentry problems probably require only a few of the many tools you have available and are easily accomplished with in-house expertise.

The use of in-house personnel builds skills that will be helpful in minimizing and resolving future problems. On the other hand, some jobs may be best handled by contractors who have specialized knowledge and experience. In the same way, diagnosing some indoor air quality problems may require equipment and skills that are complex and unfamiliar. Your knowledge of your organization and building operations will help in selecting the right tools and deciding whether in-house personnel or outside professionals should be used in responding to the specific IAQ problem.

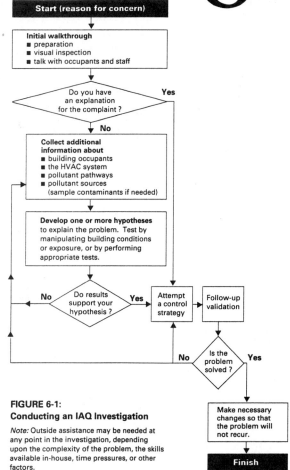

FIGURE 6-1:
Conducting an IAQ Investigation

Note: Outside assistance may be needed at any point in the investigation, depending upon the complexity of the problem, the skills available in-house, time pressures, or other factors.

(Continued)

(Figure 9.7 continued)

The detailed section of the process description begins with this overview section that provides general information about the process and refers to the flowchart on the previous page.

The IAQ investigation is often a repetitive cycle of information-gathering, hypothesis formation, and hypothesis testing.

The overview contains a number of subsections, each of which has its own heading. Note that although the writer uses the second-person pronoun— "you"—the document is not a set of instructions but a process description. A set of instructions would include detailed information about tools and materials, as well as step-by-step instructions.

OVERVIEW: CONDUCTING AN IAQ INVESTIGATION

An IAQ investigation begins with one or more reasons for concern, such as occupant complaints. Some complaints can be resolved very simply (e.g., by asking a few common sense questions of occupants and facility staff during the walkthrough). At the other extreme, some problems could require detailed testing by an experienced IAQ professional. In this section "the investigator" refers to in-house staff responsible for conducting the IAQ investigation.

The flowchart on page 45 shows that the IAQ investigation is a cycle of information-gathering, hypothesis formation, and hypothesis testing. The goal of the investigation is to understand the IAQ problem well enough so that you can solve it. Many IAQ problems have more than one cause and may respond to (or require) several corrective actions.

Initial Walkthrough

An initial walkthrough of the problem area provides information about all four of the basic factors influencing indoor air quality (occupants, HVAC system, pollutant pathways, and contaminant sources). The initial walkthrough may provide enough information to resolve the problem. At the least, it will direct further investigation. For example, if the complaint concerns an odor from an easily identified source (e.g., cooking odors from a kitchen), you may want to study pollutant pathways as a next step, rather than interviewing occupants about their patterns of discomfort.

Developing and Testing Hypotheses

As you develop an understanding of how the building functions, where pollutant sources are located, and how pollutants move within the building, you may think of many "hypotheses," potential explana-

tions of the IAQ complaint. Building occupants and operating staff are often a good source of ideas about the causes of the problem. For example, they can describe changes in the building that may have occurred shortly before the IAQ problem was noticed (e.g., relocated partitions, new furniture or equipment).

Hypothesis development is a process of identifying and narrowing down possibilities by comparing them with your observations. **Whenever a hypothesis suggests itself, it is reasonable to pause and consider it.** Is the hypothesis consistent with the facts collected so far?

You may be able to test your hypothesis by modifying the HVAC system or attempting to control the potential source or pollutant pathway to see whether you can relieve the symptoms or other conditions in the building. If your hypothesis successfully predicts the results of your manipulations, then you may be ready to take corrective action. Sometimes it is difficult or impossible to manipulate the factors you think are causing the IAQ problem; in that case, you may be able to test the hypothesis by trying to predict how building conditions will change over time (e.g., in response to extreme outdoor temperatures).

Collecting Additional Information

If your hypothesis does not seem to be a good predictor of what is happening in the building, you probably need to collect more information about the occupants, HVAC system, pollutant pathways, or contaminant sources. Under some circumstances, detailed or sophisticated measurements of pollutant concentrations or ventilation quantities may be required. Outside assistance may be needed if repeated efforts fail to produce a successful hypothesis or if the information required calls for instruments and procedures that are not available in-house.

(Figure 9.7 continued)

Results of the Investigation

Analysis of the information collected during your IAQ investigation could produce any of the following results:

The apparent cause(s) of the complaint(s) is (are) identified.

Remedial action and follow-up evaluation will confirm whether the hypothesis is correct.

Other IAQ problems are identified that are not related to the original complaints.

These problems (e.g., HVAC malfunctions, strong pollutant sources) should be corrected when appropriate.

A better understanding of potential IAQ problems is needed in order to develop a plan for corrective action.

It may be necessary to collect more detailed information and/or to expand the scope of the investigation to include building areas that were previously overlooked. Outside assistance may be needed.

The cause of the original complaint cannot be identified.

A thorough investigation has found no deficiencies in HVAC design or operation or in the control of pollutant sources, and there have been no further complaints. In the absence of new complaints, the original complaint may have been due to a single, unrepeated event or to causes not directly related to IAQ.

Using Outside Assistance

Some indoor air quality problems may be difficult or impossible for in-house investigators to resolve. Special skills or instruments may be needed. Other factors can also be important, such as the benefit of having an impartial outside opinion or the need to reduce potential liability from a serious IAQ problem. You are best able to make the judgment of when to bring in an outside consultant. See *Section 8* for a discussion of hiring professional assistance to solve an IAQ problem.

This is the main page for this device. ————

This is the most general description of the device. ————

Along with a graphic of the main component of the device are buttons linking to four related pages. ————

More information about the device. ————

The main components of the device. ————

Contact information at the bottom of the main page. ————

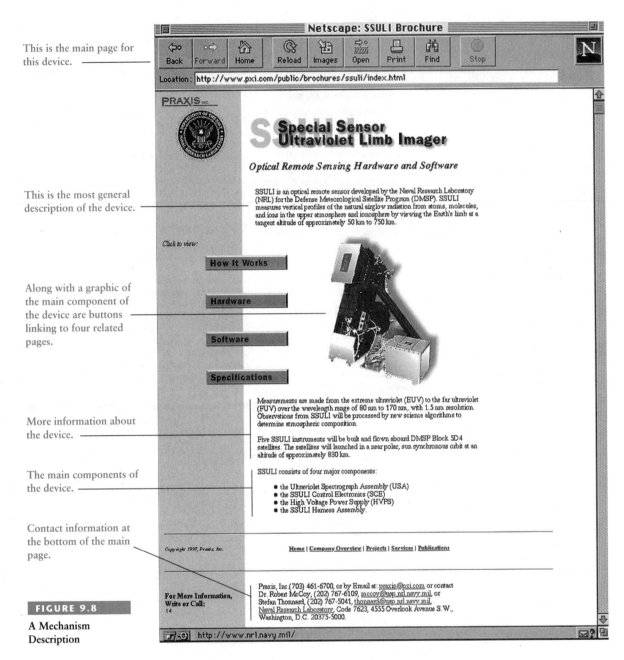

FIGURE 9.8

A Mechanism Description

CASE

Annotating a Patent Application

Form small groups for this project on patents. Your group works for your university's Office of Research Administration. One function of this office is to apply for patents for devices invented by faculty. The director of the office, Sandra Gustafson, has asked you to do some research.

"Go to the site for the United States Patent and Trademark Office (http://www.uspto.gov). It's a good site, containing many files defining and describing patents and trademarks. However, the site doesn't include an example of a successful patent application. I'd like you to get a copy of the *Official Gazette of the United States Patent and Trademark Office,* which includes the actual patents and trademarks granted by the office. Take some representative pages from a sample, then annotate them. I want you to show what the two major components of the patent application—the specification and the drawing—really look like."

"Who is this for?" you ask her.

"Sorry. It's for faculty. When they come in asking about patents, I want to be able to show them what they look like. Most people have never seen one," Sandra replies.

"Okay, you want us to mark up some sections of a patent application, pointing out key elements of the specification and the drawing."

"Yes, that's what I want."

"Would you let us look at the site and get back to you to make sure we understand what you need?"

"Fine," Sandra responds. "Can we talk tomorrow morning?"

You and your group go to the Patent and Trademark Office site (http://www.uspto.gov/web/offices/pac/doc/general/specifications.html) and find the following definition of the word *claim,* an important part of the patent application.

The specification must conclude with one or more claims particularly pointing out and distinctly claiming the subject matter which the applicant regards as the invention.

The claims are brief descriptions of the subject matter of the invention, eliminating unnecessary details and reciting all essential features necessary to distinguish the invention from what is old. The claims are the operative part of the patent. Novelty and patentability are judged by the claims, and, when a patent is granted, questions of infringement are judged by the courts on the basis of the claims.

You meet with Sandra. "Here's what we think would work. We want to take definitions, like this one on *claims,* and annotate the claims section of an actual patent application, showing how it meets the requirements of the definition."

"That's it, exactly. Thanks," Sandra replies.

Annotate a patent application, focusing on the key elements of the specification and the drawing.

This checklist covers parenthetical, sentence, and extended definitions.

1. Are all necessary terms defined?

2. Are the parenthetical definitions
   ❑ appropriate for the audience?
   ❑ clear?

3. Does each sentence definition
   ❑ contain a sufficiently specific category and distinguishing characteristics?
   ❑ avoid describing a particular item when a general class of items is intended?
   ❑ avoid circular definition?
   ❑ contain a noun or a noun phrase in the category?

4. Are the extended definitions developed logically and clearly?

5. Are the definitions placed in the most useful location for the readers?

The following questions cover descriptions of objects and mechanisms:

1. Have you clearly indicated the nature and scope of the description?

2. In introducing the description, have you answered, if appropriate, the following questions:
   ❑ What is the item?
   ❑ What does it do?
   ❑ What is its function?
   ❑ What does it look like?
   ❑ What is its principle of operation?
   ❑ What are its principal parts?

3. Have you included a graphic identifying all the principal parts?

4. In providing detailed information, have you
   ❑ chosen an appropriate organizing principle?
   ❑ included graphics for each of the components?

5. In concluding the description, have you
   ❑ summarized the major points in the part-by-part description?
   ❑ included (where appropriate) a description of the item performing its function or an attempt to motivate the reader to take action?

The following questions cover process descriptions:

1. Have you clearly indicated the nature and scope of the description?

2. In introducing the description, have you answered, if appropriate, the following questions:
   ❑ What is the process?
   ❑ What is its function?
   ❑ Where and when does the process take place?
   ❑ Who or what performs it?
   ❑ How does the process work?
   ❑ What are its principal steps?

3. Have you included a graphic identifying all the principal steps?

4. In providing detailed information, have you
   ❑ answered, for each of the major steps, the questions in item 2?
   ❑ discussed the steps in chronological order or other logical sequence?
   ❑ made clear the causal relationships among the steps?
   ❑ included graphics for each of the principal steps?

5. In concluding the description, have you
   ❑ summarized the major points in the step-by-step description?
   ❑ discussed, if appropriate, the importance or implications of the process?
   ❑ attempted (if appropriate) to motivate the reader to take action?

EXERCISES

1. Add parenthetical definitions of the italicized terms in the following sentences:
   a. Reluctantly, he decided to *drop* the physics course.
   b. Last week the computer was *down*.
   c. The department is using *shareware* in its drafting course.
   d. The tire plant's managers hope they do not have to *lay off* any more employees.
   e. Please submit your assignments *electronically* and as *hard copy*.

2. Write a sentence definition for each of the following terms:
   a. catalyst
   b. compact-disc player
   c. job interview
   d. Web site
   e. flextime
   f. automatic teller machine
   g. fax machine
   h. intranet

3. Write a 500- to 1,000-word extended definition of one of the following terms, or of a term used in your

field of study. If you do secondary research, cite your sources clearly and accurately. If you photocopy graphics from a secondary source, cite them as well. In addition, check that the graphics are appropriate for your audience and purpose. In a brief note at the start, indicate the audience and purpose for your definition.

a. flextime
b. binding arbitration
c. robotics
d. an academic major (don't focus on any particular major; define what a major is)
e. quality control
f. bioengineering
g. fetal-tissue research
h. community policing
i. software

4. Revise any of the following sentence definitions that need revision:
   a. Dropping a course is when you leave the class.
   b. A thermometer measures temperature.
   c. The spark plugs are the things that ignite the air-gas mixture in a cylinder.
   d. Double-parking is where you park next to another car.
   e. A strike is when the employees stop working.

5. Identify the techniques used to create the following extended definition:

   Holography, from the Greek *holos* (entire) and *gram* (message), is a method of photography that produces images that appear to be three-dimensional. A holographic image seems to change as the viewer moves in relation to it. For example, as the viewer moves, one object on the image appears to move in front of another object. In addition, the distances between objects on the image seem to change.

   Holographs are produced by coherent light, that is, light of the same wavelength, with the waves in phase and of the same amplitude. This light is produced by laser. Stereoscopic images are created by incoherent light—random wavelengths and amplitudes, out of phase. The incoherent light, which is natural light, is focused by a lens and records the pattern of brightness and color differences of the object being imaged.

   How are holographic images created? The laser-produced light is divided as it passes through a beam splitter. One portion of the light, called the *reference beam,* is directed to the emulsion—the "film." The other portion, the object beam, is directed to the subject and then reflected back to the emulsion. The reference beam is coherent light, whereas the object beam becomes incoherent because it is reflected off the irregular surface of the subject. The resulting dissonance between the reference beam and the object beam is encoded; it records not only the brightness of the different parts of the subject but also the different distances from the laser. This encoding creates the three-dimensional effect of holography.

6. Locate an extended definition in one of your textbooks, a journal article, or a Web site. In a memo to your instructor, describe the techniques the author uses to define the term. Then evaluate the effectiveness of these techniques. Submit a photocopy of the definition along with your assignment. For information on memos, see Chapter 16.

7. Write a 500- to 1,000-word description of one of the following items or of a piece of equipment used in your field. Include appropriate graphics. In a note preceding the description, specify your audience and indicate the type of description (general or particular) you are writing.

   a. carburetor   h. jet engine
   b. locking bicycle rack i. telescope
   c. deadbolt lock   j. ammeter
   d. folding card table   k. television set
   e. lawn mower   l. automobile jack
   f. photocopy machine   m. stereo speaker
   g. cooling tower at a   n. refrigerator
      nuclear power plant   o. personal computer

8. Write a memo to your instructor evaluating the effectiveness of the description of the personal computer that appears on pages 264–266. (See Chapter 16 for a discussion of memos.) Which aspects of the description are successful? Which parts of the description would you change? Why?

9. Rewrite the description in exercise 8.

10. Write a 500- to 1,000-word description of one of the following processes or a similar process with which you are familiar. Include appropriate graphics. In a note preceding the description, specify your audience and indicate the type of description (general or particular) you are writing. If you use secondary sources, cite them properly (see Chapter 10 for documentation systems).

The Personal Computer

Introduction

This paper will briefly describe what a personal computer system is and the major components of the system. It is intended for a general reader who has a little knowledge about computers. Most readers will be familiar with the small amount of jargon contained herein since many of these words have crossed over into other professions.

A personal computer system is made up of input devices, a central processing unit, and output devices (see Figure 1). Information is put into the computer via input devices, processed in the central processing unit, then viewed via output devices.

This particular type of computer system is called a personal computer system because it has been scaled down small enough that most people have enough space for one in their home or business. The days of warehouse-sized computers are over—personal computers have become dominant in the computer industry.

Central Processing Unit

The central processing unit (CPU) is the "brain" of the computer (see Figure 1). Input is sent to the CPU via the input devices, processed in the CPU, then

Figure 1

sent to output devices. The CPU contains components such as memory, a microprocessor, floppy disk drives, hard disk drives, graphics adapters, a power supply, modems, and so on. All of these components work together to process information. This processing can be in the form of documents such as letters, grants, and reports; number analysis such as statistics, spreadsheets, and mathematical calculations; or graphics such as slides.

Input Devices

Input devices put data and commands into the central processing unit. There are many input devices such as bar code readers, scanners, touch screens, and voice-activated devices, but the two most common input devices are the keyboard and the mouse.

Keyboard

A keyboard is an input device that roughly resembles a typewriter keyboard in both size and arrangement (see Figure 2). However, there are some important differences. One difference is that the computer keyboard has a numeric keypad that resembles an adding machine keypad in addition to the usual letter, number, and character keys. Also, the computer keyboard has directional keys and a row of function keys (programmable keys used to perform routine tasks). The computer user can use the keyboard to type in commands to an operating system (a program that converts English words such as print, copy, move, and so on into series of electrical pulses that the machine can understand) or a software package, or type letters and numbers into software programs such as word processors and spreadsheets.

Figure 2

Mouse

A mouse is an input device about the size of a deck of cards (see Figure 3). The mouse has two or three buttons on the front edge that are depressed in order to choose commands from the computer screen. The mouse is connected by a wire to the back of the CPU. The purpose of the mouse is to bypass the directional keys on the keyboard and directly point to a command or graphical icon on the computer monitor's screen. To move the cursor on the screen, the user slides the mouse around on a rubber pad or even on a table or desktop.

Figure 3

Output Devices

Output devices display information that is contained inside the computer. There are many output devices such as plotters, slide makers, and projection screens, but the two most common output devices are the monitor and the printer.

Monitor

A monitor is an output device that resembles a small television in both size and appearance (see Figure 1). Monitors, like television sets, come in many screen sizes and in both black-and-white and color. The most common computer monitor uses a 14-inch diagonal screen with color video graphics. The monitor displays (much like a television) the characters and commands that the user types into the computer. The monitor also displays the results after a software package has processed data that was input by the user.

Printer

A printer is an output device that prints output onto paper. Printers vary in size from the size of a notebook to the size of a small desk. However, most printers are about 2 feet wide by 1 foot high by 2 feet deep (about the same size as the CPU). Documents, letters, spreadsheets, and so on are printed onto paper with the same quality as a typewriter. The printed copies are called "hard copies."

Conclusion

All of the components of the personal computer system are important. If one component was taken away, the user's ability to process information would be impaired. To remember the components, the reader should remember input devices, then the central processing unit, then the output devices.

a. how steel is made
b. how a nuclear power plant works
c. how a food co-op works
d. how a suspension bridge is constructed
e. how we see
f. how a dry battery operates
g. how a baseball player becomes a free agent

11. The following topics are appropriate for both process descriptions and instructions. Write a 500- to 1,000-word process description. Later in the semester, when you are studying Chapter 20, write a set of instructions on the same topic.
   a. how to clean a fish
   b. how a fax machine operates
   c. how to study for a test
   d. how an audit is conducted
   e. how to winterize a recreational vehicle

Collaborative Exercise

12. Form small groups according to major for this project on definitions and descriptions on the Web. The Web has many descriptions of processes and mechanisms. Find a description of an item used in your field and, in a brief memo to your instructor, respond to the following questions.

   • Does the description contain definitions? If so, what kind? How clear are they? How would you revise them?

   • What is the purpose of the description, and how does that purpose determine the nature, scope, and organization of the description?

   • In what ways have the authors used the hypertext medium in this description? In other words, how is this version of the description different from one that would appear as a traditional printed document?

   • How effective is the description? Which aspects are well done? Which aspects would you do differently?

Web Exercise

13. You are a student intern working for the chamber of commerce of Seattle, Washington. To publicize the business environment in Seattle, the chamber is assembling a directory of leading high-technology manufacturers in the Seattle area. One of the major corporations to be featured in this directory is Boeing. You have been asked to write a brief description—no more than three or four pages—of one of Boeing's newest products: the F-22 Fighter. When you phone the public-affairs office at Boeing, you are told that all the information you will need about the F-22 is on their Web site (http://www.boeing.com) and that you are free to refer to the text and graphics in doing your project, so long as you cite Boeing if you quote any text or reproduce any graphics. Write the description of the F-22, addressed to a general audience.

REFERENCES

Krol, E. (1992). *The whole Internet user's guide & catalog.* Sebastopol, CA: O'Reilly & Associates.

Praxis, Inc. (1997). SSULI brochure. Http://www.pxi.com/public/brochures/ssuli/index.html. March 21.

Roblee, C. L., & McKechnie, A. J. (1981). *The investigation of fires.* Englewood Cliffs, NJ: Prentice Hall.

U.S. Congress, Office of Technology Assessment. (1995a). *Bringing health care online: The role of information technologies.* OTA-ITC-624. Washington, DC: U.S. Government Printing Office.

U.S. Congress, Office of Technology Assessment. (1995b). *Renewing our energy future.* OTA-ETI-614. Washington, DC: U.S. Government Printing Office.

U.S. Environmental Protection Agency. (1991). *Building air quality: A guide for building owners and facility managers.* Washington, DC: U.S. Environmental Protection Agency.

Wilson, J. R. (1964). *The mind.* New York: Time, Inc.

CHAPTER

10

Drafting the Front and Back Matter

Using the Word Processor 269

Writing the Front Matter 270
  The Letter of Transmittal 270
  The Cover 271
  The Title Page 271
  The Abstract 274
  The Table of Contents 276
  The List of Illustrations 279
  The Executive Summary 279

APA and MLA Documentation of Sources 283
  APA Style 284
  MLA Style 289

Writing the Back Matter 295
  The Glossary and List of Symbols 296
  References 297
  Appendices 297
  The Index 298

Case: Comparing Online Indexes 304

Writer's Checklist 304

Exercises 305
  Collaborative Exercise 307
  Web Exercise 307

Front matter consists of the elements that precede the body of a substantial document, such as title pages and tables of contents. *Back matter* consists of the elements that follow the body, such as glossaries and indexes. Front and back matter appear in various kinds of documents, including proposals, reports, and manuals. Some of the elements discussed in this chapter, such as glossaries, can appear in almost any kind of document, including brochures and flyers.

Different elements in the front and back matter play various roles:

- *To help readers find the information they seek.* The table of contents and the index exist so that readers can find what they want to read.
- *To help readers decide if they want to read the document.* The abstract—one of the summaries—enables readers to decide whether to take the time to read the whole document.
- *To substitute for the whole document.* The executive summary is written especially for managers, who often do not read anything else in the document.
- *To help readers understand the document.* The glossary—a list of definitions—helps readers who don't know the subject thoroughly.
- *To protect the document.* The cover, for instance, might contain no information at all.

Few documents have all the elements of the front matter and back matter in the order in which they are covered in this chapter; most organizations have their own format preferences. Therefore, you should study the style guide used in your company. If there is no style guide, study examples from the files to see how other writers have assembled their documents.

Using the Word Processor

A word processor simplifies assembling the front and back matter in three ways:

- *You can use word-processing software to determine the length of different elements.* Knowing the length is useful because many of the elements, such as abstracts, have length guidelines or restrictions.
- *You can use the copy function in drafting the different summaries and the transmittal letter.* You can make a copy of the body of the document and then revise it to make the elements you need. Using the copy function rather than drafting from scratch is also more accurate because you don't introduce errors.
- *You can use special features in some word processors to create tables of contents and indexes.*

For more information on using the word processor, see Chapter 4.

Writing the Front Matter

The front matter consists of the seven elements shown in Figure 10.1.

The Letter of The letter of transmittal introduces the purpose and content of the document to
Transmittal the principal reader. The letter is attached to the document, bound in with it, or
simply placed on top of it. Even though the letter might contain no information
that is not included elsewhere in the document, it is important, because it is the
first thing the reader sees. It establishes a courteous and graceful tone for the document. Letters of transmittal are customary even when the writer and the reader
both work for the same organization and ordinarily communicate by memo.

Transmittal letters generally contain the following seven elements:

- a statement of the title and the purpose of the document
- a statement of who authorized or commissioned the project and when
- a statement of the methods used in the project (if they are noteworthy) or
  of the principal results, conclusions, and recommendations

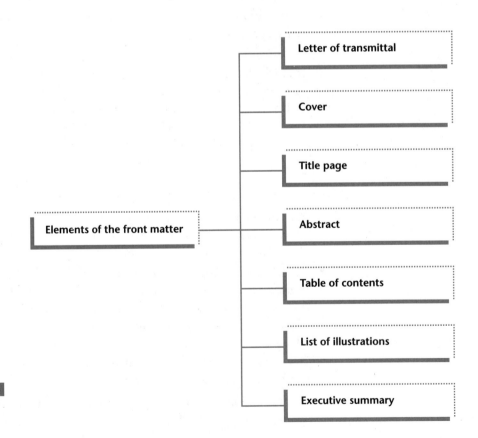

FIGURE 10.1

Elements of the Front
Matter

- a statement urging that the organization carry out an additional project as a result of the findings of the current project

- an acknowledgment of any assistance the writers received in preparing the materials

- a gracious offer to assist in interpreting the materials or in carrying out further projects

- a reference to any errors or omissions (for example, the writers might want to include information that was gathered after the document was typed or printed)

Don't use the transmittal letter to apologize to the reader or ask for the reader's patience or compassion. Present your information professionally. Don't write sentences such as the following:

I didn't know a lot about hot air ballooning before I began this report, so please excuse any errors you might see.

I realize the figures aren't as neat as they should be, but I ran out of time and didn't have the chance to fix them.

Figure 10.2 on page 272 shows a transmittal letter. (For a discussion of letter format, see Chapter 19.)

The Cover   The cover protects the document, either from normal wear and tear or from harsher environmental conditions such as salt water or grease. The cover usually contains the following information:

- the title of the document

- the name and position of the writer

- the date of submission

- the name or logo of the writer's company

Sometimes the cover includes a security notice (if it is a military document) or a statement of proprietary information.

See Chapter 11 for information about different kinds of materials used in covers and different kinds of bindings.

The Title Page   The title page includes at least the following elements:

- the title

- the date of submission

- the names and positions of the writer and the principal reader

- the name or logo of the writer's company

ALTERNATIVE ENERGY, INC.
Bar Harbor, ME 00314

April 3, 19XX

Rivers Power Company
15740 Green Tree Road
Gaithersburg, MD 20760

Attention: Mr. J. R. Hanson, Project Engineering Manager

Subject: Project #619-103-823

Gentlemen:

We are pleased to submit "A Proposal for the Riverfront Energy Project" in response to your request of February 6, 19XX.

The windmill described in the attached proposal uses the most advanced design and materials. Of particular note is the state-of-the-art storage facility described on pp. 14–17. As you know, storage limitations are a crucial factor in the performance of a generator such as this.

If you have any questions, please do not hesitate to call us.

Yours very truly,

Ruth Jeffries

Ruth Jeffries
Project Manager

Enclosures 2

FIGURE 10.2

Letter of Transmittal

A good title answers two basic questions:

- What is the subject of the document?
- What is the purpose of the document?

For example:

Choosing a Microcomputer: A Recommendation

An Analysis of the Kelly 1013 Packager

Open Sea Pollution-Control Devices: A Summary

PETROLEUM PRICES AT THE END OF THE CENTURY:
A FORECAST

Prepared for: Harold Breen, President

                      Reliance Trucking Co.

by: Adelle Byner, Manager

                      Purchasing Department

                      Reliance Trucking Co.

                      April 19, 19XX

RELIANCE TRUCKING CO.
moving with the times since 1942

FIGURE 10.3

Simple Title Page

Note that a convenient way to define the type of document is to use a generic term—such as *analysis, recommendation, summary, review, guide,* or *instructions*—in a phrase following a colon. For more information on titles, see Chapter 13.

Figure 10.3 shows a simple title page. A more complex title page might include other elements that identify the document, the different personnel associated with the project, and the distribution list.

The Abstract   An abstract is a brief technical summary—usually no more than 200 words—of the document. An abstract addresses readers who are familiar with the technical subject and who need to decide whether to read the full document. Therefore, in writing an abstract you can use technical terminology freely and refer to advanced concepts in your field.

You will see abstracts in three different places:

- *In the document itself.*
- *In different departments in the organization.* Because abstracts can be useful before and after a document is read—and can even be read in place of the document—often they are kept on file in several locations within an organization: the division in which the project originated, as well as many or all of the higher-level departments. To facilitate this wide distribution, some organizations type abstracts on special forms.
- *In abstracting services.* As discussed in Chapter 7, abstract services print abstracts of journal articles and some government publications.

Abstracts are classified into two types: descriptive and informative. You will see the *descriptive abstract* most often where space is at a premium. Some government proposals, for example, call for descriptive abstracts to be placed at the bottom of the title page. You will see the *informative abstract* most often when more comprehensive information is required. If you are ever called on to write an abstract and don't know which kind the reader wants, write an informative one, not a descriptive one.

Abstracts often contain a list of about a half-dozen *keywords*—words or phrases that are entered into electronic databases. Your job is to think of the different keywords that people might use to access your document.

The Descriptive Abstract

The *descriptive abstract,* sometimes called the *topical* or *indicative* or *table-of-contents* abstract, does only what its name implies: it describes the kinds of information in the document. It does not provide the information itself: the important results, conclusions, or recommendations. It simply lists the topics covered, giving equal emphasis to each. Thus the descriptive abstract essentially duplicates the major headings in the table of contents. Figure 10.4 is an example of a descriptive abstract.

The Informative Abstract

The informative abstract presents the major information that the document conveys. Rather than merely list topics, it states the problem, scope and methods (if appropriate), and major results, conclusions, and recommendations.

ABSTRACT
"Design of a Radio-Based System for Distribution Automation"
by Brian D. Crowe

At this time, power utilities' major techniques of monitoring their distribution systems are after-the-fact indicators such as interruption reports, meter readings, and trouble alarms. These techniques are inadequate in two ways. One, the information fails to provide the utility with an accurate picture of the dynamics of the distribution system. Two, after-the-fact indicators are expensive. Real-time load monitoring and load management would offer the utility both system reliability and long-range cost savings. This report describes the criteria we used to design the radio-based system for a pilot program of distribution automation. It then describes the hardware and software of the system.

Keywords: distribution automation, distribution systems, load, meters, radio-based systems, utilities

FIGURE 10.4

Descriptive Abstract

The informative abstract includes three elements:

- *The identifying information.* Include the title of the document and your name. Sometimes, you will want to include your position, too.
- *The problem statement.* Write one or two sentences that define the problem or opportunity that led to the project.
- *The important findings.* Devote the final three or four sentences to the findings: the crucial information contained in the document. In reports, findings usually consist of results, conclusions, recommendations, and implications for further projects. Sometimes, however, the abstract presents other information. For instance, many technical projects focus on new or unusual methods for achieving results that have already been obtained through other means. In such a case, the abstract will focus on the methods, not on the results.

Figure 10.5 on page 276 is an informative abstract.

The distinction between descriptive and informative abstracts is not absolute. Sometimes you will have to combine elements of both in a single abstract. For instance, suppose you are writing an informative abstract, but the report includes 15 recommendations, far too many to list. You might identify the major results and conclusions, as you would in any informative abstract, but add that the report contains numerous recommendations, as you would in a descriptive abstract.

ABSTRACT
"Design of a Radio-Based System for Distribution Automation"
by Brian D. Crowe

At this time, power utilities' major techniques of monitoring their distribution systems are after-the-fact indicators such as interruption reports, meter readings, and trouble alarms. This system is inadequate in that it fails to provide the utility an accurate picture of the dynamics of the distribution system, and it is expensive. This report describes a project to design a radio-based system for a pilot project. The basic system, which uses packet-switching technology, consists of a base unit (built around a personal computer), a radio link, and a remote unit. The radio-based distribution monitoring system described in this report is more accurate than the currently used after-the-fact indicators, it is small enough to replace the existing meters, and it is simple to use. We recommend installing the basic system on a trial basis.

Keywords: distribution automation, distribution systems, load, meters, radio-based systems, utilities

FIGURE 10.5

Informative Abstract

The Table of Contents

A table of contents has two main functions:

- *To help readers find the information they want.* Your table of contents is the most important guide readers have to help them navigate the document.
- *To help readers understand the scope and organization of the document.*

The table of contents uses the same headings that appear in the document itself. To create an effective table of contents, therefore, you first must make sure the document has effective headings—and that it has enough of them. If the table of contents shows no entry for five or six pages, the document could probably be divided into additional subsections. In fact, some tables of contents have an entry, or several entries, for every page in the document.

Insufficiently specific tables of contents generally result from the exclusive use of generic headings (those that describe entire classes of items) in the report. The following table of contents is relatively uninformative because it simply lists generic headings.

Table of Contents

Introduction 1

Materials 3

Methods 4

Results. 19

Recommendations 23

References. 26

Appendices 28

This table of contents is ineffective because it does not provide enough detail. The methods section, for instance, runs from page 4 to page 18; the writer should add subentries to break up the text and to help readers find the information they seek.

To make the headings more informative, combine generic and specific items, as in the following examples:

> Recommendations: Five Ways to Improve Information-Retrieval Materials Used in the Calcification Study
>
> Results of the Commuting-Time Analysis

Then build more subheadings into the report. For instance, in the "Recommendations" example, make a subheading for each of the five recommendations.

Once you have created a clear system of headings within the document, transfer them to the contents page. Use the same text attributes—capitalization, boldface, italics, and outline style (traditional or decimal)—that you use in the body. (See the discussions of headings in Chapters 11 and 13.)

A word processor makes it simple to transfer headings from the body of the document to the table of contents. Some software programs do the job automatically. But even the simplest software helps you make the table of contents quickly and easily, without introducing errors. Simply make a copy of the document, so that you have two copies back-to-back in the same file. Then scroll through one copy, erasing the text. What you have left are the headings. If they are inconsistent, you will notice immediately and be able to fix the problems.

Figure 10.6 on page 278 shows how you can combine generic and specific headings in a table of contents. The report is titled "Methods of Computing the Effects of Inflation in Corporate Financial Statements: A Recommendation."

The table of contents in Figure 10.6 works well for three reasons:

- Managers can find the executive summary quickly and easily.
- All other readers can find the information they are looking for because each substantive section is listed separately.
- The decimal numbering system used to identify the headings (such as 2.1 and 2.1.1) helps readers to understand the organization of the document. For more information on numbering systems, see Chapter 8.

Two notes about pagination:

- The table of contents page is not listed as an entry on the table of contents.

Contents

Abstract ii

List of Illustrations iv

1. **Executive Summary: Recommending a
    Method of Computing the Effects of Inflation** 1

2. **Detailed Discussion** 2

    2.1 *Users of Financial Statements* 2

         2.1.1 Stockholders and Investors 2

         2.1.2 Creditors 3

         2.1.3 Managers 3

    2.2 *Criteria of an Inflation-Accounting System* 4

         2.2.1 Convenience 4

         2.2.2 Accuracy 5

         2.2.3 Sensitivity 5

    2.3 *Advantages and Disadvantages
    of the Current Value Method* 6

         2.3.1 Advantages of the Current Value Method 7

         2.3.2 Disadvantages of the Current Value Method 8

    2.4 *Advantages and Disadvantages of
    the General Purchasing Power (GPP) Method* 10

         2.4.1 Advantages of the GPP Method 11

         2.4.2 Disadvantages of the GPP Method 14

    2.5 *Illustrations of the GPP Method* 15

         2.5.1 Longstreet Electric, Inc. 15

         2.5.2 Consolidated Mining Co. 17

    2.6 *Opinions of Professional Organizations* 19

         2.6.1 FASB 19

         2.6.2 AICPA 21

    2.7 *Implementing the GPP Method* 24

         2.7.1 Planning the GPP Method 25

         2.7.2 Starting Up the GPP Method 27

         2.7.3 Evaluating the GPP Method 29

References 31

Appendices 33

    A. *Text of the FASB Opinion* 33

    B. *Text of the AICPA Opinion* 39

FIGURE 10.6

Effective Table of
Contents

- Front matter is numbered with lowercase roman numerals (i, ii, and so
  forth). The title page of a document is not numbered, although it represents page i. The abstract is numbered page ii. The table of contents is
  usually not numbered, although it represents page iii. The roman numerals

are often centered at the bottom of the page. The body of the document is numbered with arabic numerals (1, 2, and so on), usually in the upper outside corner of the page. See Chapter 11 for more information on pagination.

The List of Illustrations

A *list of illustrations* is a table of contents for the figures and tables of a document. (See Chapter 12 for a discussion of figures and tables.) If the document contains figures but not tables, the list is called a *list of figures*. If the document contains tables but not figures, the list is called a *list of tables*. If the document contains both figures and tables, figures are listed separately, before the list of tables, and the two lists together are called a *list of illustrations*.

Some writers begin the list of illustrations on the same page as the table of contents; others prefer a separate page. If the list of illustrations begins on a separate page, it is listed in the table of contents. Figure 10.7 (U.S. Congress, 1995, p. 214) on page 280 shows a list of illustrations.

The Executive Summary

The *executive summary* (sometimes called the *epitome*, the *executive overview*, the *management summary*, or the *management overview*) is a one- or two-page condensation of the document. Its audience is managers, who rely on executive summaries to cope with the tremendous amount of paper crossing their desks every day. Managers do not need or want a detailed and deep understanding of the various projects undertaken in their organizations; this kind of understanding would be impossible for them because of limitations in time and knowledge.

What managers *do* need is a broad understanding of the projects and how they fit into a coherent whole. Consequently, a one-page (double-spaced) maximum for the executive summary has become an unwritten standard for documents of under 20 pages. For longer documents, the maximum length is often calculated as a percentage of the document, such as 5 percent.

The special needs of managers dictate a two-part structure for the executive summary:

- *Background.* Because managers are not necessarily technically competent in the writer's field, the background of the project is discussed clearly. The specific problem or opportunity is stated explicitly—what was not working, or not working effectively or efficiently; or what potential modification of a procedure or product had to be analyzed.
- *Major findings and implications.* Because managers are not interested in the details of the project, the methods rarely receive more than one or two sentences. The conclusions and recommendations, however, are discussed in a full paragraph.

For instance, if the research-and-development division at an automobile manufacturer has created a composite material that can replace steel in engine

Figures

Chapter 1

1-1: Information Technology Applications Currently Being Adopted 13

Chapter 2

2-1: Health Information Pathways to Hospital Information Systems 17
2-2: Health Information on the Internet 18
2-3: Messaging Standards for Clinical Institutions 22

Chapter 3

3-1: Health-Care Industry Trading Relationships 29
3-2: Payment-Related Transactions 34
3-3: High-Level Architecture of a Community
     Health Information Network 35
3-4: Common User Interface 39
3-5: Federated Database Approach for a Community Health
     Information Network 46

Chapter 4

4-1: Algorithm Depicting a Clinical Practice Guideline
     for Evaluation and Care of Patients with Heart Failure 54

Chapter 5

5-1: A Telemedicine Health-Care Network 62
5-2: East Carolina University Telemedicine Network 64
5-3: Rural Area Networks 65

Tables

Chapter 2

2-1: Key Information Technologies for Health Care 19
2-2: Storage Requirements for Imaging Techniques 20

Chapter 3

3-1: Estimates of Cost Savings for Payers, Providers,
     and Employers through the Use of Information Technology 32

Chapter 5

5-1: Summary of Telemedicine Application Categories 61

FIGURE 10.7

List of Illustrations

components, the technical details of the report might deal with the following kinds of questions:

- How was the composite devised?
- What are its chemical and mechanical structures?
- What are its properties?

The managerial implications, however, involve other kinds of questions:

- Why is this composite better than steel?
- How much do the raw materials cost? Are they readily available?
- How difficult is it to make the composite?
- Are there physical limitations to the amount we can make?
- Is the composite sufficiently different from similar materials to prevent legal problems?
- Does the composite have other possible uses in cars?

Executives are less concerned about chemistry than about how this project can help them make a better automobile for less money.

In drafting an executive summary, follow these six guidelines:

- *Describe the background specifically.* In describing the problem or the opportunity, provide specific evidence. For most managers, the best evidence includes costs and savings. Instead of writing that the equipment we are now using to cut metal foil is ineffective, write that the equipment jams on the average of once every 72 hours, and that every time it jams we lose $400 in materials and $2,000 in productivity because the workers have to stop the production line. Then add up these figures for a monthly or annual total.

- *In describing opportunities you researched, use the same strategy.* Your company uses thermostats to control the heating and air conditioning. Research suggests that if you had a computerized energy-management system you could cut your energy costs by 20 to 25 percent. If your energy costs last year were $300,000, you could save $60,000 to $75,000. With these figures in mind, readers have a good understanding of what motivated the study.

- *Describe the methods briefly.* In most cases, your principal reader does not care how you did what you did. He or she assumes you did it competently and professionally. However, if you think your reader is interested, include a brief description—no more than a sentence or two.

- *Describe the findings in accordance with your readers' needs.* If your readers understand your subject sufficiently and want to know your principal results, provide them. Sometimes, however, your readers are not able to understand the technical data or are not interested. If so, go directly to the conclusions—the inferences you draw from the data. Most managers want your recommendations.

- *After drafting the executive summary, have someone review it.* Give it to someone who has had nothing to do with the project—preferably someone outside the field. That person should be able to read your summary and understand what the project means to the organization.

- *Decide how to integrate the executive summary with the body of the document.* The current practice in business and industry is to place it before the body. To highlight the executive summary further, writers commonly make it equal in importance to the entire detailed discussion. They signal this strategy in the table of contents, which shows the document divided into two units: the executive summary and the detailed discussion.

Figure 10.8 shows an effective executive summary.

Executive Summary

Presently, we monitor our distribution system using after-the-fact indicators such as interruption reports, meter readings, and trouble alarms. This system is inadequate in two respects:

- It fails to give us an accurate picture of the dynamics of the distribution system. To ensure enough energy for our customers, we must overproduce. Last year we overproduced by 7 percent, for a loss of $273,000.

- It is expensive. Escalating labor costs for meter readers and the increased number of difficult-to-access residences have led to higher costs. Last year we spent $960,000 reading the meters of 12,000 such residences.

This report describes a project to design a radio-based system for a pilot project on these 12,000 homes.

The basic system, which uses packet-switching technology, consists of a base unit (built around a personal computer), a radio link, and a remote unit.

The radio-based distribution monitoring system described in this report is feasible because it is small enough to replace the existing meters and because it is simple to use. It would provide a more accurate picture of our distribution system, and it would pay for itself in 3.9 years. We recommend installing the system on a trial basis. If the trial program proves successful, radio-based distribution-monitoring techniques will provide the best long-term solution to the current problems of inaccurate and expensive data collection.

FIGURE 10.8

Executive Summary

Notice the differences between this executive summary and the informative abstract (Figure 10.5). The abstract focuses on the technical subject: whether the new radio-based system can effectively monitor the energy usage. The executive summary concentrates on whether the system can improve operations *at this company.* The executive summary describes the symptoms of the problem in financial terms.

After a one-sentence paragraph describing the system design—the results of the study—the writer describes the findings in a final paragraph. Notice that this last paragraph clarifies how the pilot program relates to the overall problem described in the first paragraph.

APA and MLA Documentation of Sources

Documentation of sources is the process of explicitly identifying the sources of the ideas and quotations in your document. Documentation serves three basic functions:

- *To help you acknowledge your debt to your sources.* Complete and accurate documentation is primarily a professional obligation—a matter of ethics. Failure to document a source, whether intentional or unintentional, is plagiarism. At most universities and colleges, plagiarism means automatic failure of the course and, in some instances, suspension or dismissal. In many companies, it is grounds for immediate dismissal.
- *To help you establish credibility.* Effective documentation helps you place your document within the general context of continuing research and to define it as a responsible contribution to knowledge in the field. Knowing how to use existing research is a mark of a professional.
- *To help your readers find the source you have relied on in case they want to read more about a particular subject.*

Three kinds of material should be cited:

- *Any quotation from a written source or an interview, even if it is only a few words.*
- *A paraphrased idea, concept, or opinion gathered from your reading.* There is one exception to this rule: an idea or concept so well known that it has become, in effect, general knowledge, such as Einstein's theory of relativity, needs no citation. If you are unsure whether an item is general knowledge, document it, just to be safe.
- *Any graphic from a written or electronic source.*

Many organizations have their own documentation style; others use published style guides, such as the *U.S. Government Printing Office Style Manual,* the American Chemical Society's *Handbook for Authors,* or the *Chicago Manual of Style.* Find out what your organization's style is and abide by it.

Two documentation systems are shown here:

- *Publication Manual of the American Psychological Association,* 4th ed. (1994). This system, often referred to as APA, is used widely in the social sciences.
- *MLA Handbook for Writers of Research Papers,* 4th ed. (Gibaldi, 1995). This system, from the Modern Language Association, is used widely in the humanities.

Naturally, the discussion of these two systems is not comprehensive. Consult the APA or MLA manual for detailed information on documenting the many kinds of sources you use. And check with your instructor to see which documentation system to use in your documents for class.

APA Style Like all documentation systems, the APA style consists of two elements: the citation in the text and the references at the end of the document.

Textual Citations

In the APA style, the citation is a parenthetical notation that includes the name of the source's author and the date of its publication immediately following the quoted or paraphrased material:

This phenomenon was identified as early as 50 years ago (Wilkinson, 1948).

But if your sentence already includes the source's name, do not repeat it in the parenthetical notation:

Wilkinson (1948) identified this phenomenon as early as 50 years ago.

If the author is an organization rather than a person, use the name of the organization:

The causes of narcolepsy are discussed in a recent booklet (Association of Sleep Disorders, 1994).

You can refer to two sources in one citation. When you do, present the sources in alphabetical order:

This phenomenon was identified as early as 50 years ago (Betts, 1949; Wilkinson, 1948).

Use first initials if two or more sources have the same last name:

This phenomenon was identified as early as 50 years ago (B. Wilkinson, 1948; R. Wilkinson, 1949).

Sometimes, particularly if the reference is to a specific fact, idea, or quotation, the page(s) from the source is also listed:

(Wilkinson, 1948, p. 36)

If two or more sources by the same author in the same year are listed in the references, add the letters *a, b, c,* and so on after the date of each to prevent confusion. (In your references list, these works are alphabetized first by year—starting with the earliest year—and then by title.)

(Wilkinson, 1948a, pp. 36–37)

The second source would be identified similarly:

(Wilkinson, 1948b, p. 11)

For a source written by two authors, cite both names. Use an ampersand in the citation itself, but the word *and* in regular text:

(Allman & Jones, 1987)

As Allman and Jones (1987) suggested, . . .

For a source written by three, four, or five authors, cite all authors the first time you use the reference; after that, include only the last name of the first author followed by *et al.*

FOR THE FIRST REFERENCE:

Bradley, Edmunds, and Soto (1995) argued . . .

FOR SUBSEQUENT REFERENCES:

Bradley et al. (1995) found . . .

For a source written by six or more authors, use the first author's name followed by *et al.*

(Smith et al., 1987)

The Reference List

A *reference list* provides the information your readers will need to find each source you have used. Note that a reference list includes only those items that you actually used in researching and preparing your document. Do not include sources used in your background reading. In addition, do not list sources (such as personal interviews, phone calls, e-mail, and messages from Usenet newsgroups and bulletin boards) that, in the APA's phrase (p. 174), "do not provide recoverable data." Cite such sources in your text, as follows:

C. Connors (personal communication, October 6, 1997) argued that . . .

The individual entries in the references list are arranged alphabetically by author (and then by date if two or more works by the same author are listed). Works by an organization are alphabetized by the first significant word in the name of the organization; works signed "Anonymous" are listed alphabetically by the word "Anonymous." Where several works by the same author are included, they are arranged by date, beginning with the earliest.

APA style calls for the references to be typed double spaced. In addition, the first line of each reference is indented five to seven spaces. Paragraph indentations are helpful for journal editors who will typeset the article. However, APA style permits students to use a hanging indent, with second and subsequent lines indented five to seven spaces. Check with your instructor to see which format you should use.

PARAGRAPH INDENT:

Chapman, D. L. (1995, June 12). Detroit makes a big comeback. *Motorist's Metronome,* 12, 17–26.

HANGING INDENT:

Chapman, D. L. (1995, June 12). Detroit makes a big comeback. *Motorist's Metro-nome,* 12, pp. 17–26.

Following are standard reference forms used with various types of sources. Note that this is only a partial list; you might need to cite many more types of sources.

The title of the book, italicized, or underlined if italics are not available. Initial uppercase letters are used only for the first word of the title and of the subtitle, as well as any proper names (such as Three Mile Island).

The author's surname is followed by the first initial or initials only.

1. A Book

Cunningham, W. (1980). *Crisis at Three Mile Island: The aftermath of a near meltdown.*

The date of publication, in parentheses followed by a period.

New York: Madison.

The location and name of the publisher come last. Note that publishers' names are often abbreviated. For instance, Prentice Hall, Inc., is often abbreviated as Prentice; consult your style guide. Include the name of the publisher's city and the state or country unless the city is well known (such as New York, Boston, or London).

2. A Book by Two or More Authors

Note the use of the comma after the first author's name.

Note the use of the ampersand (&) in place of the word *and.*

Bingham, C., & Withers, S. (1997). *Neural networks and fuzzy logic.* New York: IEEE.

3. A Book Issued by an Organization

The organization takes the position of the author. Although APA style calls for the names of organizations to be written out in full, the U.S. Department of Energy uses "U.S.," not "United States."

U.S. Department of Energy. (1996). *The energy situation in the next decade.* Technical Publication 11346-53. Washington, DC: U.S. Government Printing Office.

Note that the report number follows the title of the report.

4. A Book Compiled by an Editor or Issued under an Editor's Name

Notice that a space separates the two initials of the editor: D. E. Morgan.

Morgan, D. E. (Ed.). (1995). *Readings in alternative energies.* Boston: Smith-Howell.

The abbreviation for *editors* is (Eds.).

5. A Book in an Edition Other Than the First

Schonberg, N. (1997). *Solid state physics.* (3rd ed.). London: Paragon.

6. An Article Included in a Book

The name of the article is presented without extra punctuation or highlighting devices such as italics.

The editor's initials and name—unlike those of the author or authors—are presented in the normal order.

May, B., & Deacon, J. (1995). Amplification systems. In A. Kooper (Ed.), *Advances in electronics* (pp. 101–114). Miami, FL: Valley Press.

7. A Journal Article

The title of the journal is italicized or underlined.

The first letters of important words are capitalized.

Hastings, W. (1997). The space shuttle debate. *The Modern Inquirer, 13,* 311–318.

The volume number of the journal is given in italics.

The page numbers of the article

8. An Article in a Proceedings

In the phrase "45th International Technical Communication Conference," initial letters are uppercase because it is a proper noun.

Carlson, C. T. (1995). Advanced organizers in manuals. In K. Rainey (Ed.), *Proceedings of the 45th International Technical Communication Conference* (pp. RT56–58). Fairfax, VA: Society for Technical Communication.

9. An Unsigned Journal Article

If the author of a journal article is not indicated, alphabetize it by title. If the title begins with a grammatical article such as *the* or *a*, alphabetize it under the first word following the article (in this case, *state*).

The state of the art in microcomputers. (1995, Fall). *Newscene, 56,* 406–421.

If the journal issue is identified by a phrase such as "Fall," that phrase is included in the date.

10. A Newspaper Article

The specific date is added after the year.

The abbreviation *p.* or *pp.* is used for newspapers.

Eberstadt, A. (1995, July 31). Carpal tunnel syndrome. *Morristown Mirror and Telegraph,* p. B3.

The page number "B3" refers to page 3 of section B.

11. A Technical Report

Identifying numbers are included for technical reports. ——————

Birnest, A. J., & Hill, G. (1996). *Early identification of children with ATD* (Report No. 43-8759). State College, PA: Pennsylvania State University College of Education. (ERIC Document Reproduction Service No. ED 186 389).

——— The name of the service used to locate the item: in this case, the ERIC system.

12. World Wide Web Site

The citation styles for Web sites and the other forms of electronic communication are based on the styles used in Andrew Harnack and Eugene Kleppinger's *Online!: A reference guide to using Internet sources* (St. Martin's Press, 1997). Consult this book for a more detailed discussion of citation consistent with APA style.

The author of the Web site ———   ┌—The name of the page   ┌—The name of the site

Hendl, K. B. (n.d.). Internet resources for nursing students. *The Nursing Page.*

<http://www.carney.edu/nursing/index.html> (1997, October 21).

The URL—the address—of the site ———   └—The date you viewed the site

13. E-mail Message

Although APA style does not call for citing e-mail because it is not recoverable, here is a format that you can use if you want to cite it.

The author of the e-mail and his or her address ———   The date the mail was sent   ┌—The subject line of the mail

Nelworth, K. C. <knelwo@ttu.edu> (1997, May 8). Flat-panel displays [Personal mail].

(1997, May 10).

——— The date you accessed the mail

14. Listserv or Newsgroup Message

The author of the e-mail and his or her address ———   The date the mail was sent   ┌—The subject line of the post

Rajiv, C. V. <rajiv@comdex.com> (1997, May 25). Portable document formats.

The address of the listserv or newsgroup ———   <techwr-l@listserv.okstate.edu> (1997, June 10).

——— The date you accessed the post

15. Synchronous Communication

Synchronous discussions include MOOs (multi-user domains, object-oriented), MUDs (multi-user domains), and IRCs (Internet relay chats).

The name of the speaker or the name of the site ———   The date of the discussion   ┌—The type of event

Prisley, L. V. (1997, October 5). [Group discussion]. Telnet moo.ku.org/port=9999

(1997, November 19).

——— The address of the site hosting the conversation

The date you accessed the information ———

16. Telnet Site

The author of the site ——————— The date of publication ——— The title of the document   The title of the full work

Center for Marine Biology. (1997, December 1). Scary sea creatures. *CSU Net*. telnet

freenet.castleton.ca login as anon, go scc, press 2 (1997, December 21).

The complete telnet address, ———————————   └ Directions for accessing   └ The date of   └ The word
with no closing punctuation   the document   access   *telnet*

17. FTP Site

The date of publication ———   The title of the document   ┌ The abbreviation *ftp*

The author's name ————————— Wislow, A. B. (1997, January 10). Interplanetary travel. ftp http://www.nasa.gov

Flight/Documents/Interp.txt (1997, February 23).   └————— Address of the FTP site

The full path to the document ————————┘   The date of access

18. Computer Software

The author (if identified) ——————— Block, K. (1997). Planner (Release 3.1) [Computer software]. New York: Global Software.
of the software

(IBM-PC).

The program is identified as computer software. ┐   The location and name of the publisher

The name of the program ——————— Tools for Drafting (Version 2.3) [Computer software]. (1996). San Jose, CA: Software
is not italicized.

International. (Macintosh version).

└———————— Additional identifying information, such as the kind of
system the software runs on, is added in parentheses.

Figure 10.9 on page 290 is a sample reference list using the APA citation system.

MLA Style   Like the APA documentation system, MLA style consists of two elements: the citation in the text and the references—called a *list of works cited*—at the end of the document.

Textual Citations

In MLA style, the citation is a parenthetical notation that includes the name of the source's author immediately following the quoted or paraphrased material. If you are referring to a specific fact, idea, or quotation, include the page(s) from the source:

This phenomenon was identified as early as 50 years ago (Wilkinson 134).

Here the citation shows that you are referring to page 134 of the Wilkinson source. If you are referring to the whole Wilkinson source, not to a particular page or pages, use only the name:

This phenomenon was identified as early as 50 years ago (Wilkinson).

References

Andress, K. (1997, July 12). More on the Scanlon Plan [36 paragraphs]. *Online Journal of Accounting Studies* [On-line serial]. Available: Doc. No. 113.

Daly, P. H. (1996). Selecting and designing a group insurance plan. *Personnel Journal, 54,* 322–323.

Flanders, A. (1993). Measured daywork and collective bargaining. *British Journal of Industrial Relations, 9,* 368–392.

Goodman, R. K., Wakely, J., & Ruh, K. (1996). What employees think of the Scanlon Plan. *Personnel, 6,* 22–29.

Trencher, P., & Coughlin, C. (1993). *Recent trends in labor-management relations.* New York: Westly.

Zwicker, D. (1995, August 13). More on the Scanlon Plan: A response. *Insurance Fundamentals.* <http://www.swisu.edu/ec201/index.html> (1997, October 6).

FIGURE 10.9

References in APA Style

If your sentence already includes the source's name, do not repeat it in a parenthetical notation:

> Wilkinson identified this phenomenon as early as 50 years ago (134).

If the author is an organization rather than a person, use the name of the organization:

> The causes of narcolepsy are discussed in a recent booklet (Association of Sleep Disorders).

You can incorporate the name of the organization into the sentence:

> The Association of Sleep Disorders discusses the causes of narcolepsy in a recent booklet.

You can refer to two or more sources in one citation:

> This phenomenon was identified as early as 50 years ago (Betts 34; Wilkinson).

Use first initials if two or more sources have the same last name:

This phenomenon was identified as early as 50 years ago (B. Wilkinson; R. Wilkinson).

If you cite two or more sources by the same author in the list of works cited, add a brief form of the title after the author's name to prevent confusion:

(Wilkinson, *Particle Physics* 36–37)

(Wilkinson, "Ascent" 11)

For a source written by two or three authors, cite all names:

(Allman and Jones)

(Finn, Crenshaw, and Zander)

For a source written by four or more authors, you may list all the authors or only the first author followed by *et al.*

Bradley, Edmunds, Stipe, and Soto argued . . .

Bradley et al. argued . . .

The List of Works Cited

A list of works cited provides the information your readers will need to find each source you used. Note that the reference list includes only those items that you actually used in researching and preparing your document and that you cite in your document.

Alphabetize the entries in the list of works cited by author (and then by title if two or more works by the same author are listed). Works by an organization are alphabetized by the first significant word in the name of the organization.

Type the entries double spaced, with second and subsequent lines indented one-half inch. Leave one space between each item in the entry; for example, leave one space between the name of the author and the title of the work. Leave one space after a comma or colon.

Following are standard reference forms for various types of sources. Note that this is only a partial list; you might need to cite many more types of sources.

1. A Book

The author's full name is given, in reverse order.

The title of the book, italicized. All important words in the title are capitalized.

Cunningham, Walter. *Crisis at Three Mile Island: The Aftermath of a Near Meltdown.* New York: Madison, 1980.

The location, name, and date of the publisher come last. Note that publishers' names are often abbreviated. For instance, Prentice Hall, Inc., is often abbreviated as Prentice; consult your style guide.

2. A Book by Two or More Authors

For a book by two or three authors, present the names in the sequence in which they appear on the title page.

Only the name of the first author is presented in reverse order.

Bingham, Christine, and Stephen Withers. *Neural Networks and Fuzzy Logic.* New York: IEEE, 1995.

A comma separates the names of the authors.

For a book by four or more authors, either present all the names or use *et al.*

Foster, Glenn, et al. *The American Renaissance.* Binghamton, NY: Archive, 1995.

3. Two Books by the Same Author

For second and subsequent entries, use three hyphens followed by a period. Arrange the entries alphabetically by title of the work.

Smith, Louis. *International Standards.* New York: IEEE, 1997.

- - -. *Wave-Propagation Technologies.* Berkeley: Stallings, 1995.

4. A Book Issued by an Organization

The organization takes the position of the author.

National Commission on Corrections. *The Future of Incarceration.* Publication 11346-53. St. Louis, MO: Liberty, 1997.

5. A Book Compiled by an Editor or Issued under an Editor's Name

The abbreviation for editors is "eds."

Morgan, Donald E., ed. *Readings in Alternative Energies.* Boston: Smith-Howell, 1995.

6. A Book in an Edition Other Than the First

Schonberg, Nathan. *Solid State Physics.* 3rd ed. London: Paragon, 1996.

7. A Quotation of a Quotation

The title of Hanson's original article

The page numbers of Hanson's article

The source being quoted in the Ortiz book

Hanson, Alfred. "Multimedia for Tomorrow." *Technology Advances* 3 June 1995: 36–39, qtd. in Ortiz, Manuel. *Multimedia Systems.* Boston: Allied, 1996.

The book in which Ortiz quotes Hanson

In the parenthetical citation, give the pages in Ortiz's book on which Hanson is quoted.

According to Hanson (211), multimedia will be the dominant mode of instruction in colleges by 2005.

8. Magazine Article

Use the first three letters to abbreviate a month, but write out *May*, *June*, and *July*.

Newman, Daniel. "Passive Restraint Systems." *Car Lore* 12 Dec. 1995: 41+.

If the article has no author listed, alphabetize it by title:

"Passive Restraint Systems." *Car Lore* 12 Dec. 1995: 41+.—The plus sign signifies that the article begins on page 41 and then continues on subsequent pages interrupted by other articles or advertisements.

9. A Newspaper Article

If the newspaper has more than one edition, cite the edition.

Felder, Melissa. "Smokeless Tobacco: New Danger Signs." *New York Post* 4 May 1997, morning ed.: 13.

10. An Article Included in a Book

The name of the article is enclosed in quotation marks.

May, Bruce, and James Deacon. "Amplification Systems." *Advances in Electronics*. Ed. Alvin Kooper. Miami, FL: Valley Press, 1997.

11. An Article in a Journal with Continuous Pagination

The title of the journal is italicized.

The pages on which the article appears

Hastings, Wendy. "The Space Shuttle Debate." *The Modern Inquirer* 13 (1996): 311–18.

The volume of the journal in which the article appears

The year of publication

12. An Article in a Journal That Pages Each Issue Separately

The title of the journal is italicized.

The year of publication

The pages on which the article appears

Juneja, Gupta. "Asynchronous Transmission Techniques." *Video Quarterly* 6.2 (1997): 11–19.

The volume of the journal in which the article appears

The issue of the journal in which the article appears

13. An Article in a Proceedings

Carlson, Carl T. "Advanced Organizers in Manuals." *Proceedings of the 45th International Technical Communication Conference*. Ed. Ken Rainey. Fairfax, VA: Society for Technical Communication, 1995: RT56-58.

14. An Unsigned Journal Article

If the author of a journal article is not indicated, alphabetize it by title. If the title begins with an article (*the* or *a*), alphabetize it under the first word following the article (in this case, *state*).

"The State of the Art in Microcomputers." *Newscene* 56 (1997): 406–21.

15. An Interview That You Conducted

Gangloff, Richard. Personal interview. 24 Jan. 1995.

Abbreviate all months except May, June, and July.

16. A Lecture

Robbins, Bruce. "Trends in Secondary Education." Lecture in E382. Boise State University. 2 May 1997.

17. World Wide Web Site

The citation styles for Web sites and other forms of electronic communication are based on the styles used in Andrew Harnack and Eugene Kleppinger's *Online!: A reference guide to using Internet sources* (St. Martin's Press, 1997). Consult this book for a more detailed discussion of citation consistent with MLA style.

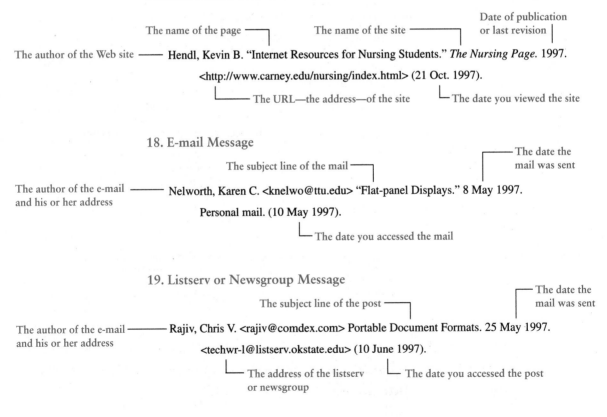

The author of the Web site — The name of the page — The name of the site — Date of publication or last revision

Hendl, Kevin B. "Internet Resources for Nursing Students." *The Nursing Page.* 1997.
<http://www.carney.edu/nursing/index.html> (21 Oct. 1997).

The URL—the address—of the site | The date you viewed the site

18. E-mail Message

The author of the e-mail and his or her address — The subject line of the mail — The date the mail was sent

Nelworth, Karen C. <knelwo@ttu.edu> "Flat-panel Displays." 8 May 1997.
Personal mail. (10 May 1997).

The date you accessed the mail

19. Listserv or Newsgroup Message

The author of the e-mail and his or her address — The subject line of the post — The date the mail was sent

Rajiv, Chris V. <rajiv@comdex.com> Portable Document Formats. 25 May 1997.
<techwr-1@listserv.okstate.edu> (10 June 1997).

The address of the listserv or newsgroup | The date you accessed the post

20. Synchronous Communication

Synchronous discussions include MOOs, MUDs, and IRCs.

The date of the discussion ── ┐   ┌ The type of event

The name of the speaker ──── Prisley, Lauren. (1997, Oct. 5). "Seminar Discussion on FTP." 5 Oct. 1997. telnet
or the name of the site

moo.ku.org/port=9999 (19 Nov. 1997). ──── The date you accessed the information

└── The address of the site hosting the conversation

21. Telnet Site

The title of ── The date of
the full work ┐   publication ┐

The title of the document ── ┐

The author of the site ──── Center for Marine Biology. "Scary Sea Creatures." *CSU Net.* (1 Dec. 1997). telnet

freenet.castleton.ca login as anon, go scc, press 2 (21 Dec. 1997).

The complete telnet address, ──┘   └ Directions for accessing   └ The date of   └ The word
with no closing punctuation   the document   access   *telnet*

22. FTP Site

The date of
publication ┐   ┌ Address of the *ftp* site

The title of the document ── ┐

The author's name ──── Wislow, Ann B. "Interplanetary Travel." 10 Jan. 1997. ftp http://www.nasa.gov

Flight/Documents/Interp.txt (23 Feb. 1997).   └ The abbreviation *ftp*

The full path to the document ──┘   └ The date of access

23. Computer Software

The distributor ──┐
of the software

The author (if identified) ──── Block, Karen. *Planner.* Version 3.1. Computer software. Global Software, 1997. IBM-PC.
of the software

At the end of the entry add any ──┘
other relevant information,
such as the operating system on
which the software runs.

Figure 10.10 on page 296 is a sample list of works cited using the MLA citation system.

Writing the Back Matter

The back matter usually consists of several or all of the four components shown in Figure 10.11 on page 297.

Works Cited

Daly, Peter H. "Selecting and Designing a Group Insurance Plan." *Personnel Journal* 54 (Nov. 1996): 322–23.

Flanders, Andrew. "Measured Daywork and Collective Bargaining." *British Journal of Industrial Relations* 9.4 (1997): 368–92.

Goodman, Raymond K., Janice Wakely, and Kathleen Ruh. "What Employees Think of the Scanlon Plan." *Personnel* 6 (1995): 22–29.

Trencher, Patricia, and Christopher Coughlin. *Recent Trends in Labor-Management Relations.* New York: Westly, 1995.

Zwicker, David. "More on the Scanlon Plan: A Response." *Insurance Fundamentals.* <http://www.swisu.edu/ec201/index.html> (6 Oct. 1997).

FIGURE 10.10

References in MLA Style

The Glossary and List of Symbols

A *glossary* is an alphabetical list of definitions. It is particularly useful if your audience includes readers unfamiliar with the technical vocabulary in your document.

Instead of slowing down the detailed discussion by defining technical terms as they appear, you can use boldface or another method of highlighting words to inform your readers that the term is defined in the glossary. A footnote at the bottom of the page on which the first boldfaced term appears serves to clarify this system for readers. For example, the first use of a term defined in the glossary might occur in the following sentence in the detailed discussion of the document: "Thus the **positron*** acts as the. . . ." At the bottom of the page, add a comment such as the following:

*This and all subsequent terms in boldface are defined in the Glossary, page 26.

Although the glossary is usually placed near the end of the document, right before the appendices, it can also be placed right after the table of contents. This placement is appropriate if the glossary is brief (less than a page) and defines terms that are essential for managers likely to read the body of the document. Figure 10.12 on page 298, based on Ruiu (1994), provides an example of a glossary.

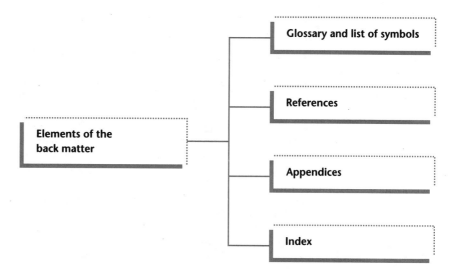

FIGURE 10.11

Elements of the Back
Matter

A *list of symbols* is structured like a glossary, but it defines not words and phrases, but the symbols and abbreviations used in the document. Don't hesitate to include a list of symbols if some of your readers might not know what your symbols and abbreviations mean or might misinterpret them. Like the glossary, the list of symbols may be placed before the appendices or after the table of contents. Figure 10.13 on page 299 shows a list of symbols (in this case, including abbreviations).

References   As explained earlier in this chapter, documentation of sources consists of textual citations and a list of works cited (in the MLA system) or a reference list (in the APA system).

Appendices   An *appendix* is any section that follows the body of the document (and the list of references or bibliography, glossary, or list of symbols). Appendices (or *appendixes*) conveniently convey information that is too bulky to be presented in the body or that will interest only a few readers. Appendices might include maps, large technical diagrams or charts, computations, computer printouts, test data, and texts of supporting documents.

Appendices, which are usually lettered rather than numbered (Appendix A, Appendix B, and so on), are listed in the table of contents and are referred to at the appropriate points in the body of the document. Therefore, they are accessible to any reader who wants to consult them.

Remember that an appendix including or consisting of a graphic is titled "Appendix," not "Figure" or "Table," even if it would have been so designated had it appeared in the body.

> **Glossary**
>
> **ISDN:** integrated-services digital network. A completely digital public telecommunications network that can accommodate both voice and data traffic.
>
> **Sonet:** synchronous optical network. A family of interfaces that defines the transportation and management of digital traffic on an optical-fiber network in North America. Outside North America, a similar system, called synchronous digital hierarchy (SDH), is used.
>
> **Switching fabric:** the circuitry in a communications system that routes transmissions from their source to their destination.
>
> **Virtual channel:** a defined connection through a network in which bandwidth is not allocated permanently, but only when a user's cells require transport.

FIGURE 10.12

Glossary

The Index An *index* is an alphabetical list of the major items—topics, ideas, functions, and tasks—described and explained in a document. For long documents—100 pages or longer—the index is the most important accessing tool. Readers consult indexes to locate specific pieces of information; when they can't find what they are looking for, or when it takes them a long time to search for it, they become confused and frustrated. According to Bonura (1994, p. 4), the most common complaint about technical documents is that the index is poor or that the document lacks one altogether.

Who creates an index? In an organization with only a few technical communicators, the author of a document often creates its index. In larger technical-publications departments, there might be a specialist who does only indexes, or a technical editor might be responsible for indexing. Organizations can also turn to freelance indexers.

Each approach has its advantages and disadvantages. Writers who index their own documents have a deeper understanding of the subject and might be able to work faster than anyone else, but they also lack the objectivity of an outsider. For instance, an author/indexer might use the term *compact disc* throughout the document and the index, without realizing that many readers will be looking for the term *CD-ROM*. If this omission is not noticed and corrected, the index will be less useful than it should be.

List of Symbols

β	beta
CRT	cathode-ray tube
γ	gamma
Hz	hertz
rcvr	receiver
SNR	signal-to-noise ratio
uhf	ultra high frequency
vhf	very high frequency

FIGURE 10.13

List of Symbols

Characteristics of an Effective Index

An effective index is accurate, comprehensive, precise, and easy to use.

- *Accuracy.* Because readers turn to the index to find where a topic is discussed, the page numbers listed must be accurate.

- *Comprehensiveness.* The index should include an entry for every term the reader is likely to seek. The average index might have three or four items listed for each page of text in the document, but that number is only a rough guide; the actual number might vary from one to more than ten.

- *Precision.* An index that includes entries such as "computers" followed by 30 page numbers is ineffective. Useful indexes have subentries and, when appropriate, sub-subentries. For instance, an index to a manual about a word-processing program should have numerous subentries for *tables*, including *creating, editing,* and *formatting,* and the subentry *formatting* should have sub-subentries such as *justifying, sizing,* and *shading.*

- *Ease of use.* The reader should not have to struggle to find an item. In creating the index, use two kinds of cross-references to help the reader navigate: *see* and *see also.*

  – *See* is used to direct readers to the proper entry, for example: "Drive, *see* disk drive." Use *see* only if the entry—*disk drive*, in this case—contains subentries. Otherwise, it is just as easy to list the page numbers for *drive* as well as for *disk drive.* Remember, your job is to help readers find what they are looking for; so if you can help them find it on the first try, do so, even if their choice of what to call the item is different from yours.

– *See also* is used to direct readers to related entries. For example: "Search-and-replace, 17. *See also* Global searching; Editing text." This entry tells readers that two other entries—one on global searching and one on editing text—contain information related to the search-and-replace function.

Note that *see* and *see also* are usually italicized. An obvious point: readers are not amused when the index does a double reverse: "floppy, *see* floppy disk," then "floppy disk, *see* disk drive." And documents actually go airborne when there is an entry "floppy disk, *see* disk drive" but there is no entry for *disk drive*.

Sometimes it is useful to add an explanatory statement at the start of an index. If, for example, you include names of people in a separate index, or you choose to group all the function keys in an entry called *function keys*, rather than alphabetizing them separately by the name of the individual function, alert readers to your scheme in a brief headnote. When an index covers several documents, especially in software documentation, tell your readers in an explanatory note how you will be abbreviating the names of the different documents.

Planning, Drafting, and Revising an Index

Creating an index is like creating any other part of a document. You have to plan, draft, and revise.

- *Planning*. Think of your audience and purpose. How much do your readers know about the subject? If you are addressing novices, include many entries for basic items. And include numerous cross-references to help them get oriented and find related entries. Think about why they are reading. What kinds of information do they want? As with all kinds of technical communication, think in terms of tasks (what they want to accomplish) in addition to functions (what the item does). If time permits, interview people who fit the profile of the reader.

  In the planning stage, you might also want to create a list of items to index. For instance, if you are indexing a manual for a word-processing program, you might include items such as the following:

  – commands
  – control keys
  – tasks (adding text, deleting text, printing)
  – text attributes (boldface, italics, and so on)

  Such a list is particularly helpful when someone other than the writer is doing the indexing; it helps the indexer understand what the writer thinks is important to include.

- *Drafting.* As with writing, people draft their indexes differently. The most popular method is the oldest: with index cards and a pencil. (Indexing software will be discussed later in this section.) First you read a section, such as a chapter or a part of one, circling the key terms. Don't forget to take advantage of the thinking you have already done: the key terms are likely to appear in chapter titles, section headings, and the topic sentences of the paragraphs.

  After finishing the section, go back and make a card for each entry. Some indexers like to create subentries at this point; others like to wait and then create all the subentries and sub-subentries later, after they have all the page references to the item. Many indexers like to check their page references as soon as they enter them on the card because checking will be much more difficult later. Some indexers like to alphabetize their cards as they go along; others wait because they don't want to break their concentration. Waiting can take longer, however, because you might end up making several different cards for the same entry, which you will have to consolidate later. At the end of this process, all the entries have to be word processed or typed.

  Some writers work directly on the computer, typing in their entries (either in the order in which they occur or alphabetically) during their first pass through the document.

- *Revising.* As you revise the index, check for six different items:

  – Have you entered all the items and are the entries easy to understand? For instance, in revising, you might want to change *location of files* to *files, location of,* to emphasize the key term.

  – Have you used terms consistently throughout the index? If you use *personal computer* in one place, don't use *microcomputer* in another.

  – Do you need to create more subentries or sub-subentries? The *Chicago Manual of Style* (1993, p. 722) recommends creating subentries when an entry contains five or six page numbers.

  – Have you used singular nouns in some places and plural nouns in others? Don't use *printer* in one location and *disk drives* in another; use all singular nouns (*printer, disk drive*) or all plural nouns.

  – Do you need to change the way you indicated page numbers? For instance, in your first draft, you might have the entry *macros, 19–24.* In looking again at those pages, however, you see they should be listed as *19–22, 23, 24,* because the main explanation is really on pages 19–22, and macros are merely mentioned on 23 and 24.

  – Do the page numbers make sense? For example, numbers might be inadvertently transposed (14–12 instead of 12–14) or the hyphen between numbers might be missing (1214 instead of 12–14).

In revising an index, you will find yourself going back and forth a lot; it simply is not possible to get it right the first time.

Computerizing the Indexing Process

If the index is part of a word-processed or typeset document, as it almost certainly will be, isn't there some way to take advantage of the power of the computer? Yes and no.

Indexing software automates the process of finding terms that you designate and then formatting them as entries, subentries, or sub-subentries. These programs let you change the format of the entries easily. Desktop-publishing software contains indexing functions as well, and even word-processing software lets you create an index by locating the items you designate. All indexing functions are basically a variation on the search function of standard word-processing programs.

But finding each reference to an item is not the hard part of indexing. The real challenge is to determine hierarchy (what is an entry and what is a subentry) and to create the *see* and *see also* cross-references. For this reason, indexing will always require an intelligent and informed person.

Guidelines for Indexing

Here are eight practical guidelines for indexing.

- Use either the run-in or indented format.

    Run-in format:

    Tables, 419–429
    converting text to, 423; converting to text, 424; copying, 426; creating, 419; importing spreadsheets to, 423; lines in, 428

    Indented format:

    Tables, 419–429
    converting text to, 423
    converting to text, 424
    copying, 426
    creating, 419
    importing spreadsheets to, 423
    lines in, 428

    If a main entry takes up more than one line in the indented format, indent the second line more than the indent for the subentries:

    Macros, 206–209, 233, 408. *See also* Keyboard map;
        Templates
    chaining of, 208
    definition of, 206

    As you can see, the run-in format uses less space, but the indented form is easier to read.

- Format the index in two columns, in a type size two points smaller than the body text. See Chapter 11 for a discussion of columns and type sizes.
- Alphabetize either letter by letter or word by word. In the *letter-by-letter format,* ignore spaces between words:

box drawing
boxes
box shading

  In the *word-by-word format,* alphabetize all entries that begin with the same word, then go on to the next word:

box drawing
box shading
boxes

- When you use acronyms and abbreviations, consider how most readers will search for the term. For instance, use "CD-ROM," not "Compact Disk-Read Only Memory." But you might want to include the spelled-out version in parentheses:

CD-ROM (Compact Disk-Read Only Memory), 18

  You might also want to include a *see* entry under the spelled-out version.

- Alphabetize symbols and numbers as they are spoken. For instance, "586 microprocessor" is alphabetized under *f.*
- If the document has a glossary, make an entry for each term listed in the glossary. If the term also appears in the body of the document, list the glossary citation as a definition. For instance, if *curly braces* is defined in the glossary on page 234 and mentioned in a discussion of math on page 135 of the body, here is how to make the index entry:

Curly braces
   definition of, 234
   in mathematical equations, 135

- Consider using boldface or italics to indicate the pages on which the main discussion of an item appears:

dot leaders, **171–172,** 211, 306

  Be sure to add a note at the start of the index to explain your use of boldface.

- Find out the appropriate way to express inclusive numbers. The easiest and clearest way is to use all the digits: 289–293. Writing all the digits is important when you use section numbering (page 3-6 is the sixth page in section 3). To express inclusive numbers for a document with section numbering, use the following form: 3-6 to 3-9.

Some organizations abbreviate the second number: 289–93. But abbreviating inclusive numbers can get complicated. For example, most style guides call for writing "100–104," rather than "100–04" or "100–4." Check your style guide to see how your organization expresses inclusive page numbers.

CASE

Comparing Online Indexes

You work in the technical communication group at United Software. Your group is about to begin writing the online documentation for the first version of your company's newest product: a word-processing program. Your supervisor, Andrea Bonnano, wants to be sure that the online documentation is high quality. "Let's take a look at how our competitors do their indexes. Would you and the other members of your group study a couple of word processors and write me a memo about their indexes?" As a group, meet and determine the appropriate criteria by which to describe and evaluate the online indexes. Once you have agreed on criteria, split up and study the indexes. What are the main similarities and differences? Which is more complete? Which is easier to use? What suggestions would you offer to the indexers? Write the memo to Andrea Bonnano, presenting your findings and recommendations on how to index your online documentation. See Chapter 16 for a discussion of memos.

WRITER'S CHECKLIST

1. Does the transmittal letter
   - ❑ clearly state the title and, if necessary, the subject and purpose of the document?
   - ❑ state who authorized or commissioned the document?
   - ❑ briefly state the methods you used?
   - ❑ summarize your major results, conclusions, or recommendations?
   - ❑ acknowledge any assistance you received?
   - ❑ courteously offer further assistance?

2. Does the cover include
   - ❑ the title of the document?
   - ❑ your name and position?
   - ❑ the date of submission?
   - ❑ the company name or logo?

3. Does the title page
   - ❑ include a title that clearly states the subject and purpose of the document?

   - ❑ list the names and positions of both you and your principal reader?
   - ❑ include the date of submission of the document and any other identifying information?

4. Does the abstract
   - ❑ list your name, the document title, and any other identifying information?
   - ❑ clearly define the problem or opportunity that led to the project?
   - ❑ briefly describe (if appropriate) the research methods?
   - ❑ summarize the major results, conclusions, or recommendations?

5. Does the table of contents
   - ❑ clearly identify the executive summary?
   - ❑ contain a sufficiently detailed breakdown of the major sections of the body of the document?
   - ❑ reproduce the headings as they appear in your document?
   - ❑ include page numbers?

6. Does the list of illustrations (or the list of tables or the list of figures) include all the graphics found in the body of the document?

7. Does the executive summary
   - ❏ clearly state the problem or opportunity that led to the project?
   - ❏ explain the major results, conclusions, recommendations, and managerial implications of your document?
   - ❏ avoid technical vocabulary and concepts that the managerial audience is not likely to know?

8. Does the documentation system conform to the guidelines of the APA system, the MLA system, or another appropriate system?

9. Does the glossary include definitions of all the technical terms your readers might not know?

10. Does the list of symbols include all the symbols and abbreviations your readers might not know?

11. Do the appendices include the supporting materials that are too bulky to present in the document body or that will interest only a small number of your readers?

12. Does the index
   - ❏ use either the run-in or indented format?
   - ❏ appear in a two-column format in a type size two points smaller than the body text?
   - ❏ use either letter-by-letter or word-by-word alphabetization?
   - ❏ alphabetize acronyms and abbreviations according to how most readers will search for the term?
   - ❏ alphabetize symbols and numbers as they are spoken?
   - ❏ include an entry for each item in the glossary?
   - ❏ express inclusive numbers appropriately?

EXERCISES

Several of the following exercises call for you to write memos. See Chapter 16 for a discussion of memos.

1. The following letter of transmittal is from a report written by an industrial engineer to his company president. Write a one-paragraph evaluation that focuses on the clarity, comprehensiveness, and tone of the letter.

Dear Mr. Smith:

The enclosed report, "Robot and Machine Tools," discusses the relationship between robots and machine tools.

Although loading and unloading machine tools was one of the first uses for industrial robots, this task has only recently become commonly feasible. Discussed in this report are concepts that are crucial to remember in using robots.

If at any time you need help understanding this report, please let me know.

Sincerely yours,

2. The following informative abstract is from a report by an electrical engineer to her manager. Write a one-paragraph evaluation. How well does the abstract define the problem, methods, and important results, conclusions, and recommendations?

"Design of a New Computer Testing Device"

The modular design of our new computer system warrants the development of a new type of testing device. The term *modular design* indicates that the overall computer system can be broken down into parts or modules, each of which performs a specific function. It would be both difficult and time consuming to test the complete system as a whole, for it consists of sixteen different modules. A more effective testing method would check out each module individually for design or construction errors prior to its installation into the system.

(Continued)

(Continued)

This individual testing process can be accomplished by the use of our newly designed testing device.

The testing device can selectively call or "address" any of the logic modules. To test each module individually, the device can transmit data or command words to the module. Also, the device can display the status or condition of the module on a set of LED displays located on the front panel of the device. In addition, the device has been designed so that it can indicate when the module being tested has produced an error.

3. The following table of contents is from a report titled "An Analysis of Corporations v. Sole Proprietorships." Write a one-paragraph evaluation. How effective is the table of contents in highlighting the executive summary, defining the overall structure of the report, and providing a detailed guide to the location of particular items?

CONTENTS

EXECUTIVE SUMMARY 2
INTRODUCTION 4
ANALYSES 5
I.   START-UP COSTS 5
II.   PERSONAL LIABILITY 6
III. INCOME TAX EFFECTS 7
IV.   FRINGE BENEFITS
       AVAILABLE 10
V.   RETIREMENT OPTIONS 12
VI.   TRANSFER OF OWNERSHIP 14
VII. AVAILABILITY OF FUNDS
       FROM BANKS 15
CONCLUSION 17
LIST OF WORKS CITED 18

4. The following executive summary is from a report titled "Analysis of Large-Scale Municipal Sludge Composting as an Alternative to Ocean Sludge-Dumping." Write a one-paragraph evaluation. How well does the executive summary present concise and useful information to the managerial audience?

Coastal municipalities currently involved with ocean sludge-dumping face a complex and growing sludge management problem. Estimates suggest that treatment plants will have to handle 65 percent more sludge in 1998 than in 1988, or approximately seven thousand additional tons of sludge per day. As the volume of sludge is increasing, traditional disposal methods are encountering severe economic and environmental restrictions. The EPA has banned all ocean sludge-dumping as of next January 1. For these reasons, we are considering sludge composting as a cost-effective sludge management alternative.

Sludge composting is a 21-day biological process in which waste-water sludge is converted into organic fertilizer that is aesthetically acceptable, essentially pathogen-free, and easy to handle. Composted sludge can be used to improve soil structure, increase water retention, and provide nutrients for plant growth. At $150 per dry ton, composting is currently almost three times as expensive as ocean dumping, but effective marketing of the resulting fertilizer could dramatically reduce the difference.

5. Find a document, such as a textbook or a manual, that contains an index. Study a page of the index; then write a memo to your instructor describing and evaluating the index. For instance, what form of alphabetization is used? Are the entries clear and easy to use? How effectively are *see* and *see also* used? Does the index appear comprehensive and accurate? See Chapter 16 for a discussion of memos.

6. Choose a journal article of some 8 to 15 pages and then write an index for it. Attach a note describing the audience and indicating what indexing style you are using. For instance, are you alphabetizing letter-by-letter or word-by-word? What style of pagination are you using? Attach a photocopy of the article.

Collaborative Exercise

7. Form small groups for this project comparing the indexing in an online help system and a printed manual. Have two people in the group study the indexing principles used in the online help system of one piece of software, such as a word-processing program, a spreadsheet, or a presentations-graphics package. Have two other people in the group study the index in a printed manual for the same product. Meet as a group to compare the two indexes. What are the main similarities and differences? Which is more thorough? Which is easier to use? What suggestions would you offer to the indexers? Present your findings in a memo to your instructor.

Web Exercise

8. Organizations that publish style sheets for documentation, such as the Modern Language Association and the American Psychological Association, are trying to keep up with advances in electronic communication tools. Find two sites on the Internet that offer suggestions on how to document electronic sources.

Download the sections that explain how to cite Web sites, e-mail messages, and FTP documents. Compare and contrast the guidelines on these three types of citations. What are the basic differences between the two documentation styles? Overall, which is easier to understand? Which would be more effective in helping the reader find the source? Present your findings in a memo to your instructor.

REFERENCES

Bonura, L. S. (1994). *The art of indexing.* New York: Wiley.

Chicago manual of style (14th ed.). (1993). Chicago: University of Chicago Press.

Crowe, B. (1985). *Design of a radio-based system for distribution automation.* Unpublished manuscript, Drexel University.

Gibaldi, J. (1995). *MLA handbook for writers of research papers* (4th ed.). New York: Modern Language Association.

Publication manual of the American Psychological Association (4th ed.). (1994). Washington, DC: American Psychological Association.

Ruiu, D. (1994). Testing ATM systems. *IEEE Spectrum, 31*(6), 25.

U.S. Congress, Office of Technology Assessment. (1995). *Bringing health care online: The role of information technologies.* OTA-ITC-624. Washington, DC: U.S. Government Printing Office.

Integrating Design Elements

Goals of Document Design and Page Design 309

Planning the Design of the Document and the Page 311
  Analyzing Your Audience's Needs and Expectations 311
  Determining Your Resources 311

Designing the Document 312
  Size 312
  Paper 314
  Bindings 315
  Accessing Tools 316

Designing the Page 319
  Principles of Effective Page Design 320
  Learning Theory and Its Relation to Page Design 321
  Page Grids, White Space, and Columns 323
  Typography 330
  Designing Titles and Headings 337
  Other Design Techniques 338

Cases: The Brochure with the Inconsistent Design 340
          Updating an Old Design 340

Writer's Checklist 340

Exercises 340
      Collaborative Exercise 340
      Web Exercise 343

To a large extent, the effectiveness of a document depends on how well it is designed. Of course, the information it contains is critical, but before your readers actually read and understand a document, they *see* it. In less than a second, the document has started to make an impression on them, and that impression might determine how effectively they read it. (A first impression might even determine *whether* they read it at all.) Of course, your goal is not to entertain your readers with flashy colors and outrageous designs, like those in mass-market magazines. Still, an understanding of effective design principles can help you make a document that your readers will want to read—and that will help them understand and remember what you say.

Design principles are not new, but we are becoming more and more conscious of how powerful they are, largely because of the word processor. Modern word-processing programs are sophisticated design tools that let us use different typefaces and sizes, include graphics, and make multicolumn layouts. More powerful still are desktop-publishing programs, which provide extraordinarily sophisticated design options.

Where does this leave technical communicators and technical professionals? Today's technical communicators are expected to be familiar not only with word processing but also with desktop publishing, for they are likely to be responsible for producing a monthly newsletter, an internal procedures manual, or even a user's guide for customers. Technical professionals such as scientists are not expected (yet) to have desktop-publishing skills and experience, but the more they know about design, the more effective their documents will be.

This chapter covers the basics of designing documents and pages. The chapter first describes the goals of document and page design. It then discusses planning the design of the document and page. Next, the chapter explains how to design a document and how to design the individual pages.

Most of the concepts discussed in this chapter apply to the design of Web pages. For instance, in both paper and electronic media you need to make titles and headings prominent, and you need to use headers and footers. A big difference between the two media, however, is that with paper you control exactly what your reader sees. With Web pages, however, the viewer's hardware and software can greatly influence what appears on the screen. For a discussion of Web page design, see Chapter 4.

Goals of Document Design and Page Design

In designing a document, you have five major goals, as shown in Figure 11.1 on page 310.

- *To make a good impression on readers.* All documents should look professional, if only to show your readers that you take your work seriously. If you are writing to people outside your organization, the document

FIGURE 11.1

Goals of Document
Design

should reflect not only your own professional standards but also those of
your organization.

- *To help readers understand the structure and hierarchy of the information.* Regardless of what kind of document they are reading, from a one-
  page data sheet to a multivolume proposal, your readers should know
  where they are in the document. And they should be able to see the hierarchy of the information; that is, they need to know whether the portion
  they are reading is more important or less important than the portion
  that preceded it. Design helps communicate these relationships.

- *To help readers find the information they need.* One obvious characteristic of technical communication is that, most of the time, people don't
  read documents from cover to cover; they refer to documents to find
  answers to particular questions. Therefore, a well-designed document
  enables readers to find information quickly and easily. A table of contents and an index help in some kinds of documents, but design elements
  such as tabs, icons, and color can be extremely effective too. In addition,
  page design and choice of type are critical to a reader's ability to find
  information.

- *To help readers understand the information.* If you design the document
  and the page carefully, you help readers understand the information

quickly and easily. For instance, if you design a set of instructions so that the text is next to its accompanying graphic, you have made it easier for readers to understand the instructions.

- *To help readers retain the information.* An effective design enables readers to create a visual image of the information, enhancing retention. Design elements such as text boxes and pull quotes help readers remember important explanations and quotations.

Planning the Design of the Document and the Page

The first step in creating a design for any kind of technical document is to plan. Start by analyzing your audience's needs and expectations, and then consider your resources.

Analyzing Your Audience's Needs and Expectations

Use the techniques described in Chapter 5 for analyzing your audience. Consider such factors as their knowledge of the subject, their attitudes, their reasons for reading, the way they will be using the document, and the kinds of tasks they will be performing. For instance, if the document is a benefits manual for employees, you know that few people will read it like a book, but many people will refer to it often. Therefore, you will want to build in as many accessing tools as you can: a table of contents and index, of course, but perhaps tabs to separate the sections or a different color paper or type for each section. You might create a separate section containing the numerous forms the readers might have to fill out.

Think too about your audience's expectations. Readers expect to see certain kinds of information presented in certain ways, and unless you have a good reason to present information in other ways, you should fulfill their expectations. For instance, if you have worked with the documentation sets that come with word-processing programs, you have learned that certain kinds of information—tutorial instructions, installation guides, quick-start guides, and reference material—come in documents of different sizes and shapes. Tutorial information, for instance, is often presented in a small-format book, bound so that it lies flat on the table next to the keyboard.

Determining Your Resources

Once you have considered what your readers need and expect, think about your major resources: time, money, and equipment. Small, informal documents are generally made in-house, but for more ambitious projects, you need to decide how much of the job should be subcontracted to professionals. If your organization has a technical-publications department, you should consult the professionals there for information on scheduling and budgeting.

- *Time.* What is your schedule? A sophisticated design might require the expertise of professionals at service bureaus and print shops, and their services can require weeks. Working up even a relatively simple design for a newsletter can require many hours.

- *Money.* What is your budget for the project? Can you afford the services of professional designers and print shops? An in-house newsletter should look professional and attractive, but most managers would be unwilling to authorize thousands of dollars for a sophisticated design. They *would* authorize many thousands of dollars for the design of an annual report, however.

- *Equipment.* With a pencil and a straightedge you can create a terrific design, but to make that design come to life in the document itself, you need equipment. Although a word-processing program by itself can fulfill most of the routine design needs in most organizations, more complicated designs also require graphics and desktop-publishing programs. A good laser printer, less expensive than the average personal computer, will produce attractive documents in black and white. For high-resolution color, however, a more expensive printer is required. One common communication decision an organization faces is whether to invest more money in equipment and hire people with design and production skills, or to subcontract the design and production to professionals outside the company.

With an understanding of your readers' needs and expectations and of your own resources, you can begin to plan the design for the document and for the individual pages.

Designing the Document

Before you begin to design individual pages, think through the design of the whole document—how you want the different elements to work together to accomplish your objectives.

There are four major elements to consider in designing the whole document: size, paper, accessing tools, and bindings. Table 11.1 shows how these elements appear in typical documents produced in the United States. In other countries, conventions vary.

Size   Size refers to two aspects of document design: page size and page count.

- *Page size.* Your readers' needs are critical. Think about the best page size for communicating your information, and about how the document will be handled. For a procedures manual that will sit on a shelf most of the day, standard 8.5 × 11-inch paper, 3-hole-punched to fit in a ring binder, might be the obvious choice. But for a software tutorial, you will probably want to create relatively narrow columns in a document that fits easily on a desk while the reader works at the keyboard. Therefore, you might choose a 5.5 × 8-inch size. The physical dimensions of the

TABLE 11.1	Document-Design Elements in Typical Documents

Typical Documents	Document-Design Elements			
	Size	Paper	Accessing Tools	Binding
MEMOS	8.5 × 11 inches, as well as smaller sizes.	Printer paper; bond.	Icons, headers and footers, page numbering.	Paper clip or staple.
LETTERS	8.5 × 11 inches.	Letterhead stationery for the first page; bond or printer paper for subsequent pages.	Icons, headers and footers, page numbering, color (sometimes).	Usually not bound.
INSTRUCTIONS	Varies, depending on the complexity of the subject and the size of the packaging.	Usually printer paper, but sometimes higher grade if environmental needs call for greater strength.	Icons, headers and footers, dividers and tabs, cross-reference tables, page numbering, color (sometimes).	Often, instructions are on a single sheet. Otherwise, saddle binding with staples, or (occasionally) ring binding.
MANUALS	Varies, up to 8.5 × 11 inches.	Varies, depending on environmental needs.	Icons, headers and footers, dividers and tabs, cross-reference tables, page numbering, color (sometimes).	Varies, but usually loose-leaf, ring, or saddle. Large manuals are often perfect bound.
PROPOSALS	Usually 8.5 × 11 inches.	Usually printer paper or bond.	Icons, headers and footers, dividers and tabs, cross-reference tables, page numbering, color (sometimes).	Varies, but usually loose-leaf, ring, or saddle. Often, proposals are bound like a book, with text on the left-hand page and graphics on the right-hand page.
REPORTS	Usually 8.5 × 11 inches.	Usually printer paper or bond.	Icons, headers and footers, dividers and tabs, cross-reference tables, page numbering, color (sometimes).	For brief, internal reports, usually ring binding. For longer or external reports, varies, including loose-leaf, ring, saddle, and perfect.
NEWSLETTERS	Usually 8.5 × 11 inches, but sometimes larger.	Usually newsprint or printer paper.	Icons, headers and footers, page numbering, color (sometimes).	Usually folded like a newspaper; sometimes saddle bound.
BROCHURES	Usually 8.5 × 11 inches folded into three panels of 8.5 × 3.67 inches. Sometimes larger.	Usually a heavier stock than printer paper, such as text paper.	Icons and colors.	None.

document can be important if it is to be used in a cramped area or has to fit in a standard-size compartment, such as in the cockpit of an airplane.

Paper comes precut in a number of standard sizes in addition to 8.5 × 11 inches, such as 4.5 × 6 inches and 6 × 9 inches. Of course, paper can be cut to whatever size you want, but costs increase substantially when you use a nonstandard size. Check with your technical-publications department or a print shop for current prices and availability of paper sizes.

- *Page count.* Although page count can be important because a document that is too thick might not fit where it must be stored, for the most part page count is a cost factor and a psychological factor.

  The cost factor is simple: paper is expensive and heavy, so you want to reduce the number of pages as much as you can, especially if you are printing large numbers of the document and mailing it. Professional journals usually have a very cramped design because mailing costs make it necessary to fit as much information as possible on every page. In the software industry, many companies now use the minimalist design, replacing 600-page manuals with 100-page versions—in part because of printing, warehousing, and mailing costs.

  The psychological factor is also easy to understand: everyone wants to spend as little time as possible reading technical communication. Therefore, if you can figure out a way to design the document so that it is 15 pages long rather than 30—but still easy to read—your readers will appreciate it.

Paper   Paper is made not only in different standard sizes but also in different weights and with different coatings.

The cheapest paper is newsprint. It is extremely porous, so it not only allows inks to bleed through to the other side but also picks up smudges and oil easily. In addition, it can turn yellow in as little as a few weeks. For these reasons, newsprint is generally used only for newspapers, informal newsletters, and other types of quick, inexpensive communication to large numbers of people.

The most common types of paper used in technical communication are bond (for typing letters and memos), book paper (a higher grade that permits better print resolution), and text paper (an even higher grade used for more formal documents such as announcements and brochures). Heavier paper is also available for cover pages and for cards, such as quick-reference cards. But probably the most common kind of paper is the relatively inexpensive paper used in photocopy machines and laser printers; this paper is made to increase the resolution of the type.

Most types of paper are available uncoated or coated. The coating increases strength and durability, and it produces the best print resolution. However, some glossy coated papers produce an annoying glare. To deal with this problem, designers recommend choosing a paper with a slight tint; a bone white, for instance, produces less glare than a bright white.

The different types of papers are available in different grades. For small jobs—say, under 5,000 copies of a brief document—you probably want high-grade paper. For bigger jobs, the grade of paper will have a larger effect on the total cost.

Unless you have worked in the printing industry, you probably have not had a chance to learn much about paper. Work closely with printing professionals; they know, for example, about UV-coated paper, which greatly reduces the extent to which the sun fades the ink. And they know the current costs of recycled paper, which is constantly improving in quality and becoming less expensive.

Bindings Many technical documents are one sheet, or folded brochures, or held together with a paper clip or a staple. But you might soon be producing documents that require more sophisticated means of binding.

In technical communication, the four types of bindings shown in Figure 11.2 are used frequently.

a. Loose-leaf

b. Ring or spiral

c. Saddle

d. Perfect

FIGURE 11.2

Common Types of
Bindings

- *Loose-leaf binders.* Loose-leaf binders are reusable, come in various sizes, and are convenient when pages must be added and deleted frequently. One drawback, however, is that most people don't keep their binders up-to-date, especially if there are frequent updates, so after a while few people in an organization have up-to-date binders. Another drawback is that a high-quality loose-leaf binder can cost several dollars.

- *Ring or spiral binders.* Similar to loose-leaf binders, ring binders use wire or plastic coils or combs to hold the pages together. Ring binders let you lay the document open on a desk or even fold it over, so that it takes up no more space than one page. This feature makes the document easy to hold in one hand. Neighborhood print shops can put plastic coils or combs on documents of almost any size for about a dollar each.

- *Saddle binding.* In saddle binding, a set of large staples is inserted from the outside of the document, when the document is opened to its middle pages. Saddle binding is not practical for large documents.

- *Perfect binding.* In perfect binding, the pages are glued together and a cover is attached. Perfect binding, which is used in book publishing, produces the most formal appearance, but it is relatively fragile, and the open document usually does not lie flat.

Accessing Tools In a well-designed document, readers know how to move around easily and find the information they seek. Figure 11.3 shows six kinds of accessing tools that you can build into most documents.

- *Icons.* Icons are pictures that symbolize actions or ideas. A garbage can on your computer screen symbolizes the task of erasing a file. An hourglass or a clock tells you to wait for the computer to perform a task. Perhaps the most common icon in technical communication is the stop sign, which often alerts the reader to a warning.

  Icons enable your readers to create a visual link between the icon and a particular kind of information. For instance, the icon of the pointing hand often introduces hints or suggestions in software manuals. Once readers learn that the icon has this function, they seek it out when they need help. For you as the writer, icons save you the trouble of writing out repetitive comments such as "Following are some hints that might help you with this task."

  But beware of being too cute in thinking up icons. One computer manual uses a cocktail glass about to fall over to symbolize "tip." Not a good idea. And don't use too many different icons, or your readers will forget what they represent. Figure 11.4 presents common icons.

- *Color.* Color is perhaps the strongest visual attribute you can use (Keyes, 1993). If you have access to colors other than black and shades of gray,

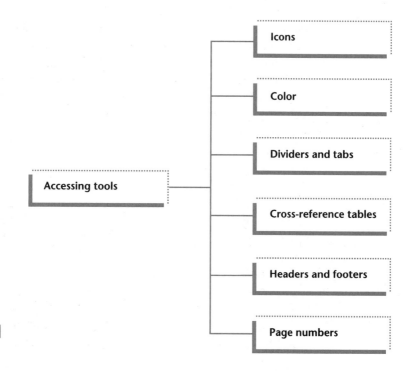

FIGURE 11.3

Accessing Tools

use them sparingly, for they can easily overpower everything else in your document.

Just as people *see* documents before they read them, they see colors before they see shapes. Therefore, don't send mixed signals to your readers. Your third-level headings should not be in color, for example, if your first- and second-level headings are black; headings in color will stand out more, regardless of other visual elements, such as boldface or size

FIGURE 11.4

Icons

variations. (See Chapter 13 for a discussion of heading levels.) Use colors to draw readers' attention to important features of the document, such as warnings, hints, major headings, and section tabs.

Colored paper is used frequently now as an accessing tool. Just as a telephone book has white pages, yellow pages, and blue pages, many technical documents now use colored paper to represent different sections. As always, however, use restraint; avoid using more than two or three colors, and avoid garish neons. See Chapter 12 for a detailed discussion of color.

- *Dividers and tabs.* You know dividers and tabs—the heavy-paper sheets with the paper or plastic extensions—from your loose-leaf notebooks. You may also have seen them in some textbooks. The tabs provide a place to add a label, and they enable readers to flip to a particular section. Often, a color scheme is used with dividers and tabs.

- *Cross-reference tables.* In a complicated document, you might want to refer readers to a related discussion. A cross-reference table, such as the one excerpted in Figure 11.5 (Fischer, 1992, p. 6), is an effective way to do so. As Figure 11.6 shows, a cross-reference table lets you direct readers with different backgrounds to various sections of the document.

- *Headers and footers.* A header is at the top of the page; a footer is at the bottom of the page. Headers and footers enable readers to see, at a glance, where they are in the document. In this book, for example, the headers on the left-hand pages indicate the chapter of the book. Those on the right-hand pages indicate the most recent first-level heading. Note that, like page numbers, headers and footers are usually near the outside edge of the page, not near the spine of the document.

  Sometimes writers build other identifying information into their headers. For example, your instructor might ask you to identify your assignments with the following information in a header: "Smith, Progress Report, English 302, page 6."

- *Page numbers.* Page numbering can be complicated. For memos, use arabic numbers in the upper right corner of the page. (Note that the first page of most documents is not numbered.) For documents printed on both sides of the page, such as this book, put the page numbers in the outside corners.

  Reports and other complex documents often use two sets of numbers: lowercase roman numerals (i, ii, and so on) for front matter and arabic numerals for the body. The title page is not numbered; the page after it is numbered ii.

  Appendices are often numbered and lettered. That is, Appendix A begins with page A-1, followed by A-2, and so on. Appendix B starts with page B-1.

Table I.1
Where to Start

Topic	See Chapter/Appendix
Adding Commands or Command Names	20
Advanced 2D commands	8–13
Changing Command Names or Keystrokes	20
Changing or Customizing Menus	20
Changing the Coordinate System	1, 3, 7, 11
Controlling Drawing Accuracy	3

FIGURE 11.5

Cross-Reference Table

If you have never used the Internet:
Read . . .   *To learn how to . . .*
   Chapter 1   connect to the Internet
   Chapter 2   use basic e-mail functions
etc.

FIGURE 11.6

Cross-Reference Table Addressed to Readers with Different Backgrounds

If you are familiar with the Internet, but not a user:
Read . . .   *To learn how to . . .*
   Chapter 5   join newsgroups
etc.

You should know about two other aspects of page numbering:

– Sometimes documents that are photocopied and distributed in a meeting, through interoffice mail, or (especially) faxed include in the page numbering system the total number of pages in the document. The second page is numbered "2 of 17," and the third page is "3 of 17." This approach looks awkward, but at least readers are sure they have received the complete document.
– Sometimes documents that will be updated use section numbering. Section 3 begins with page 3-1, followed by 3-2. Section 4 begins with 4-1. This way, a complete revision of one section does not affect the pagination of the subsequent sections.

Designing the Page

Not long ago, page design was a simple matter for technical communicators: text was double spaced, margins were one inch. Graphics pages followed the text pages. When readers needed to find the graphic that went with the text, they flipped forward through the document to find the table or chart. Now that sophisticated desktop-publishing software is available, we are studying the best ways to design the page so that it is attractive and functional.

This section begins with a discussion of page-design principles and of what learning theory can teach us about designing better pages. The section also discusses page grids, the visual plan for the page; typography; the design of titles and headings; and special design techniques (such as text boxes).

Principles of Effective Page Design

A page of technical communication is effectively designed if the reader recognizes a pattern—where to look for certain kinds of information, for instance—but isn't put to sleep by a monotonous, mechanical appearance. As shown in Figure 11.7, designers often refer to three qualities when they discuss page design: balance, consistency, and simplicity.

- *Balance.* Balance refers to the visual stability of the page; it should not look as if it is tipping to one side or is top-heavy or bottom-heavy. In documents that open to a two-page spread, the concept of balance also applies to the two pages. On a balanced page, a graphic on one side can be offset by a similar size graphic on the other side, or one large graphic can be balanced by two smaller ones. Balance can be *symmetrical* or *asymmetrical*. Symmetrical balance is achieved when the two halves of the page are similar in appearance: a graphic balances a graphic, or white space balances white space. Asymmetrical balance is achieved when the two halves of the page are dissimilar but still stable. For example, an asymmetrically balanced page might have a black-and-white graphic in the upper left balanced by an icon in color in the lower right. Figure 11.8 shows some common designs for balanced pages.

  But don't be a slave to symmetry. Readers can be bored by a balance that is too regular; slight variations add visual interest to the page, as seen in Figure 11.9 on page 322.

- *Consistency.* A well-designed page uses consistent patterns, just as a well-designed document does. Margins, for example, should be consistent from page to page. Indentation should follow a regular pattern too:

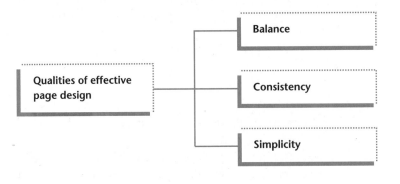

FIGURE 11.7

Qualities of Effective Page Design

On the left-hand page, symmetrical balance is achieved by centering key elements.

The page on the right is asymmetrically balanced because the graphic in the upper-left corner balances the two-column text on the right side of the page.

The Falcon *The Falcon*

FIGURE 11.8

Symmetrical and Asymmetrical Balance

second-level text, for instance, might be indented five characters. Type-faces and sizes help you create a consistent pattern; first-level heads might be printed in 18-point Helvetica, for example, with body copy in 12-point Century Schoolbook. Other elements that can create consistent visual patterns are colors, icons, rules, and screens. See the discussion of page grids later in this chapter for more information on how to create and sustain consistent patterns.

- *Simplicity.* Keep in mind that the primary purpose of technical communication is to inform the reader. In general, a simple design is more effective than a complicated one. Don't be tempted to use all the design tricks loaded on the CD-ROM disk that came with your software; the philosophy of some manufacturers is "If the feature fits on the disk, throw it in." For instance, typeface designers offer outlined and shadowed typefaces, but they are harder to read than regular typefaces. A page with four or five typefaces and sizes is likely to confuse the reader.

Learning Theory and Its Relation to Page Design

Your job in designing the page is to create visual patterns so that readers can easily find, understand, and remember the information they need. A basic understanding of three principles of learning theory—chunking, queuing, and filtering—can help you design effective pages. (See Keyes [1993] for a summary of the research on learning theory.)

- *Chunking.* People understand information best if it is delivered to them in chunks—small units—rather than all at once. In a business letter,

Each page of this two-page spread is asymmetrical, but viewed together the two pages are well balanced. Notice that white space can be an important element: the white space at the top of the right page balances the title at the top of the left page.

In this two-page spread, the angled placement of the art adds visual interest.

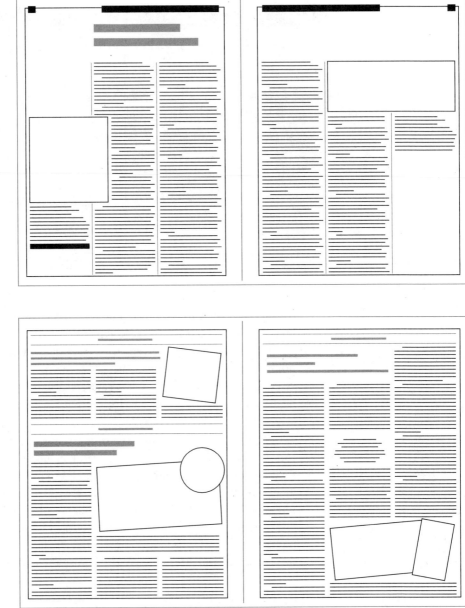

Variations on Balanced Pages

which is typed single spaced, chunking involves double spacing between paragraphs. Figure 11.10 shows how chunking helps readers see the units of related information.

- *Queuing.* Queuing refers to creating visual patterns that show levels of importance. In a traditional outline, the "I" heading is more important than the "1" heading. In page design, you want to use more emphatic

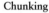 **FIGURE 11.10**

Chunking   a. Without chunking   b. With chunking

elements—such as bigger type or boldface type—to suggest importance.

Another visual element of queuing is indentation. Designers start the more important information closer to the left margin and indent the less important information. (An exception is titles, which are often centered in reports in the United States.) The page in Figure 11.11(a) on page 324, for example, shows queuing by size: the first-level head at the top is bigger than the second-level heads. The page in (b) shows queuing by size and by indentation: each of the three headings is a different size, and the headings and their accompanying text are aligned on different left margins.

• *Filtering.* Filtering is the use of visual patterns to distinguish types of information so that readers can decide what they want to read. For instance, a stop sign often signals safety information in a set of instructions. Designers also use typography to create filters. Introductory material might be displayed in larger type, and notes might be displayed in italics or in a different typeface. Figure 11.12(a) on page 325 shows a design that uses text boxes to set off notes. Figure 11.12(b) shows a text box with an icon used as a filtering device.

Page Grids, White Space, and Columns   Every page includes two kinds of space: white space and space for text and graphics. The best way to start designing a page is to make a grid—a picture of what it will look like. In making a grid, you will decide how to use white space and determine how many columns to have.

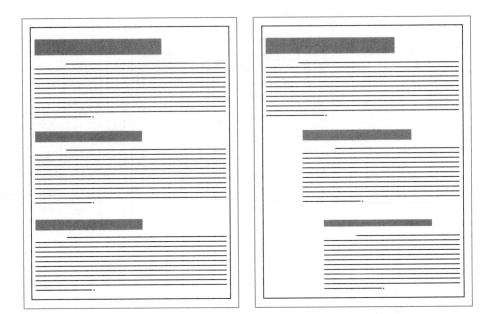

FIGURE 11.11

Queuing by Size and by
Indentation

a. Queuing by size

b. Queuing by size and indentation

Table 11.2 on page 326 shows how the page-design elements discussed in this section are used in the United States. In other countries, the conventions vary.

Page Grids

As the phrase suggests, a page grid is like a map on which you plan where the white space, the text, and the graphics will go. To create an effective grid, you must first understand your audience, their purpose in reading, and their reading behavior.

Many writers like to begin with a thumbnail sketch. A *thumbnail sketch* is a rough drawing that helps you see how the different elements—the text and the graphics—will be arranged on the page. Figure 11.13 shows two thumbnail sketches for a page from the body of a manual.

Keep experimenting with thumbnail sketches of the different kinds of pages your document will have: body pages, front matter, and so on. When you are satisfied with your thumbnail sketches, make page grids.

You can use either a computer or a pencil and paper to create page grids. Or you can combine the two techniques. On a computer, you can use a word-processing program or a draw program or paint program (see Chapter 12) to create the grid. Many word-processing programs come with templates that you can use as is or modify. Desktop-publishing software offers the greatest variety of templates, many quite sophisticated.

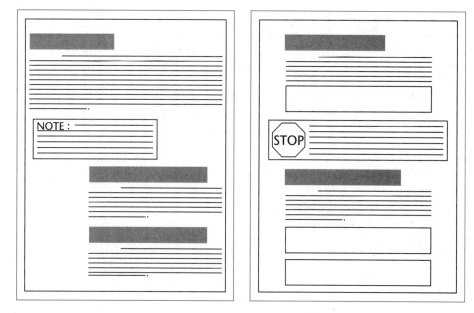

FIGURE 11.12

Filtering

a. Filtering using a text box

b. Filtering using a text box and an icon

You don't need sophisticated software to design a grid, however. With a pencil, you can draw the grid on a sheet of paper the size you will be using. Some writers like to use graph paper because the grid lines are convenient. You can use inches or the metric system for your measurements, or you can use the unit that

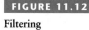

FIGURE 11.13

Thumbnail Sketches

TABLE 11.2	Page-Design Elements in Typical Documents				
	Page-Design Elements				
Typical Documents	Columns	Typeface	Type Size (in points)	Line Spacing	Justification
MEMOS	One.	Varies, but usually serif.	Usually 10–12.	Single spaced.	Left-justified or justified.
LETTERS	One.	Varies, but usually serif.	Usually 10–12.	Single spaced.	Left-justified or justified.
INSTRUCTIONS	Varies, but usually a multi-column design.	Varies.	Varies.	Varies.	Left-justified or justified.
MANUALS	Varies.	Varies, but usually serif.	Usually 10–12.	Varies, but usually single spaced.	Left-justified or justified.
PROPOSALS	Varies.	Usually serif.	Usually 10–12.	Varies.	Left-justified or justified.
REPORTS	Often one.	Usually serif.	Usually 10–12.	Often double spaced.	Left-justified or justified.
NEWSLETTERS	Usually multicolumn.	Serif for body text; sans serif for display text.	Usually 10–12 for body text.	Usually single.	Left-justified or justified.
BROCHURES	Usually one column per panel.	Varies.	Varies.	Varies.	Varies.

printing professionals use—the pica—which equals one-sixth of an inch. Some writers like to experiment with pencil and paper, then transfer their grid to a computer. Figure 11.14 on page 328 shows two simple grids: one using picas and one using inches.

Experiment with different grids until you think you have a page design that is attractive, able to meet the needs of your readers, and appropriate for the information you are conveying. Figure 11.15 on page 329 shows the variety of grids you can create.

White Space

Sometimes called *negative space,* white space is the portion of the paper that does not have writing or graphics on it: the space between two lines of text, the space between text and graphics, and, most obvious, the margins. White space directs readers' eyes to a particular element, emphasizing it; in addition, white space helps readers see relationships among elements on the page.

Most people are surprised to learn that almost half the area of a typical page is devoted to margins. Why so much? Margins serve four main purposes:

- To cut down on the amount of information on the page, making the page seem easier to read and use.

- To provide enough space that the document can be bound or that the reader can easily hold the page without covering up the text.

- To provide a neat frame. Pages on which the type extends all the way out to the edge of the paper look sloppy.

- To provide space for marginal glosses. (Marginal glosses are discussed later in this chapter.)

Figure 11.16 on page 330 shows the common margins for left-hand and right-hand pages. The actual sizes of the margins are determined by several factors:

- *The size of the page.* In general, the smaller the page, the smaller the margins.

- *The amount of text that needs to fit on the page.* For an article in a journal that will be mailed, the margins tend to be small, to reduce the page count and, therefore, postage costs.

- *The difficulty of the text.* In general, the more technical the text, the larger the margins. This strategy makes the pages look less intimidating.

- *The background of the audience.* In general, the less knowledgeable the readers, the larger the margins.

White space other than margins also serves as a border that sets off and emphasizes an element on the page. For instance, white space around a graphic separates it from the text and draws the reader's eye to it. White space between columns helps the reader to read the text easily, without mistaking two adjacent columns for one. And white space between sections of text helps the reader see that one section is ending and another is beginning.

Columns

Most documents you write on the job will look much like the one-column documents you write as a student. But things become a lot more complicated when you go beyond the world of paper clips and staples. For example, you will have to leave bigger inside margins to accommodate the binding. But the greatest difference between the design of the pages you write as a student and those you will write on the job is in the use of multiple columns. A multicolumn design can have four major advantages for the technical communicator:

a. Grids using picas (left) and inches (right)

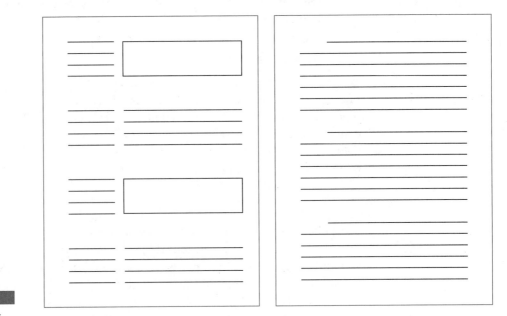

FIGURE 11.14

Sample Grids Using
Picas and Inches

b. Pages based on the grids

a. Double-column grid   b. Two-page grid, with narrow outside columns for notes

c. Three-panel brochure

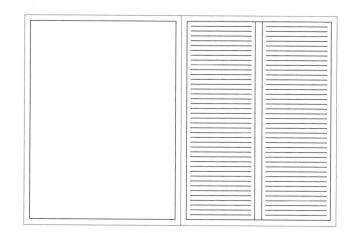

d. Two-page grid, with graphics on the left page and double-column text on the right page.

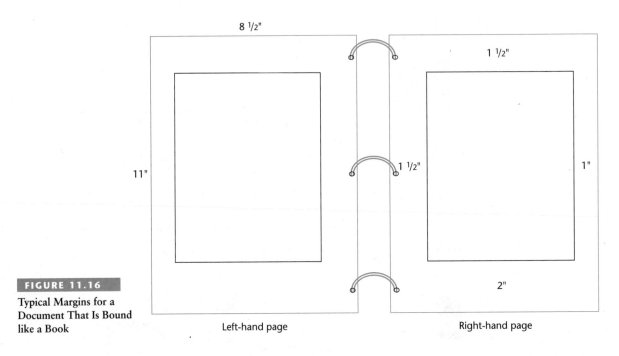

FIGURE 11.16

Typical Margins for a
Document That Is Bound
like a Book

- Text in columns is generally easier to read because the lines are shorter.
- The use of columns lets you fit more information on the page because many of your graphics can fit in one column. Wider graphics can extend across two or more columns. In addition, you can fit more words on a page with a multicolumn design than with a single-column design.
- The use of columns allows you to create a visual pattern in which text is in one column and the accompanying graphic in an adjacent column.
- The use of columns gives you greater flexibility in sizing graphics.

Typography *Typography* is the study of type and the way people read it. When designers and technical communicators study typography, they learn about typefaces, families, case, and sizes. They also consider the white space of typography: line spacing, line length, and justification.

Typefaces

A *typeface* is a set of letters, numbers, punctuation marks, and other symbols, all bearing a characteristic design. When the first movable, reusable metal type was invented in the Western world in the fifteenth century, the typefaces resembled ornate hand lettering. Today there are thousands of typefaces, generally bearing the name of their designer or some historical name (such as Times Roman, Avant

This paragraph, for example, was typed on a word processor using the Kaufman typeface. You are not likely to see this style of font in any technical communication because it is too ornate and too hard to read. It is better suited to wedding invitations and other formal announcements.

This paragraph was typed with the Times Roman typeface. It looks like the kind of type used by newspapers such as the *New York Times* in the nineteenth century. It is an effective typeface for body text in technical documents.

This paragraph was typed with the Univers typeface, which has a very modern, high-tech look. It is best suited for headings and titles in technical documents.

FIGURE 11.17

Typefaces

Garde, Helvetica, Garamond, or Zapf Chancery). More are designed every year. Figure 11.17 shows three typefaces.

Typefaces are generally classified into two categories: *serif* and *sans serif*. A serif typeface, such as Times Roman, has short lines—serifs—at the ends of the major strokes in most of the letters. For instance, look at the serifs in the letter N in *New York Times* in the second paragraph in Figure 11.17. Sans serif typefaces, such as Univers, do not have these lines. Serif typefaces are generally considered easier to read because the short lines encourage the movement of the reader's eyes along the line. Sans serif typefaces are harder on our eyes than serif typefaces because the letters are less distinct from one another than they are in a serif typeface. For this reason, sans serif typefaces are used mostly for short documents and for headings. However, sans serif typefaces are easier to read on the screen and when printed on dot-matrix printers, because the letters are simpler; serif fonts can look smudged.

Cultural factors can be important, though. For instance, Europeans find sans serif typefaces easier to read than serif typefaces because sans serif is more common in Europe.

Most of the time, you will use a standard font, such as Times Roman, that your applications software includes and that your printer can reproduce. Just remember that different typefaces create different impressions—from highly formal and ceremonial to very modern and high tech—and that some typefaces cause less eye fatigue than others.

Type Families

Each typeface is actually a family of typefaces, consisting of a number of variations on the basic style. Figure 11.18 on page 332, for example, is the Helvetica family.

As with typefaces, avoid the temptation to overdo the use of different members of a type family. Used sparingly and consistently, the different members of a

Helvetica Light	**_Helvetica Bold Italic_**
Helvetica Light Italic	**Helvetica Heavy**
Helvetica Regular	**_Helvetica Heavy Italic_**
Helvetica Regular Italic	Helvetica Regular Condensed
Helvetica Bold	_Helvetica Regular Condensed Italic_

FIGURE 11.18

Helvetica Family of Type

type family help you direct the reader's attention to different kinds of text, such as warnings, notes, and characters that you want your readers to type (Felker, Pickering, Charrow, Holland, & Redish, 1981). You will need italics for book titles and other elements, and you might use bold type for emphasis and headings, but you can go a long time without needing condensed and expanded versions of typefaces. And you can live a full, rewarding life without ever using outlined or shadowed versions of typefaces.

Case

To make your technical communication easy to read, use uppercase and lowercase letters just as you would in any other kind of writing. Some writers mistakenly think that text printed in uppercase letters is easier to read or more emphatic than text printed in a mix of uppercase and lowercase letters. It isn't. Lowercase letters are actually easier to read because the individual variations from one letter to another are greater for lowercase than for uppercase, as shown in Figure 11.19.

The average person requires 10 to 25 percent more time to read all-uppercase writing than writing using both cases (Haley, 1991). In addition, using lowercase and uppercase letters provides the visual cues that signal new sentences (Poulton, 1968). One more reason to use both cases: uppercase letters require up to 35 percent more space than do lowercase letters (Haley, 1991).

Type Sizes

Using different sizes of one typeface can create other visual cues. Sizes are measured in a unit called a _point_. There are 12 points in a pica, or 72 points in an inch. Figure 11.20 shows the basic range of type sizes.

Software and printers with scalable typefaces can produce letters and other characters of almost any size, ranging from 0.25 to 999.75 points, in increments

FIGURE 11.19

Individual Variations in Lowercase and Uppercase Type

Individual variations are greater in lowercase words

THAN THEY ARE IN UPPERCASE WORDS.

6	ABCDEFGHIJKLMNOPQRSTUVWXYZABCDEFGHIJKLMNOPQRSTUVWXYZABC
8	ABCDEFGHIJKLMNOPQRSTUVWXYZABCDEFGHIJKLMNO
10	ABCDEFGHIJKLMNOPQRSTUVWXYZABCDEFG
12	ABCDEFGHIJKLMNOPQRSTUVWXYZAB
14	ABCDEFGHIJKLMNOPQRSTUVW
18	ABCDEFGHIJKLMNOPQR
24	ABCDEFGHIJKLMN
30	ABCDEFGHIJKL
42	ABCDEFG
48	ABCDEF
60	ABCDE

FIGURE 11.20

Helvetica Medium
in 11 Sizes

of 0.25 points. In most technical communication, body text is presented in 10-, 11-, or 12-point type.

This paragraph is printed in 10-point type. This size is easy to read, provided it is reproduced on a letter-quality impact printer or laser printer. On most impact printers, however, the resolution isn't high enough for type this small.

This paragraph is printed in 12-point type. If you are using a dot-matrix printer, 12-point type is the best size to use because it balances readability and economy.

This paragraph is printed in 14-point size. This size is appropriate for titles or headings.

Some other sizes that are used occasionally in technical communication:

footnotes	8- or 9-point type
indexes	2 points smaller than the body text
slides or transparencies	24- to 36-point type

In general, you want a 4 to 6 point difference between headings and body text, and between headings and titles.

Don't go overboard with size variations, or your document will look like a sweepstakes advertisement in the mail.

Line Length

Ironically, the line length that we use most often—about 80 characters on an 8.5 × 11-inch page—is somewhat difficult to read. In general, a short line—of perhaps 50 to 60 characters—is less tiring to read than a longer line, especially for long documents (Biggs, 1980). With a multicolumn format, you can choose a line length that is appropriate for the size of the type and the paper, as well as for the degree of difficulty of the text and the knowledge level of the readers.

Line Spacing

Sometimes called *leading* (pronounced *ledding*), *line spacing* refers to the white space between lines, or between one line of text and a graphic. If the line spacing is too big, the page looks diffuse, the text loses its coherence, and the reader tires quickly. If the line spacing is too small, the page looks crowded and is difficult to read; your eye sees the lines above and below the one you are reading, causing fatigue. Some research suggests that smaller type requires greater line spacing, that longer lines benefit from additional line spacing, and that sans serif typefaces need a little more line spacing than serif typefaces. Figure 11.21 shows the same text with three amounts of line spacing.

Line spacing is usually determined by the kind of document you are writing. Memos are single spaced; other documents, such as reports and proposals, are double spaced or one-and-a-half spaced.

Line spacing can also separate one section of text from another. Breaks between single-spaced paragraphs are usually two spaces. Double-spaced and one-and-a-half-spaced text has no extra line spacing between paragraphs, but the first line of each paragraph is indented, as in handwritten documents.

For breaks between sections, the line spacing is usually greater than that between paragraphs. For instance, if you are typing a single-spaced letter, double space between paragraphs and triple space between sections. In other words, the line spacing *between* two sections should be greater than the line spacing *within* a section. Figure 11.22 on page 336 shows the effective use of line spacing to distinguish one section from another and to separate text from graphics.

Line spacing is also used to set off block quotations. If the text is single spaced, set off the quotation above and below with double spacing; if the text is one-and-a-half or double spaced, use triple spacing.

a.   Excessive line spacing

Aronomink Systems has been contracted by Cecil Electric Cooperative, Inc. (CECI)

to design a solid waste management system for the Cecil County plant, Units 1 and 2,

to be built in Cranston, Maryland. The system will consist of two 600 MW pulverized

coal-burning units fitted with high-efficiency electrostatic precipitators and limestone

reagent FGD systems. The coal will contain an estimated 3% sulfur and 10% ash. The

station will output approximately 64 TPH (DWB) of FGD sludge and 24 TPH fly ash at

100% load.

b.   Appropriate line spacing

Aronomink Systems has been contracted by Cecil Electric Cooperative, Inc. (CECI)
to design a solid waste management system for the Cecil County plant, Units 1 and 2,
to be built in Cranston, Maryland. The system will consist of two 600 MW pulverized
coal-burning units fitted with high-efficiency electrostatic precipitators and limestone
reagent FGD systems. The coal will contain an estimated 3% sulfur and 10% ash. The
station will output approximately 64 TPH (DWB) of FGD sludge and 24 TPH fly ash at
100% load.

c.   Inadequate line spacing

Aronomink Systems has been contracted by Cecil Electric Cooperative, Inc. (CECI)
to design a solid waste management system for the Cecil County plant, Units 1 and 2
to be built in Cranston, Maryland. The system will consist of two 600 MW pulverized
coal-burning units fitted with high-efficiency electrostatic precipitators and limestone
reagent FGD systems. The coal will contain an estimated 3% sulfur and 10% ash. The
station will output approximately 64 TPH (DWB) of FGD sludge and 24 TPH fly ash at
100% load.

FIGURE 11.21

Line Spacing

Justification

Justification refers to the alignment of words along the left and right margins. In technical communication, text is often *left-justified;* that is, except for paragraph indentations, the lines begin along a uniform left margin but end on an irregular right border. This irregular right border is often called *ragged right.* Ragged right is most common in typewritten and word-processed text (even though word-processing programs can justify the right margin).

In *justified text,* also called *full-justified text,* both the left and right margins are justified. Justified text is seen most often in typeset, formal documents. However, left-justified text offers an advantage even when you are typesetting: if you have to make minor revisions after the text is set, your changes won't affect as many lines because the irregular right margin gives you more room to make

FIGURE 11.22

Line Spacing Used to Distinguish One Section from Another

changes. For example, if you add a word to a short line, the change might not affect any subsequent lines.

The following passage (Berry, Mobley, & Turk, 1994, p. 647) is presented first in left-justified form and then in justified form:

Notice that the space between words is uniform in left-justified text.

> Until recently, most software development has adhered to a linear development process, often referred to as the "waterfall" process. Rimbaugh (1992) describes this as several phases performed in sequence, one after another. Using this approach, a product is first designed, then developed, tested, and delivered. Nothing is assembled until the design is complete, and the design is not changed once assembly starts. The car being built moves straight down the assembly line and isn't complete until it reaches the end.

In justified text, the spacing between words is irregular, slowing down the reader. Because a big space suggests a break between sentences, not a break between words, the reader can become confused, frustrated, and fatigued.

Notice that the irregular spacing not only slows down reading but also creates "rivers" of white space. Readers are tempted to concentrate on the rivers running south rather than on the information itself.

> Until recently, most software development has adhered to a linear development process, often referred to as the "waterfall" process. Rimbaugh (1992) describes this as several phases performed in sequence, one after another. Using this approach, a product is first designed, then developed, tested, and delivered. Nothing is assembled until the design is complete, and the design is not changed once assembly starts. The car being built moves straight down the assembly line and isn't complete until it reaches the end.

Should you justify your technical communication? If you are using a standard word-processing program rather than a sophisticated desktop-publishing system, probably not.

Justification can make the text harder to read in one more way. Some word-processing programs and typesetting systems automatically hyphenate words that do not fit on the line and sometimes hyphenate incorrectly. Hyphenation slows down the reader and can be distracting. Left-justified text does not require hyphenation the way justified text sometimes does.

Designing Titles and Headings   Chapters 13 and 14 discuss the strategy of creating effective titles and headings. This section, however, explains how to integrate titles and headings into the page design. The principle is that you want titles and headings to stand out visually because they announce and communicate new information; your reader needs to see that you are beginning a new idea.

Designing Titles

Because a title is the most important heading in a document, it should be displayed clearly and prominently. If it is on a cover page or a title page, you might present it in boldface in a large size, such as 18 points. If it also appears at the top of the first page, you might make it slightly bigger than the rest of the text—perhaps 18-point type for a document printed in 12-point type—but not as big as on the cover page or title page. Titles are often centered on a page horizontally. For more information on titles, see Chapter 13.

Designing Headings

Readers should be able to tell when you are beginning a new idea. The most effective way to distinguish one level of headings from another is to use size variations (Williams & Spyridakis, 1992). For example, a 20 percent difference in size between a first-level head and a second-level head will be clear for most readers. Boldface also sets off headings effectively. The least effective way to set off headings is by underlining them, perhaps because in many word-processing programs the underline obscures the descenders, the portions of letters such as *p* and *y* that extend below the body of the letters.

Indenting the headings can also help readers. In general, the more important the level, the closer to the left margin it appears. First-level headings, therefore, usually begin at the left margin. Second-level headings are often indented four or five characters; third-level headings, seven or eight. Another common design for third-level headings is to run the text in on the same line as the heading. When you indent a heading, consider indenting the text that follows it. This design accomplishes two goals:

- The indented heading remains clear and emphatic; it does not get swallowed up by the text that follows it.

- The indented text appears appropriately subordinate. In some documents, writers even decrease the type size for lower-level text. For instance, in a document that is printed in 12-point type, all third-level and fourth-level text is printed in 10-point type.

As mentioned in the discussion of margins, writers sometimes highlight their headings by using hanging indentation: extending the heading further out into the left margin than the accompanying text.

In designing headings, you also need to use line spacing well. A noticeable distance between a heading and the text increases the impact of the heading. Consider the three examples in Figure 11.23.

Other Design Techniques Technical communicators are adapting the design features of newspapers and magazines. Five design techniques that are appearing more frequently in technical communication are rules, boxes, screens, marginal glosses, and pull quotes.

- *Rules.* The designer's term for a straight line is *rule*. Using word-processing software, you can easily create horizontal or vertical rules for various purposes. Horizontal rules are often used to separate headers and footers from the body of the page, or to distinguish two sections. Vertical rules can separate columns in a multicolumn page or identify revised text in a manual (see Chapter 20).

- *Boxes.* Create boxes by adding rules on all four sides of an item. Boxes can enclose graphics or special sections of text; sometimes they are used as a border for the whole page. Boxed text is often placed so that it extends into a margin, giving it increased emphasis.

- *Screens.* The shading that appears behind text or a graphic to give it emphasis is known as a *screen*. The degree of screen ranges from 1 percent to 100 percent. Usually, a 5 to 10 percent screen provides the desired emphasis, while leaving the boxed text legible. You can use screens with or without boxes. Figure 11.24 shows a box with a screen behind it.

- *Marginal glosses.* A marginal gloss is a brief text, usually a summary statement, placed in the margin of the document. Usually, marginal

---

Summary

In this example, the writer has skipped a line between the heading and the text that follows it. The heading stands out clearly.

Summary
In this example, the writer has not skipped a line between the heading and the text that follows it. The heading stands out, but not as emphatically.

FIGURE 11.23

The Effect of Different Line Spacing on the Visibility of Headings

Summary. In this example, the writer has begun the text on the same line as the heading. This run-in style makes the heading stand out the least.

3. Slide the old toner cartridge out.
4. Prepare the new toner cartridge.

Important: Do not remove the flexible plastic sheets. They need to remain in place.

FIGURE 11.24

Box with a Screen

glosses are set in a different typeface—and sometimes a different color—from the main discussion, and often they are separated from the main discussion by a vertical rule. Marginal glosses can be helpful in providing an overview of the longer explanation; however, they can also compete with the main discussion for the reader's attention.

- *Pull quotes.* You might not have heard the term *pull quotes,* but you see them all the time, particularly in newspapers and magazines. A pull quote is simply a quotation—usually just a sentence—pulled from the text and displayed in a larger size and generally in a different typeface. Often a pull quote is boxed. The purpose of the pull quote is to attract readers' attention and make them want to read the document. Pull quotes are not appropriate for some kinds of documents, such as most reports, because they look too informal. But they are increasingly popular in newsletters. Figure 11.25 shows pull quotes.

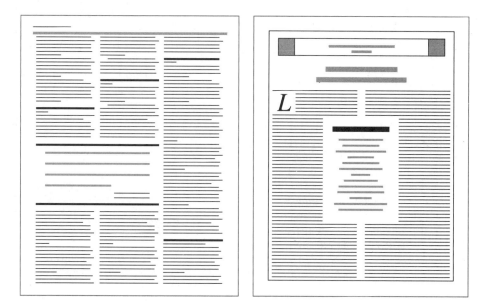

FIGURE 11.25

Pull Quotes

CASES

1. The Brochure with the Inconsistent Design

You work in the corporate communications department at Milton Bradley Company, which produces Scrabble® and many other board games. Your boss, Louise Willis, approaches you one day. "We've got a problem. The redesign of the brochure for the next printing of Scrabble® isn't going well. The designer quit halfway through the job. John was in a real hurry to meet our production deadline, so he hired another firm to finish the design. Would you mind taking a look at the mockup and letting me know whether you think it looks okay or still needs work?" Study the copy of the four-page brochure. Does it seem appropriate for the intended audience? Is it clear and attractive? Is there any evidence that two different sets of designers worked on it? If so, would it be easy to fix or should the company start over? Present your analysis and recommendations in a memo to Louise Willis. See Chapter 16 for a discussion of memos.

2. Updating an Old Design

You are a student intern at the Bonneville Power Administration. Your department is creating a new design for its reports (see page 342). Your supervisor, Bill Irwin, e-mails you, asking for your advice on redesigning the body pages. "The people in technical publications tell me this design is out of date, but I don't see anything wrong with it. I put a lot of work into it in 1989, and everyone said it was fine then. What do you think? Get back to me." Draft a response to him.

WRITER'S CHECKLIST

Have you
- ❏ analyzed your audience—knowledge of the subject, attitudes, reasons for reading, and the kinds of tasks they will be carrying out?
- ❏ thought about what your readers will expect to see when they pick up the document?
- ❏ determined your resources: time, money, and equipment?
- ❏ considered the best size for the document?
- ❏ considered the best paper for the document?
- ❏ considered the best binding for the document?
- ❏ thought about the most appropriate accessing tools to include, such as icons, color, dividers, tabs, and cross-reference tables?
- ❏ devised a style for headers and footers?
- ❏ devised a style for page numbers?
- ❏ drawn thumbnail sketches and page grids that define columns and white space?
- ❏ chosen typefaces that are appropriate to your subject?
- ❏ used appropriate members of the type families?
- ❏ used type sizes that are appropriate for your subject and audience?
- ❏ decided on whether to use left-justified text or justified text?
- ❏ chosen a line length that is appropriate for your subject and audience?
- ❏ chosen line spacing that is appropriate for your line length, subject, and audience?
- ❏ designed your title for clarity and emphasis?
- ❏ worked out a logical, consistent style for each heading level?
- ❏ used rules, boxes, screens, marginal glosses, and pull quotes, where appropriate?
- ❏ used color, if you have access to it, appropriately to highlight items such as warnings?

EXERCISES

Collaborative Exercise

1. Form small groups for this collaborative exercise in analyzing design. Photocopy a page from a book or

CASE 1

The Brochure with the Inconsistencies

Note: These pages are reduced 50% from the original.

Back cover

Special Features

The Official SCRABBLE® Players Dictionary (Second Edition) offers lots of features to help you play faster, better, and more confidently:
• No other dictionary includes all the two-to-eight letter words you'll find here.
• Only words that are permissible in SCRABBLE® can be found here. Proper names, words requiring hyphens or apostrophes, words considered foreign, and abbreviations have been omitted.
• All main entries are given a part-of-speech label, followed by appropriate inflected forms.
• All acceptable variant forms are shown alphabetically.
• All noun plurals are shown, as well as comparatives and superlatives of adjectives and adverbs.

Where to Get It

The Official SCRABBLE® Players Dictionary (Second Edition) costs only $14.95* at your local bookstore.

How to Order It

If the dictionary is out of stock, you can order it directly from the publishers. Send a check or money order in U.S. funds for $15.95,* which includes postage and handling, to:

MERRIAM-WEBSTER INC.
47 Federal Street
P.O. Box 281
Springfield, MA 01102
Attn: Consumer Sales

*Prices subject to change.
©1991, 1992 Milton Bradley Company. All Rights Reserved.
The gameboard shown in the book jacket photograph is
©1948 by Milton Bradley Company.

SCRABBLE®
CROSSWORD GAME
by Milton Bradley

Front cover

JOIN THE CLUB!

•

The National
SCRABBLE®
Association

TURN THE PAGE TO FIND OUT HOW

Also Inside:

• OFFICIAL SCRABBLE® PLAYERS DICTIONARY OFFER
• FREE TILE REPLACEMENT

Page 2

The National
SCRABBLE®
Association

As a member of the National SCRABBLE® Association, you'll receive a membership card; a roster of SCRABBLE® Players Clubs in the U.S. and Canada; 3 special word lists to help improve your play; and a 1-year subscription to the SCRABBLE® News. That's 8 exciting issues with...
• News about clubs and tournaments in your area
• Official National SCRABBLE® Association Tournament Rules
• Special word lists and helpful playing hints
• The latest in strategy
• Challenging quizzes & puzzles
Fill out the form below (print clearly) and send to: NATIONAL SCRABBLE® ASSOCIATION, P.O. Box 700, Greenport, NY 11944

YES! I'd like to join the National SCRABBLE® Association. I enclose a check or money order for $15.00 ($20.00 for Canadian membership; $25.00 outside the 50 states and Canada). Only U.S. funds will be accepted.
☐ *I am interested in starting a club in my area.*

NAME

STREET

CITY

STATE   ZIP

Are any of your tiles missing or defective?

If so, we want to replace them free of charge. Please list them on the form below and send it, with any defective tiles, to: MILTON BRADLEY COMPANY, Attn. Customer Service Department, 443 Shaker Road, East Longmeadow, MA 01028

TILE REPLACEMENT FORM
Please print clearly.
NAME   STREET

CITY   STATE   ZIP
GAME:
☐ SCRABBLE ☐ DELUXE SCRABBLE ☐ TRAVEL SCRABBLE   ITEM # (from package):

MISSING TILES: DEFECTIVE TILES (please send with form):

Page 3

IT'S OFFICIAL!

Have you ever had a word challenged that you knew was acceptable—yet it couldn't be found in your dictionary? Now there's a dictionary that was created especially for SCRABBLE® players...

The Official
Players
Dictionary
Second Edition

T.H.E   B.E.S.T

The Official SCRABBLE® Players Dictionary

Second Edition!
Over 5,000 new entries!

Any standard abridged dictionary can be used to settle a SCRABBLE® challenge. But this easy-to-use, 683-page hardcover volume includes well over 100,000 words—and they're *all* acceptable SCRABBLE® words! That's why it's the official dictionary of first reference for all SCRABBLE® game tournaments sponsored by the National SCRABBLE® Association.

More on next page

2.2.5 Anadromous Fish
Columbia River Basin

The Pacific Northwest supports a large number of anadromous fish (species that migrate downriver to the ocean to mature, then return upstream to spawn). The principal anadromous fish runs in the Columbia Basin are chinook, coho, sockeye salmon, and steelhead trout.

These fish are an important resource to the Pacific Northwest, both for their economic value to the sport and commercial fisheries, and for their cultural and religious value to the region's Indian tribes and others.

The development of dam and reservoir projects on the Columbia and Snake Rivers and tributaries has reshaped the natural flows of these rivers. The use of storage reservoirs to capture runoff for later release results in reduced flows during the spring and early summer, when juvenile salmon and steelhead are migrating downstream to the ocean. Water velocities have also been reduced as a result of the increased cross-sectional area of the river due to run-of-river projects. These changes have slowed juvenile fish migration, exposing juvenile salmon and steelhead to predation and disease and impairing their ability to adapt to salt water when they reach the ocean. Additional mortality occurs as fish attempt to pass each dam on their downstream migration to the ocean.

Flow: Flow plays an important part in moving juvenile fish downstream to the ocean. In 1982, the Northwest Power Planning Council established a specific volume of water, known as the water budget, to increase river flows during the April 15 through June 15 period. This coincides with the peak out-migration of spring fish (predominately yearling chinook, steelhead, and sockeye), which depend on adequate river flow, particularly velocity, for a successful migration. The federal hydro system is operated to provide this water each year.

Not only is flow important for moving juvenile fish downstream past the dams, but flow is an important component of ensuring successful spawning and emergence of fall chinook on Vernita Bar (a gravel bar used by spawning fall chinook, located in the Hanford Reach of the Columbia River, downstream from Priest Rapids Dam). In 1988, BPA and the mid-Columbia operators signed a long-term Vernita Bar Agreement, which specifies protection requirements for fall chinook spawning, incubation and emergence on Vernita Bar.

Spill: Until adequate bypass systems are installed at all the dams, spill remains a necessary means of moving juvenile fish past dams. Planned fish spill now includes the negotiated Spill Agreement, as well as a restricted operation at Bonneville Dam by the Corps. Planned spill also includes spill levels specified by the federal Energy Regulatory Commission for non-federal projects. Planned spill does not include overgeneration spill (water which is spilled because there is no market for the energy it would produce) and is not changed as a result of the resource additions. Planned fish spills are met under all water conditions.

a magazine. Have each person work separately for the first part of this project:

- One person will describe the design elements.
- One person will evaluate the design. Which aspects of the design are effective, and which could be improved?
- One person will create a new design, using thumbnail sketches.

Then, the group members will meet and compare notes. Do all members of the group agree with the first member's description of the design? With the second member's analysis of the design? Do all members like the third member's redesign? What have your discussions taught you about design? Write a memo to your instructor presenting your findings. Include the photocopy with your memo. See Chapter 16 for a discussion of memos.

Web Exercise

2. The design of an organization's Web site is often far different from the design of its printed materials. One reason for this difference is that some design features cannot be achieved in Web pages, and some cannot be achieved in print. A second reason is that the audiences for the two media are often quite different; people visiting Web sites tend to be younger and less conservative than people reading print documents. Still, it makes sense for an organization to try to achieve some consistency in the design of its printed materials and its Web site. Choose an organization that has a Web site as well as printed information for the public, such as IBM. To what extent is the design of the two kinds of communication consistent? How is consistency achieved? Are the inconsistencies effective, or do they appear haphazard? Do the differences in the audiences and purposes of the

two kinds of communication justify the inconsistencies? Present your findings in a memo to your instructor. Include the photocopy with your memo. See Chapter 16 for a discussion of memos.

REFERENCES

Berry, R., Mobley, K., & Turk, K. (1994). Preparing to document an object-oriented project. *Technical Communication, 41,* 643–652.

Biggs, J. R. (1980). *Basic typography.* New York: Watson-Guptill.

Bonneville Power Administration. (1993). *Resource programs: Final environmental impact statement, volume 1: Environmental analysis.* Washington, DC: U.S. Department of Energy.

Felker, D. B., Pickering, F., Charrow, V. R., Holland, V. M., & Redish, J. C. (1981). *Guidelines for document designers.* Washington, DC: American Institutes for Research.

Fischer, J. M. (1992). *Inside DesignCAD.* Carmel, IN: New Riders Publishing.

Haley, A. (1991). All caps: A typographic oxymoron. *U&lc, 18*(3), 14–15.

Keyes, E. (1993). Typography, color, and information structure. *Technical Communication, 40,* 638–654.

Poulton, E. (1968). Rate of comprehension of an existing teleprinter output and of possible alternatives. *Journal of Applied Psychology, 52,* 16–21.

White, J. V. (1990). *Great pages: A common-sense approach to effective desktop design.* El Segundo, CA: Serif Publishing.

Williams, T., & Spyridakis, J. (1992). Visual discriminability of headings in text. *IEEE Transactions on Professional Communication, 35,* 64–70.

chapter 12

Integrating Graphics

Characteristics of an Effective Graphic 347

Planning, Creating, and Revising the Graphics 350
   Planning the Graphics *350*
   Creating the Graphics *351*
   Revising the Graphics *355*

Using Color Effectively 356

Choosing the Appropriate Kind of Graphic 361
   Illustrating Numerical Values *361*
   Illustrating Logical Relationships *376*
   Illustrating Instructions and Process Descriptions *378*
   Illustrating Visual and Spatial Characteristics *381*

Showing Motion in Your Graphics 387

Creating Graphics for International Readers 388

Understanding Graphics Software 391
   Spreadsheet Business Graphics *391*
   Paint Programs and Draw Programs *391*

Case: Adding Graphics to a Web Site 394

Writer's Checklist 394

Exercises 394
   Collaborative Exercise *397*
   Web Exercise *397*

Graphics are the "pictures" in technical communication: drawings, maps, photographs, diagrams, charts, graphs, and tables. In terms of appearance, graphics range from realistic, such as photographs, to highly abstract, such as organization charts. In terms of function, graphics range from the decorative, such as clip art that shows a group of people seated at a conference table, to highly informational, such as a table or a schematic diagram of an electrical device.

Graphics are important in technical communication because they can

- communicate some kinds of information that are difficult to communicate with words
- help you clarify and emphasize information
- interest the reader
- help nonnative speakers of English understand the document
- help communicate information to multiple audiences with different interests, aptitudes, and reading habits

This chapter begins by explaining the characteristics of effective graphics and describing the process of planning, creating, and evaluating them. Next it discusses the use of color and then covers the basic kinds of graphics used in technical communication. The chapter then explains how to show motion in graphics and how to create graphics for international readers. It concludes with a discussion of graphics software.

The most obvious reason to use graphics is that they motivate people to study the document. According to Gatlin (1988), 83 percent of what people learn derives from visual stimuli, whereas only 11 percent derives from our hearing. Because as a species we are attuned to acquiring information through sight, communication that includes a visual element beyond words on the page is more effective than communication that doesn't. People studying text with graphics learn about one-third more than people reading text without graphics (Levie & Lentz, 1982). And graphics help readers retain what they learn; people remember some 43 percent more when the document includes graphics (Morrison & Jimmerson, 1989). Readers want graphics. One survey indicates that readers of computer documentation consistently desire more graphics and fewer words (Brockmann, 1990, p. 203).

Graphics offer several benefits that words alone cannot provide.

- *Graphics are almost indispensable in demonstrating logical and numerical relationships.* For example, the organization chart used in most businesses is an effective way to represent the lines of authority in the organization. And graphics are useful in showing numerical relationships of all kinds. If you want to show the trends in the number of nuclear power plants completed each year over the last decade, a line graph would be much easier to understand than a paragraph full of numbers. Graphics can also show the relationships among several variables over time, such as the numbers of 4-cylinder, 6-cylinder, and 8-cylinder cars manufactured in the United States during each of the last five years.

- *Graphics can communicate spatial information more effectively than words alone.* Try to convey in words what a simple hammer looks like. It's not easy. But in 10 seconds you can draw a simple diagram that clearly shows the hammer. If you want to show your readers the location of the San Andreas fault, you can shade the fault area on a standard map of the United States. If you want to show the details of the derailleur mechanism on a bicycle, you can present a diagram of the bicycle with a close-up of the derailleur.

- *Graphics can communicate steps in a process more effectively than words alone.* A troubleshooter's guide, a common kind of table, explains how to understand what might be causing a problem in a process and what you might do to fix it. A diagram can explain clearly how acid rain forms.

- *Graphics can save space.* Consider the following paragraph:

  In the Wilmington area, some 90 percent of the population aged 18–24 watches movies or tapes on a VCR. They watch an average of 2.86 tapes a week. Among 35- to 49-year-olds, the percentage is 82, and the average number of movies or tapes is 2.19. Among the 50–64 age group, the percentage is 67, and the number of movies and tapes watched averages 2.5. Finally, among those people 65 years old or older, the percentage is 48, and the average number of movies and tapes watched weekly is 2.71.

  Presented as a paragraph, this information is uneconomical and hard to remember. Presented as a table, however, the information is more concise and more memorable.

Age	Percentage Watching Tapes/Movies	Number of Tapes Watched per Week
18–24	90	2.86
35–49	82	2.19
50–64	67	2.50
65+	48	2.71

- *Graphics can reduce the cost of documents intended for international readers.* Translation costs can reach 60¢ per word; used effectively, graphics can reduce the number of words you have to translate (Horton, 1993).

As you draft your document, think of opportunities to use graphics to clarify, emphasize, summarize, and organize information. McGuire and Brighton

(1990) recommend that you be alert to certain words and phrases that often signal the opportunity to create a graphic:

categories	features	numbers	routines
components	fields	phases	sequence
composed of	functions	procedures	shares
configured	if and then	process	structured
consists of	layers	related to	summary of
defines			

If you find yourself writing a sentence such as "The three categories of input modules are . . . ," consider creating a diagram to reinforce your idea. If you write "The first step in the procedure is . . . ," think in terms of a flowchart or a logic box. If you write "This structure is related to . . . ," consider a diagram.

Characteristics of an Effective Graphic

An effective graphic must be clear and understandable, and meaningfully related to the larger discussion. Six characteristics make a graphic clear and understandable:

- *A graphic should have a purpose.* Don't put a graphic in your document unless it will help your reader understand or remember the information. Beware of decorative, content-free clip art, the images that you can buy in hard copy or electronic form: drawings of businesspersons standing with clipboards, shaking hands, and so on. The novelty of meaningless clip art has worn thin.

- *A graphic should be honest.* Graphics can be dishonest, just as words can. You are responsible for making sure the graphic does not lie or mislead the reader. Following are some common ethical concerns to keep in mind as you create graphics:

  - If you did not create the graphic or generate the data, you must cite your source.
  - You must include all the relevant data in your graphic. For example, if you have a data point that you cannot explain, it is unethical to change the scale as a way to eliminate it.
  - You must show items as they really are. If the electrical appliance shown in your photograph requires an electrical cord, show it.
  - You should begin the axes in your graphs at zero—or mark them clearly—so that quantities are represented honestly.
  - You should not use a table to hide a data point that would be obvious in a graph.

– You should not use color or shading to misrepresent the importance of an item. A dark-shaded bar in a bar graph, for example, appears larger and nearer than a light-shaded bar of the same size.

Common problem areas are pointed out in the discussions of the different kinds of graphics throughout this chapter.

- *A graphic should be simple and uncluttered.* Three-dimensional bar graphs are easy to make with modern software, but they are harder to understand than two-dimensional ones, as shown in Figure 12.1. Graphics scholar Edward Tufte (1983) uses the term *chartjunk* to describe the ornamentation that clutters up the graphic, distracting the reader from the message. Unnecessary 3-D is one example of chartjunk. In Figure 12.1 the two-dimensional bar graph is clean and uncluttered, but the three-dimensional graph is more difficult to understand because the third dimension obscures the main data points. The number of uninsured emergency-room visits in February, for example, is very difficult to see in the three-dimensional graph.

- *A graphic should present a manageable amount of information.* If you present too much information in a graphic, you risk confusing your readers. Although some kinds of graphics can accommodate much more information than others, the most important factors to consider are your audience and purpose: what kinds of graphics are your readers familiar with, how much do they already know about the subject, and what do you want the document to do? In general, readers learn best if you present information in small chunks. Therefore, it is better to create several simple graphics than one complicated one.

- *A graphic should meet the reader's format expectations.* Through experience, readers learn how to read different kinds of graphics. Follow the conventions—for instance, by using diamonds to represent decision points in a flowchart—unless you have a good reason not to.

- *A graphic should be clearly labeled.* Every graphic (except a brief, informal one) should have a unique, clear, and informative title. The columns of a table and the axes of a graph should be labeled fully, complete with the units of measurement. The lines on a line graph should also be labeled. Readers should not have to guess whether you are using meters or yards, or whether you are including statistics from last year or just this year.

In addition, five characteristics of an effective graphic concern its integration with the surrounding text:

- *A graphic should be placed in an appropriate location.* If your readers need the information to understand the discussion, put the graphic directly after the pertinent point in the text—or as soon after that point as

FIGURE 1
Insured and Uninsured Emergency Room Visits
Parkland Memorial Hospital
January–March 1994

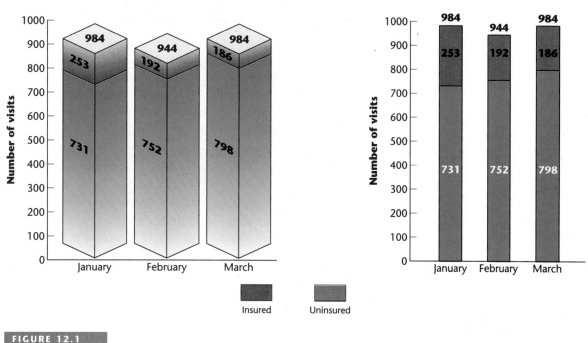

FIGURE 12.1

Chartjunk and
Clear Art

possible. If the information functions merely as support for a point that is
already clear, or as elaboration of it, include it as an appendix.

- *A graphic should be introduced in the text.* Whenever possible, refer to a
  graphic before it appears. Ideally, place the graphic on the same page as
  the reference. Refer to the graphic by number (such as *Figure 7*). Do not
  refer to "the figure above" or "the figure below"; the graphic might be
  moved during the production process. If the graphic is in an appendix,
  tell your readers where to find it: "For the complete details of the operating characteristics, see Appendix B."

- *A graphic should be explained in the text.* State what you want your
  readers to learn from looking at it (Dean & Kulhavy, 1981). When introducing a graphic, ask yourself whether a mere paraphrase of its title will
  communicate its significance: "Figure 2 is a comparison of the costs of
  the three major types of coal gasification plants." Sometimes you need to

explain why the graphic is important or how to interpret it. If you want the graphic to make a point, don't just hope your readers will understand it. Make the point explicitly:

As Figure 2 shows, a high-sulfur bituminous coal gasification plant is more expensive than either a low-sulfur bituminous or anthracite plant, but more than half of its cost is cleanup equipment. If these expenses could be eliminated, high-sulfur bituminous would be the least expensive of the three types of plants.

Often graphics are accompanied by captions—explanations ranging in length from a sentence to several paragraphs. Captions are useful for readers who are skimming the document; if the captioned material were included in the body text, they would not see it. Captioned text is usually smaller than the body text and presented in a different typeface.

- *A graphic should be clearly visible in the text.* Distinguish the graphic from the surrounding text by adding white space or rules—lines—or by enclosing it in a box. See Chapter 11 for a discussion of these design techniques.

- *A graphic should be accessible to your readers.* If the document is more than a few pages long and contains more than four or five graphics, consider including a list of illustrations—a table of contents for the graphics—so that readers can find them easily. See Chapter 10 for a discussion of lists of illustrations.

Planning, Creating, and Revising the Graphics

Using graphics involves planning, creating, and revising them. The following sections discuss these three steps.

Planning the Graphics

Some people, particularly those with a strong writing background, create documents by writing the words and then figuring out what kinds of graphics to add. Others, especially scientists, engineers, and technicians, think graphically first. They sketch their graphics and then write the text. Some people combine the two approaches, and some use a text-first process for some kinds of documents and a graphics-first process for others.

Regardless of which process you use, however, at some point you must think about the role of the graphics in the document. In planning your graphics, consider the following four aspects of the document, shown in Figure 12.2.

- *Audience.* Some types of graphics, such as pie charts, are easy for any reader to understand. Others, such as scatter graphs, are incomprehensible to the general reader. Do your readers understand certain kinds of graphics, such as schematics or computer flowcharts? Are they familiar with standard icons used in your field? Are they already motivated to

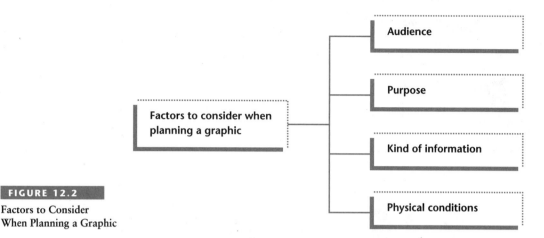

FIGURE 12.2

Factors to Consider
When Planning a Graphic

read your document or do you need to enliven the text with graphics to hold their attention, or perhaps even use color for emphasis?

- *Purpose.* What point are you trying to make with the graphic? Figure 12.3 on page 352 shows that even a few simple facts can yield a number of different points. Your responsibility is to determine what you want to show and then to figure out how best to show it. Don't rely on your software; it doesn't think.

  Each of these four graphs emphasizes a different point. Graph (a) focuses on the total number of cars disabled each month, classifying the problem by cause; graph (b) focuses on the three rail lines during one month; and so forth.

- *Kind of information.* The subject will help you determine what type of graphic to create. If you are writing about the number of languages spoken by citizens in your state, you will probably use tables for the mass of statistical data, maps to show the patterns of language use, and graphs to show trends in the statistics over time. If you are discussing how the sales of two products have changed over several years, a grouped bar chart might be the best type of graphic.

- *Physical conditions.* The physical conditions in which the document will be used—amount of lighting, amount of space available, exposure to water or grease, and so forth—might affect not only what kind of graphic to create but also the size and shape, the thickness of lines and size of type, and the color.

Creating the Graphics You have thought about what kinds of graphics would help you communicate your information, and you might even have made some quick sketches. Now it is

TABLE 1
Number of Rail Cars Disabled by Electrical Problems November–January

Rail Line	November		December		January	
	Disabled by electrical problems	Total disabled	Disabled by electrical problems	Total disabled	Disabled by electrical problems	Total disabled
Bryn Mawr	19	27	17	28	20	26
Swarthmore	12	16	9	17	13	16
Manayunk	22	34	26	31	24	33

a. Number of rail cars disabled, November–January

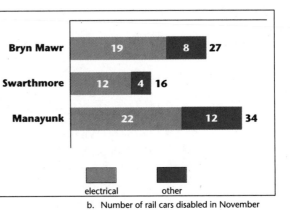

b. Number of rail cars disabled in November

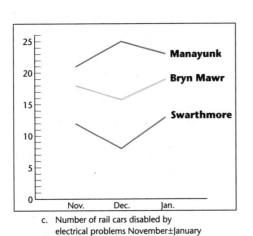

c. Number of rail cars disabled by electrical problems November±January

d. Range in percent of rail cars, by line, disabled by electrical problems, November±January

FIGURE 12.3

Different Graphics Emphasizing Different Points

time to consider how you are going to create them. As you plan, consider four important factors:

- *Time.* Creating a complicated graphic can take a lot of time, so you need a schedule. See Chapter 6 for a discussion of planning.
- *Money.* A high-quality graphic can be expensive. How big is the budget for the project? How can you use that money effectively? See Chapter 6 for a discussion of budgeting.
- *Equipment.* Although some graphics can be made with a straightedge and a pencil, most require other tools, including word-processing software for tables, spreadsheets for tables and graphs, and graphics software for diagrams.
- *Expertise.* How much do you know about creating graphics? How much expertise do you have access to?

Time, money, equipment, and expertise are overlapping categories, of course; if you have a large budget, you might be able to hire experts with the latest equipment and take the time to revise the graphics several times. Except for special projects, however, such as annual reports or extremely important proposals, people usually don't have all the resources they would like. Therefore, it is necessary to choose from among a range of different approaches to creating graphics, as shown in Figure 12.4.

- *Using existing graphics.* First, it is necessary to distinguish between two kinds of documents: unpublished and published.

    For a student paper that will not be published, some instructors permit the use of photocopies of existing graphics; other instructors do not. If you are permitted to photocopy existing graphics, you are ethically obligated to cite them, just as you would any other information you borrow.

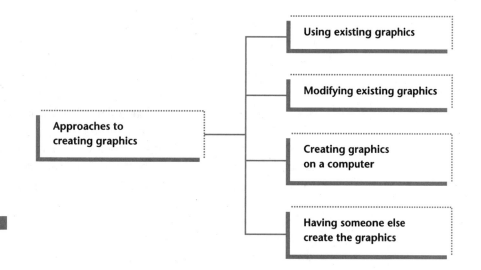

FIGURE 12.4

Approaches to
Creating Graphics

For a published document, written by a student or a nonstudent, the use of an existing graphic is permissible if the graphic is the property of the organization or if the organization has obtained permission to use it. If the graphic was created under work-for-hire guidelines (see Chapter 2), the organization has an unlimited right to use it. If, however, the graphic was not created under work-for-hire guidelines, the organization must acquire the right to multiple use; that is, the contract under which the graphic was originally created for the organization must grant the organization multiple use, not merely one-time use.

Be particularly careful about graphics that you find on the Web. Many people mistakenly think that anything they download from the Web can be used without permission. This is not true. The same copyright laws that apply to printed material apply to Web-based material, both words and graphics. However, many Web-based companies would be happy to sell you rights to use their graphics in your documents. Search for "computer graphics," "computer art," and related terms. In addition, there are a number of sources of free clip art, as well as many sources of free art to use in Web pages, such as buttons, icons, and backgrounds.

Aside from the issue of copyright, think carefully before using existing graphics, for three reasons:

- The style of the graphic might not match that of the other graphics you want to use.
- The graphic might lack some features you want or contain some you don't.
- If you are photocopying a printed graphic, the lines in the photocopy might lack crispness, and the shading in the original—the shades of gray or the colors—might "void out"; that is, they might not reproduce sharply, creating a blotchy appearance.

If you use an existing graphic, create your own number and title for it; the existing number will not match your numbering system, and the title is likely to be written in a different style.

- *Modifying existing graphics.* You can modify existing graphics in several ways: redraw an existing graphic, revise as you go, or use a scanner to digitize the graphic and then manipulate it electronically with graphics software. In either case, you are under the same ethical and legal obligations described in the previous section. Your source statement should indicate that your graphic is "based on" or "adapted from" your source.
- *Creating graphics on a computer.* Although simple tables and diagrams can be done by hand, in most cases you will want to use a computer to create professional-quality graphics.

Computer graphics offer two main advantages over noncomputer graphics: variety and reusability. The variety and complexity of new computer graphics is considerable. With each passing month, new graphics software comes on the market with powerful capabilities. Three-dimensional effects are routine now, and animation is becoming common. Because they are digital, computer graphics can be stored, reused, and revised. After you create a graphic showing this month's budget, you can copy it and easily revise it next month. Remember, however, that simpler graphics are usually preferable to complicated ones.

In addition to the general kinds of graphics software, there are specialized graphics packages for different applications. For instance, architecture students routinely create three-dimensional images using computer-assisted design (CAD) software. Computer scientists use flowcharting software to create flowcharts quickly. Proposal writers use project-management software to create network diagrams such as Gantt charts and PERT charts (see Chapter 15). Artists and designers use animation graphics that show movement over time. Physicians practice surgery using computer-generated graphics instead of cadavers, and aircraft designers use computer simulations rather than wind-tunnel experiments.

Computer graphics are not simple to create, however, and an unprofessional-looking graphic can hurt your credibility, leading the reader to question the professionalism of the whole document.

Visit the computer labs at your college or university to test some of the many excellent graphics software packages available. Also, consult the bibliography (Appendix D) for a list of books about computers and technical communication. But beware: any book about computer graphics is probably out of date as soon as it is published.

For more information on computer graphics, see the discussion later in this chapter.

- *Having someone else create the graphics.* If graphics software is so versatile, why would technical communicators choose not to make the graphics themselves on their own computers? One obvious reason: professional-level software can cost hundreds of dollars. Another less obvious reason: some people don't have the artistic ability to create professional-quality graphics. And even if the person does have the ability, proficiency with a sophisticated graphics package can require dozens or even hundreds of hours.

  In most cases, your choices for having someone else create your graphics are constrained by your budget and policies. Some companies have technical-publications departments with graphics experts. Others subcontract this work. Many print shops and service bureaus have graphics experts on staff or can direct you to them.

Revising the Graphics   As with any other aspect of technical communication, build in enough time and money to revise the graphics.

Create a checklist and evaluate each graphic to test its effectiveness. The Writer's Checklist at the end of this chapter is a good starting point.

Show the graphics to people with backgrounds like the intended readers' and ask them for suggestions. Revise the graphics and solicit more reactions.

Using Color Effectively

Color draws the reader's eyes to information you want to emphasize, helps you establish visual patterns to increase the reader's understanding, and adds interest. Color has been used for many decades in magazines and some books. But now that the price of color printers is dropping, color is becoming more common in technical communication. Color is a powerful tool that is easy to misuse. The following discussion is based on Jan V. White's excellent text *Color for the Electronic Age* (1990).

In using color in graphics and page design, keep in mind seven guidelines:

- *Don't overdo it.* Even if your computer monitor is capable of displaying 16 million colors, readers can understand only two or three at a time (White, 1990, p. 19). Use colors for small items, such as portions of graphics and important words. And don't use colors where black will work better. Use black for the main text; black is easier for the reader to interpret.

- *Use color to emphasize particular items.* Horton (1991) notes that we interpret color before we interpret shape, size, or placement on the page. Therefore, color effectively draws the reader's attention to a particular item or class of items on a page. Figure 12.5 shows how color emphasizes different kinds of items.

- *Use colors to create patterns.* We all recognize the pattern of black writing against a yellow background as a traffic sign. The same principle applies in graphics and document design. Choose a warm color such as red as the background or the text color for all your safety comments; choose a cool color such as green for your tips. Use different-colored paper for different sections of a manual. In creating patterns, consider using shapes too. For instance, use red for the safety comments, placing each one in an octagon resembling a stop sign. This way, you give your readers two visual characteristics to help them recognize the pattern. Figure 12.6 (Curtis & Barnes, 1989, p. 532) on page 358 shows how the page of a biology textbook uses color to establish patterns.

  Color can also be used effectively to emphasize other kinds of particular items, such as text boxes, rules, screens, and headers and footers. These design elements are discussed in Chapter 11.

Color draws the reader's
attention to the line.

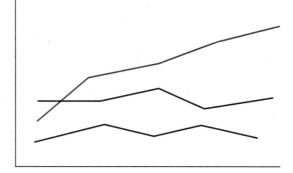

Color is useful for emphasizing short phrases. Used
for longer passages, color
loses its impact.

**Pixel Driver Renders 600,000 Polygons
per Second**

The colored frame focuses
the reader's attention on the
information in the graph.

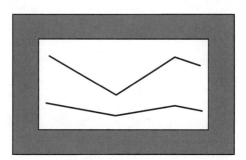

Here the color emphasizes a
row; it could also be used to
emphasize a column or a single
data cell.

Table 2
Israeli Power Plant Sites

Site	Generating Units	Installed Capacity (MW)
Haifa	4	430
Hadera	4	1400
Tel Aviv	4	530
Ashdod	9	1210
Total	21	3570

FIGURE 12.5

Color Used for Emphasis

Source: B. Golany, Y. Roll, & D. Rybak. (1994, August). Measuring efficiency of power plants in Israel by data envelopment analysis, *IEEE Transactions on Engineering Management 41*, 3, p. 292.

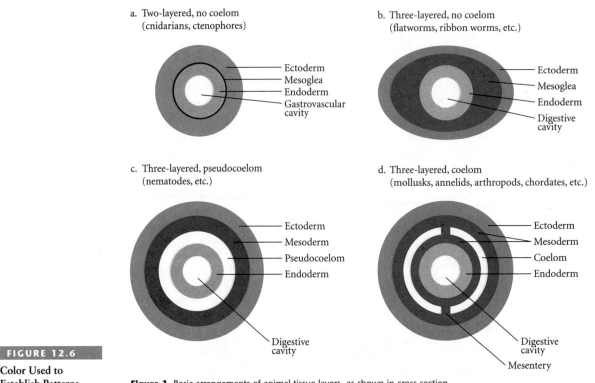

a. Two-layered, no coelom
(cnidarians, ctenophores)

— Ectoderm
— Mesoglea
— Endoderm
— Gastrovascular cavity

b. Three-layered, no coelom
(flatworms, ribbon worms, etc.)

— Ectoderm
— Mesoglea
— Endoderm
— Digestive cavity

c. Three-layered, pseudocoelom
(nematodes, etc.)

— Ectoderm
— Mesoderm
— Pseudocoelom
— Endoderm
— Digestive cavity

d. Three-layered, coelom
(mollusks, annelids, arthropods, chordates, etc.)

— Ectoderm
— Mesoderm
— Coelom
— Endoderm
— Digestive cavity
— Mesentery

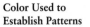

FIGURE 12.6

Color Used to
Establish Patterns

Figure 1. Basic arrangements of animal tissue layers, as shown in cross section.

- *Use contrast effectively.* The visibility of a color is a function of the background against which it appears. The strongest contrasts are between black and white and black and yellow. In Figure 12.7, notice that a color washes out if the background color is too similar.

  The need to provide effective contrast applies not only in text but in graphics used in presentations (see Chapter 21). You want to provide a natural-looking background against which the text or graphics provide a vivid contrast. Transparencies are used with the room lights on; therefore, the most popular background is the white provided by the screen,

FIGURE 12.7

The Effect of Background
in Creating Contrast

and the most popular color for the text or graphics is black or dark blue. For slides, however, the room lights are either dim or off. Therefore, the most popular background color for slides is dark blue, with yellow or white text and graphics. Figure 12.8 shows the use of contrast in a transparency and a slide.

- *Take advantage of the meanings that colors already have.* In our culture, red signals danger, heat, or electricity. Yellow signals caution; orange signals warning. If you use any of these warm colors in ways that depart from these cultural meanings, you can confuse the reader. The cooler colors—blues and greens—are more conservative and subtle. Keep in mind, however, that different cultures interpret colors differently. See Chapter 5 for a discussion of cultural differences. In the drawing of the electric car shown in Figure 12.9 on page 360, the batteries are red. The warm red contrasts effectively with the cool green of the car body.

- *Realize that color swallows up text.* If you are printing against a colored background, you might need to make the text a little bigger, because the color will make text look smaller. Figure 12.10 on page 360 shows that the text printed against a white background looks bigger than the same size text printed against a colored background. White type counteracts this effect, as shown in Figure 12.11 on page 360.

- *Use light colors to make objects look bigger.* In Figure 12.12 on page 360, the yellow box looks bigger than the blue box, even though they are the same size.

New Video Interfaces

- Video Scope
- Video Space Icon
- Video Space Monitor
- Paper Video

New Video Interfaces

- Video Scope
- Video Space Icon
- Video Space Monitor
- Paper Video

FIGURE 12.8

Contrast in a Transparency (left) and a Slide (right)

Safety system
Motor controller
Energy management system
Battery monitor system
Regenerative braking system
Variable-temperature seat
Audio navigation system
Battery pack
Brushless dc motor
Inductive charging system

FIGURE 12.9

Colors Already Have Clear Associations for Readers

FIGURE 12.10

A Colored Background Minimizes the Appearance of Text

This text looks bigger because of the white background.

This text looks smaller, even though it is the same size, because of the colored background.

FIGURE 12.11

White and Black Type against a Colored Background

This line of type appears to reach out to the reader.
This line of type appears to recede into the background.

FIGURE 12.12

Light-Colored Objects Appear Bigger Than Dark-Colored Objects

Choosing the Appropriate Kind of Graphic

The graphics used in technical documents are classified into two basic categories: tables and figures. *Tables* are lists of data—usually numbers—arranged in columns. *Figures* are everything else: graphs, charts, diagrams, photographs, and the like. Generally, tables and figures have their own sets of numbers: the first table in a document is Table 1; the first figure is Figure 1. In documents of more than one chapter (as in this book), the graphics are usually numbered chapter by chapter. Figure 3.2, for example, would be the second figure in Chapter 3.

But this broad distinction between tables and figures offers little guidance about what kind of graphic to use in a particular situation. There is no simple system for choosing the type of graphic, because for many kinds of situations several different types would do the job. In general, however, graphics can be categorized according to the kind of information they communicate. The discussion that follows is based on a classification system created by William Horton in his chapter "Pictures Please—Presenting Information Visually," in *Techniques for Technical Communicators* (Barnum & Carliner, 1993). I strongly recommend Horton's guide to graphics.

Technical information can be classified into four categories, as shown in Figure 12.13. Note that some kinds of graphics can convey several kinds of information. For instance, tables can communicate both numerical values and procedures.

Figure 12.14 on page 362 summarizes the following discussion.

Illustrating Numerical Values

In technical communication, you will often have to display numerical information. Sometimes you will communicate precise numerical data; sometimes, relative values. Sometimes you will communicate numerical values at one point in

FIGURE 12.13

Four Categories of Technical Information Communicated by Graphics

Categories of technical information

- Numerical values
- Logical relationships
- Instructions and process descriptions
- Visual and spatial characteristics

Purpose	Type of Graphic	What the Graphic Does Best
ILLUSTRATING NUMERICAL INFORMATION	Table	• Shows large amounts of numerical data, especially when there are several variables for a number of items.
	Bar graph	• Shows the relative values of two or more items.
	Pictograph	• Enlivens statistical information for the general reader.
	Line graph	• Shows how the quantity of an item changes over time. A line graph can accommodate much more data than a bar graph can.
	Pie chart	• Shows the relative size of the parts of a whole. Pie charts are instantly familiar to most readers.
ILLUSTRATING LOGICAL RELATIONSHIPS	Diagram	• Represents items or properties of items.
	Organization chart	• Shows the lines of authority and responsibility in an organization.
ILLUSTRATING INSTRUCTIONS AND PROCESS DESCRIPTIONS	Checklist	• Lists or shows what equipment or materials to gather, or describes an action.
	Table	• Shows numbers of items or indicates the state (on/off) of an item.
	Flowchart	• Shows the stages of a procedure or a process.
	Logic box	• Shows which of two or more paths to follow.
	Logic tree	• Shows which of two or more paths to follow.
ILLUSTRATING VISUAL AND SPATIAL CHARACTERISTICS	Drawing	• Shows simplified representations of objects.
	Map	• Shows geographical areas.
	Photograph	• Shows precisely the external surface of objects.
	Screen shot	• Shows what appears on a computer screen.

FIGURE 12.14

Choosing the Appropriate Kind of Graphic (Based on Horton [1993])

time; sometimes, these values as they change over time. The basic kinds of graphics used to communicate numerical values are tables, bar graphs, pictographs, line graphs, and pie charts.

Tables

Tables easily convey large amounts of numerical data, and often provide the only means of showing several variables for a number of items. For example, if you want to show the numbers of people employed in six industries in 10 states, a table would probably be the best graphic. Although tables lack the visual appeal of other kinds of graphics, they can handle much more information with complete precision.

Keep in mind the needs and aptitudes of your audience. You can convey much more information in a table than your readers need or can understand easily. If you choose to create a large table—one with, say, 20 columns and 30

Table 6. Test Results for Valves #1 and #2 ——— Title

Valve Readings	Maximum Bypass Cv	Minimum Recirculation Flow (GPM)	Pilot Threads Exposed	Main ΔP at Rated Flow (psid)
Valve #1				
Initial	43.1	955	+3	4.4
Final	43.1	955	+3	. . .
Valve #2				
Initial	48.1	930	+3	4.5
Final	48.2	950	+2	. . .

Number
Stub head
Row head
Column head

Source statement ——— Source: "Third Progress Report: Anderson Machine Tools Reconfiguration Project."

FIGURE 12.15

Parts of a Table

rows—consider placing it in an appendix and making a summary table for the body of the document.

Figure 12.15 shows the standard parts of a table. Tables are identified with both a number ("Table 1") and a substantive, informative title, one that encompasses the items being compared, as well as the basis (or bases) of comparison.

The basis of comparison is time: each year from 1990 to 1994.

Mallard Population in Rangeley, 1990–1994

The basis of comparison is the two nations: Japan and the United States.

The Growth of the Robotics Industry in Japan and the United States, 1997

Note that most tables are numbered and titled above the data. The number and title are centered horizontally or left justified.

Here are 10 guidelines for constructing tables:

- *Indicate the units in which you are presenting the data.* If all the data are expressed in the same unit, indicate that unit in the title:

  Farm Size in the Midwestern States (in Hectares)

  If the data in one column are expressed in a different unit from the data in another column, indicate the units in the column headings:

Population (in millions)	Per Capita Income (in thousands of U.S. dollars)

  If all the data cells in a column use the same unit, indicate that unit in the column head, not in each data cell:

Speed (knots)
15
18
14

You can express data in both real numbers and percentages. A column heading and the first data cell under it might read as follows:

Number of Students (Percentage)
53 (83)

- *In the stub—the left-hand column—list the items being compared.* Arrange the items in the stub in some logical order: big to small, important to unimportant, alphabetical, chronological, geographical, and so forth. If the items fall into several categories, you can include the names of the categories in the stub:

Snow Belt States ....
    Connecticut............................
    New York...............................
    Vermont ....

Sun Belt States ....
    Arizona..................................
    California...............................
    New Mexico...........................

If the items in the stub are not grouped in logical categories, skip a line every five rows to help the reader follow the rows across the table. Or use a shaded or colored background for every other set of five rows. *Dot leaders,* a row of dots that links the stub and the next column, are also useful.

- *In the columns, arrange the data clearly and logically.* Line up the numbers consistently by using the decimal tab function:

3,147.4
  365.7
46,803.5

In general, don't change units. If you use meters for one of your quantities, don't use feet for another, unless the quantities are so dissimilar that your readers would have a difficult time understanding them as expressed in the same units.

3.4 hr
12.7 min
4.3 sec

This list would probably be easier for most readers to understand than one in which all quantities were expressed in the same unit.

- *Perform the computations your readers will need.* If your readers will need to know the totals for the columns or the rows, provide them. If your readers will need to know percentage changes from one column to the next, present them:

Number of Students (Percentage Change from Previous Year)

1991	1992	1993
619	644 (+4.0)	614 (−4.7)

- *Use dot leaders if a column contains a "blank" spot: a place where there are no appropriate data:*

3,147

. . .

46,803

However, don't substitute dot leaders for a quantity of zero:

3,147

0

46,803

- *Don't make the table wider than it needs to be.* The reader should be able to scan across a row easily. As Jan White (1984) points out, there is no reason to make the table as wide as the text column in the document. If a column heading is long—say, more than five or six words—stack the words rather than make the column wide:

Computers Sold
Without a CD-ROM

- *Minimize the use of lines.* Grimstead (1987) recommends using lines only when necessary: to separate the title and the headings, the headings and the body, and the body and the notes. When you use lines, make them thin rather than thick.

- *Try to keep the table on one page.* You can tell your software not to split a table between pages. If you have to split it, write *continued* at the bottom; on the next page, include the full title, followed by *continued*. Repeat the column headings.

- *Provide footnotes for any information that needs to be explained.* You should provide all the information your readers need to understand the graphic.

- *Indicate the source of your information if you did not generate it yourself.* A typical source statement is: "Source: Data from *IEEE Spectrum*, vol. 30, no. 8 (1994), p. 34."

Tables can also convey textual information effectively, as shown in Figure 12.16 (U.S. Congress, *Making*, 1993: 41) on page 366.

Bar Graphs

Like tables, bar graphs can communicate numerical values, but they are better at showing the relative values of two or more items. In considering whether to make the bars vertical or horizontal, think about how people envision information.

Table 2.3 Types of Electronic Kiosks: Key Characteristics and Selected Applications

Type of Kiosk	Key Characteristics	Selected Applications
Off-line: Stand-alone	For information that does not need updating: no telecommunications costs	GSA's Central Office Building directory
Off-line: Polled	Can update information, and retrieve queries and survey results over a telephone line and modem at night	USPS's "Postal Buddy"; "24-Hour City Hall"
Online	Can process information immediately; can update rules and software in central computer; requires dedicated telephone line and central computer capacity.	Tulare County, CA's "Tulare Touch"; State of California's "InfoCalifornia"
Online: Transactional	Online, but can also collect money via credit or debit cards for bills and services	Long Beach, CA's "Auto Clerk"; State of California's "InfoCalifornia"

KEY: GSA = General Services Administration; USPS = U.S. Postal Service.
Source: Office of Technology Assessment, 1993.

FIGURE 12.16

Text Table

Vertical bars are best for showing quantities such as height, size, and amount. Horizontal bars are best for showing quantities such as speed and distance. However, these distinctions are not ironclad; as long as the axes are labeled carefully, your readers should have no trouble understanding the graph.

Figure 12.17 shows the structure of typical horizontal and vertical bar graphs.

When you construct bar graphs, follow six basic guidelines.

FIGURE 12.17

Structures of Horizontal and Vertical Bar Graphs

Figure 1. Horizontal graph

Figure 1. Vertical graph

- *Make the proportions fair.* For all bar graphs, number the axes at regular intervals. If you are drawing the graph (rather than using software), use a ruler or graph paper.

  For vertical bar graphs, choose intervals that make your vertical axis about three-quarters the length of the horizontal axis. If the vertical axis is much longer than that, the differences in the height of the bars will be exaggerated. If the horizontal axis is too long, the differences will be unfairly flattened. Make all bars equally wide, and make the amount of space between them about half the width of a bar. Here are two poorly proportioned graphs.

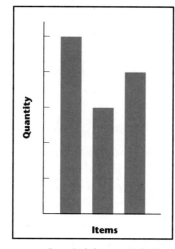

a. Excessively long vertical axis

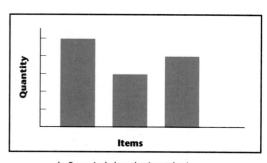

b. Excessively long horizontal axis

- *If possible, begin the quantity scale at zero.* Doing so will ensure that the bars accurately represent the quantities. Notice how misleading a graph is if the scale doesn't begin at zero.

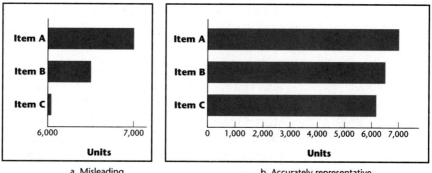

a. Misleading

b. Accurately representative

Version (a) exaggerates the differences among items a, b, and c. If it is not practical to start the quantity scale at zero, break the quantity axis clearly.

- *Use tick marks or grid lines to signal the amounts.* Ticks are the little marks drawn across the axis:

Grid lines are ticks extended through the bars. In most cases, grid lines are necessary only if you have several bars (some of which would be too far away from tick marks for readers to gauge the quantity easily) and if the actual quantities are not indicated near the bars.

- *Arrange the bars in a logical sequence.* For a vertical bar graph, use chronology for the sequence when you can. For a horizontal bar graph, arrange the bars in order of descending size, beginning at the top of the graph, unless some other logical sequence seems more appropriate.

- *Place the title underneath the figure.* Unlike tables, which are usually read from top to bottom, figures are usually read from the bottom up.

- *Indicate the source of your information if you did not generate it yourself.*

Figure 12.18 shows an effective bar graph. Notice the use of grid lines. Four variations on the basic bar graph can help you accommodate different communication needs.

- The *grouped bar graph* lets you compare two or three quantities for each item you are representing. Grouped bar graphs are useful for showing information such as the numbers of full-time and part-time students at several universities. One kind of bar represents the full-time students; the other, the part-time. To distinguish the bars from each other, use hatching (striping), shading, or color, and either label one set of bars or provide a key.

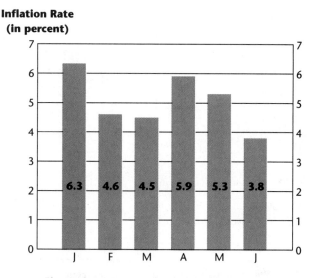

**Inflation Rate
(in percent)**

FIGURE 12.18

Effective Bar Graph

Figure 1. Tri-County Inflation Rate This Year to Date

- In the *subdivided bar graph,* Aspect I and Aspect II are stacked, just as wooden blocks are placed on top of one another. Although the totals are easy to compare in a subdivided bar graph, the individual quantities (except those that begin on the horizontal axis) are not.
- The *100-percent bar graph,* which shows the relative proportions of the elements that make up several items, is useful in portraying, for example, the proportion of full-scholarship, partial-scholarship, and no-scholarship students at a number of colleges.
- The *deviation bar graph* shows how various quantities deviate from a norm. Deviation bar graphs are often used when the information contains both positive and negative values, as with profits and losses. Bars on the positive side of the norm line represent profits; on the negative side, losses.

Figure 12.19 on page 370 shows a grouped bar graph, a subdivided bar graph, a 100-percent bar graph, and a deviation bar graph.

Pictographs

Pictographs are bar graphs in which the bars are replaced by a series of symbols that represent the items, as shown in Figure 12.20 on page 371. Pictographs are used primarily to present statistical information for the general reader. The quantity scale is usually replaced by a statement that indicates the numerical value of each symbol. Thousands of clip-art symbols and pictures are available for creating pictographs.

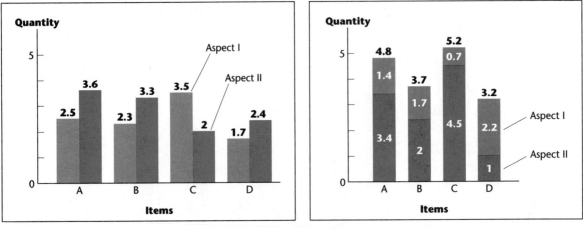

a. Grouped bar graph

b. Subdivided bar graph

c. 100-Percent bar graph

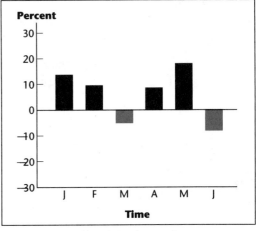

d. Deviation bar graph

FIGURE 12.19

Variations on the
Bar Graph

Follow these two guidelines in creating pictographs:

- *In most cases, arrange pictographs horizontally rather than vertically.*
  Pictures of houses balanced on top of each other can look foolish.
- *Represent quantities honestly.* Figure 12.21 shows that a picture drawn
  to scale can appear many times larger than it should. The reader sees the
  total area of the symbol rather than its height.

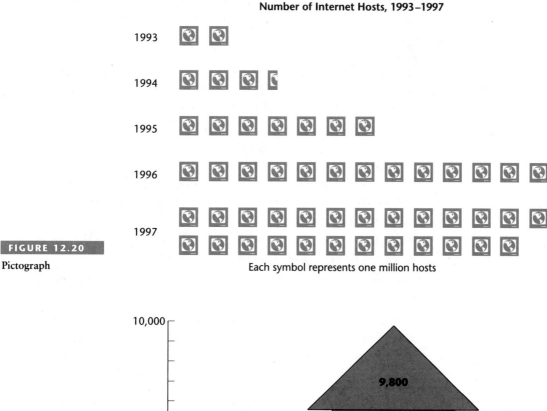

FIGURE 12.20

Pictograph

Each symbol represents one million hosts

FIGURE 12.21

Misleading Pictograph

Figure 3. Housing Starts in the Tri-State Area, 1995–1996

Line Graphs

In line graphs, the quantities are represented by points linked by a line. This line traces a pattern that in a bar graph would be formed by the highest point of each bar. Line graphs are used almost exclusively to show how the quantity of an item changes over time, such as the month-by-month production figures for a product. A line graph focuses the reader's attention on the change in quantity, whereas a bar graph emphasizes the quantities themselves.

An advantage of the line graph for demonstrating change is that it can convey much more data than a bar graph. Because you can plot three or four lines on one graph, you can compare trends conveniently. If the lines intersect often, however, the graph will be unclear. If this is the case, draw separate graphs. Figure 12.22 shows a line graph. You can use a different color or a different pattern—or both—to distinguish the lines.

Use the following three guidelines in constructing line graphs:

- *If possible, begin the quantity scale at zero.* Just as in a bar graph, beginning at zero portrays the information most honestly. If you cannot begin at zero, clearly indicate a break in the axis.

- *Use reasonable proportions for the vertical and horizontal axes.* As with bar graphs, try to make the vertical axis about three-quarters the length of the horizontal axis.

- *Use grid lines—horizontal, vertical, or both—rather than tick marks when your readers need to read the quantities precisely.*

Two common variations on the line graph are the stratum graph and the ratio graph.

- A *stratum graph*, sometimes called an *area graph*, shows an overall change while partitioning that change into its parts. Figure 12.23 (U.S. Department of Commerce, 1994, p. 20-4) is an example of a stratum

Million Barrels per Day

Source: Coglin Associates

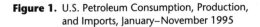

Figure 1. U.S. Petroleum Consumption, Production, and Imports, January–November 1995

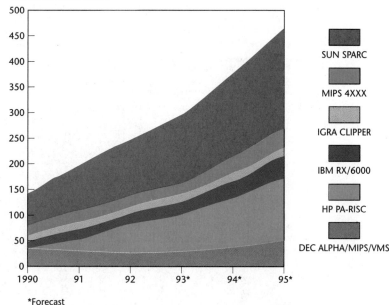

Figure 26-10. U.S. RISC Workstations Shipments

Thousands of Units

*Forecast
Note: RISC = reduced instruction set computing.
Source: InfoCorp.

FIGURE 12.23

Stratum Graph

graph. The figure shows that in 1990, DEC shipped some 30,000 Alpha/MIPS/VMS workstations; Sun shipped some 70,000 workstations. Together, the six companies represented in the stratum graph shipped 145,000 RISC workstations in 1990; by 1995, that figure was expected to rise to over 450,000.

- A *ratio graph* is a line graph used to emphasize percentages of change rather than the change in real numbers. A ratio graph charts data that could not be represented fairly on a standard line graph. For example, you might wish to compare the month-by-month sales of a large corporation with those of a small one. You would have great trouble making a vertical axis that would accommodate a company with monthly sales of $200,000 and one with monthly sales of $2,000,000. Even if you had a giant piece of paper, the graph could not reflect a true relation between the companies. If both companies increased their sales at the same rate (such as 1 percent per month), the small company's line would appear relatively flat, whereas the big company's line would shoot upward, just because of the large quantities involved.

   To solve this problem, the ratio graph compresses the vertical axis more and more as the quantities increase. Figure 12.24 (Council, 1994,

Sources of Personal Income

Personal income increased $71.3 billion (annual rate) in February, following a decline of $18.8 billion in January. The changes were affected by a number of special factors, the most significant of which was uninsured losses to residential and business property from the California earthquake in January. Excluding the special factors, personal income increased $17.1 billion in February and $34.1 billion in January.

FIGURE 12.24

Ratio Graph

*Seasonally adjusted annual rates
Source: Department of Commerce

Council of Economic Advisors

p. 5) shows how ratio graphs work. The bottom line—transfer payments—increases from about $520 billion in 1986 to about $920 billion in 1993. This increase is about 75 percent. During the same period, the top line, total personal income, increased from about $3.6 trillion to about $5.5 trillion, an increase of about 52 percent. Even though these quantities differ greatly, the ratio graph accurately represents how each one changes.

Pie Charts

The *pie chart* is a simple but limited design used for showing the relative size of the parts of a whole. Pie charts can be instantly recognized and understood by the untrained reader. Figure 12.25 (Mankiw, 1997, p. 37) shows a typical pie chart. Here are eight guidelines for creating pie charts:

- *Restrict the number of slices to six or seven.* Readers cannot easily handle more than seven slices because, as the slices get smaller, judging their sizes becomes more difficult.

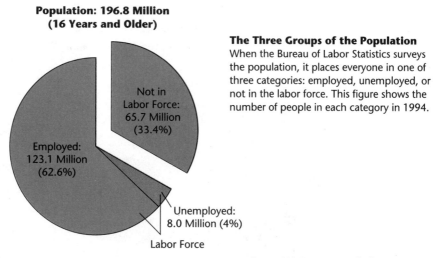

**Population: 196.8 Million
(16 Years and Older)**

Not in
Labor Force:
65.7 Million
(33.4%)

Employed:
123.1 Million
(62.6%)

Unemployed:
8.0 Million (4%)

Labor Force

The Three Groups of the Population
When the Bureau of Labor Statistics surveys
the population, it places everyone in one of
three categories: employed, unemployed, or
not in the labor force. This figure shows the
number of people in each category in 1994.

FIGURE 12.25

Pie Chart

Source: U.S. Department of Labor.

- *Begin with the largest slice at the top of the pie (noon position) and work clockwise in decreasing-size order, unless you have a good reason for arranging the slices in a different order.*

- *Create a miscellaneous slice for very small quantities that would make the chart unclear.* Explain the contents of this slice in a footnote. This miscellaneous slice, sometimes called "other," appears after the other sections as you work in a clockwise direction.

- *Label the slices (horizontally, not radially) inside the slice, if space permits.* It is customary to include the percentage that each slice represents. Sometimes the absolute quantity is added.

- *To emphasize one of the slices, use a brightly contrasting color or separate the slice from the pie.* Do this, for example, when you introduce a discussion of the item represented by that slice.

- *Check to see that your software follows appropriate guidelines for pie charts.* Some software violates basic guidelines for creating pie charts. Many software programs violate the principle of simplicity. They add fancy visual effects, which can actually hurt comprehension. For instance, many programs portray the pie in three dimensions, as in Figure 12.26 (Health, 1990, p. 7) on page 376.

  The "shadows" misrepresent the slices to which they are attached by making them appear larger than they should be. The three-dimensional pie chart is misleading in another way: the slices at the bottom of the pie appear larger than they should because they seem to be closer to the reader. To communicate clearly, make your pies two dimensional, not three.

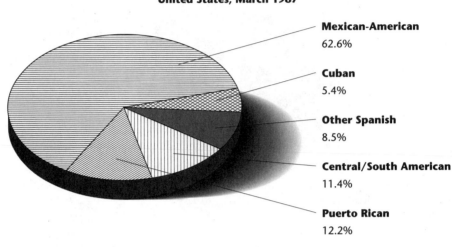

Chart 2.

**Population of Spanish origin by type of Spanish origin:
United States, March 1987**

Mexican-American
62.6%

Cuban
5.4%

Other Spanish
8.5%

Central/South American
11.4%

Puerto Rican
12.2%

Source: Bureau of the Census: Current Population Reports. Series P-20, No. 434, Page 12.

- *Don't overdo fill patterns.* Fill patterns are the designs, shades, or colors that distinguish one slice from another. In general, use understated, simple patterns, or no patterns; instead, include information about the slice, such as raw numbers, percentages, or labels. This tactic reduces the need for legends.
- *Check that your percentages add up to 100.* If you are doing the calculations yourself, rather than using software, check your math.

**Illustrating
Logical
Relationships**

Graphics can help you communicate the logical relationships among different items. For instance, in describing a piece of hardware you might want to show its major components. In writing a proposal, you might want to show alternative courses of action and explain the advantages and disadvantages of each. The two kinds of graphics most useful for showing logical relationships are diagrams and organization charts.

Diagrams

A *diagram* is a visual metaphor, a drawing in which you use symbols to represent items or properties of items. Common kinds of diagrams in technical communication are blueprints, wiring diagrams, and schematics. Figure 12.27 (U.S. Environmental, 1991, p. 124) is an example of a schematic diagram.

In this diagram, the artist has used visual symbols, such as the fan and the filter, to represent components and how they work together. Note that diagrams

FIGURE B-1: Typical HVAC System Components

Schematic Diagram

are not representational. Although the silhouette of the person looks realistic, the HVAC system itself is drawn to show its function, not its appearance. The clip art of the person helps the reader see the location of the HVAC equipment.

Diagrams can also clarify difficult concepts, as shown in Figure 12.28 (Labuz, 1984, p. 184). Here the writer has effectively arranged a series of simple calculations to show how to convert binary numbers into decimal numbers.

A popular form of diagram is the *block diagram*, which uses simple geometric shapes—usually rectangles—to suggest logical relationships. In Figure 12.29 on page 378, notice how much clearer the block diagram is than the prose version of the same information.

Diagram Used to Clarify
a Difficult Concept

Converting a Binary Number into a Decimal Number							
Binary Number:	1	1	1	0			
	8	4	2	1			
	x1	x1	x1	x0			
	8	4	2	0			
Total:	8 +	4 +	2 +	0	=	**14 Decimal Number**	

Prose Version

DFC is composed of two modules: the ADM and the COT. The ADM microcode consists of three components: the 120 protocol, the ADM control, and the TCG module.

Block-Diagram Version

FIGURE 12.29

Block Diagram and Prose Description

Organization Charts

An *organization chart* is a form of block diagram that portrays the lines of authority and responsibility in an organization. In most cases, the positions are represented by rectangles. The more important positions can be emphasized through the size of the boxes, the width of the lines that form the boxes, the typeface, or color. If space permits, the boxes can contain brief descriptions of the positions, duties, or responsibilities. Figure 12.30 is a typical organization chart. Unlike most other figures, the title of an organization chart generally appears above the chart because the chart is read from the top down.

Illustrating Instructions and Process Descriptions

Graphics often accompany instructions (Chapter 20) and process descriptions (Chapter 9). The following discussion covers some of the kinds of graphics used in writing about actions: checklists, tables, flowcharts, logic boxes, and logic trees.

Checklists

In explaining how to carry out a task, you often need to show the reader what equipment or materials to gather, or describe an action or series of actions to

FIGURE 12.30

Organization Chart

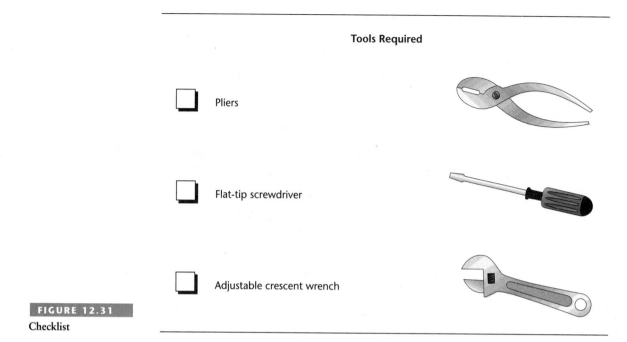

Tools Required

☐ Pliers

☐ Flat-tip screwdriver

☐ Adjustable crescent wrench

FIGURE 12.31

Checklist

take. A *checklist* is simply a list of items, with check boxes before each one. If your reader might be unfamiliar with the items you are listing, it's a good idea to include drawings of the items, as shown in Figure 12.31.

Often you need to indicate that your reader is to carry out certain tasks at certain intervals. A table is a useful graphic for this kind of information, as illustrated in Figure 12.32 (McComb, 1991, p. 133) on page 380. In this figure, the darkened check boxes indicate when an action is to be performed. For example, a VCR subjected to heavy use should have its exterior and interior cleaned and dusted at 6 months, 12 months, 18 months, 24 months, 30 months, and 36 months.

Flowcharts

A flowchart, as its name suggests, shows the stages of a procedure or a process. A flowchart might be used, for example, to show the steps involved in transforming lumber into paper or in synthesizing an antibody. Flowcharts are useful, too, for summarizing instructions. The basic flowchart portrays stages with labeled rectangles or circles. If you are addressing general readers, consider pictorial symbols instead of geometric shapes. If the process involves quantities (for example, paper manufacturing might "waste" 30 percent of the lumber), they can be listed or merely suggested by the thickness of the line used to connect the stages. Flowcharts can portray *open systems* (those that have a "start" and a "finish") or *closed systems* (those that end where they began).

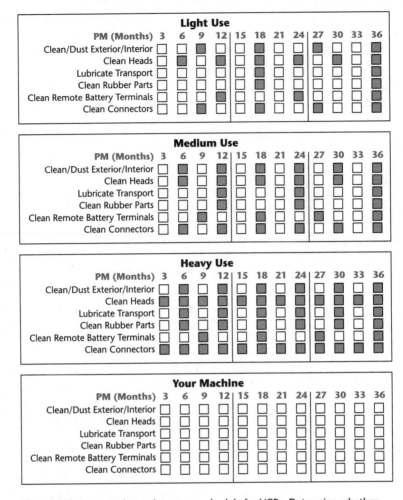

Figure 5-1. Preventative maintenance schedule for VCRs. Determine whether your machine gets light, medium, or heavy use. Then, perform the task indicated by each green box. Record the maintenance in the bottom portion, "Your Machine."

Figure 12.33 (Chernicoff & Venkatakrishnan, 1995, p. 159) shows an open-system flowchart. Figure 12.34 (Curtis & Barnes, 1989, p. 251) on page 382 shows a closed-system flowchart.

Logic Boxes

Logic boxes show the reader which of two or more paths to follow. Figure 12.35 on page 383 shows a typical example.

Logic Trees

Logic trees are like logic boxes, but they use the metaphor of branches on a tree. Figure 12.36 on page 383 shows a logic tree that helps students think through the process of registering for a course.

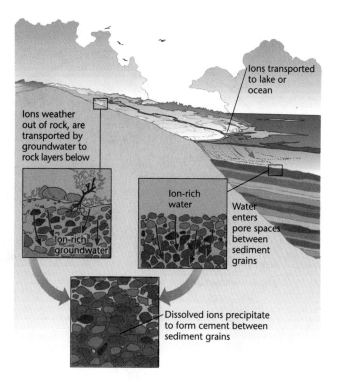

Ions transported
to lake or
ocean

Ions weather
out of rock, are
transported by
groundwater to
rock layers below

Ion-rich
water

Water
enters
pore spaces
between
sediment
grains

Ion-rich
groundwater

Dissolved ions precipitate
to form cement between
sediment grains

FIGURE 12.33

Open-System Flowchart

Logic trees are probably somewhat easier to understand than logic boxes because the tree metaphor is visually more distinct than the box metaphor. However, logic boxes can accommodate more textual information than logic trees can.

**Illustrating
Visual and Spatial
Characteristics**

To show visual and spatial characteristics, use photographs, screen shots, line drawings, and maps.

Photographs

Photographs are unmatched for reproducing visual detail. If you want to show realistically the kinds of tire-tread wear caused by various alignment problems, a photograph is best. If you want your readers to recognize a new product such as a new automobile model, you will probably use a photograph. And recent advances in specialized kinds of photography—especially in internal medicine and biology—are expanding the possibilities of the art.

Ironically, however, sometimes a photograph can provide too much information. In an advertising brochure for an automobile, a glossy photograph of the dashboard might be very effective. But if you are creating an owner's manual and you want to show how to find the trip odometer, a diagram will probably

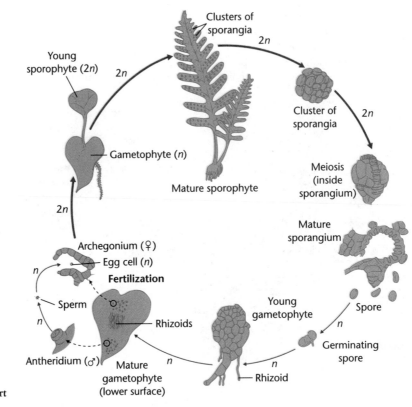

Young
sporophyte (2*n*)

Clusters of
sporangia

2*n*

2*n*

Gametophyte (*n*)

Cluster of
sporangia

2*n*

Mature sporophyte

Meiosis
(inside
sporangium)

2*n*

Mature
sporangium

Archegonium (♀)

Egg cell (*n*)

Fertilization

n

Sperm

n

Antheridium (♂)

Rhizoids

Mature
gametophyte
(lower surface)

Young
gametophyte

Spore

n

n

n

Germinating
spore

Rhizoid

FIGURE 12.34

Closed-System Flowchart

work better because it lets you focus on the item you want to emphasize. Sometimes a photograph can provide too little information; the item you want to show can be inside the mechanism or obscured by some other component.

Taking pictures to be used in publications is best left to professionals, because including photographs in a document involves many complicated technical details. On occasion you will want to use existing photographs in a document; if so, follow these five basic guidelines:

- *Do not electronically manipulate the photograph.* Today it is routine to digitize a photograph, remove blemishes, and crop the photograph (cut it down to a desired size). There is nothing unethical about doing so. Figure 12.37 on page 384 shows examples of cropped and uncropped photographs. However, manipulating a photograph—for example, enlarging the size of the monitor that comes with a computer system shown in a photograph—is unethical. As consumers we have all been angered and frustrated when we take a product out of the box and realize that it bears only a faint resemblance to the photograph on the box.

- *Help the reader understand the perspective.* Most objects portrayed in magazines and journals are photographed from an angled view; this per-

Do you want to replace all occurrences of the first word with the second word?		
YES	NO	
Press Y. The computer replaces all occurrences of the first word with the second word.	Press N. The cursor moves to the next occurrence of the word.	
	Do you want to replace that occurrence of the word?	
	YES	NO
	Press SHIFT/F2.	Press F2.
	Press F4 to move the cursor to the next occurrence of the word. Then press SHIFT/F2 or F2, as you did above, to replace that occurrence of the word or to leave it as is.	
Do you want to use the search function again?		
YES	NO	
Press F8.	Press F10.	

FIGURE 12.35

Logic Boxes

spective provides more information because it shows the depth of the object as well as its height and width.

- *If appropriate, include in the photograph some common object, such as a coin or a ruler, to give a sense of scale.*
- *Eliminate extraneous background clutter that can distract your reader.* Do this by cropping the photograph or by digitizing it and deleting unnecessary detail.
- *If appropriate, label components or important features.*

FIGURE 12.36

Logic Trees

<figure>FIGURE 12.37

Cropping a Photograph
Emphasizes the
Important Information
</figure>

Screen Shots

A *screen shot* is a computer graphic of what appears on the computer monitor. Screen shots are often used in software manuals to show the reader what the screen looks like at various points during the use of a program. The reader is reassured to see that the image on the screen looks like the screen shot in the documentation. Screen shots also help the reader locate information on the screen. Figure 12.38 (Hewlett Packard, 1997) is an example of a screen shot.

Line Drawings

Line drawings are simplified visual representations of objects. Line drawings can have three advantages over photographs:

- Line drawings can focus the reader's attention on desired information better than a photograph can.
- Line drawings can highlight information that might be obscured by bad lighting or a bad angle in a photograph.
- Line drawings are sometimes easier for readers to understand than photographs are.

A line drawing such as the one in Figure 12.39 (Briggs & Stratton, 1989) on page 386 shows readers enough of the lawnmower to orient them and highlights the

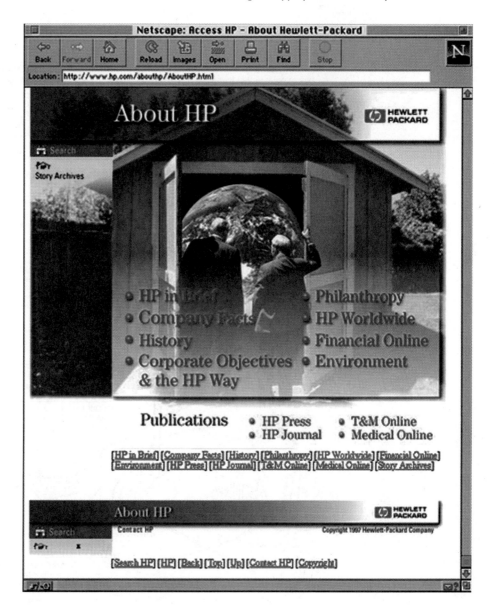

FIGURE 12.38

Screen Shot

important information. Notice how close-ups show details that could not be seen from an ordinary perspective.

You have probably seen three variations on the common drawing. *Phantom drawings* show parts that are hidden from view by outlining external items that obscure their view. *Cutaway drawings* "remove" a part of the surface to expose what is underneath. *Exploded drawings* separate components while maintaining their physical relationship. Figure 12.40 on page 387 shows phantom, cutaway, and exploded views.

STARTERS
Nicad System

ELECTRIC STARTER KEY SWITCH AND WIRING RECOMMENDATIONS
FOR 12 VOLT NICKEL CADMIUM BATTERY STARTING SYSTEM — SERIES 92000 AND 110900 ENGINES

STARTING SWITCH — Lettering around key should be as indicated when the standard switch case is not used.

CONNECTOR RETAINER — When the standard switch case of the key switch is not used, a retainer is required to prevent an unintentional disconnect of the cord where it attaches to the switch. If the plug becomes disconnected, turning the key to "off" position will not stop the engine.

6" MIN.
(152.4 mm)
←—CLIP

CLIPS should be used to direct wires toward battery plug. This will minimize the hazard of shrubs, etc., pulling out the leads and disengaging the connector.

BATTERY CLIPS should be used to guide and retain harness in a neat installation.

SUFFICIENT SLACK in the harness should be provided to allow full movement of the handle.

FIGURE 12.39

Line Drawing

Maps

Maps are readily available as clip art that can be modified with a graphics program. Include a scale and a legend if the map is not thoroughly familiar to your readers. Also, use conventional colors, such as blue for water. Figure 12.41 shows a map made from clip art.

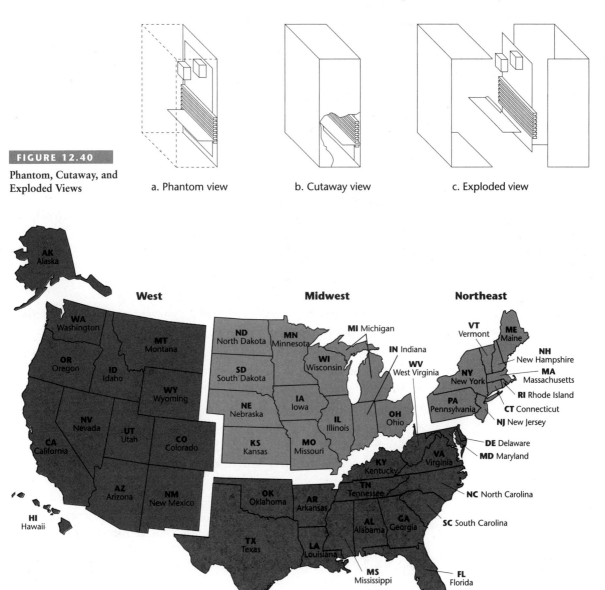

FIGURE 12.40

Phantom, Cutaway, and Exploded Views

a. Phantom view   b. Cutaway view   c. Exploded view

FIGURE 12.41

Map

Showing Motion in Your Graphics

Often in technical communication you want to show motion. For instance, in an instruction manual for military helicopter technicians, you might want to show the process of removing an oil dipstick or tightening a bolt. Or you might want to show a warning light flashing. Although animation or video is often used

Showing the Action from
the Reader's Perspective

today to show these sorts of processes, printed graphics are still needed to communicate this information. Following are two guidelines for showing motion:

- *If the reader is to perform the action, show the action from the reader's point of view.* In most cases, you need to show only the person's hands, not the whole body, as in Figure 12.42.

- *Use arrows or other symbols to show actions.* Arrows show the direction in which something is moving or should be moved; shake lines suggest vibration; starburst lines suggest a blinking light. Another way to suggest motion is to show an object both before and after the motion. Figure 12.43 shows these different symbols.

  Note that these symbols are conventional, not universal. If you are addressing international readers, consult a qualified person from your target culture to make sure the symbols you want to use are clear and inoffensive.

Creating Graphics for International Readers

Whether you are writing to people inside or outside your organization, consider the needs of readers who do not speak English as their primary language. If you communicate with people in other countries, be aware that many do not speak English at all. In 1990, almost 32 million Americans spoke a language other than English at home (U.S. Bureau, 1996, p. 53). With the rate of immigration rising each year, that figure will surely increase.

Graphics are excellent tools for communicating with an international reader, but use them carefully. Like words, graphics have cultural meanings, and if you

a. Arrow

c. Shake lines

b. Starburst lines

d. Before and after

FIGURE 12.43

Showing Motion

are unaware of them, you can communicate something very different from what you had intended. The following discussion, based on William Horton's article "The Almost Universal Language: Graphics for International Documents" (1993), presents six guidelines for using graphics when addressing an international audience:

- *Understand that there are different reading patterns.* In most countries, people read from left to right, but in some they read right to left or top to bottom. In some cultures, direction signifies value: the right-hand side is superior to the left, or the reverse. Therefore, you need to think about how to sequence graphics that show action, or where you put "before" and "after" graphics. If you want to show a direction, as in an informal flowchart, consider using arrows to indicate the direction in which the chart should be read.

- *Understand that there are different cultural attitudes toward giving instruction.* You might have noticed that instructions for products made in Japan are highly polite and deferential: "Please insert the batteries at this time." Some cultures favor spelling out general principles but leaving the reader to infer the details. In instructions written for people from these cultures, a detailed close-up of how to carry out a task might appear insulting. An instructional table with the headings "When You See This . . ." and "Do This . . ." might be inappropriate.

- *Deemphasize trivial details.* Common objects such as plugs on the ends of power cords can take any number of shapes around the world; therefore, draw them so that they look generic rather than specific to one country.

- *Avoid culture-specific language, symbols, and references.* Don't use a picture of a mouse to symbolize a computer mouse; the device is not known by that name everywhere. Avoid the casual use of national symbols (such as the maple leaf or national flags); you might make an error in a detail and unknowingly insult your readers. Use colors carefully: red means danger to most people from a Western culture, but it is a celebratory color to the Chinese. Don't portray an executive sitting in an office with a big window; the window is a sign of prestige in most cultures, but in Japan a "window watcher" is an employee who has disappointed the organization and is being paid to gaze out the window, not to do any productive work.

- *Portray people very carefully.* Every aspect of people's appearance, from their clothing to their hair styles to their features, is culture-specific or race-specific. A photograph of a woman in casual Western attire seated at a workstation would not be effective in Saudi Arabia; in Islamic cultures only the hands and eyes of a woman may be shown. Horton (1993) recommends using stick figures or silhouettes that do not reveal culture, race, or gender.

- *Be particularly careful in portraying hand gestures.* Many Western hand gestures, such as the "okay" sign, or the palm-forward gesture used by a traffic officer to stop cars, are obscene in other cultures. And long red fingernails would not be appropriate in many cultures. For this reason, use hands only when necessary, for example, when they are carrying out a task. Disguise the person's sex and race. When communicating with Arabic readers, try not to display the left hand, which they reserve for personal hygiene.

Cultural differences are many and subtle, and sometimes it seems as if everything you want to portray will offend someone. The best advice is to learn as much as possible about your readers and try to consider their habits and outlook. For more information on this subject, consult these three books:

Edward Hall. (1990). *Understanding cultural differences.* Yarmouth, ME: Intercultural Press.

Roger Axtel. (1991). *Gestures: The do's and taboos of body language around the world.* New York: John Wiley.

Scott Jones, Cynthia Kennelly, Claudia Mueller, et al. (1992). *Developing international user information.* Bedford, MA: Digital Press.

Also see Chapters 5 and 14 for additional discussions of writing to international readers.

Understanding Graphics Software

Two basic categories of graphics software are available for the personal computer:

- spreadsheet business graphics
- paint programs and draw programs

Spreadsheet Business Graphics

Spreadsheet programs—software created to help businesspeople calculate budgets and evaluate hypothetical business scenarios—offer business graphics capabilities. After you type in the numerical data you want to portray, such as the profits and losses of the four divisions of your company, you can ask the software to portray the data in various kinds of graphics and charts.

Spreadsheets make different kinds of tables, bar graphs, line graphs, and pie charts, in both two-dimensional and three-dimensional formats. You then add labels and a title and customize the graphic to suit your needs. Figure 12.44 on page 392 shows how data entered on a spreadsheet can be displayed as a table and then as a bar chart. Most spreadsheets can display the same numerical data in dozens of styles of graphics.

Some spreadsheets also let you produce slide shows; that is, the software presents the different graphics you have created, one after the other. You can choose the length of time that each graphic remains on the screen (such as 10 seconds) or you can advance to the next graphic by clicking the mouse.

Be careful in using spreadsheet business graphics, because the software doesn't provide advice on what kind of graphic to make or how to modify the basic presentation; it just draws whatever you ask it to. The software often makes poor choices: unnecessary 3-D, useless clip art, and lurid color combinations. To use the graphics capabilities effectively, you need to understand the basics of the different kinds of graphics.

Paint Programs and Draw Programs

Paint programs and draw programs let you create and then modify freehand drawings in a number of ways. For instance, you can

- modify the width of the lines
- modify the size of the shapes

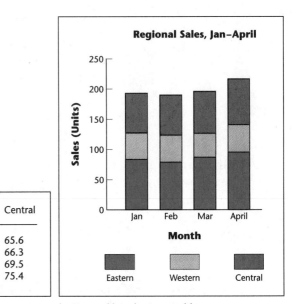

	Eastern	Western	Central
Jan	83.5	43.7	65.6
Feb	78.9	44.6	66.3
Mar	86.7	39.5	69.5
April	95.3	45.7	75.4

FIGURE 12.44

Creating a Graphic from a Spreadsheet

a. Spreadsheet data

b. Grouped bar chart created from the spreadsheet data

- copy, rotate, and flip images
- fill in shapes with different colors and textures
- add text with a simple text editor

With both kinds of programs, you can start in three different ways:

- By drawing on a blank screen.
- By importing an image from a clip-art library.
- By scanning an image. Scanning an image involves using a special piece of hardware—a scanner—and a special software program to translate a graphic on a piece of paper into a computerized image.

What is the difference between paint programs and draw programs?

Paint programs use pixel-based technology (also called *bit-map technology*). A pixel, short for *picture element,* is the smallest unit of space on the computer screen that can be treated as an individual element. In a paint program, each pixel is dark or lit with a different color.

One disadvantage of paint programs is that an enlarged image will look washed out and blurry, like a photograph that has been enlarged too much in a newspaper. In the larger image, you will notice the little dots that make up the picture. One way to avoid this problem is to create the image larger than you want it to appear and reduce it on a photocopy machine.

Draw programs use vector technology. In a draw program, you work with preexisting shapes, called *graphics primitives,* as shown here.

To create an image, manipulate the size and shape of the graphics primitives. The computer remembers the shapes you create by storing in memory the beginning and ending points of the lines that make the shapes. These beginning and ending points are called *vectors*.

Figure 12.45 shows how paint programs and draw programs represent the letter *I*.

Draw programs are considered more sophisticated than paint programs for two reasons:

- *Vector information is independent of the display device.* You can output the image in any size or display it on any monitor without affecting its resolution.

- *Draw programs allow you to create more sophisticated three-dimensional effects.* If you draw a circle and a square, you can place the circle in front of the square. The computer recognizes and remembers the circle and the square as different shapes. With a paint program, the two shapes would merge; the pixels where the two original shapes overlap would assume whichever pattern is placed on them more recently, and you could not separate the two original shapes.

For these reasons, graphic artists and others use vector technology in sophisticated draw programs such as those used in CAD applications. As Figure 12.46 on page 394 illustrates, you can create the same basic graphic with either a paint program or a draw program, but with a draw program you can manipulate the elements.

As Figure 12.46 on page 394 illustrates

FIGURE 12.45

How an Image Is Represented in a Paint Program and a Draw Program

Paint program

Draw program

FIGURE 12.46

Images Created with a Paint Program and a Draw Program

Paint program   **Draw program**

CASE

Adding Graphics to a Web Site

You are a graphic artist working for Ford Motor Company. A member of the Web site development group, Chien Su, has come to you for your thoughts on the graphics on the Ford site (www.ford.com). "We've got a lot of photographs of our cars," she tells you, "but I'm concerned we don't have enough other kinds of graphics. Would you mind taking a look at the site and giving me some ideas on how we might improve it visually?" Study the site, thinking about the kinds of people who might be viewing it and about their needs. Evaluate the current integration of text and graphics, as well as the variety and quality of the graphics. Can you suggest other kinds of graphics that might be added to provide other kinds of information for the different audiences that might visit the site? Write a memo to Chien Su, evaluating the graphics and giving your recommendations. See Chapter 16 for a discussion of memos.

WRITER'S CHECKLIST

This checklist focuses on the characteristics of an effective graphic.

1. Does the graphic have a purpose?

2. Is the graphic honest?

3. For an existing graphic, do you have the legal right to use it? If so, have you cited it appropriately?

4. Does the graphic present a manageable amount of information?

5. Does the graphic meet the reader's format expectations?

6. Is the graphic simple and uncluttered?

7. Is the graphic clearly labeled?

8. Is the graphic in an appropriate location in the document?

9. Is the graphic introduced clearly in the text?

10. Is the graphic explained in the text?

11. Is the graphic clearly visible in the text?

12. Is the graphic easily accessible to your readers?

EXERCISES

1. Find out from the admissions department of your college or university the numbers of students from the different states, or from the different counties in your state. Communicate this information using four different graphics:

   a. a map   c. a bar graph
   b. a table   d. a pie chart

   In three or four paragraphs, explain how the different graphics would be appropriate for different audiences and purposes, and how they emphasize different aspects of the information.

2. Create a flowchart for a process you are familiar with, such as applying for a summer job, studying for a test, preparing a paper, or performing some

task at work. Your audience is someone who will be carrying out the process portrayed in the graphic.

3. Create an organizational chart for an organization you belong to or are familiar with: a department at work, a fraternity or sorority, a campus organization, or student government.

4. In two or three paragraphs, define an audience and purpose and then make a drawing of some object you are familiar with, such as a tennis racquet, a stereo speaker, or a weight bench. Be certain that the object and the drawing are appropriate to your audience and purpose.

5. In two or three paragraphs, define an audience and purpose and then draw a diagram of some concept, such as the effects of smoking or the advantages of belonging to the student chapter of a professional organization. Be certain that the concept and the diagram are appropriate to your audience and purpose.

6. Create a graphic for one of the following scenarios:
   a. In a user's manual for a computer, how to insert the floppy disk
   b. In the owner's manual for a bicycle, how to locate its serial number
   c. In a training manual for new waitpersons, a guide to explaining the menu to customers
   d. In a set of instructions on how to change the oil in a car, a checklist for the equipment and materials you will need
   e. In a student handbook at your college or university, the process of applying for graduation

7. Create a set of graphics to complement a set of instructions on how to use an automated-teller machine. Your readers are nonnative speakers of English who are unfamiliar with automated-teller machines. For each graphic you create, write a one- or two-sentence explanation of what the text would be describing.

8. In the *Statistical Abstract of the United States* or an issue of *Forbes* or *Business Week,* find and photocopy a bar graph, a line graph, a pictograph, a pie chart, a flowchart, a photograph, and a diagram. For each graphic, write a brief discussion that responds to the following questions:
   a. Is the graphic necessary?
   b. Is it ethical, or does it misrepresent the information?
   c. Is it professional in appearance?
   d. Does it conform to the guidelines for that kind of graphic?

e. Is it effectively integrated into the discussion?

9. The table on page 396 (U.S. Bureau, 1996, p. 99) provides statistics on AIDS in the United States. Study the table, then perform the following tasks:
   a. Create two different graphics, each of which communicates information about the incidence of AIDS among different age groups.
   b. Create two different graphics, each of which communicates information about the incidence of AIDS in women and in men.
   c. Create two different graphics, each of which communicates information about the incidence of AIDS among different racial/ethnic groups.

10. For each of the following graphics, write a paragraph evaluating its effectiveness and describing how you would revise it.

a.

	1995	1996	1997
Civil Engineering	236	231	253
Chemical Engineering	126	134	142
Comparative Literature	97	86	74
Electrical Engineering	317	326	401
English	714	623	592
Fine Arts	112	96	72
Foreign Languages	608	584	566
Materials Engineering	213	227	241
Mechanical Engineering	196	203	201
Other	46	42	51
Philosophy	211	142	151
Religion	86	91	72

b.

Merchant Marines

(Exercise 9, continued)

No. 134. Acquired Immunodeficiency Syndrome (AIDS) Deaths, by Selected Characteristics: Thru 1995
[Data are shown by year of death and are subject to substantial retrospective changes.
For data on AIDS cases reported, see table 217. Based on reporting by state health departments]

CHARACTERISTIC	NUMBER									PERCENT DISTRIBUTION	
	Total cases	1985 and before[1]	1989	1990	1991	1992	1993	1994	1995	1995	1993–1995
Total[2]	**305,843**	**12,493**	**26,355**	**29,934**	**34,651**	**38,813**	**41,077**	**43,975**	**31,256**	**100.0**	**100.0**
Age:											
13 to 29 years old	53,632	2,558	4,924	5,427	6,059	6,481	6,815	6,930	5,105	16.3	16.2
30 to 39 years old	138,660	5,714	12,056	13,695	15,582	17,437	18,548	20,037	14,083	45.1	45.3
40 to 49 years old	77,425	2,726	6,205	7,329	8,943	10,339	10,975	11,921	8,543	27.3	27.0
50 to 59 years old	25,383	1,104	2,218	2,433	2,822	3,175	3,389	3,721	2,572	8.2	8.3
60 years old and over	10,743	391	952	1,050	1,245	1,381	1,350	1,366	953	3.0	3.2
Sex:											
Male	268,987	11,553	23,742	26,752	30,725	34,072	35,551	37,360	26,375	84.4	85.4
Female	36,856	940	2,613	3,182	3,926	4,741	5,526	6,615	4,881	15.6	14.6
Race/ethnicity:											
White	159,292	7,267	14,481	16,517	18,693	19,982	20,376	20,733	14,389	46.0	47.7
Black	100,167	3,424	7,950	9,010	10,722	12,839	14,420	16,294	11,879	38.0	36.6
Hispanic	43,154	1,719	3,656	4,120	4,851	5,578	5,820	6,421	4,574	14.6	14.5
Other/unknown	3,230	83	268	287	385	414	461	527	414	1.3	1.2

Z Less than 0.5 percent.   [1]Includes deaths prior to 1982.   [2]Includes other race/ethnicity groups not shown separately.
Source: U.S. Centers for Disease Control, *Surveillance Report,* annual.

(Exercise 10, continued)

c.

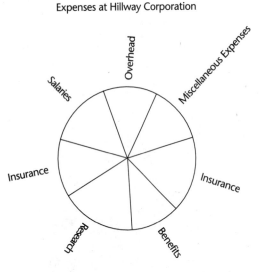

Expenses at Hillway Corporation

d.

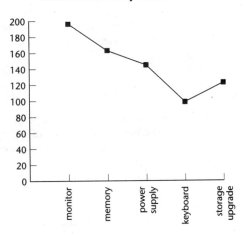

Costs of the Components of a PC

11. In each of the following exercises, translate the written information into at least two kinds of graphics. For each exercise, write a two- or three-paragraph statement about which kinds work best. If one kind works well for one audience but not so well for another, explain.

   a. Following are the profit and loss figures for Pauley, Inc., in early 1994: January, a profit of 6.3 percent; February, a profit of 4.7 percent; March, a loss of 0.3 percent; April, a loss of 2.3 percent; May, a profit of 0.6 percent.

   b. The prime interest rate had a major effect on our sales. In January, the rate was 4.5 percent. It went up a full point in February, and another half point in March. In April, it leveled off, and it dropped one full point each in May and June. Our sales figures were as follows for the Crusader 1: January, 5,700; February, 4,900; March, 4,650; April, 4,720; May, 6,200; June, 8,425.

   c. This year, our membership can be broken down as follows: 45 percent, electrical engineering; 15 percent, mechanical engineering; 20 percent, materials engineering; 10 percent, chemical engineering; and 10 percent, other.

   d. In January of this year we sold 50,000 units of the BG-1, of which 20,000 were purchased by the army. In February, the army purchased 15,000 of our 60,000 units sold. In March, it purchased 12,000 of the 65,000 we sold.

   e. The normal rainfall figures for this region are as follows: January, 1.5 in.; February, 1.7 in.; March, 1.9 in.; April, 2.1 in.; May, 1.8 in.; June, 1.2 in.; July, 0.9 in.; August, 0.7 in.; September, 1.3 in.; October, 1.1 in.; November, 1.0 in.; December, 1.2 in. The following rainfall was recorded in this region: January, 2.3 in.; February, 2.6 in.; March, 2.9 in.; April, 2.0 in.; May, 1.6 in.; June, 0.7 in.; July, 0.1 in.; August, 0.4 in.; September 1.3 in.; October, 1.2 in.; November, 1.4 in.; December, 1.8 in.

   f. You can access the Rightfile from programs written in six languages. Rightfile classifies these languages as two groups (A and B) and provides separate files for each group. The A group includes C, PL/1, and Fortran. The B group includes COBOL and RPGII. The object module for the A group is GRTP. The object module for the B group is GRAP.

12. Interview three people: an engineer or scientist, a technical communicator, and a graphic artist or illustrator who works in technical communication. Find out about their approaches to creating graphics for technical documents. Specifically, ask about their education in graphics, their use of tools or software, the way they integrate graphics into the process of writing a document, and their views of the importance of graphics in the work they do. Write your results in a memo to your instructor. See Chapter 16 for a discussion of memos.

Collaborative Exercise

13. Form small groups. Go to one of your campus' computer centers and study a popular piece of software, such as a spreadsheet or graphics package. Have each group member, working separately, print out three of the basic kinds of graphics, such as graphs, tables, and pie charts, and take notes on the process. Come together as a group and share your experiences in learning to use the software. What techniques did you use to learn the software: using a tutorial, reading a manual, trial-and-error, asking someone for help? If appropriate, have the group go back to the lab together and work through some of the questions that arose in the group meeting. Write a memo to your instructor, describing how easy or difficult it was to learn how to create the graphics and evaluating the quality of the graphics produced. How well do the graphics conform to the basic guidelines presented in this chapter? Would you recommend the software to other members of your class? Why or why not? See Chapter 16 for a discussion of memos.

Web Exercise

14. Locate a graphic on the Web that you feel would be inappropriate for an international audience because it might be offensive or unclear to some cultures. Choose an intended audience, such as people from the Middle East, and write a brief statement explaining the potential problem. Finally, revise the graphic so that it would be appropriate for its intended audience.

REFERENCES

Barnum, C. M., & Carliner, S. (1993). *Techniques for technical communicators.* New York: Macmillan.

Briggs & Stratton Corporation. (1989). *Briggs & Stratton service and repair instructions.* Part #270962.

Brockmann, R. J. (1990). *Writing better computer user documentation: From paper to hypertext.* New York: Wiley.

Chernicoff, S., & Venkatakrishnan, R. (1995). *Geology: An introduction to physical geology.* New York: Worth.

Council of Economic Advisers. (1994, March). *Economic indicators.* Washington, DC: U.S. Government Printing Office.

Curtis, H., & Barnes, N. S. (1989). *Biology* (5th ed.). New York: Worth.

Dean, R. S., & Kulhavy, R. W. (1981). Influence of spatial organization in prose learning. *Journal of Educational Psychology, 73,* 57–64.

Gatlin, P. L. (1988). Visuals and prose in manuals: The effective combination. In *Proceedings of the 35th International Technical Communication Conference* (pp. RET 113–115). Arlington, VA: Society for Technical Communication.

Grimstead, D. (1987). Quality graphics: Writers draw the line. *Proceedings of the 34th International Technical Communication Conference* (pp. VC 66–69). Arlington, VA: Society for Technical Communication.

Health status of the disadvantaged chartbook. (1990). U.S. Department of Health and Human Services. DHHS Publication HRS-P-DV-90-1, p. 7.

Hewlett Packard. (1997). About HP. *Hewlett Packard.* <http:\\www.hp.com> (1997, May 12).

Horton, W. (1991). *Illustrating computer documentation: The art of presenting information graphically on paper and online.* New York: Wiley.

Horton, W. (1993). The almost universal language: Graphics for international documents. *Technical Communication, 40,* 682–693.

Labuz, R. (1984). *How to typeset from a word processor: An interfacing guide.* New York: R. R. Bowker.

Levie, W. H., & Lentz, R. (1982). Effects of text illustrations: A review of research. *Journal of Educational Psychology, 73,* 195–232.

Mankiw, N. G. (1997). *Macroeconomics* (3rd ed.). New York: Worth.

McComb, G. (1991). *Troubleshooting and repairing VCRs* (2nd ed.). Blue Ridge Summit, PA: Tab Books.

McGuire, M., & Brighton, P. (1990). Translating text into graphics. Session at the 37th International Technical Communication Conference. Dallas, TX.

Morrison, C., & Jimmerson, W. (1989, July). Business presentations for the 1990s. *Video Manager, 4,* 18.

Tufte, E. R. (1983). *The visual display of quantitative information.* Cheshire, CT: Graphics Press.

U.S. Bureau of the Census. (1996). *Statistical abstract of the United States: 1996.* Washington, DC: U.S. Government Printing Office.

U.S. Congress, Office of Technology Assessment. (1993). *Making government work: Electronic delivery of federal services,* OTA-TCT-578. Washington, DC: U.S. Government Printing Office.

U.S. Department of Commerce. (1994). *U.S. industrial outlook 1994: Forecasts for selected manufacturing and service industries.* Washington, DC: U.S. Government Printing Office.

U.S. Environmental Protection Agency. (1991). *Building air quality: A guide for building owners and facility managers.* Washington, DC: U.S. Government Printing Office.

White, J. V. (1984). *Using charts and graphs: 1000 ideas for visual persuasion.* New York: R. R. Bowker.

White, J. V. (1990). *Color for the electronic age.* New York: Watson-Guptill.

CHAPTER

13

Revising for Coherence

Understanding the Given/New Strategy 400

Techniques Used in Revising 401
   Revising by Yourself *401*
   Obtaining Help from Someone Else *403*
   Testing Readers as They Use the Document *404*

Revising the Whole Document for Coherence 405

Revising Titles for Coherence 406

Revising Headings for Coherence 407

Revising Lists for Coherence 408

Revising Paragraphs for Coherence 410
   Structure Paragraphs Clearly *411*
   Keep Paragraphs to a Manageable Length *413*
   Use Coherence Devices within and between Paragraphs *415*
   Turning a "Writer's Paragraph" into a "Reader's Paragraph" *418*

Case: Writing Guidelines about Coherence *419*

Writer's Checklist 421

Exercises 421
   Collaborative Exercise *422*
   Web Exercise *423*

Coherence is a measure of how effectively a document hangs together. A coherent document seems to flow smoothly from one part to the next; the reader can concentrate on understanding the technical information it contains. An incoherent document is harder to read; the reader can misunderstand the information or become confused, wondering how a particular point relates to one that preceded it. This chapter explains how to revise your drafts to make them coherent.

Perhaps the most durable axiom about writing is only three words long: *Writing is revising.* The chapter begins by describing the basic techniques of revising, then discusses how to revise the whole document for coherence. Next, the chapter discusses revising titles, headings, lists, and paragraphs.

Chapter 14 concentrates on revising for sentence-level effectiveness.

Understanding the Given/New Strategy

The process of understanding text is much more complicated than we used to think. Once we believed that word length and sentence length were the major factors in understanding text, but now we think that other factors are more important. Because understanding occurs within a person, not within a document, it involves the reader's knowledge, attitude, and reading behavior. In reading, a person extracts new information and tries to associate it with existing information.

Every sentence that we read or hear contains both given and new information. *Given information* is what the listener or reader is expected to know already. *New information,* obviously, is what the person is not expected to know already. We try to understand what we hear or read by searching our memories for information that matches the given information in the sentence. Once we find a match, we can attach the new information to it, thereby changing our original information. The given/new strategy is shown in Figure 13.1.

Given information:
What we already know about the subject.

New information:
We link the new information and the given information, changing or expanding our understanding.

New information:
We link additional new information and the given information, etc.

FIGURE 13.1

The Given/New Strategy

For example, the sentence "The proposal we submitted last month won the contract" consists of two pieces of information: (1) we submitted a proposal last month and (2) that proposal won the contract. Readers will first try to understand the first piece by searching for a memory of the company's having submitted that proposal. Then, they will add the second piece—that proposal won the contract—to the first piece. Thereafter, their memories will contain the updated information.

However, if you wrote that same sentence, "The proposal we submitted last month won the contract," but your reader did not know that the company had submitted a proposal last month, communication would break down, at least momentarily, as the reader tried (unsuccessfully) to remember the proposal you are referring to. That reader could continue to try to remember, give up in frustration, or create a meaning: "We must have submitted a proposal last month, and it apparently won the contract." If the created meaning is accurate, the only cost has been the reader's labor; if the created meaning is inaccurate, the price could be much higher.

What does the given/new strategy have to do with technical communication? It supports a simple, fundamental idea: when you want to transmit information to people, you will be most successful if you deliver the new information gradually, building on what they already know. This idea applies to all units of discourse, from large to small. The rest of this chapter will show how this idea applies to revising the different elements of a document.

Techniques Used in Revising

Revising is the process of making sure the document says what you want it to say—and that it says it professionally. As is the case with other phases of the writing process, everyone uses a different technique for revising. But the important point is that you must have a technique. You cannot hope to simply read through your document, waiting for the problems to leap off the page. Some of them might; most won't.

There are three basic ways to revise a document:

- Revise it by yourself.
- Obtain help from someone else.
- Test readers as they use the document.

Revising by Yourself   Before you give a document to someone else for help in revision, and before you test the document with real users, revise it yourself. Although this process won't help you find and fix all the problems, you'll make a lot of progress. The four guidelines shown in Figure 13.2 on page 402 can help you evaluate your own writing.

FIGURE 13.2

Revising by Yourself

- *Let it sit.* You cannot revise your document effectively right after you finish writing it. Although you can immediately identify some of the smaller writing problems—such as errors in spelling or grammar—you cannot objectively assess the quality of what you have said: whether the information is clear, comprehensive, and coherent.

  To revise your draft effectively, set it aside for a time. Give yourself at least a night's rest. If possible, work on something else for a few days. This will give you time to "forget" the draft and approach it more as your readers will. In a few minutes, you will see problems that would have escaped your attention even if you had spent hours revising it right after completing the draft.

  You're probably thinking that you don't have time to set it aside for a night. There is no easy solution to the time problem. You just have to try to plan your schedule so that you are done drafting a few days before the document is due.

- *Read it aloud.* Reading a document aloud gives you a better sense of what it sounds like than reading it silently. Any teacher of writing has seen this effect a hundred times. A student submits a paper in which the first sentence is missing a necessary word. The teacher asks the student to look at the sentence and fix the problem, and she replies that she can't see any problem. When asked to read the sentence out loud, the student immediately "hears" the mistake. Big problems, too, such as illogical reasoning or missing evidence, also seem more obvious when you hear your document. Reading your draft out loud to identify areas for improvement will make you feel foolish. Welcome to the club.

- *Use checklists.* Before airplane pilots take off, they perform a series of safety checks written out in checklists; they can't count on remembering all the different safety checks they need to carry out. The same applies to writing.

  This text is full of checklists that summarize the major points you should look for in revising different kinds of documents. However, any checklist that someone else has written for you has a big drawback: it doesn't necessarily cover the things you need to check most. For example, you might often confuse the use of *its* and *it's* in your writing. Don't count on a generic checklist to include this issue. Instead, keep a list of the points your instructors make about your writing and then customize the checklists you use.

Many writers construct checklists that help them remember to review the document three times: once for overall meaning and clarity, once for organization and development, and once for correctness. Of course, each of the three checklists includes numerous items.

- *Use software.* Many software programs can help you spot different kinds of style problems. Like a checklist, however, a program is a generalized tool that cannot understand your context—the purpose and audience of your communication—and therefore cannot be relied on to offer relevant and wise advice all the time. Revision software—spelling checkers, style programs, and thesaurus programs—is discussed in Chapter 4.

Obtaining Help from Someone Else

Once you have done what you can by yourself, enlist a colleague to help you revise. In general, you can turn to two kinds of people for help:

- *Subject-matter experts.* If, for instance, you have written an analysis of alternative fuels for automobiles, you could ask an automotive expert to review it. In the working world, important documents are routinely reviewed by technical experts, lawyers, and marketing specialists before being released to the public.
- *People who are like the eventual readers.* People who fit the profile of the eventual readers will help you see problems you didn't notice in revising by yourself. People more knowledgeable about the subject than your eventual readers will understand the document even if it isn't clear; therefore, they might not be sufficiently critical.

If possible, show the document to both kinds of people. But think twice before asking a personal friend or a spouse. You want objective advice. You don't want that advice to be affected by your relationship—and you don't want the relationship to be affected by the advice.

When you ask someone to help you, provide specific instructions. Tell the person as much as you can about your audience and purpose and about the tasks you still have to complete before you finish the document. Then tell the person—or, better yet, write out—the kinds of problems you want him or her to look for.

After reading the draft, your colleagues may be able to point out its strong points, unclear passages, and sections that need to be added, deleted, or revised.

Testing Readers as They Use the Document

Analyzing a document can provide only limited information; it cannot tell you how well the document will work when it gets into the hands of the people who will use it. To find out that information, you must test readers as they use the document.

Testing readers entails four commonsense steps, as shown in Figure 13.3.

- *Determining what you want to test.* The questions you want to answer are likely to vary from one kind of document to the next. For a sales brochure, you might want to determine whether it makes a positive impression on potential customers. For a set of instructions, you might want to see whether a general reader can carry out the task safely, easily, and effectively.

- *Choosing representative test participants.* A basic principle of testing is that the test participants must match the real users of the document as closely as possible. If the real users will be mostly men in their twenties, test men in their twenties. Try to match the test participants to the eventual users in terms of education, experience, and all the other factors involved in analyzing audience as discussed in Chapter 5.

- *Making the tests realistic.* After recruiting representative test participants, make the tests themselves realistic. If you are testing a user's manual for a word-processing program, ask the test subject to use the manual to carry out representative word-processing tasks on the appropriate computer.

  For documents that communicate information but don't require that the reader actually do anything, you have to be more inventive. You can ask participants to do the following:

  – Tell you what they are thinking as they read the document.
  – Answer questions after they read the document.

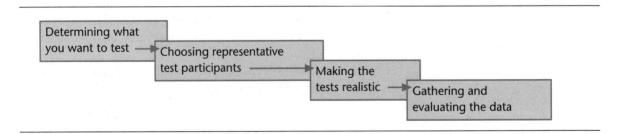

FIGURE 13.3

Testing Readers as They
Use the Document

- Write a paraphrase of a particular piece of information after they read the document.
- Read the document, then read a second document, also written by the test team, and state whether the second document accurately reflects what was said in the first document.

- *Gathering and evaluating the data.* The data you collect from testing the documents must be evaluated as carefully as the data from any experiment. Basic principles of experimentation apply, such as having enough test subjects to ensure valid results and making sure that the process of gathering data does not unintentionally influence the test subjects' behavior. If you interview the test subjects, ask each the same questions in the same way, and record the information accurately and precisely.

Testing is discussed in more detail in Chapter 20.

Revising the Whole Document for Coherence

In looking for problems to fix, most writers prefer a top-down pattern. That is, they look first for the largest and most important problems and proceed to the smaller, less important ones so that they don't waste time fixing awkward paragraphs they might eventually throw out. For this reason, many writers begin revising by considering the whole document.

In revising the whole document, answer the following five questions:

- *Have you left out anything in turning your outline into a draft?* In the process of drafting, it is easy to leave out a topic. Check the document against the outline to see that all the topics in the outline are presented in the document itself.

- *Have you included all the elements your readers expect to see?* In planning the document, you thought about what the readers would expect to see. For instance, readers of a report expect to see certain elements, such as a transmittal letter, a table of contents, and an abstract. If your document is missing elements the readers expect, they might be distracted and unable to understand the document easily.

- *Are all the elements presented consistently?* In drafting quickly, you probably were not paying close attention to consistency. Check to see that parallel items are presented consistently. For instance, if you are using 18-point boldface text for your first-level headings, make sure all your first-level headings are presented that way. And check for grammatical parallelism, particularly in lists.

- *Is the emphasis appropriate throughout the document?* If the document has a table of contents, study it carefully. If a relatively minor topic seems to be treated at great length, with numerous subheadings, check the text itself. Maybe you created more headings at that point than you did in treating some of the other topics. But if your treatment of the minor topic is in fact excessive, mark passages for possible condensing.

- *Are the cross-references accurate?* In a technical document you often refer to other elements in the document. For instance, you will cross-reference an appendix or refer to a table or a figure that appears on a different page. Check to see that items you refer to are where you say they are.

Revising Titles for Coherence

The title is a key element in a document because it becomes the readers' first piece of "given" information. If the document is to cohere, its various topics have to relate clearly to the title.

Readers look for the title to provide two important pieces of information: your *subject* and *purpose*. Don't revise your title until you complete the document. The reason is simple: until you finish the document, you cannot be sure of the extent to which your subject and purpose have changed since you began to plan the document.

Precision is the key to a good title. If you are writing a feasibility study on the subject of offering free cholesterol screening at your company, make sure the title contains the key terms *cholesterol screening* and *feasibility*. The following title would be effective:

Offering Free Cholesterol Screening at Thrall Associates: A Feasibility Study

If your document is an internal report discussing company business, you might not need to identify the organization. The following would be clear:

Offering Free Cholesterol Screening: A Feasibility Study

Of course, you could present the purpose before the subject:

A Feasibility Study of Offering Free Cholesterol Screening

Do not substitute general terms, such as *health screening* for *cholesterol screening*. Keep in mind that key terms from your title might be used in various kinds of indexes; the more precise your terms, the more useful your readers will find the title.

Your readers should be able to translate your title into a clear and meaningful sentence. For instance, the title "A Feasibility Study of Offering Free Cholesterol Screening" could be translated into the following sentence: "This document reports on whether it is feasible to offer free cholesterol screening for our employees." Notice, however, what happens when the title is incomplete: "Free Cholesterol Screening." All the reader knows is that the document has something to do with free cholesterol screening. But is the writer recommending that free cholesterol screening be instituted—or discontinued? Or is the writer reporting on how well an existing program is working?

Following are more examples of effective titles:

Choosing a Laptop: A Recommendation

An Analysis of the Kelly 1013 Packager

Open Sea Pollution-Control Devices: A Summary

A Forecast of Smoking Habits in the United States in the Coming Decade

Revising Headings for Coherence

A heading is a lower-level title inside a document. A clear and informative heading is vital in technical communication because it announces the subject of the discussion that follows it. This announcement helps readers understand what they will be reading or, in some cases, helps them decide whether to read it at all. For the writer, a heading eliminates the need to announce the subject in a sentence such as "Let us now turn to the advantages of the mandatory enrollment process."

In revising your headings, follow these six guidelines:

- *Avoid long noun strings.* They are hard to understand:

  Production Enhancement Proposal Analysis Techniques

  Instead, put some prepositions into the phrase to make the title clearer.

  Techniques for Analyzing the Proposal for Enhancing Production

  This version says, more clearly than the noun-string version, that the writer is going to describe the techniques. See Chapter 14 for a detailed discussion of noun strings.

- *Be informative.* You could add information to the preceding example: some indication of how many techniques will be described.

  Three Techniques for Analyzing the Proposal for Enhancing Production

  And you can go one step further, by indicating what you wish to say about the three techniques:

  Advantages and Disadvantages of the Three Techniques for Analyzing the Proposal for Enhancing Production

  Don't worry if the heading seems too long; clarity is more important than brevity.

- *Use the question form for less knowledgeable readers.*

  What Are the Three Techniques for Analyzing the Proposal for Enhancing Production?

The question form is easier for less knowledgeable readers to understand (Benson 1985).

- *Use the "how to" form in instructional material, such as manuals.*

  How to Analyze the Proposal for Enhancing Production

- *Use the verbal form (-ing) to suggest a process.*

  Analyzing the Proposal for Enhancing Production

- *Avoid back-to-back headings.* Headings should be followed by text, not by lower-level headings. In other words, avoid the following:

  3. Approaches to Neighborhood Policing

      3.1 Community Policing

Text should separate the two headings, as in this example:

3. Approaches to Neighborhood Policing
   The past decade has seen unprecedented strains on the social institutions of society, as well as a further weakening of the family structure. One of the results of these twin problems is the increase in violent crime. Police departments throughout the country have struggled to keep up with the increase, especially the rise in violence among young people. In this section, we discuss two new approaches to neighborhood policing that appear to be working successfully in cities of different sizes and with different problems: community policing and police-sponsored activities for youth. For each of the two approaches, we will . . .

   3.1 Community Policing

The text after the heading "3. Approaches to Neighborhood Policing," called an *advance organizer,* introduces the material in section 3. Like any other kind of introduction, the advance organizer indicates the background, purpose, scope, and organization of the discussion that follows it. By using an introduction, you improve coherence by giving readers an overview of the discussion before they encounter the details.

For information on how to format the headings, see Chapter 11.

Revising Lists for Coherence

Lists are fundamental to technical communication because they add a visual dimension to the text, making it easier for readers to understand the discussion. As you revise your text, look for opportunities to turn traditional paragraphs into lists.

Not every paragraph should be a list, of course. But lists work especially well in conveying any kind of information that can be enumerated or expressed in a sequence. For instance, the following sentences introduce material that would be well suited to a list:

Contractors typically use one of four methods to shore the walls of an excavation.

Solar-energy research stalled in the 1980s for three major reasons.

When you arrive at the site, please make the following arrangements.

The next chapter discusses the use of lists in individual sentences, but lists can also enhance coherence in longer passages. For readers, the chief advantage of a list format is that the information is easier to read and to remember. The reason is that the logic of the discussion is evident on the page. The key terms in the list are set off with bullets or numbers, enabling readers to see them at a glance. Thus the coherence of the discussion is enhanced: readers see the overall structure before reading the detailed discussion.

For you as the writer, turning paragraphs into lists has four advantages:

- *It forces you to look at the big picture.* As you draft, it is easy to lose sight of the information outside the paragraph you are working on. But when you revise, looking for opportunities to create lists, you force yourself to study the key idea in each paragraph. This practice increases your chances of noticing that an important item is missing or that an item needs to be clarified.

- *It forces you to examine the sequence of the items.* As you transform some of your paragraphs into bulleted or numbered lists with key phrases, you get a clearer look at the sequence of information. If you want to make a change, the list format makes it easy.

- *It forces you to create a clear lead-in.* In the lead-in, you can add a number:

  Auto imports declined last year because of four major factors:

  Of course, you can add the same number in a traditional paragraph, but you are less likely to be thinking in those terms because you will not be focusing on the bulleted items.

- *It forces you to tighten and clarify the prose.* When you see an opportunity to make a list, look for a word, phrase, or sentence to identify each item. Your focus shifts from weaving sentences together in a paragraph to highlighting key ideas. And once you have formatted the list, look at it critically and revise it until it is as clear and concise as you can make it.

Figure 13.4 on page 410 (based on Cohen & Grace, 1994, p. 15) shows the same passage displayed in a paragraph form and in a list form. The authors are discussing the idea that engineers have a special social responsibility. Notice that turning the paragraph into a list accomplishes two things:

- It forces the writer to create headings that provide a sharp focus for each bulleted entry.

Currently, there are three conceptions of the relation between engineering as a profession and society as a whole.

The first conception is that there is no relation. Engineering's proper regard is properly instrumental, with no constraints at all. Its task is to provide purely technical solutions to problems.

The second conception is that engineering's role is to protect. It must be concerned, as a profession, with minimizing the risk to the public. The profession is to operate on projects as presented to it, as an instrument; but the profession is to operate in accordance with important safety constraints, which are integral to its performing as a profession.

The third conception is that engineering has a positive social responsibility to try to promote the public good, not merely to perform the tasks that are set for it, and not merely to perform those tasks such that risk is minimized or avoided in performing them. Rather, engineering's purpose as a profession is to promote the social good.

Currently, there are three conceptions of the relation between engineering as a profession and society as a whole.

- *There is no relation.* Engineering's proper regard is properly instrumental, with no constraints at all. Its task is to provide purely technical solutions to problems.
- *The engineer's role is to protect society.* Engineering is concerned, as a profession, with minimizing the risk to the public. The profession is to operate on projects as presented to it, as an instrument; but the profession is to operate in accordance with important safety constraints, which are integral to its performing as a profession.
- *The engineer's role is to promote social responsibility.* Engineering has a positive social responsibility to try to promote the public good, not merely to perform the tasks that are set for it, and not merely to perform those tasks such that risk is minimized or avoided in performing them. Rather, engineering's purpose as a profession is to promote the social good.

FIGURE 13.4

Paragraph Format
and List Format

- It enables the writer to delete the wordy topic sentences from the paragraph version. By deleting these sentences, the writer saves space so that the list version of the passage is not significantly longer than the paragraph version, despite the indentations and extra white space.

Revising Paragraphs for Coherence

The preceding discussion on the virtues of listing information is not meant to suggest that you could or should eliminate paragraphs. In fact, you will often write paragraphs within listed passages, as you saw in Figure 13.4. But regardless of whether you are writing traditional paragraphs or paragraphs within lists, try to enhance their coherence as you revise.

The two kinds of paragraphs are body paragraphs and transitional paragraphs.

A *body paragraph* is the basic structural unit for communicating the technical information. A body paragraph could be defined as a group of sentences (or sometimes a single sentence) that is complete and self-sufficient but that also contributes to a larger discussion. The challenge of creating an effective paragraph in technical communication is to make sure, first, that all the sentences clearly and directly substantiate one main point, and second, that the whole paragraph follows logically from the material that precedes it.

Readers tend to pause between paragraphs (not between sentences) to digest the information given in one paragraph and link it with that given in the previous paragraphs. For this reason, the paragraph is the key unit of composition. Readers might forgive or at least overlook a slightly fuzzy sentence. But if they can't figure out what a paragraph says or why it appears where it does, communication is likely to break down.

A *transitional paragraph* helps readers to progress from one major point to another. In most cases, it summarizes the previous point and helps readers understand how it relates to the next one.

The following example of a transitional paragraph is taken from a manual explaining how to write television scripts. The writer has described six principles of writing for an episodic program, including introducing characters, pursuing the plot, and resolving the action at the end of the episode.

> The six basic principles of writing for episodic television, then, are the following:
> - Reintroduce the characters.
> - Make the extra characters episode-specific.
> - Present that week's plot swiftly.
> - Make the characters react according to their personalities.
> - Resolve the plot neatly.
> - Provide a denouement that hints at further developments.
>
> But how do you put these six principles into action? The following section provides specific how-to instructions.

Notice that the first sentence contains the word "then" to signal that it is introducing a summary. Note, too, that the final sentence of the transitional paragraph clearly indicates the relationship between what precedes it and what follows it.

Structure Paragraphs Clearly

If you draft quickly—as you should—your paragraphs will need a lot of revision before they are easy to read and understand. A hasty paragraph often starts off with a number of details: about who worked on the problem before and what equipment or procedure they used; about the ups and downs of the project, the successes and setbacks; about specifications, dimensions, and computations. The paragraph winds its way down the page until, finally, the writer concludes: "No problems were found."

This structure—moving from the particular details to the general statement—reflects the way we think: we accumulate the details and then draw conclusions from them. But the paragraph will be easier to read if the main point is presented first and then supported in the remainder of the paragraph.

The Topic Sentence

Help your readers. Put the point—the *topic sentence*—up front. Technical communication should be clear and easy to read, not full of suspense. If a paragraph describes a test you performed on a piece of equipment, include the result in your first sentence:

> The point-to-point continuity test on Cabinet 3 revealed no problems.

Then go on to explain the details. If the paragraph describes a complicated idea, start with an overview:

> Mitosis occurs in five stages: (1) interphase, (2) prophase, (3) metaphase, (4) anaphase, and (5) telophase.

Then describe each phase. In other words, put the "bottom line" on top.

Notice, for instance, how difficult the following paragraph is to read because the writer structured the discussion in the same order that she performed her calculations:

> Our estimates are based on our generating power during eight months of the year and purchasing it the other four. Based on the 1997 purchased power rate of $0.034/KW (January through April cost data) inflating at 8 percent annually, and a constant coal cost of $45–$50, the projected 1998 savings resulting from a conversion to coal would be $225,000.

Putting the bottom line on top makes the paragraph much easier to read. Notice how the writer adds a numbered list after the topic sentence:

> The projected 1998 savings resulting from a conversion to coal are $225,000. This estimate is based on three assumptions: (1) that we will be generating power during eight months of the year and purchasing it the other four, (2) that power rates inflate at 8 percent from the 1997 figure of $0.034/KW (January through April cost data), and (3) that coal costs remain constant at $45–$50.

The topic sentence in technical communication functions just as it does in any other kind of writing: it summarizes or forecasts the main point of the paragraph.

The Support

The support follows the topic sentence. The purpose of the support is to make the topic sentence clear and convincing. Sometimes a few explanatory details can provide all the support needed. In the paragraph about estimated fuel savings presented earlier, for example, the writer simply fills in the assumptions used in making the calculation: the current energy rates, the inflation rate, and so forth. Sometimes, however, the support must carry a heavier load: it has to clarify a difficult thought or defend a controversial one.

Because every paragraph is unique, it is impossible to generalize about the exact function of the support. Usually, however, the support fulfills one of the following five roles:

- to define a key term or idea included in the topic sentence
- to provide examples or illustrations of the situation described in the topic sentence
- to identify causes: factors that led to the situation
- to define effects: implications of the situation
- to defend the assertion made in the topic sentence

The techniques that writers use in developing the support include those used in most nonfiction writing, including definition, comparison and contrast, classification and partition, and causal analysis. Another key to supporting a paragraph is the use of specific, concrete details. These techniques are described in detail in Part Three.

Keep Paragraphs to a Manageable Length

How long should a paragraph of technical communication be? In general, 75 to 125 words will provide enough length for a topic sentence and four or five supporting sentences. Long paragraphs are more difficult to read than short paragraphs simply because readers have to concentrate longer. Long, unbroken stretches of type can so intimidate some readers that they actually skip them.

But don't let an arbitrary guideline about length take precedence over your analysis of the audience and purpose. Often you will need to write very brief paragraphs. You might need only one or two sentences—to introduce a graphic, for example. A transitional paragraph also is likely to be quite short. If a brief paragraph fulfills its function, let it be. Do not combine two ideas in one paragraph to achieve a minimum word count.

Although it is confusing to include more than one basic idea in a paragraph, often you will need to break up one idea into two or more paragraphs. An idea that would require 200 or 300 words to develop probably should not be squeezed into one paragraph.

The following example shows how a writer addressing a general audience divided one long paragraph into two:

> High-tech companies have been moving their operations to the suburbs for two main reasons: cheaper, more modern space and a better labor pool. A new office complex in the suburbs will charge anywhere from half to two-thirds of the rent charged for the same square footage in the city. And that money goes a lot further, too. The new office complexes are bright and airy, with picture windows looking out on lush landscaping. New office space is already wired for the computers; and exercise clubs, shopping centers, and even libraries are often on-site.

The second major factor attracting high-tech companies to the suburbs is the availability of experienced labor. Office workers and middle managers are abundant; many suburbanites, especially women returning to the labor force after their children start school, are highly trained and willing to make the short trip to the office complex. In addition, the engineers and executives, who tend to live in the suburbs anyway, are happy to forego the commuting, the city wage taxes, and the noise and stress of city life.

A strict approach to paragraphing would have required one paragraph, not two, because all the information presented supports the topic sentence that opens the first paragraph. Many readers, in fact, could easily understand a one-paragraph version. However, the writer found a logical place to create a second paragraph and thereby improved the communication.

Following is another approach: making each "reason for moving to the suburbs" a separate paragraph.

High-tech companies have been moving their operations to the suburbs for two main reasons: cheaper, more modern space and a better labor pool.

First, office space is a bargain in the suburbs. A new office complex will charge anywhere from half to two-thirds of the rent charged for the same square footage in the city. And that money goes a lot further, too. The new office complexes are bright and airy, with picture windows looking out on lush landscaping. New office space is already wired for the computers; and exercise clubs, shopping centers, and even libraries are often on-site.

Second, experienced labor is plentiful. Office workers and middle managers are abundant; many suburbanites, especially women returning to the labor force after their children start school, are highly trained and willing to make the short trip to the office complex. In addition, the engineers and executives, who tend to live in the suburbs anyway, are happy to forego the commuting, the city wage taxes, and the noise and stress of city life.

The original topic sentence becomes a transitional paragraph that leads clearly and logically into the two body paragraphs.

Of course, the writer could also use a list format:

High-tech companies have been moving their operations to the suburbs for two main reasons:

- *Cheaper, more modern space.* Office space is a bargain in the suburbs. A new office complex will charge anywhere from half to two-thirds of the rent charged for the same square footage in the city. And that money goes a lot further, too. The new office complexes are bright and airy, with picture windows looking out on lush landscaping. New office space is already wired for the computers; and exercise clubs, shopping centers, and even libraries are often on-site.
- *A better labor pool.* Office workers and middle managers are abundant; many suburbanites, especially women returning to the labor force after their children start school, are highly trained and willing to make the short trip to the office complex. In addition, the engineers and executives, who tend to live in the suburbs anyway, are happy to forego the commuting, the city wage taxes, and the noise and stress of city life.

FIGURE 13.5
Coherence Devices

Use Coherence Devices within and between Paragraphs After you have revised the main structure of the paragraph—the topic sentence and the support—make sure that the paragraph is coherent. In a coherent paragraph, thoughts are linked together logically and clearly. Parallel ideas are expressed in parallel grammatical constructions. Even if the paragraph already moves smoothly from sentence to sentence, emphasize the coherence in three ways, as shown in Figure 13.5.

Transitional Words and Phrases

Transitional words and phrases help the reader understand a discussion by pointing out the direction the thoughts are following. Table 13.1 lists the most common logical relationships between two thoughts and some of the common transitions that express those relationships.

In the following examples, the first versions contain no transitional words and phrases. Notice how much clearer the second versions are.

TABLE 13.1	Transitional Words and Phrases
Relationship	**Transitions**
addition	also, and, finally, first (second, etc.), furthermore, in addition, likewise, moreover, similarly
comparison	in the same way, likewise, similarly
contrast	although, but, however, in contrast, nevertheless, on the other hand, yet
illustration	for example, for instance, in other words, to illustrate
cause-effect	as a result, because, consequently, hence, so, therefore, thus
time or space	above, around, earlier, later, next, to the right (left, west, etc.), soon, then
summary or conclusion	at last, finally, in conclusion, to conclude, to summarize

WEAK: Neurons are not the only kind of cell in the brain. Blood cells supply oxygen and nutrients.

IMPROVED: Neurons are not the only kind of cell in the brain. For instance, blood cells supply oxygen and nutrients.

WEAK: The project was originally expected to cost $300,000. The final cost was $450,000.

IMPROVED: The project was originally expected to cost $300,000. *However,* the final cost was $450,000.

WEAK: The manatee population of Florida has been stricken by an unknown disease. Marine biologists from across the nation have come to Florida to assist in manatee-disease research.

IMPROVED: The manatee population of Florida has been stricken by an unknown disease. *As a result,* marine biologists from across the nation have come to Florida to assist in manatee-disease research.

Use transitional words to maintain coherence *between* paragraphs just as you use them to maintain coherence *within* paragraphs. Place the transitions as close as possible to the beginning of the second element. For example, the link between two sentences within a paragraph should be near the start of the second sentence:

The new embossing machine was found to be defective. *However,* the warranty on the machine will cover replacement costs.

The link between two paragraphs should be near the start of the second paragraph:

The complete system would be too expensive for us to purchase now _____

_____

In addition, a more advanced system is expected on the market within six months

_____

_____

Key Words

Repeating key words—usually nouns—helps readers follow the discussion. Notice in the following example how the first version can be confusing:

UNCLEAR: For months the project leaders carefully planned their research. The cost of the work was estimated to be over $200,000. (What is the work: the planning or the research?)

CLEAR: For months the project leaders carefully planned their *research.* The cost of the *research* was estimated to be over $200,000.

Out of a misguided desire to be interesting, some writers keep changing their important terms. *Plankton* becomes *miniature seaweed,* then *the ocean's fast food.*

Leave this kind of word game to TV sportscasters; technical communication must be clear and precise.

Of course, too much repetition can be boring. You can vary nonessential terms, so long as you don't sacrifice clarity.

SLUGGISH: The purpose of the new plan is to *reduce the problems* we are seeing in our accounting operations. We hope to see a *reduction in the problems* by early next quarter.

BETTER: The purpose of the new plan is to *reduce the problems* we are seeing in our accounting operations. We hope to see an *improvement* by early next quarter.

Demonstrative Pronouns Followed by Nouns

In addition to using transitional words and phrases and repeating key phrases, using demonstrative pronouns—*this, that, these,* and *those*—can help you maintain the coherence of a discussion by linking ideas securely. In almost all cases, demonstrative pronouns should serve as adjectives rather than as pronouns; that is, they should be followed by nouns. In the following examples, notice that a demonstrative pronoun by itself can be confusing.

UNCLEAR: New screening techniques are being developed to combat viral infections. These are the subject of a new research effort in California. (What is being studied in California: new screening techniques or viral infections?)

CLEAR: New screening techniques are being developed to combat viral infections. *These techniques* are the subject of a new research effort in California.

UNCLEAR: The task force could not complete its study of the mine accident. This was the subject of a scathing editorial in the union newsletter. (What was the subject of the editorial: the mine accident or the task force's inability to complete its study of the accident?)

CLEAR: The task force failed to complete its study of the mine accident. *This failure* was the subject of a scathing editorial in the union newsletter.

Even when the context is clear, a demonstrative pronoun used without a noun refers readers backward to an earlier idea and therefore interrupts readers' progress.

INTERRUPTIVE: The law firm advised that the company initiate proceedings. This caused the company to search for a second legal opinion.

FLUID: The law firm advised that the company initiate proceedings. *This advice* caused the company to search for a second legal opinion.

Transitional words and phrases, the repetition of key words, and demonstratives cannot *give* your writing coherence: they can only help readers appreciate the coherence that already exists. Your job is, first, to make sure your writing is coherent and, second, to highlight that coherence.

Turning a "Writer's Paragraph" into a "Reader's Paragraph"

To demonstrate the strategies that improve coherence, the following examples show how these techniques can improve a weak paragraph. The paragraph is from a status report written by a branch manager of a utility company. In it the writer explains how he decided on a method to increase the company's business within his particular branch. (The sentences are numbered to clarify the marginal comments.)

Sentence 1 focuses on the two alternatives, not on the final decision the writer made. Sentences 2 and 3 describe one alternative, then the other. Sentence 4 explains why the first alternative was undesirable. Sentences 5 and 6 explain why he chose the second alternative. Throughout the paragraph, the focus is on the process of the study, not on the results.

(1) There were two principal alternatives considered for improving the Montana Branch. (2) The first alternative was to drill and equip additional sources of supply with sufficient capacity to provide for the present and projected system deficiencies. (3) The second alternative was to provide for said deficiencies through a combination of additional sources of supply and a storage facility. (4) Unfortunately, groundwater studies which were conducted in the Southeast Montana area by the consulting firm of Smith and Jones indicated that although groundwater is available within this general area of our system, it is limited as to quantity, and there is considerable separation between said sources. (5) This being the case, it becomes necessary to utilize the sources that are available or that can be developed in the most efficient manner, which means operating them in conjunction with a storage facility. (6) In this way, the sources only have to be capable of providing for the average demand on a maximum day, and the storage facility can be utilized to provide for the peaking requirements plus fire protection. (7) Consequently, the second alternative as mentioned hereinabove was determined to be the more desirable alternative.

First, let's be fair. The paragraph has been taken out of its context—a 17-page report—and was never meant to stand alone on a page. Also, it was written not for the general reader but for an executive of the water company—someone who, in this case, is technically knowledgeable in the writer's field. Still, an outsider's analysis of an essentially private communication can at least isolate the weaknesses.

Following is the writer's paragraph translated into a reader's paragraph. (The sentences are numbered to clarify the marginal comments.)

The topic sentence clearly states the main point of the paragraph. Sentence 2 justifies the assertion from sentence 1. Sentence 3 goes back to describe the other alternative they considered. Sentence 4 explains why that alternative was rejected.

(1) We found that the best way to improve the Montana branch would be to add a storage facility to our existing supply sources. (2) Currently, we can handle the average demand on a maximum day; the storage facility will enable us to meet peaking requirements and fire-protection needs. (3) In conducting our investigation, we considered developing new supply sources with sufficient capacity to meet current and future needs. (4) This alternative was rejected, however, when our consultants (Smith and Jones) did groundwater studies that revealed that insufficient groundwater is available and that the new wells would have to be located too far apart if they were not to interfere with each other.

This revision is superior to the original in that it is shorter and more direct, and therefore easier to read and understand. Notice the use of transitional words ("currently" in sentence 2 and "however" in sentence 4); key words ("storage facility," "sources," and "needs"); and demonstrative pronouns with nouns ("this alternative").

The only possible objection to the streamlined version is that it is *too* clear, that it leaves the writer vulnerable in case his decision turns out to have been wrong. But if the decision doesn't work out, the writer will be responsible anyway, and poor writing will not endear him to his supervisor. Good writing is the best bet under any circumstances.

CASE

Writing Guidelines about Coherence

You are a public-information officer recently hired by the Agency for Health Care Policy and Research. One of your responsibilities is to make sure that your agency's public information on the Web is clear and accurate. Your supervisor, José Martinez, has asked you to write guidelines for physicians and other researchers who write the articles you put on your site. You ask him why he thinks they need the guidelines. "Their writing is factually correct," José replies, "but because they are taking excerpts from longer, more scientific studies, their documents can be choppy. They need to be smoothed out." You ask your supervisor to point you in the right direction by identifying a sample that shows the qualities he wants you to describe. He directs you to the following brief report (U.S. Department of Health and Human Services, 1997) about therapies for treating pneumonia. "This is a good sample of how to write to the general reader," José tells you. Study this report, noting the different coherence techniques the writer has used. Focus on the title, the headings, and the paragraphs. Write a brief set of guidelines for cohesive writing, using excerpts from this sample to illustrate your advice.

PNEUMONIA: MORE PATIENTS MAY BE TREATED AT HOME
Research Findings for Consumers

Overview

A promising new model for doctors developed by U.S. and Canadian researchers could lead to more patients being treated for pneumonia in the comfort of their own homes as safely and as effectively as if they were hospitalized, according to new research supported by the federal government's Agency for Health Care Policy and Research (AHCPR). The model—a clinical algorithm—can help doctors quickly and easily determine a patient's risk level, which is essential for deciding where treatment should take place and the type of therapy to be used.

Although pneumonia—a disease characterized by a bacterial or viral infection of the lungs—can be deadly, research has shown that the majority of patients are

(Case, continued)

low-risk, meaning that they are in little danger of dying from the disease or of suffering serious consequences because of it.

Some patients already being treated in the hospital develop pneumonia. However, this research addresses the vast majority of pneumonia cases, patients not already hospitalized. About 4 million Americans a year develop this "community-acquired" pneumonia and 600,000 of these, or 15 percent, are hospitalized.

Although about 85 percent of pneumonia cases currently are treated outside the hospital, medical experts believe that an even higher percentage are eligible for outpatient care, and others could be hospitalized just for short periods. Treating more pneumonia patients at home also could help lower the cost of care. Inpatient treatment of pneumonia costs an estimated 10 to 15 times as much as outpatient care.

The problem is that doctors do not have science-based criteria to guide their decisions for admitting pneumonia patients. Part of the solution, researchers believe, may lie in this new model, which helps physicians identify pneumonia patients who do not need intensive treatment by accurately estimating their progress, or prognosis, from basic medical information.

To ensure the accuracy of the algorithm, the researchers tested it using data on thousands of pneumonia patients, including roughly 2,300 individuals in Pittsburgh, Boston, and Halifax, Nova Scotia, who were treated at home or in the hospital. If the prediction model had been available to doctors in those three cities, roughly a quarter to nearly one-third of the hospitalized patients could have been assigned outpatient care, and slightly over a tenth to almost one-fifth could have been kept only briefly for observation instead of having a longer stay.

What Patients Prefer

Most of the low-risk patients in the study who were surveyed, including those hospitalized for initial treatment, said they generally preferred home-based care. But the researchers found that patients usually are not asked where and how they would like their pneumonia to be treated. Of the doctors from the three-city study who were surveyed, 83 percent said that they alone made the decisions for outpatients, and 72 percent said they did so for the inpatients. A number of factors weigh against home care, such as the lack of a family caregiver, limited availability of home nursing services, inability to drink fluids and take medication by mouth, and certain severe medical conditions.

About the Study

These findings are from a recently completed, 5-year study of variations and outcomes in pneumonia care. The research is part of a series of studies on the quality, effectiveness, and cost-effectiveness of current therapies for treating some of the most common and costly medical conditions in the United States. The project is supported by AHCPR—the U.S. Department of Health and Human Services agency spearheading federal efforts to improve the quality of American medical care. The study was directed by Wishwa N. Kapoor, M.D., M.P.H., from the University of Pittsburgh School of Medicine.

WRITER'S CHECKLIST

1. Have you revised the document yourself, by
   - ❑ letting it sit?
   - ❑ reading it aloud?
   - ❑ using checklists?
   - ❑ using software?

2. Have you revised the document by obtaining help from
   - ❑ appropriate subject-matter experts?
   - ❑ people similar to the eventual readers?

3. If possible and appropriate, have you tested readers as they use the document?

4. Have you revised the whole document by checking to make sure that
   - ❑ you haven't left out anything in turning your outline into a draft?
   - ❑ you have included all the elements your readers expect to see?
   - ❑ all the elements are presented consistently?
   - ❑ all the cross-references are accurate?
   - ❑ the emphasis is appropriate throughout the document?

5. Have you revised the title so that it
   - ❑ clearly states the audience and purpose of the document?
   - ❑ is sufficiently precise and informative?

6. Have you revised the headings to
   - ❑ avoid long noun strings?
   - ❑ be informative?
   - ❑ use the question form for less knowledgeable readers?
   - ❑ use the "how to" form in instructional material, such as manuals?
   - ❑ use the verbal form (-ing) to suggest a process?
   - ❑ avoid back-to-back headings by including an advance organizer?

7. Have you looked for opportunities to turn traditional paragraphs into lists?

8. Have you revised your paragraphs so that each one
   - ❑ begins with a clear topic sentence?
   - ❑ has adequate and appropriate support?
   - ❑ is not too long for readers?
   - ❑ uses coherence devices such as transitional words and phrases, repetition of key words, and demonstratives followed by nouns?

EXERCISES

1. Write a one-paragraph evaluation for each of the following titles. How clearly does the title indicate the subject and purpose of the document? On the basis of your analysis, rewrite each title.
   a. Recommended Forecasting Techniques for Haldane Company
   b. Robotics in Japanese Manufacturing
   c. A Study of Disc Cameras
   d. Agriculture in the West: A 10-Year View
   e. Synfuels—Fact or Hoax?

2. Write a one-paragraph evaluation for each of the following headings. How clearly does the heading indicate the subject of the text that will follow it? On the basis of your analysis, rewrite each heading to make it more clear and informative. Invent any necessary details.
   a. Multigroup Processing Technique Review Board Report Findings
   b. The Great Depression of 1929
   c. Low-Level Radiation and Animals
   d. Minimize Down Time
   e. Intensive Care Nursing

3. Turn the following passage (based on Snyder, 1993) into a list format. The subject is bioremediation: the process of using microorganisms to restore natural environmental conditions.

   Scientists are now working on several new research areas. One area involves using microorganisms to make some compounds less dangerous to the environment. Although coal may be our most plentiful fossil fuel, most of the nation's vast Eastern reserve cannot meet air-pollution standards because it emits too much sulfur when it is burned. The problem is that the aromatic compound dibenzothiothene (DBT) attaches itself to hydrocarbon molecules, producing sulfur dioxide. But the Chicago-based Institute of Gas Technology last year patented a bacterial strain that consumes the DBT (at least 90 percent, in recent lab trials) while leaving the hydrocarbon molecules intact.

   A second research area is the genetic engineering of microbes in an attempt to reduce the need for toxic chemicals. In 1991, the EPA approved the first genetically engineered pesticide. Called Cellcap, it incorporates a gene from one microbe that produces a toxin deadly to potato beetles and corn borers into a thick-skinned microbe that is hardier. Even then, the engineered bacteria are dead when applied to the crops.

A third research area is the use of microorganisms to attack stubborn metals and radioactive waste. Microbes have been used for decades to concentrate copper and nickel in low-grade ores. Now researchers are exploiting the fact that if certain bacteria are given special foods, they excrete enzymes that break down metals and minerals. For example, researchers at the U.S. Geological Survey found that two types of bacteria turn uranium from its usual form—one that easily dissolves in water—into another—one that turns to a solid that can be easily removed from water. They are now working on doing the same for other radioactive waste.

4. Provide a topic sentence for each of the following paragraphs:
   a. _____. All service centers that provide gas and electric services in the tri-county area must register with the TUC, which is empowered to carry out unannounced inspections periodically. Additionally, all service centers must adhere to the TUC's Fair Deal Regulations, a set of standards that encompasses every phase of the service-center operations. The Fair Deal Regulations are meant to guarantee that all centers adhere to the same standards of prompt, courteous, and safe work at a fair price.
   b. _____. The reason for this difference is that a larger percentage of engineers working in small firms may be expected to hold high-level positions. In firms with fewer than 20 engineers, for example, the median income was $42,200. In firms of 20 to 200 engineers, the median income was $40,345. For the largest firms, the median was $38,600.

5. Develop the following topic sentences into full paragraphs:
   a. Job candidates should not automatically choose the company that offers the highest salary.
   b. Every college student should learn at least the fundamentals of computer science.
   c. The one college course I most regret not having taken is _____.

   d. Sometimes two instructors offer contradictory advice about how to solve the same kind of problem.

6. In the following paragraph, transitional words and phrases have been removed. Add an appropriate transition in each blank space. Where necessary, add punctuation.

   As you know, the current regulation requires the use of conduit for all cable extending more than 18″ from the cable tray to the piece of equipment. _____ conduit is becoming increasingly expensive: up 17% in the last year alone. _____ we would like to determine whether the NRC would grant us any flexibility in its conduit regulations. Could we _____ run cable without conduit for lengths up to 3′ in low-risk situations such as wall-mounted cable or low-traffic areas? We realize _____ that conduit will always remain necessary in high-risk situations. The cable specifications for the Unit Two report to the NRC are due in less than two months; _____ we would appreciate a quick reply to our request, because this matter will seriously affect our materials budget.

7. In each of the following exercises, the second sentence begins with a demonstrative pronoun. Add a noun after the demonstrative to enhance the coherence.
   a. The Zoning Commission has scheduled an open hearing for March 14. This _____ will enable concerned citizens to voice their opinions on the proposed construction.
   b. The university has increased the number of parking spaces, instituted a shuttle system, and increased parking fees. These _____ are expected to ease the parking problems.
   c. Congress's decision to withdraw support for the supercollider in 1994 was a shock to the U.S. particle-physics community. This _____ is seen as instrumental in the revival of the European research community.

Collaborative Exercise

8. Form small groups. Have each person in the group find a multipage document he or she has written

recently, either in this class or in another. Make copies of this document for each group member. Have each member annotate the document according to the principles of coherence discussed in this chapter, then write a summary statement at the end of the document, highlighting those techniques of coherence that are well done and those that could be improved. Meet as a group and study these annotated documents. Write a memo to your instructor describing those aspects of the annotations on your document that more than one group member cited, as well as those aspects that only one group member cited. Overall, what are the basic differences among group members between the annotations as well as the summary statement? Do you think that, as a general practice, it would be worthwhile to have a draft reviewed and annotated by more than one person? What have you learned about the usefulness of peer reviewing of a document?

 Web Exercise

9. Find a report on the Web for this project on coherence. The report should be at least several pages long and contain a number of the elements discussed in this chapter, such as titles, headings, and lists. You might use a search engine to find the sites of companies and professional organizations in your field (see Chapter 7). In a memo to your instructor, describe

the use of coherence principles in the site. Have the authors of the site presented information differently than they would have presented it in printed text? For instance, do headings appear as they would in a printed document, or are they links to other files or sites? Does the paragraphing style seem tailored to the Web, or would a printed version of the information look essentially the same? Would you make any changes to the writing to improve its coherence? Present your findings in a memo to your instructor. Include copies of representative pages from the site. See Chapter 16 for a discussion of memos.

REFERENCES

Benson, P. (1985). Writing visually: Design considerations in technical publications. *Technical Communication, 32*(4), pp. 35–39.

Cohen, S., & Grace, D. (1994). Engineers and social responsibility: An obligation to do good. *IEEE Technology and Society, 13*(3), pp. 12–19.

Snyder, J. D. (1993). Off-the-shelf bugs hungrily gobble our nastiest pollutants. *Smithsonian, 24*(1), pp. 66+.

U.S. Department of Health and Human Services. (1997). Pneumonia: More patients may be treated at home: Research findings for consumers. *Agency for Health Care Policy and Research.* <http://www.ahcpr.gov/consumer/pneucons.htm> AHCPR Publication No. 97-R030 (1997, March 31).

Revising for Sentence Effectiveness

Structuring Effective Sentences 425

    Use Lists *425*

    Put New and Important Information at the End of the Sentence *430*

    Choose an Appropriate Sentence Length *430*

    Focus on the "Real" Subject *432*

    Focus on the "Real" Verb *433*

    Express Parallel Elements in Parallel Structures *434*

    Use Modifiers Effectively *435*

Choosing an Appropriate Level of Formality 437

Choosing the Right Words and Phrases 438

    Be Clear and Specific *438*

    Be Concise *444*

    Use Inoffensive Language *448*

Understanding Simplified English for Nonnative Speakers 449

Understanding Structured Writing 450

Case: Revising a Draft for Sentence Effectiveness *453*

Writer's Checklist 454

Exercises 454

    Collaborative Exercise *458*

    Web Exercise *458*

The previous chapter discussed techniques for improving the coherence of the whole document, of long passages, and of paragraphs. This chapter concentrates on sentences and the clauses, phrases, and words they contain.

Technical communication is workplace writing; it is meant to get a job done, not to show off the writer's skill. Therefore, the advice offered in this chapter is based on a simple idea: the sentences and words of technical communication should be clear, simple, concise, and easy to understand.

Your readers should not be aware of your presence. They should not notice that you have a wonderful vocabulary or that your sentences flow beautifully—even if those things are true. In the best technical communication, the writer stays in the background.

This is as it should be. Few people read technical communication for pleasure. Most readers either must read it as part of their work or to keep abreast of developments. People read technical communication to gather information, not to appreciate the writer's flair. For this reason, experienced writers do not try to be fancy. The old saying has never been more appropriate: *Write to express, not to impress.*

Structuring Effective Sentences

Good technical communication is characterized by clear, correct, and graceful sentences that convey information without calling attention to themselves. This section consists of seven guidelines for structuring effective sentences, as shown in Figure 14.1 on page 426.

Use Lists

list

Chapter 13 contains a discussion of lists as a technique for organizing paragraph-length passages. This section discusses lists as a technique for communicating shorter passages.

Many sentences in technical communication are long and complicated:

> We recommend that more work on heat-exchanger performance be done with a larger variety of different fuels at the same temperature, with similar fuels at different temperatures, and with special fuels such as diesel fuel and shale-oil-derived fuels.

Here readers cannot concentrate fully on the information because they are trying to remember all the *with* phrases following "done." If they could "see" how many phrases they had to remember, their job would be easier.

Revised as a list, the sentence is easier to follow:

> We recommend that more work on heat-exchanger performance be done:
> - with a larger variety of different fuels at the same temperature
> - with similar fuels at different temperatures
> - with special fuels such as diesel fuels and shale-oil-derived fuels

FIGURE 14.1

Structuring
Effective Sentences

In this version, the arrangement of the words on the page reinforces the meaning. The bullets direct the reader's eyes to the three items in a series. And the fact that each item begins at the same left margin helps, too.

If you don't have enough space to list the items vertically, or if you are not permitted to do so, number the items within the sentence:

> We recommend that more work on heat-exchanger performance be done (1) with a larger variety of different fuels at the same temperature, (2) with similar fuels at different temperatures, and (3) with special fuels such as diesel fuels and shale-oil-derived fuels.

As you create lists, follow these six guidelines:

- *Indent the items in the list.* The amount of indentation depends on the length of the items. Single words or short phrases might be indented so that the list appears centered; longer items might be indented only two or three characters.

- *Set off each listed item with a number, a letter, or a symbol (usually a bullet).*

  – Use numbered lists to suggest sequence (as in the numbered steps in a set of instructions) or priority (the first item is the most important). Sometimes writers number a list to emphasize the total number of items (as in the "Seven Warning Signals of Cancer" from the American Cancer Society). For sublists, use lowercase letters:

    1. Item
        a. subitem
        b. subitem

    2. Item
        a. subitem
        b. subitem

  – Use bullets when you do not wish to suggest sequence or priority. Make sure you don't use numbers for lists of people; everyone except number 1 gets offended. For sublists, use hyphens.

    - Item
        – subitem
        – subitem

  – Use an open (unshaded) box (□) for checklists. See Chapter 12 for more information on checklists.

- *Break up long lists.* Because most people can remember only 5 to 9 items, break up lists that contain 10 or more items. Figure 14.2 shows how to break up a long list.

Original List	Revised List
Tool Kit:	Tool Kit:
• hand saw	• Saws
• coping saw	– hand saw
• hacksaw	– coping saw
• compass saw	– hacksaw
• adjustable wrench	– compass saw
• box wrench	
• Stillson wrench	• Wrenches
• socket wrench	– adjustable wrench
• open-end wrench	– box wrench
• Allen wrench	– Stillson wrench
	– socket wrench
	– open-end wrench
	– Allen wrench

FIGURE 14.2

Breaking Up a
Long List

Carliner (1987) recommends breaking up a bulleted list of more than 5 items and a numbered list of more than 10. (The numbers give readers extra cues to help them remember the information.)

- *Present the items in a parallel structure.* Parallelism is discussed later in this chapter. Figure 14.3 shows how to revise a nonparallel list. In Figure 14.3, the nonparallel list is sloppy, a mixture of noun phrases (items 1, 3, 4, and 5), a verb phrase (item 2), and a participial phrase preceded by a dependent clause (item 6). The revision uses parallel verb phrases and deemphasizes the dependent clause in item 6 by placing it after the verb phrase.

- *Structure and punctuate the lead-in correctly.* Although standards vary from organization to organization, the most common is to use a grammatically complete lead-in followed by a colon, as shown in the following examples:

Following are the three main assets:

The three main assets are as follows:

The three main assets are the following:

If you cannot use a grammatically complete lead-in, use a colon, a dash, or no punctuation at all:

The committee found that the employee
- did not cause the accident
- acted properly immediately after the accident
- reported the accident properly

- *Punctuate the list correctly.* Rules for punctuating lists vary. Like most stylistic issues in technical communication, there is no accepted standard, so you should find out whether your organization has a preference. If it does not, you can generally punctuate lists as follows:

– If the items are sentence fragments, use a lowercase letter at the start and do not use a period or a comma at the end.

Nonparallel	**Parallel**
Here is the sequence we plan to follow:	Here is the sequence we plan to follow:
1. construction of the preliminary proposal	1. write the preliminary proposal
2. do library research	2. do library research
3. interview with the Bemco vice president	3. interview the Bemco vice president
4. first draft	4. write the first draft
5. revision of the first draft	5. revise the first draft
6. after we get your approval, typing of the final draft	6. type the final draft, after we receive your approval

FIGURE 14.3

Parallelism in Lists

> The new facility will offer three advantages:
> - lower leasing costs
> - easier commuting distance
> - a greater pool of potential workers

The last item in a list of fragments is generally *not* followed by a period. The white space to the right of the last listed item, as well as the white space that separates the list from the next line, clearly indicates the end of the list. Some writers, however, prefer to add a period after the last item.

– If the items are complete sentences, use an uppercase letter at the start and a period at the end.

> The new facility will offer three advantages:
> - The leasing costs will be lower.
> - The commuting distance for most of the employees will be shorter.
> - The pool of potential workers will be larger.

– If the items are fragments followed by complete sentences, begin the fragment with an uppercase letter and end with a period. Then begin the complete sentences with uppercase letters and end them with periods.

> The new facility will offer three advantages:
> - Lower leasing costs. The lease will cost $1,800 per month; currently we pay $2,300.
> - Easier commuting distance. According to a recent questionnaire, our workers now spend an average of 18 minutes traveling to work. At the new location, the average would drop to 14 minutes.
> - A greater pool of potential workers. In the last decade, the population has begun to shift westward, to the area of the new facility. We would be able to increase our potential workforce, especially in the semiskilled and managerial categories.

– If the list consists of two kinds of items—some fragments, some fragments followed by complete sentences—punctuate all the items the same way: with uppercase letters and periods.

> The new facility will offer three advantages:
> - Lower leasing costs.
> - Easier commuting distance. According to a recent questionnaire, our workers now spend an average of 18 minutes traveling to work. At the new location, the average would drop to 14 minutes.
> - A greater pool of potential workers. In the last decade, the population has begun to shift westward, to the area of the new facility. We would be able to increase our potential workforce, especially in the semiskilled and managerial categories.

Note that in the design of lists, the second and subsequent lines in an item, called *turnovers*, are indented. Turnovers are aligned under the first letter of the first line. In other words, the bullet or the number at the start of the list extends out to the left of the text. This form, called *hanging indentation,* highlights the bullet or the number and thereby helps the reader see the organization of the passage.

Put New and Important Information at the End of the Sentence

`new`

Chapter 13 discussed the given/new concept: we learn by adding new information to the information we already know. This principle applies not only to large elements in a text but also to individual sentences. In general, your sentences will be easier to understand and more emphatic if you put the new information at the end of the sentence.

For instance, your company is experiencing labor problems, and you want to describe the possible results. Structure the sentence like this:

> Because of the labor problems, we anticipate a three-week delay.

In this case, the *three-week delay* is the new information.

If, however, your readers know that there will be a three-week delay but don't know the reason for it, structure the sentence this way:

> We anticipate the three-week delay in the production schedule because of labor problems.

Here, *labor problems* is the new information. Notice how the articles *the* and *a* signal the old and new information. In the second sentence, the definite article *the* is used with given information; in the first sentence, the indefinite article *a* is used with new information.

Try not to end the sentence with qualifying information that blunts the impact of the new information.

> WEAK: The joint could fail under special circumstances.
>
> IMPROVED: Under special circumstances, the joint could fail.

Also, put new or difficult terms at the end of the sentence.

> WEAK: You use a wired glove to point to objects.
>
> IMPROVED: To point to objects, you use a wired glove.

Choose an Appropriate Sentence Length

`len`

Sometimes sentence length affects the quality of writing. In revising a draft, you might want to compute the average sentence length of a representative passage. (Many software programs do this for you.)

No firm guidelines for appropriate sentence length exist. In general, however, an average length of 15 to 20 words is effective for most technical communication. A succession of 10-word sentences would be choppy; a series of 35-word sentences would probably be too demanding. And a succession of sentences of approximately the same length can distract the reader.

Avoid Overly Long Sentences

How long is too long? There is no simple answer, because the ease of reading a sentence depends on its length, vocabulary, and structure; the reader's knowledge of the topic; and the purpose of the communication.

Yet often your draft will include sentences such as the following:

> The construction of the new facility is scheduled to begin in March, but it might be delayed by one or even two months by winter weather conditions, which can make it impossible or nearly impossible to begin excavating the foundation.

This 40-word sentence can be tiring to read. To make it more readable, divide it into two sentences:

> The construction of the new facility is scheduled to begin in March. However, construction might be delayed until April or even May by winter weather conditions, which can make it impossible or nearly impossible to begin excavating the foundation.

Sometimes an overly long sentence can be fixed by creating a list:

WEAK: To connect the CD player to the amplifier, first be sure that the power is off on both units, then insert the plugs firmly into the jacks (the red plug into the right-channel jack and the black plug into the left-channel jack), making sure that you leave a little slack in the connecting cord to prevent shock or vibration.

IMPROVED: To connect the CD player to the amplifier, follow these steps:

1. Be sure that the power is off on both units.
2. Insert the plugs firmly into the jacks. The red plug goes into the right-channel jack, and the black plug into the left-channel jack.

Make sure that you leave a little slack in the connecting cord to prevent shock or vibration.

As this revision suggests, sometimes the best way to communicate technical information is to switch to a more visually oriented format by using lists or graphics.

Avoid Overly Short Sentences

Just as sentences can be too long, they can also be too short, as in the following example:

> The fan does not oscillate. It is stationary. The blade is made of plastic. This is done to increase safety. Safety is especially important because this design does not include a guard around the blade. A person could be seriously injured by putting his or her hand into the turning blade.

The problem is not that the word count of these sentences is too low, but rather that the sentences are choppy and contain too little information. The best way to revise a passage such as this is to combine sentences, as in the following example:

> The fan is stationary, not oscillating, with a plastic blade. Because the fan does not have a blade guard, the plastic blade is intended to prevent injury if a person accidentally touches it when it is moving.

One symptom of excessively short sentences is that they needlessly repeat key terms. Consider combining sentences, as in the following examples:

SLUGGISH: Computronics, a medium-sized consulting firm, consists of many diverse groups. Each group handles and develops its own contracts.

BETTER: Computronics, a medium-sized consulting firm, consists of many diverse groups, each of which handles and develops its own contracts.

SLUGGISH: I have experience working with various microprocessor-based systems. Some of these systems include the T90, RCA 9600, and the AIM 7600.

BETTER: I have experience working with various microprocessor-based systems, including the T90, RCA 9600, and the AIM 7600.

Focus on the "Real" Subject

rs

The conceptual or "real" subject of the sentence should also be the grammatical subject, and it should be prominent. Don't bury the real subject in a prepositional phrase following a useless or phantom grammatical subject. In the following examples, notice how the limp subjects disguise the real subjects. (The grammatical subjects are italicized.)

WEAK: The *use* of this method would eliminate the problem of motor damage.

STRONG: This *method* would eliminate the problem of motor damage.

WEAK: The *presence* of a six-membered lactone ring was detected.

STRONG: A six-membered lactone *ring* was detected.

Another way to make the subject of the sentence prominent is to reduce the number of grammatical *expletives: it is, there is,* and *there are.* In most cases, these constructions just waste space.

WEAK: There is no alternative for us except to withdraw the product.

STRONG: We have no alternative except to withdraw the product.

WEAK: It is hoped that testing the evaluation copies of the software will help us make this decision.

STRONG: I hope that testing the evaluation copies of the software will help us make this decision.

Often, as in this second example, the expletive *it is* is used along with the passive voice. See the discussion of the passive voice later in this chapter.

Expletives are not errors. Rather, they are conversational expressions that sometimes help the reader by emphasizing the information that follows them.

WITH THE
EXPLETIVE: It is hard to say whether the recession will last more than a few months.

WITHOUT THE
EXPLETIVE: Whether the recession will last more than a few months is hard to say.

The version without the expletive is a little more difficult to understand because the reader has to read and remember a long subject—"Whether the recession will last more than a few months"—before getting to the verb—"is." However, the sentence could also be rewritten in other ways to make it easy to understand and to eliminate the expletive.

I don't know whether the recession will last for more than a few months.

Nobody really knows whether the recession will last more than a few months.

Use the search function of your word-processing program to find most weak subjects: usually they're right before the word *of*. Expletives are also easy to find.

Focus on the "Real" Verb

rv

A "real" verb, like a "real" subject, should stand out in every sentence. Few stylistic problems weaken a sentence more than nominalizing verbs. Writers nominalize verbs by changing them into nouns and then adding another verb, usually a weaker one, to clarify the meaning. *To install* becomes *to effect an installation; to analyze* becomes *to conduct an analysis*. Notice how nominalizing the verbs makes the following sentences both awkward and unnecessarily long (the nominalized verbs are italicized).

WEAK: Each *preparation* of the solution is done twice.
STRONG: Each solution is prepared twice.

WEAK: An *investigation* of all possible alternatives was undertaken.
STRONG: All possible alternatives were investigated.

WEAK: *Consideration* should be given to an *acquisition* of the properties.
STRONG: We should consider acquiring the properties.

Like expletives, nominalizations are not errors. Many common nouns are nominalizations; in addition, writers can often use nominalizations effectively to summarize an idea from a previous sentence. The nominalizations in the following example are italicized.

The new *legislation* could delay our *entry* into the HDTV market. This *delay* could cost us millions.

Some software programs search for the most common nominalizations. With any word-processing program, you can identify most of the nominalizations by searching for character strings such as *tion, ment, sis, ence, ing,* and *ance.* If you search for the word *of,* you will also find many nominalizations.

Express Parallel Elements in Parallel Structures

//

A sentence is parallel if its coordinate elements are expressed in the same grammatical form: that is, its clauses are either passive or active, its verbs are either infinitives or participles, and so on. Creating and sustaining a recognizable pattern makes the sentence easier to follow.

Notice how faulty parallelism weakens the following sentences.

NONPARALLEL: Our present system *is costing* us profits and *reduces* our productivity. (unparallel verbs)

PARALLEL: Our present system is *costing* us profits and *reducing* our productivity.

NONPARALLEL: The dignitaries *watched* the launch, and the crew *was applauded.* (unparallel voice [see the discussion of voice later in this chapter])

PARALLEL: The dignitaries *watched* the launch and *applauded* the crew.

NONPARALLEL: *The typist should follow* the printed directions; *do not change* the originator's work. (unparallel mood: the first clause is subjunctive; the second, imperative)

PARALLEL: *The typist should* follow the printed directions and *not change* the originator's work.

When creating parallel constructions, make sure that parallel items in a series do not overlap, thus changing or confusing the meaning of the sentence:

CONFUSING: The speakers will include partners of law firms, businesspeople, and civic leaders.

CLEAR: The speakers will include businesspeople, civic leaders, and partners of law firms.

The problem with the original sentence is that "partners" appears to apply to "businesspeople" and "civic leaders." The revision solves the problem by rearranging the items so that "partners" cannot apply to the other two groups in the series.

CONFUSING: We need to buy more lumber, hardware, tools, and hire the subcontractors.

CLEAR: We need to buy more lumber, hardware, and tools, and we need to hire the subcontractors.

The problem with the confusing sentence is that the writer has two ideas joined inappropriately. The first idea is that we need to buy three things: lumber, hardware, and tools. The second idea is that we need to hire the subcontractor. In the clear revision, each of these ideas has its own subject and verb.

Use Modifiers Effectively

Technical communication is full of *modifiers*—words, phrases, and clauses that describe other elements in the sentence. To make your meaning clear, you must communicate to your readers whether a modifier provides necessary information about the word or phrase it refers to (its *referent*) or simply provides additional information. Further, you must make sure that the referent itself is always clearly identified.

Distinguish between Restrictive and Nonrestrictive Modifiers

mod

A *restrictive modifier,* as the term implies, restricts the meaning of its referent; that is, it provides information necessary to identify the referent. In the following examples, the restrictive modifiers are italicized:

> The aircraft *used in the exhibitions* are slightly modified.
>
> Please disregard the notice *you just received from us.*

In most cases, the restrictive modifier doesn't require a relative pronoun, such as *that* or *which.* If you choose to use a pronoun, however, use *that:*

> The aircraft *that* are used in the exhibits are slightly modified.

(If the pronoun refers to a person or persons, use *who.*) Notice that restrictive modifiers are not set off by commas.

A *nonrestrictive modifier* does not restrict the meaning of its referent: the information it provides is not necessary to identify the referent. In the following examples, the nonrestrictive modifiers are italicized:

> The Hubble telescope, *intended to answer fundamental questions about the origin of the universe,* was repaired in 1997.
>
> When you arrive, go to the Registration Area, *which is located on the second floor.*

Like the restrictive modifier, the nonrestrictive modifier usually does not require a relative pronoun. If you use one, however, choose *which* (*who* or *whom* when referring to a person). Note that nonrestrictive modifiers are separated from the rest of the sentence by commas.

Avoid Misplaced Modifiers

mn

The placement of the modifier often determines the meaning of the sentence. Notice, for instance, how the placement of *only* changes the meaning of the following sentences.

> *Only* Turner received a cost-of-living increase last year.
> (Meaning: Nobody else received one.)

Turner received *only* a cost-of-living increase last year.
   (Meaning: He didn't receive a merit increase.)

Turner received a cost-of-living increase *only* last year.
   (Meaning: He received a cost-of-living increase as recently as last year.)

Turner received a cost-of-living increase last year *only.*
   (Meaning: He received a cost-of-living increase in no other year.)

Misplaced modifiers—those that appear to modify the wrong referent—are common in technical communication. In general, the solution is to place the modifier as close as possible to its intended referent. Frequently, the misplaced modifier is a phrase or a clause:

MISPLACED: The subject of the meeting is the future of geothermal energy *in the downtown Webster Hotel.*

CORRECT: The subject of the meeting *in the downtown Webster Hotel* is the future of geothermal energy.

MISPLACED: *Jumping around nervously in their cages,* the researchers speculated on the health of the mice.

CORRECT: The researchers speculated on the health of the mice *jumping around nervously in their cages.*

A *squinting modifier* is a special kind of misplaced modifier. A squinting modifier is placed ambiguously between two potential referents, so that the reader cannot tell which one is being modified:

UNCLEAR: We decided *immediately* to purchase the new system.
   (Did we decide immediately, or did we decide to make the purchase immediately?)

CLEAR: We *immediately* decided to purchase the new system.

CLEAR: We decided to purchase the new system *immediately.*

UNCLEAR: The people who worked on the Eagle assembly line *reluctantly* picked up their last paychecks.
   (Did they work reluctantly, or did they pick up their last checks reluctantly?)

CLEAR: The people who worked *reluctantly* on the Eagle assembly line picked up their last paychecks.

CLEAR: The people who worked on the Eagle assembly line picked up their last paychecks *reluctantly.*

A subtle form of misplaced modification often occurs with the *correlative constructions,* such as *either . . . or, neither . . . nor,* and *not only . . . but also:*

NONPARALLEL: The new refrigerant not only decreases energy costs but also spoilage losses.

PARALLEL: The new refrigerant decreases not only energy costs but also spoilage losses.

In this example, "decreases" applies to both "energy costs" and "spoilage losses." Therefore, the first half of the correlative construction should follow "decreases." Note that if the sentence contains two different verbs, the first half of the correlative construction should precede the verb:

> The new refrigerant not only decreases energy costs but also reduces spoilage losses.

Avoid Dangling Modifiers

dm A *dangling modifier* has no referent in the sentence:

> DANGLING: Trying to solve the problem, the instructions seemed unclear.

In this sentence, the writer has not identified the person doing the trying. To correct the problem, rewrite the sentence to put the clarifying information either *within* the modifier or *next* to the modifier:

> CORRECT: *As I was trying to solve the problem,* the instructions seemed unclear.

> CORRECT: Trying to solve the problem, *I thought* the instructions seemed unclear.

Sometimes you can correct a dangling modifier by switching from the *indicative mood* (a statement of fact) to the *imperative mood* (a request or command):

> DANGLING: To initiate the procedure, the BEGIN button should be pushed.

> CORRECT: To initiate the procedure, push the BEGIN button.

In the imperative, the referent—in this case, *you*—is understood.

Choosing an Appropriate Level of Formality

lev Although no standard definition of levels of formality exists, most experts agree that there are three levels:

> INFORMAL: The Acorn 560 is a real screamer. With 200 meg of pure computing power, it slashes through even the thickest spreadsheets before you can say 2 + 2 = 4.

> MODERATELY FORMAL: With its 200 MHz microprocessor, the Acorn 560 can handle even the most complicated spreadsheet problems quickly.

> HIGHLY FORMAL: With a 200 MHz microprocessor, the Acorn 560 is a high-speed personal computer designed for computation-intensive applications such as large spreadsheets.

In general, technical communication is written in either a moderately formal or a formal style.

As you revise your writing to achieve an appropriate tone, think about your audience, subject, and purpose:

- *Audience.* You would use a more formal style in writing to a group of retired executives than to a group of college students; you would use a more formal style in writing to the company vice president than to your subordinates.
- *Subject.* You would use a more formal style in writing about a serious subject—safety regulations, important projects—than about preparations for the office Christmas party.
- *Purpose.* You would use a more formal style in a report to shareholders than in a company newsletter. Instructions tend to be relatively informal, often using the second person and the imperative mood. As discussed in Chapter 20, sometimes instructions are quite informal; they use contractions, the second person, and (occasionally) humor.

In general, it is better to err on the side of more formality rather than less. Avoid an informal style in any writing that you do at the office, for two reasons:

- *Informal writing tends to be imprecise.* In the example about the computer, what exactly is a *screamer,* and what does it mean to *slash through* a problem?
- *Informal writing can be embarrassing.* If your boss unexpectedly sees an e-mail you wrote to a colleague, you might wish that it didn't begin, "How ya doin', loser?"

Choosing the Right Words and Phrases

Effective technical communication consists of the right words and phrases in the right places. The following section includes the three guidelines shown in Figure 14.4.

Be Clear and Specific

Follow these seven guidelines to make your writing clear and specific:

- Use active and passive voice appropriately.
- Be specific.
- Avoid unnecessary jargon.

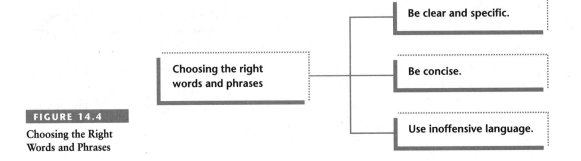

FIGURE 14.4

Choosing the Right Words and Phrases

- Use positive constructions.
- Avoid long noun strings.
- Avoid clichés.
- Avoid euphemisms.

Use Active and Passive Voice Appropriately

> voice

In the active voice, the subject of the sentence performs the action expressed by the verb. In the passive voice, the subject receives the action. (In the following examples, the subjects are italicized.)

ACTIVE: *Brushaw* drove the launch vehicle.

PASSIVE: The launch *vehicle* was driven by Brushaw.

ACTIVE: Many *physicists* support the big-bang theory.

PASSIVE: The big-bang *theory* is supported by many physicists.

In most cases, the active voice works better than the passive voice. A sentence in the active voice more clearly emphasizes the agent. In addition, the active-voice sentence is shorter, because it does not require a form of the *to be* verb and the past participle, as the passive-voice sentence does. In the active version of the second example, for instance, the verb is "support," rather than "is supported," and "by" is unnecessary.

The passive voice, however, is generally more appropriate in four cases:

- The agent is clear from the context.

  Students are required to take both writing courses.

  The context makes it clear that the college requires that students take both writing courses.

- The agent is unknown.

  The comet was first referred to in an ancient Egyptian text.

  We don't know *who* referred to the comet.

- The agent is less important than the action.

  The documents were hand-delivered this morning.

  It doesn't matter *who* the messenger was.

- A reference to the agent is embarrassing, dangerous, or in some other way inappropriate.

  Incorrect data were recorded for the flow rate.

  It might be inappropriate to say *who* recorded the incorrect data; it might be tactful to avoid pointing a finger. However, it is unethical to use the passive voice to avoid responsibility for an action. In addition, the passive voice can help you maintain the focus of your paragraph.

  LANs have three major advantages. First, they are inexpensive to run. Second, they can be expanded easily. . . .

Many people who otherwise take little interest in grammar have strong feelings about voice. Some believe that the active voice is inappropriate because it emphasizes the person who does the work rather than the work itself, robbing the writing of objectivity. In many cases, this idea is valid. Why write "I analyzed the sample for traces of iodine"? If there is no ambiguity about who did the analysis, or if it is not necessary to identify who did the analysis, a focus on the action being performed is appropriate. But be conservative; if in doubt, use the active voice.

Supporters of the active voice argue that the passive voice creates a double ambiguity. When you write "The sample was analyzed for traces of iodine," your reader is not quite sure who did the analysis (you or someone else) or when it was done (as part of the project being described or sometime previously). Even though a passive-voice sentence can contain all the information found in its active-voice counterpart, often the writer omits the actor.

The best approach to the active-passive problem is to recognize how the two voices differ and use them appropriately. In the following examples, the writer mixes active and passive voice for no good reason.

AWKWARD: He lifted the cage door, and a white mouse was seen.

BETTER: He lifted the cage door and saw a white mouse.

AWKWARD: The new catalyst produced good-quality foam, and a flatter mold was caused by the new chute-opening size.

BETTER: The new catalyst produced good-quality foam, and the new chute-opening size resulted in a flatter mold.

A number of style programs can help you find the passive voice in your writing. Beware, however, of the advice some style programs offer about the passive voice: that it is undesirable, almost an error. This advice is wrongheaded. Use passive voice when it works better than the active voice; don't be bullied into changing to the active voice.

Any word-processing program allows you to search for *is, are, was,* and *were,* the forms of the verb *to be* that are most commonly used in passive-voice expressions. In addition, searching for *ed* will isolate the past participles, which also appear in most passive-voice expressions.

Be Specific

spec   Being specific involves using precise words, providing adequate detail, and avoiding ambiguity.

- *Use precise words.* A Ford Taurus is an automobile, but it is also a vehicle, a machine, and a thing. In describing the Ford Taurus, *automobile* is better than *vehicle,* because the less specific word also refers to pickup trucks, trains, hot-air balloons, and other means of transport. As words become more abstract—from *machine* to *thing,* for instance—the chances for misunderstanding increase.

- *Provide adequate detail.* Readers probably know less than you do about your subject. What might be perfectly clear to you might be too vague for them.

  VAGUE: An engine on the plane experienced some difficulties. *(Which engine? What plane? What kinds of difficulties?)*

  CLEAR: The left engine on the Jetson 411 unaccountably lost power during flight.

- *Avoid ambiguity.* Don't let readers wonder which of two meanings you are trying to convey.

  AMBIGUOUS: After stirring by hand for 10 seconds, add three drops of the iodine mixture to the solution. *(Stir the iodine mixture or the solution?)*

  CLEAR: Stir the iodine mixture by hand for 10 seconds. Then add three drops to the solution.

  CLEAR: Stir the solution by hand for 10 seconds. Then add three drops of the iodine mixture.

What should you do if you don't have the specific data? You have two options: to approximate—and clearly tell readers you are doing so—or to explain why the specific data are unavailable and indicate when they will become available.

  VAGUE: The leakage in the fuel system is much greater than we had anticipated.

  CLEAR: The leakage in the fuel system is much greater than we had anticipated; we estimate it to be at least 5 gallons per minute, not 2.

Several style programs isolate common vague terms and suggest more precise alternatives.

Avoid Unnecessary Jargon

jarg

Jargon is shoptalk. To a banker, *CD* means certificate of deposit; to an audiophile, it means compact disc. To the general reader, *ATM* means automated teller machine; to an electrical engineer, it means asynchronous transfer mode. Although jargon is often ridiculed, it is a useful and natural kind of communication in its proper sphere. Two software designers would find it hard to talk to each other about their craft without terms such as *SCSI port* and *WYSIWYG.*

However, using unnecessary jargon is inappropriate in four ways.

- *It can be imprecise.* The best current example is the degree to which electronics terminology has crept into everyday English. In offices, we ask employees to provide *feedback* after considering a proposal. What are we asking for: a facial expression, some body language, a phone call, or a written evaluation?

- *It can be confusing.* If we ask a computer novice to boot the system, he or she might have no idea what we're talking about.

- *It is often seen as condescending.* Many people react as if the writer is showing off—displaying a level of expertise that excludes most readers. While the readers are concentrating on how much they dislike the writer, they are missing the message.

- *It is often intimidating.* People feel somehow inadequate or stupid because they do not know what the writer is talking about. Obviously, this feeling hurts communication.

If you are addressing a technically knowledgeable audience, feel free to use appropriate jargon. (Sometimes, however, a technical term can have very different meanings, even when used by technical professionals with similar backgrounds.) If you are addressing an audience that includes managers or the general public, avoid jargon. If your document has separate sections for different audiences—as in the case of a technical report with an executive summary—use jargon accordingly. A glossary (list of definitions) is a useful addition if you suspect that managers will read the technical sections. See Chapter 10 for a discussion of these elements.

Use Positive Constructions

pos

The term *positive construction* has nothing to do with cheerfulness or an optimistic outlook on life. Rather, it indicates that the writer is describing what something is instead of what it is not. In the sentence "I was sad to see this project completed," "sad" is a positive construction. The negative construction would be "not happy."

Here are a few other examples of positive and negative constructions:

Positive Construction	*Negative Construction*
most	not all
few	not many
on time	not late, not delayed
positive	not negative
inefficient	not efficient
reject	cannot accept
impossible	not possible

Why should you try to use positive rather than negative constructions? Because readers understand positive constructions more quickly and more easily. And when a writer uses several negative constructions in the same sentence, readers must work much harder to untangle the meaning. Consider the following examples:

| DIFFICULT: | Because the team did not have sufficient time to complete the project, it was not unsurprising that it was unable to prepare a satisfactory report. |
| SIMPLER: | Because the team had too little time to complete the project, it produced an unsatisfactory report. |

| DIFFICULT: | Without an adequate population of krill, the entire food chain of the Antarctic region would be unable to sustain itself. |
| SIMPLER: | If the krill population were too low, the entire food chain of the Antarctic region would be destroyed. |

Avoid Long Noun Strings

str
A *noun string* is a phrase consisting of a series of nouns (or nouns and adjectives and adverbs), all of which modify the last noun. For example, in the phrase *parking-garage regulations,* the first two words modify *regulations.* Noun strings save time, and if your readers understand them, they are fine. A *passive-restraint system* is easier to write than *a restraint system that is passive,* and it won't confuse readers.

Hyphens help clarify some noun strings by linking two or more words that go together. For example, in the phrase *self-locking washer,* the hyphen links *self* and *locking;* together they modify *washer.* In other words, it is not a self washer, or a locking washer, but a self-locking washer. For more information on hyphens, see Appendix A.

However, noun strings are sometimes so long or so complex that hyphens won't ensure clarity. Consider untangling the phrases and restoring prepositions, as in the following examples:

| UNCLEAR: | preregistration procedures instruction sheet update |
| CLEAR: | an update of the instruction sheet for preregistration procedures |

| UNCLEAR: | operator-initiated default-prevention technique |
| CLEAR: | a technique for preventing defaults that are initiated by the operator |

An additional danger is that sometimes noun strings are merely pompous. If you are writing about a smoke detector, there is no reason to call it a *smoke-detection device.*

Avoid Clichés

cl
Good writing is original. Rather than use a cliché, just say what you want to say in plain English. Instead of "It's a whole new ball game," write "The situation has changed completely." Don't write "I am sure the new manager can cut the mustard"; write "I am sure the new manager can do his job effectively." Newer clichés have joined these hallowed examples: *movers and shakers, couch potato, fast lane,* and *paradigm shift.* But the advice is the same: if you're used to hearing and reading a phrase, avoid it.

Sometimes, writers further embarrass themselves by getting their clichés wrong: expressions become so timeworn that users forget what the words mean. The phrase "I could care less" often is used when the writer means just the opposite. The best solution to this problem is, of course, not to use clichés.

Following are a cliché-filled sentence and a translation into plain English:

TRITE: Afraid that we were *between a rock and a hard place,* we decided to *throw caution to the winds* with a *grandstand play* that would *catch* our competition *with its pants down.*

PLAIN: Afraid that we were in a dilemma, we decided on a risky, aggressive move that would surprise our competition.

Avoid Euphemisms

| *eu* | A euphemism is a polite way to say something that makes people uncomfortable—the more uncomfortable the subject, the more euphemisms. Dozens of euphemisms deal with drinking, bathrooms, sex, and death. David Lord (as quoted in Fuchsberg, 1990) lists 48 euphemisms for firing someone, including the following:

personnel surplus reduction   dehiring

workforce imbalance correction   decruiting

degrowing   redundancy elimination

indefinite idling   career-change-opportunity creation

corporate downsizing

It's fun to think of euphemisms like these, but don't write them. It's an ethical issue. People suffer when they get fired. Don't use language to cloud reality.

Be Concise   These six guidelines promote concise technical communication:

- Avoid obvious statements.
- Avoid meaningless modifiers.
- Avoid unnecessary prepositional phrases.
- Avoid wordy phrases.
- Avoid redundant expressions.
- Avoid pompous words.

Avoid Obvious Statements

| *ob* | Writing can become sluggish if it overexplains. The italicized words in the example below are sluggish:

SLUGGISH: The market for *the sale of* flash memory chips is dominated by *two chip manufacturers:* Intel and Advanced Micro Systems. These two *chip manufacturers* are responsible for 76 percent of the $1.3 billion market *in flash memory chips* last year.

IMPROVED: The market for flash memory chips is dominated by Intel and
Advanced Micro Systems, two companies that claimed 76 percent
of the $1.3 billion industry last year.

You don't need to state that the chips were made by chip manufacturers, nor
do you need to explain that the $1.3 billion market is in flash memory chips.
If it were in something else, the writer wouldn't be mentioning it in this context.

Avoid Meaningless Modifiers

| w | Sometimes we include meaningless modifiers, such as the following:

basically	kind of
certain	sort of
essentially	various

BLOATED: *I think that, basically,* the board felt *sort of* betrayed, *in a sense,* by the
kind of behavior the president displayed.

BETTER: The board felt betrayed by the president's behavior.

Of course, modifiers are not necessarily meaningless. For instance, it is appropriate to use *I think* or *it seems to me* to suggest an awareness that not everyone shares your view.

BLUNT: Next year we will face unprecedented challenges to our dominance of
the market.

LESS BLUNT: In my view, next year we will face unprecedented challenges to our
dominance of the market.

Of course, a sentence that is blunt to one reader is self-confident to another. You
have to think about your audience's preferences.

Avoid Unnecessary Prepositional Phrases

| prep | A prepositional phrase consists of a preposition followed by a noun or a noun
equivalent. Here are some examples of prepositional phrases:

in the summary

on the engine

under the heading

Unnecessary prepositional phrases, often used along with abstract nouns and
nominalizations, make writing long and boring. In the following examples, the
prepositions are italicized:

LONG: The increase *in* the number *of* students enrolled *in* the materials engineering program *at* Lehigh University is a testament *to* the regard *in*
which that program is held *by* the university's new students.

SHORTER: The growth of Lehigh University's materials engineering program suggests that the university's new students think it is a good program.

Avoid Wordy Phrases

| *w* | Wordy phrases make technical communication long and boring. Sometimes people write phrases such as *on a weekly basis* rather than writing *weekly*. The long phrase rolls off the tongue easily and appears to carry the weight of scientific truth. But the humble *weekly* says the same thing more concisely.

Figure 14.5 lists common wordy phrases and their concise equivalents. Notice that almost all the wordy phrases contain prepositions, while many others contain nominalizations.

Following are a wordy sentence and a concise translation:

WORDY: I am of the opinion that, in regard to profit achievement, the statistics pertaining to this month will appear to indicate an upward tendency.

CONCISE: I think this month's statistics will show an increase in profits.

Avoid Redundant Expressions

| *redun* | Avoid redundant expressions, such as *collaborate together, past history, end result, any and all, each and every, still remain, completely eliminate,* and *very unique.* Be content to say something once.

Wordy Phrase	Concise Phrase	Wordy Phrase	Concise Phrase
a majority of	most	in view of the fact that	because
a number of	some, many	it is often the case that	often
at an early date	soon	it is our opinion that	we think that
at the conclusion of	after, following	it is our recommendation that	we recommend that
at the present time	now	it is our understanding that	we understand that
at this point in time	now	make reference to	refer to
based on the fact that	because	of the opinion that	think that
despite the fact that	although	on a daily basis	daily
due to the fact that	because	on the grounds that	because
during the course of	during	prior to	before
during the time that	during, while	relative to	regarding, about
have the capability to	can	so as to	to
in connection with	about, concerning	subsequent to	after
in order to	to	take into consideration	consider
in regard to	regarding, about	until such time as	until
in the event that	if		

FIGURE 14.5

Wordy Phrases and Their
Concise Equivalents

REDUNDANT: We initially began our investigative analysis with a sample that was spherical in shape and heavy in weight.

BETTER: We began our analysis with a heavy, spherical sample.

Avoid Pompous Words

pomp | Writers sometimes try to impress their readers by using pompous words, such as *initiate* for *begin, perform* for *do,* and *prioritize* for *rank.* When asked why they use big words where small ones will do, writers say they want to make sure their readers know they have a strong vocabulary, that they are well educated. In technical communication, plain talk is best. If you know what you're talking about, be direct and simple. Even if you're not so sure of what you're talking about, say it plainly; big words won't fool anyone for more than a few seconds.

Following are a few pompous sentences translated into plain English:

POMPOUS: The purchase of a database program will enhance our record-maintenance capabilities.

PLAIN: Buying a database program will help us maintain our records.

POMPOUS: It is the belief of the Accounting Department that the predicament was precipitated by a computational inaccuracy.

PLAIN: The Accounting Department thinks a math error caused the problem.

Figure 14.6 lists commonly used fancy words and their plain equivalents.

Fancy Word	Plain Word	Fancy Word	Plain Word
advise	tell	impact (verb)	affect
ascertain	learn, find out	initiate	begin
attempt (verb)	try	manifest (verb)	show
commence	start, begin	parameters	variables, conditions
demonstrate	show	perform	do
employ (verb)	use	prioritize	rank
endeavor (verb)	try	procure	get, buy
eventuate (verb)	happen	quantify	measure
evidence (verb)	show	terminate	end, stop
finalize	end, settle, agree, finish	utilize	use
furnish	provide, give		

FIGURE 14.6

Fancy Words and Their
Plain Word Equivalents

Several style programs isolate pompous words and expressions. Of course, you can use any word-processing program to search for terms you tend to misuse.

In the long run, your readers will be more impressed with clarity and accuracy than with fancy language.

Use Inoffensive Language

Writing inoffensively is not merely a matter of politeness. It is a matter of perception. If language not only reflects people's attitudes but also helps form them, writing inoffensively helps to break down stereotypes, so that we see people as individuals.

Use Nonsexist Language

sexist

Sexist language favors one sex at the expense of the other. Although sexist language can shortchange men—as in some writing about female-dominated professions such as nursing—in most cases it victimizes women. Common examples include nouns such as *workman* and *chairman*. When writers use male pronouns to represent both males and females, as in the sentence "Each worker is responsible for his work area," they are using sexist language. Some writers sidestep the problem of sexist language by simply claiming their innocence: "The use of the pronoun *he* does not in any way suggest a male bias." Many readers find this argument unpersuasive or even insincere.

In most organizations, people try to avoid sexist language. Although eliminating gender bias from writing is not easy, many organizations have formulated guidelines to reduce sexist language.

Six major techniques are used to eliminate sexist writing:

- *Replacing the male-gender words with non–gender-specific words. Chairman,* for instance, is replaced by *chairperson* or *chair. Firemen* are *firefighters, policemen* are *police officers.*
- *Switching to a different form of the verb.*

SEXIST: The operator must pass a rigorous series of tests before *he is promoted* to supervisor.

NONSEXIST: The operator must pass a rigorous series of tests before *being promoted* to supervisor.

- *Switching to the plural.*

SEXIST: The operator must pass a rigorous series of tests before *he* is promoted to supervisor.

NONSEXIST: Operators must pass a rigorous series of tests before *they* are promoted to supervisors.

In some organizations, you will see plural pronouns with singular nouns, particularly in informal documents such as memos:

If *an employee* wishes to apply for tuition reimbursement, *they* should consult Section 14.5 of the *Employee Manual.*

This construction is still considered a grammar error by careful writers and readers.

- *Switching to* he or she, he/she, s/he, *or* his or her. Sometimes, switching to the plural can make the sentence unclear:

  UNCLEAR: Operators are responsible for their operating manuals.
  (Does each operator have one operating manual or more than one?)

  CLEAR: Each operator is responsible for *his or her* operating manual.

  He or she, his or her, and the related constructions are awkward, especially if overused, but at least they are clear and inoffensive.

- *Addressing the reader directly.* If you use *you* and *your* or the understood *you,* you have avoided the problem.

- *Alternating* he *and* she. Joseph Williams (1997) and many other language authorities recommend alternating *he* and *she*.

If you use a word processor, search for *he, man,* and *men,* the words and parts of words most associated with sexist writing. Some style programs search out the most common sexist terms and suggest nonsexist alternatives. But use your common sense. I have read the following sentence: "Every employee is responsible for the cost of his or her gynecological examination."

For a full discussion of nonsexist writing, see *The Bias-Free Word Finder: A Dictionary of Nondiscriminatory Language* (Maggio, 1992).

Use Inoffensive Language When Referring to People with Disabilities

| dis | According to the National Organization on Disability, one in six Americans—more than 40 million—has a physical, sensory, emotional, or mental impairment that interferes with daily life (Tyler, 1990).

In writing about people with disabilities, use the "people first" approach. This approach calls for writing about the person as someone with a disability, not as someone defined by that disability. The disability is a condition the person has, not what the person is.

Figure 14.7 on page 450, based on Tyler (1990, p. 65), lists the basic tenets of the people-first approach.

Understanding Simplified English for Nonnative Speakers

Because English is the language of more than half of the world's scientific and technical communication, millions of nonnative speakers of English read technical communication in English (Peterson, 1990).

- Refer to the person first, the disability second. Write "people with mental retardation," not "the mentally retarded."

- Don't confuse *handicap* with *disability*. *Disability* refers to the impairment or condition; *handicap* refers to the interaction between the person and his or her environment. A person can have a disability without being handicapped.

- Don't refer to victimization. Write "a person with AIDS," not "an AIDS victim" or "an AIDS sufferer."

- Don't refer to a person as "wheelchair bound" or "confined to a wheelchair." People who use wheelchairs to get around are not confined.

- Don't refer to people with disabilities as abnormal. They are atypical, not abnormal.

FIGURE 14.7

Principles of the "People-First" Approach

Efforts to create a simplified language for these readers began as early as 1932, when Charles K. Ogden created an 850-word Basic English dictionary, complete with a set of simple grammar rules. The language used the active voice only and contained 18 "operators": verbs such as *put* and *get*.

Since that time, other companies and associations have created versions of *Simplified English,* as this general category of languages is known. All Simplified English languages share four characteristics:

- Each consists of a basic vocabulary of about 1,000 words.
- Each is based on the premise that a word has only one meaning. There are no synonyms.
- Each consists of a set of basic grammar rules.
- Each is geared to a specific discipline. For example, "AECMA Simplified English" is intended for aerospace workers.

Here is a sample of text translated into Simplified English.

BEFORE TRANSLATION INTO SIMPLIFIED ENGLISH:
Before filling the gas tank, it is necessary to turn off the propane line to the refrigerator. Failure to do so significantly increases the risk of explosion.

AFTER TRANSLATION INTO SIMPLIFIED ENGLISH:
Before you pump gasoline into the gas tank, turn off the propane line to the refrigerator. If you do not turn off the propane tank, it could explode.

For background information on Simplified English, see Peterson (1990). For information on a project to create a software program that assists in translating standard English into Simplified English, see Thomas, Jaffe, Kincaid, and Stees (1992).

Understanding Structured Writing

Structured writing is a general term referring to different strategies of imposing a visible structure on writing. For example, in one structure, used by many companies that write proposals, graphics are displayed on one page and the accompanying text is on the facing page. (See Chapter 15.)

Like Simplified English, structured writing is intended to place restrictions on the use of the language in order to simplify and standardize it. Structured writing is not meant exclusively for nonnative speakers of English, although to the extent that it simplifies the language it assists them as well as native speakers.

A leading example of structured writing is Information Mapping®, a technique developed by Robert Horn in the mid-1960s. Figure 14.8, from an article by Horn (1985), shows the same information displayed first in traditional paragraphs and then using his technique.

TRANSFERS

There are more and more requests for transfers as the company expands and the key workforce adopts a more flexible lifestyle. The company supervisor is a key person in facilitating such transfers and in determining whether they would be in the best interest of the company and the employees. This memorandum covers company policy that has been in effect for the past year and continues to be our policy. It outlines each supervisor's responsibilities when an employee under his or her supervision requests a transfer.

First, it is in the company's interest to retain employees who are performing satisfactorily; therefore, we will try to help employees to move to an area or job that they find more desirable. This is what you should do. Provide an employee who comes to talk about a transfer or to request one with Form 742, Application for Transfer, and tell him or her to fill it out as soon as possible.

If the employee is applying for a new job and not just a new location and if there are any parts of the new job that you as the supervisor consider may disqualify the employee, then you should discuss with the employee those areas immediately. Remember, it is the company policy that if an employee wishes to be transferred, the company will make every effort to find a job acceptable to the employee. So you should not discourage any request for transfer, even if it would disturb the completion of projects or goals in your department.

At the bottom of the form, you should fill out the supervisor's comment. Be brief and to the point. When you have finished that, you should make a photocopy of the employee's latest Performance Evaluation and attach it to the form.

If the employee's current performance rating is unsatisfactory, then your signature and your immediate supervisor's signature on Part C of the form are required. If the current performance rating is outstanding, then attach a copy of any letters of commendation. If the current performance rating is satisfactory, you do not have to attach anything.

Send a copy of the blue copy of the form to the company Placement Bureau and a pink copy to your Department File. The yellow copy should be given to the employee.

FIGURE 14.8

Information Mapping®

(continued)

(Figure 14.8 continued)

How to Handle Transfer Requests

Introduction This procedure outlines each supervisor's responsibilities when an employee under your supervision requests a transfer.

Procedure table

STEP	ACTION
1	When employees request transfers, provide them with form 742, Application for Transfer.
2	Discuss with the employee any areas in the new job that you consider may disqualify the employee. Remember, it is our policy that if an employee wishes to be transferred, the company will make every effort to find a job acceptable to the employee. So you may in *no* way discourage a request for transfer.
3	At the bottom of the form, fill out the supervisor's comments.
4	Attach a copy of the latest Performance Evaluation.
5	<table><tr><td>IF the current performance rating is . . .</td><td>THEN</td></tr><tr><td>Unsatisfactory</td><td>Your signature and your immediate supervisor's signature on Part C of the form are required.</td></tr><tr><td>Outstanding</td><td>Attach a copy of the letter of recommendation.</td></tr></table>
6	<table><tr><td>Send this copy of the form</td><td>TO</td></tr><tr><td>Blue Copy</td><td>Placement Bureau</td></tr><tr><td>Pink Copy</td><td>Departmental File</td></tr><tr><td>Yellow Copy</td><td>Employee</td></tr></table>

CASE

Revising a Draft for Sentence Effectiveness

For three years you have been employed as a work-study student in your university's advising office. Your office wishes to distribute a new pamphlet for incoming students that describes the services your office provides. A new work-study student, Kim Vavrick, has written the following brief introduction to advising. You have been asked to help her with it. Write a memo to her, evaluating the writing in her draft according to the material presented in this chapter. See Chapter 16 for a discussion of memos.

Academic advising is counseling by a university representative, usually a faculty member, to assist the student achieve their goals for their education. The counseling's character, and the relationship that exists between the advisor and the student, change as the student's career in the academic setting progresses.

In the student's freshman and sophomore years, academic advising assists the student to identify, comprehend, and finalizing the sequence of university core requirements; that is, common classes such as English composition and basic science courses. It is also the case that academic advising may also serve to help the student clarify his academic strengths and interests in order to establish a major.

During these first two years, the interpersonal relationship between the student and the advisor are usually general and impersonal. The academic advisor may very well be someone with whom the student has little or no contact beyond obtaining a signature as a formality on paperwork. Similarly, the student may well never be enrolled in a course taught by the advisor, or otherwise become involved in the advisor's activities or academic interests.

This rarely succeeds in giving the student the optimal possible guidance for progressing in their academic career, however it is very economical and usually suitable. Faculty time and resources are expensive, limited commodities. Except in small, private institutions, there is rarely a large enough faculty to provide close and individual attention to each student who needs it. Student attrition rates are high in these first two years, many students flunk out of school before they have an opportunity to benefit from detailed, personal advice. Even among those who stay there is a high percentage of changes in academic majors. The emphasis on ensuring students understand and complete the core requirements ultimately ensures that those who do remain as students are able to progress along their degree path in a relatively smooth fashion.

In the student's junior and senior years, there is a shift in the emphasis. The goal of academic advising now is more to assist the student finish fulfilling their individual educational needs, and less to help the student meet the needs of the university. Academic advising helps the student make the best choices of the remaining options and requirements.

The relationship between the student and the advisor is closer in the last two years as well. The advisor is more personally acquainted with the student; he (or she) has seen the student periodically over a substantial period of time, and may even have instructed the student in one or more classes. The advisor is also more familiar with the major department, the courses it offers, and the colleagues who teach them, and can offer the student personal recommendations regarding many important and critical issues. Owing to the fact that the student is pursuing academic interests related to the advisor's, there is likely to be more interaction between them in academic projects and programs.

Academic advising also helps the student look beyond their undergraduate years. As the student comes close to concluding a degree program, they may be considering the possibility of entering a professional career, for example, or at extending their education in a graduate program. Academic advising serves to assist students again in making the educational choices, which will be most productive in meeting those goals.

WRITER'S CHECKLIST

Lists

1. Is each list of the appropriate kind: bulleted, numbered, or checklist?

2. Does each list contain an appropriate number of items?

3. Are all the items in each list grammatically parallel?

4. Is the lead-in to each list structured and punctuated properly?

5. Are the items in each list punctuated properly?

Sentences

6. Are the sentences structured so that the new or important information comes near the end?

7. Are the sentences the appropriate length: neither short and choppy nor long and difficult to understand?

8. Does each sentence focus on the "real" subject?

9. Have you reduced the number of expletives used as sentence openers?

10. Does each sentence focus on the "real" verb, without unnecessary nominalizations?

11. Have you eliminated nonparallelism from your sentences?

12. Have you used restrictive and nonrestrictive modifiers appropriately?

13. Have you eliminated misplaced modifiers, squinting modifiers, and dangling modifiers?

Level of Formality

14. Have you chosen an appropriate level of formality for your audience, subject, and purpose?

Words and Phrases

15. Have you
    - ☐ used active and passive voice appropriately?
    - ☐ used precise words?
    - ☐ provided adequate detail?
    - ☐ avoided ambiguity?
    - ☐ avoided unnecessary jargon?
    - ☐ used positive rather than negative constructions?
    - ☐ avoided long noun strings?
    - ☐ avoided clichés?
    - ☐ avoided stating the obvious?
    - ☐ avoided meaningless modifiers?
    - ☐ avoided unnecessary prepositions?
    - ☐ used the most concise phrases?
    - ☐ avoided redundancy?
    - ☐ avoided pompous words?
    - ☐ avoided euphemisms?
    - ☐ used nonsexist language?
    - ☐ used people-first vocabulary in referring to people with disabilities?

EXERCISES

1. The information contained in the following sentences could be conveyed better in a list. Rewrite each sentence in the form of a list.
   a. The causes of burnout can be studied from three perspectives: physiological—the roles of sleep, diet, and physical fatigue; psychological—the roles of guilt, fear, jealousy, and frustration; environmental—the role of the physical surroundings at home and at work.
   b. There are many problems with the online registration system used at Dickerson. First, lists of closed sections cannot be updated as often as necessary. Second, students who want

to register in a closed section must be assigned to a special terminal. Third, the computer staff is not trained to handle the student problems. Fourth, the Computer Center's own terminals cannot be used on the system; therefore, the university has to rent 15 extra terminals to handle registration.

2. In the following exercises, determine whether the new or important information is emphasized appropriately. If it is not, revise the sentence.

   a. Asynchronous transmission mode is likely to become the dominant communication mode over the next few years.

   b. Multimedia interfaces are most successful when they are designed by engineers, computer scientists, psychologists, and applications experts, all working together.

   c. The program must be carefully designed if it is to succeed.

3. The following sentences might be too long for some readers. Break each sentence into two or more sentences.

   a. If we get the contract, we must be ready by June 1 with the necessary personnel and equipment to get the job done, so with this in mind a staff meeting, which all group managers are expected to attend, is scheduled for February 12.

   b. Once we get the results of the stress tests on the 125-Z fiberglass mix, we will have a better idea of where we stand in terms of our time constraints, because if the mix isn't suitable we will really have to hurry to find and test a replacement by the Phase 1 deadline.

   c. Although we had a frank discussion with Backer's legal staff, we were unable to get them to discuss specifics on what they would be looking for in an out-of-court settlement, but they gave us a strong impression that they would rather settle out of court.

4. The following examples contain choppy, abrupt sentences. Combine sentences to create a smoother prose style.

   a. I need a figure on the surrender value of a policy. The policy number is A4399827. Can you get me this figure by tomorrow?

   b. The program obviously contains an error. We didn't get the results we anticipated. Please ask Paul Davis to go through the program.

   c. The supervisor is responsible for processing the outgoing mail. He is also responsible for maintaining and operating the equipment.

5. In the following sentences, the real subjects are buried in prepositional phrases or obscured by expletives. Revise the sentences so that the real subjects appear prominently.

   a. There has been a decrease in the number of students enrolled in our training sessions.

   b. It is on the basis of recent research that I recommend the new CAD system.

   c. The use of in-store demonstrations has resulted in a dramatic increase in business.

6. In the following sentences, unnecessary nominalization obscures the real verb. Revise the sentences to focus on the real verb.

   a. Pollution constitutes a threat to the Wilson Wildlife Preserve.

   b. Evaluation of the gumming tendency of the four tire types will be accomplished by comparing the amount of rubber that can be scraped from the tires.

   c. Reduction of the size of the tear-gas generator has already been completed.

7. Revise the following sentences to eliminate nonparallelism.

   a. The next two sections of the manual discuss how to analyze the data, the conclusions that can be drawn from your analysis, and how to decide what further steps are needed before establishing a journal list.

   b. With our new product line, you would not only expand your tax practice, but your other accounting areas as well.

   c. Sections 1 and 2 will introduce the entire system, while Sections 3 and 4 describe the automatic application and step-by-step instructions.

8. The following sentences contain punctuation or pronoun errors related to the use of modifiers. Revise the sentences to eliminate the errors.

   a. The Greeting-Record Button records the greeting which is stored on a microchip inside the machine.

   b. This problem that has been traced to manufacturing delays, has resulted in our losing four major contracts.

   c. Please get in touch with Tom Harvey who is updating the instructions.

9. Revise the following sentences to eliminate the misplaced modifiers.
   a. Over the past three years it has been estimated that an average of eight hours per week are spent on this problem.
   b. Information provided by this program is displayed at the close of the business day on the information board.
   c. The computer provides a printout for the management team that shows the likely effects of the action.

10. Revise the following sentences to eliminate the dangling modifiers.
   a. By following these instructions, your computer should provide good service for many years.
   b. To examine the chemical homogeneity of the plaque sample, one plaque was cut into nine sections.
   c. The boats in production could be modified in time for the February debut by choosing this method.

11. Revise the following informal sentences to make them moderately formal.
   a. The learning modules were put together by a couple of professors in the department.
   b. The biggest problem faced by multimedia designers is that users freak if they don't see a button—or, heaven forbid, if they have to make up their own buttons!
   c. If the University of Arizona can't figure out where to dump its low-level radioactive waste, Uncle Sam could pull the plug on millions of dollars of research grants.

12. Rewrite the following sentences to remove the inappropriate usages of the passive voice.
   a. Most of the information you need will be gathered as you document the history of the journals.
   b. When choosing multiple programs to record, be sure that the proper tape speed has been chosen.
   c. During this time I also cowrote a manual on the Roadway Management System. Frequent trips were also made to the field.
   d. Mistakes were made.
   e. Come to the reception desk when you arrive. A packet with your name on it can be picked up there.

13. Revise the following sentences by replacing the vague elements with specific information. Make up any reasonable details.
   a. The results won't be available for a while.
   b. The fire in the lab caused extensive damage.
   c. A soil analysis of the land beneath the new stadium revealed an interesting fact.

14. The following sentences addressed to general readers contain unnecessary jargon. Revise the sentences to remove the jargon.
   a. Please submit your research assignment in hard-copy mode.
   b. The perpetrator was apprehended and placed under arrest directly adjacent to the scene of the incident.
   c. The new computer lab supports both platforms.

15. In the following sentences, convert the negative constructions to positive constructions.
   a. Williams was accused by management of making predictions that were not accurate.
   b. We must make sure that all our representatives do not act unprofessionally to potential clients.
   c. The shipment will not be delayed if Quality Control does not disapprove any of the latest revisions.

16. Rewrite the following sentences to eliminate the long noun strings that the general reader might find awkward or difficult to understand.
   a. The corporate-relations committee meeting has been scheduled for next Thursday.
   b. The research team discovered a glycerin-initiated, alkylene-oxide-based, long-chain polyether.
   c. We are considering purchasing a digital-imaging capable, diffusion-pump equipped, tungsten-gun SEM.

17. Revise the following sentences to eliminate the clichés.
   a. We hope the new program will positively impact all our branches.
   b. If we are to survive this difficult period, we are going to have to keep our ears to the ground and our noses to the grindstone.
   c. DataRight will be especially useful for those personnel tasked with maintaining the new system.

18. Revise the following sentences to eliminate the euphemisms.

a. Downsizing will enable our division to achieve a more favorable cash-flow profile.

b. Of course, accident statistics can be expected to show a moderate increase in response to a streamlining of the training schedule.

c. Should the aircraft experience a water landing, your seat cushion will function as a personal flotation device.

19. The following sentences are verbose because they state the obvious. Revise the sentences to eliminate the obvious material.

a. To register to take a course offered by the university, you must first determine whether the university will be offering that course that semester.

b. The starting date of the project had to be postponed for a certain period of time due to a delay in the necessary authorization from the Project Oversight Committee.

c. After you have installed DataQuick, please spend a few minutes responding to the questions about the process, then take the card to a post office and mail it to us.

20. Revise the following sentences to remove the meaningless modifiers.

a. It would seem to me that the indications are that the project has been essentially unsuccessful.

b. For all intents and purposes, our company's long-term success depends to a certain degree on various factors that are in general difficult to foresee.

c. This aspect of the presentation was quite well received overall, despite the fact that the meter readings were rather small.

21. Revise the following sentences to eliminate unnecessary prepositional phrases.

a. Another advantage of the approach used by the Alpha team is that interfaces of different kinds can be combined.

b. The complexity of the module will hamper the ability of the operator in the diagnosis of problems in equipment configuration.

c. The purpose of this test of your aptitudes is to help you with the question of the decision of which major to enroll in.

22. Revise the following sentences to make them more concise.

a. The instruction manual for the new copier is lacking in clarity and completeness.

b. The software packages enable the user to create graphic displays with a minimum of effort.

c. We remain in communication with our sales staff on a daily basis.

23. Revise the following sentences to remove the redundancies.

a. In grateful appreciation of your patronage, we are pleased to offer you this free gift as a token gesture of our gratitude.

b. An anticipated major breakthrough in storage technology will allow us to proceed ahead in the continuing evolution of our products.

c. During the course of the next two hours, you will see a demonstration of our new speech-recognition system that will be introduced for the first time in November.

24. Revise the following sentences to eliminate the pomposity.

a. This state-of-the-art soda-dispensing module is to be utilized by the personnel associated with the Marketing Department.

b. It is indeed a not unsupportable inference that we have been unsuccessful in our attempt to forward the proposal to the proper agency in advance of the mandated date by which such proposals must be in receipt.

c. Deposit your newspapers and other debris in the trash receptacles located on the station platform.

25. Revise the following sentences to eliminate the sexism.

a. Each doctor is asked to make sure he follows the standard procedure for handling Medicare forms.

b. Policemen are required to live in the city in which they work.

c. Professor Harry Larson and Ms. Anita Sebastian—two of the university's distinguished professors—have been elected to the editorial board of *Modern Chemistry*.

26. Revise the following sentences to eliminate the offensive language.

a. This year, the number of female lung-cancer victims is expected to rise because of increased smoking.

b. Mentally retarded people are finding greater opportunities in the service sector of the economy.

c. This bus is specially equipped to accommodate the wheelchair-bound.

Collaborative Exercise

27. Form small groups. Have one person in the group find a multipage document he or she has written recently, either in this class or in another. Have each member annotate a copy of this document according to the principles of sentence effectiveness discussed in this chapter, then write a summary statement of the document, highlighting those techniques of sentence construction that are well done and those that could be improved. Meet as a group and study these annotated documents. As a group, write a memo to your instructor, describing those aspects of the annotations that more than one group member cited, as well as those aspects that only one group member cited. Overall, what are the basic differences between the annotations and the summary statement from one group member and those from another? Do you think that, as a general practice, it would be worthwhile to have a draft reviewed and annotated by more than one person? What have you learned about the usefulness of peer reviewing of a document? See Chapter 16 for a discussion of memos.

Web Exercise

28. You work in the corporate communications department of your company. Management has decided to create a Web site for the company. Your supervisor has asked you to find out what sort of writing style your competitors use on their sites. She asks you to write her a memo comparing and contrasting the writing styles in the Web sites for three or four of your competitors. Consider such factors as sentence structure, level of formality, jargon, and clichés. Because your company has customers around the world, she also would like you to consider the issue of comprehension by nonnative speakers. Overall, which site is most effective? See Chapter 16 for a discussion of memos.

REFERENCES

Carliner, S. (1987). Lists: The ultimate organizer for engineering writing. *IEEE Transactions on Professional Communication, 30*(4), 218–221.

Fuchsberg, G. (1990, December 7). Well, at least "terminated with extreme prejudice" wasn't cited. *Wall Street Journal*, p. B1.

Horn, R. E. (1985). Results with structured writing using the Information Mapping® writing service standards. In T. M. Duffy and R. Waller (Eds.), *Designing Usable Texts*. Orlando, FL: Academic Press.

Maggio, R. (1992). *The bias-free word finder: A dictionary of nondiscriminatory language*. Boston: Beacon.

Peterson, D. A. T. (1990). Developing a Simplified English vocabulary. *Technical Communication, 37*(2), 130–133.

Thomas, M., Jaffe, G., Kincaid, J. P., & Stees, Y. (1992). Learning to use Simplified English: A preliminary study. *Technical Communication, 39*(1), 69–73.

Tyler, L. (1990). Communicating about people with disabilities: Does the language we use make a difference? *Bulletin of the Association of Business Communication, 53*(3), 65–67.

Williams, J. (1997). *Style: Ten lessons in clarity & grace* (5th ed.). New York: HarperCollins.

Writing Proposals

The Logistics of Proposals 460
  External and Internal Proposals 460
  Solicited and Unsolicited Proposals 461

The "Deliverables" of Proposals 463
  Research Proposals 463
  Goods-and-Services Proposals 464

Persuasion and Proposals 464
  Understanding the Readers' Needs 465
  Describing What You Plan to Do 466
  Demonstrating Your Professionalism 467

Writing a Proposal 468
  The Role of Resource Planning 468
  The Role of Collaboration 468

The Structure of the Proposal 469
  Summary 470
  Introduction 471
  Proposed Program 472
  Qualifications and Experience 473
  Budget 476
  Appendices 476

The Format of the Proposal 481

Sample Internal Proposal 481

Cases: Ethics and Proposals 488
        Selecting Project-Management Software 488

Writer's Checklist 488

Exercises 489
  Collaborative Exercise 489
  Web Exercise 489

Most projects undertaken by organizations, as well as most major changes made within organizations, begin with a proposal. A *proposal* is an offer to carry out research or to provide a product or service.

This chapter begins by describing the logistics of proposals, concentrating on external and internal proposals, as well as solicited and unsolicited proposals. Then it explains the "deliverables" of proposals: what you deliver at the end of the project. The chapter then discusses the need for persuasion in proposal writing. Finally, it discusses the components of a proposal and presents an example of an internal proposal.

The Logistics of Proposals

Proposals can be classified as either external or internal and as solicited or unsolicited. Figure 15.1 shows the relationship among these four terms.

External and Internal Proposals
Proposals are either *external* (if submitted to another organization) or *internal* (if submitted to the writer's own organization).

External Proposals

An *external proposal* is a proposal submitted to an outside organization. No organization produces all the products or provides all the services it needs. Paper clips and company cars have to be purchased. Offices must be cleaned and maintained. Sometimes projects that require unusual expertise, such as sophisticated market analyses, have to be carried out. Any number of companies would love to provide the paper clips or the cars, and a few dozen consulting organizations would happily conduct the studies. For this reason, it is almost always a buyer's market. To get the best deal, most organizations require that their prospective suppliers compete for the business by submitting proposals, documents by each supplier to make the case that it deserves the contract.

Internal Proposals

One day, while you're working on a project in the laboratory, you realize that if you had a new centrifuge you could do your job better and faster. The increased

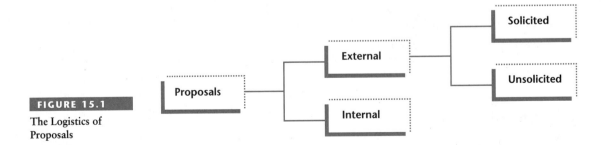

FIGURE 15.1

The Logistics of Proposals

productivity would save your company the cost of the equipment in a few months. You call your supervisor, who tells you to send a memo describing what you want, why you want it, what you're going to do with it, and what it costs; if your request seems reasonable and the money is available, you'll likely get the new centrifuge.

Your memo is an *internal proposal*—a persuasive argument, submitted within an organization, for carrying out an activity that will benefit the organization, generally by increasing productivity or quality or by reducing costs. An internal proposal can recommend that the organization conduct research, purchase a product, or change some aspect of its policies or procedures.

The scope of the proposal determines its format. A simple request might be conveyed orally, either in person or on the phone. A more ambitious request might require a brief memo. The most ambitious requests are generally conveyed in formal proposals. Organizations often use dollar figures to determine the format of the proposal. For instance, employees use a brief form to communicate proposals that would cost less than $1,000, whereas they use a report to communicate proposals that would cost more than $1,000.

Solicited and Unsolicited Proposals

External proposals are either solicited or unsolicited. A *solicited proposal* originates with a request from a customer. An *unsolicited proposal* originates with the prospective supplier.

Solicited Proposals

When an organization wants to purchase a product or service, it publishes one of two basic kinds of statements:

- An IFB—*information for bid*—is used for standard products. When an agency of the federal government needs office equipment, for instance, it lets suppliers know that it wants to purchase, say, 100 office chairs of a particular type. The supplier that offers the lowest bid wins the contract.
- An RFP—*request for proposal*—is issued for customized products or services. Police cars are likely to differ from the standard consumer model: they might have different engines, cooling systems, suspensions, and upholstery. The police department's RFP might be a long and detailed set of technical specifications. The supplier that can provide the automobile most closely resembling the specifications—at a reasonable price—will probably win the contract. Sometimes the RFP is a more general statement of goals. The customer is in effect asking the suppliers to create their own designs or describe how they will achieve the specified goals. The supplier that offers the most persuasive proposal will probably win the contract.

Most organizations issue RFPs and IFBs in newspapers or send them in the mail to past suppliers. Government RFPs and IFBs are published in the journal

Commerce Business Daily, which is available online. Figure 15.2 (CBDNET, 1997) shows a sample entry from that journal.

Unsolicited Proposals

An unsolicited proposal looks essentially like a solicited proposal, except, of course, that it does not refer to an RFP. Even though the potential customer never formally requested the proposal, in almost all cases the supplier was invited to submit the proposal after people from the two organizations met and discussed the project informally. Because proposals are expensive to write, suppliers are reluctant to submit them without assurances that the potential customer will study them carefully. Thus, the word *unsolicited* is only partially accurate.

External proposals—both solicited and unsolicited—can culminate in contracts of several types: a flat fee for a product or a one-time service; a leasing agreement; or a "cost-plus" contract, under which the supplier is reimbursed for the actual cost plus a profit set at a fixed percentage of the costs.

PART: U.S. GOVERNMENT PROCUREMENTS
SUBPART: SUPPLIES, EQUIPMENT, AND MATERIAL
CLASSCOD: 17—Aircraft Launching, Landing, and Ground Handling Equipment
OFFADD: Defense Supply Center, Columbus, PO BX 16595, dscc-pbab Columbus, OH 43216-6595
SUBJECT: 17—GUIDE, SHUTTLE ASSEMBLY. EXCEPTIONS TO DWG.516404
REV.F: SOL SP0760-97-Q-0717 DUE 051997
POC Contact: copy of solicitation. FAX 614-692-2262 or mail request to Defense Supply Center, Columbus, PO Box 16595, DS cc-pbab Columbus, OH 43216-6595, for information contact contracting officer (name/phone) LT. John Groeschel, (614) 692-1240.
DESC: PR-NO: YPC97013001791 NSN 1720-01-047-3828, CAGE 80020 Part No 516404-7 guide, shuttle assembly. MFG.I/A/W Navy dwg.80020 516404 to latest Rev. P/N 516404-7 79 -EA Del to Norfolk, VA 23512-0001 Del by 150 days after award. 58 -EA Del to Mechanicsburg, PA 17055 Del by 150 days after award. 37 -EA Del to Tracy, CA 95376-5000 Del by 140 days after award. See NOTES 1, 9, and 23. Small business size standard is 1000 employees. Solicitation not available until 15 days after published in the CBD. Written requests for drawings are to be submitted to Defense Supply Center Columbus, Attn: DSCC-VTRD, P.O. Box 3990, Columbus, OH 43216-5000, by FAX (614) 692-2998 or call (614) 692-2344. Requests should include the IFB/RFP/RFQ number, opening/closing, NSN, YPC/ZPC or NPC number; buyer's name and your complete name and address. Federal, military, and commercial specifications cannot be provided by DSCC. Automated Best Value Model (ABVM) and Delivery Evaluation Factor (DEF) apply. While price is a significant factor in the evaluation of offers, the final award decision will be based upon a combination of price, delivery, and past performance as described in the solicitation, this request will be issued via a written RFQ. The estimated award date will be 97 Jun 18. All responsible sources may submit offer which DCSC shall consider.
CITE: (I-087 SN052321)

FIGURE 15.2

Extract from *Commerce Business Daily*

The "Deliverables" of Proposals

When people talk about *deliverables,* they are referring to what the supplier will deliver at the end of the project. Deliverables can be classified into two major categories, as shown in Figure 15.3.

Research Proposals

In a research proposal, you are promising to provide a research report of some sort. Here are a few examples:

- A biologist working for a state bureau of land management writes a proposal to the National Science Foundation asking for resources to build a window-lined tunnel in the forest to study tree and plant roots and the growth of fungi. The biologist also wishes to understand better the relationship between plant growth and the activity of insects and worms. The deliverable will be a report submitted to the National Science Foundation and, perhaps, an article published in a professional journal.

- A manager of the technical-publications department at a manufacturing company writes a proposal to her supervisor asking for resources to study whether the company should convert its internal documents from paper to online. The deliverable will be a completion report (see Chapter 17) that contains her recommendations.

- A university sociologist writes to his state board of education proposing to study a nearby migrant community to determine the extent to which this community understands and uses the municipal services available to it. The deliverable will be a report, submitted to the state board of education and the municipal government, that presents his findings. If the findings are of interest to people beyond his immediate geographical area, he will also write a journal article.

A research proposal often leads to two other kinds of documents: progress reports and completion reports.

After the proposal has been approved and the researchers have begun work, they often submit one or more progress reports. A *progress report* tells the sponsor of the project how the work is proceeding: Is it following the plan of work outlined in the proposal? Is it going according to schedule? Is it staying within budget? Progress reports are discussed in Chapter 16.

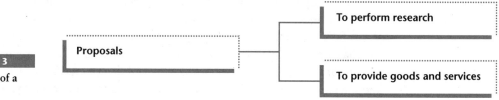

FIGURE 15.3

"Deliverables" of a Proposal

At the end of the project, researchers prepare a *completion report,* often called a *final report,* a *project report,* or simply a *report.* A completion report tells the readers the whole story of the research project, beginning with the problem or opportunity that motivated it, the methods that were used in carrying it out, and the findings: the results, conclusions, and recommendations. Completion reports are discussed in Chapter 17.

People perform research projects for three basic reasons:

- *Curiosity.* Researchers want to understand as much as they can about their field.
- *Professional advancement.* Many organizations require that their professional employees carry out research and publish in appropriate reports, journals, or books. Government researchers and university professors, for instance, are expected to remain active. Writing proposals is one way to get the resources—time as well as money for travel, equipment, and assistants—to carry out the research.
- *Personal satisfaction.* Conducting research, writing, and publishing are fun.

Goods-and-Services Proposals

Whereas a research proposal leads to a report of some kind, a goods-and-services proposal leads to a tangible product (a fleet of automobiles), a service (building maintenance), or some combination of the two (the building of a house).

A vast network of goods-and-services contracts spans the working world. The U.S. government, the world's biggest customer, spent more than $42 billion in 1997 buying military equipment from organizations that submitted proposals. But goods-and-services contracts are by no means limited to government contractors. One auto manufacturer buys engines from another, and a company that makes spark plugs buys its steel from another company. In fact, most products and services are purchased by contract. The world of work depends on goods-and-services proposals.

Persuasion and Proposals

Regardless of whether the supplier is a professor applying for a research grant, an employee proposing a project at the office, or an electronics company seeking a government contract to build a sophisticated radar device for a new jet aircraft, the proposals will be analyzed carefully and skeptically.

The agency reviewing the professor's proposal will want to see that the professor fully understands the important research questions pertaining to the subject, has a feasible plan for carrying out the project, and has a good track record in such projects. Many other professors will be competing for the same grant, and the agency wants to make sure it is spending its money wisely.

The supervisor of the employee writing the internal proposal will want to be satisfied that the employee has isolated a real problem and devised a feasible strategy for solving it, and that the employee has a good record of carrying through on similar projects.

The government officials reviewing the radar-device proposals will want to be satisfied that the supplier will live up to its promise: to build, on schedule, the best radar device at the best price. With perhaps a dozen suppliers competing for the contract, the government officials will know only that a number of companies want the work; they can never be sure—not even after the contract has been awarded—that they have made the best choice.

A proposal, then, is an argument. To be successful, it must be persuasive. The writers must convince the readers that the *future benefits* will outweigh the *immediate and projected costs*. Basically, proposal writers must clearly demonstrate that they

- understand the readers' needs
- are able to fulfill their own promises
- are committed to fulfilling their own promises

Understanding the Readers' Needs

The most crucial element of the proposal is the definition of the problem or opportunity to which the project is intended to respond. This would seem to be mere common sense: how can you expect to write a successful proposal if you don't show that you understand the readers' needs? Yet the people who evaluate proposals—whether government readers, private foundation officials, or managers in small corporations—agree that an inadequate or inaccurate understanding of the problem or opportunity is the most common weakness of the proposals they see.

Readers' Needs in an External Proposal

Sometimes the RFP fails to convey the problem or opportunity. More often, however, the writers of the proposal are at fault. The suppliers might not read the RFP carefully and simply assume that they understand the client's needs. Or perhaps the suppliers, in response to a request, know they cannot satisfy a client's needs but nonetheless prepare proposals detailing a project they can complete, hoping either that the readers won't notice or that no other supplier will come closer to responding to the real problem. Suppliers can easily concentrate on the details of what *they* want to do rather than on what the customers need.

But most readers will toss the proposal aside as soon as they realize that it does not respond to their needs. When you respond to an RFP, study it thoroughly. If you don't understand something in it, contact the organization that issued it; they will be happy to clarify it, because a bad proposal wastes everyone's time. Your first job as a proposal writer is to demonstrate your grasp of the problem.

When you write an unsolicited proposal, analyze your audience carefully. How can you define the problem or opportunity so that your readers will understand it? Keep in mind the readers' needs (even if the readers are oblivious to them) and, if possible, the readers' backgrounds. Concentrate on how the problem has decreased productivity or quality or on how your ideas would create new opportunities. When you submit an unsolicited proposal, your task in many cases is to convince your readers that a need exists. Even when you have reached an understanding with some of your customer's representatives, your proposal will still have to persuade other officials.

Readers' Needs in an Internal Proposal

An internal proposal also must respond to readers' needs. If you propose that your organization hire a new person, you have to make the case that the person is needed and would save or bring in more money than he or she costs. In addition, you have to make sure there is a place for the new person in your current facilities. Writing an internal proposal is both more simple and more complicated than writing an external proposal. It is simpler because you have more access to your readers than you would to external readers. And you can get more information more easily.

However, you might find it more difficult to get a true sense of the situation in your organization. Some coworkers might not be willing to tell you directly if your proposal is a long shot. Another danger is that when you identify the problem to solve, you are often criticizing—directly or indirectly—the person in your organization who instituted the system that needs to be revised or who failed to take necessary actions earlier. Therefore, before you write an internal proposal, discuss your ideas thoroughly with as many potential readers as you can. In this way you will more likely find out what the organization really thinks of your idea before you commit it to paper.

Describing What You Plan to Do

Once you have shown that you understand why something needs to be done, describe what you plan to do. Convince your readers that you can respond to the situation you have just described. Discuss your approach to the subject: indicate the procedures and equipment you would use. If appropriate, justify your choices. For example, after you say that you want to do ultrasonic testing on a structure, explain why, unless the reason is obvious.

Create a complete picture of how you would progress from the first day of the project to the last. Many inexperienced writers believe that they need only convince the reader of their enthusiasm and good faith. Unfortunately, few readers are satisfied with assurances, no matter how well intentioned. Most look for a detailed plan that shows that the writer has actually started to do the work.

No proposal can anticipate all of your readers' questions about what you plan to do, of course. But the more planning you have done before you submit the proposal, the greater the chances you will be able to do the work successfully if you get the go-ahead. A full discussion of your plan suggests to your readers that you are interested in the project itself, not just in winning the contract or in receiving authorization.

Demonstrating Your Professionalism

After showing that you understand the readers' needs and have a well-conceived plan, demonstrate that you are the kind of person—or yours is the kind of organization—that is committed to delivering what is promised. Many other people or organizations could probably carry out the project. You want to convince your readers that you have the pride, ingenuity, and perseverance to solve the problems that inevitably occur in any big undertaking. In short, you want to show that you are a professional.

Demonstrating your professionalism usually involves providing four kinds of information:

- *Credentials and work history.* Who are the people in your organization with the qualifications and experience to carry out the project? What equipment and facilities do you have that will enable you to do the work? What management structure will you use to maintain coordination and keep all the different activities running smoothly? What similar projects have you completed successfully? In short, make the case that you know how to make this project work because you have made similar projects work.

- *Work schedule.* Sometimes called a *task schedule,* this schedule—which usually takes the form of a graph or chart— shows when the various phases of the project will be carried out. In one sense, the work schedule is a straightforward piece of information that enables your readers to see how you would apportion your time. But in another sense, it reveals more about your attitudes toward your work than about what you will actually be doing on any given day. Events rarely proceed according to plan: some tasks take more time than anticipated, some take less. A careful and detailed work schedule is therefore really another way of showing that you have done your homework, that you have attempted to foresee any problems that might jeopardize the success of the project.

- *Quality control.* Your readers will want to see what procedures you have established to evaluate the effectiveness and efficiency of your work on the project. Sometimes quality-control procedures consist of technical evaluations carried out periodically by the project staff. Sometimes the writer will build into the proposal provisions for on-site evaluation by recognized authorities in the field or by representatives of the potential client. Quality control is also measured by progress reports (see Chapter 16).

- *Budget.* Most proposals conclude with a budget, a formal statement of how much the project will cost.

Writing a Proposal

In writing a proposal, you use the same basic techniques of prewriting, drafting, and revising that you use in any other kind of writing. However, a proposal can be such a big project that two aspects of the writing process—resource planning and collaboration—assume greater importance than they do in smaller documents.

The Role of Resource Planning

As discussed in Chapter 6, planning a project requires a lot of work. You need to see whether your organization can devote resources to two activities:

- writing the proposal
- carrying out the project if the proposal is successful

Sometimes an organization writes a proposal, wins the contract, and then loses money because it doesn't have the resources to do the project and must subcontract major portions of it.

The resources you need fall into three basic categories:

- *Personnel.* Will the necessary technical personnel, managers, and support people be available?
- *Facilities.* Will facilities be available to carry out the research and production, or can they be leased? Can you profitably subcontract portions of the job to companies that have the appropriate facilities?
- *Equipment.* Do you have the right equipment? If not, can you buy it or lease it or subcontract the work? Some contracts provide for the purchase of equipment, but others don't.

Don't write the proposal unless you are confident that you can carry out the project if the proposal is successful.

The Role of Collaboration

Collaboration is critical in most proposals of more than a few pages because no one person has the time and expertise to do all the work. Writing major proposals calls for the expertise of technical personnel, writers, editors, graphic artists, managers, lawyers, and document-production specialists. For this reason, you need to determine whether these people are available.

Usually, a project manager coordinates the writing of a large proposal. In most cases, the parts of the proposal are written at different times by different people. These components might not come together into the final package until a few days—or a few hours—before the proposal deadline.

For more information on collaboration, see Chapter 3.

INTERVIEW: PROPOSAL WRITER JIM HILL

on the importance of proposals

They are absolutely essential if you're going to do any government contract work or indeed any kind of contract work. You certainly hope your win ratio will be good; companies that don't have a good win ratio just go under.

on the team members who write proposals

The critical ones are the people who know the technology. Then you usually have a few people on a professional proposal team who are writers or editors or project managers who bring people together and outline the proposal. They assign it to the technical specialists to write their parts. And there are marketing people and legal staff involved in looking at a proposal before it goes out.

on the importance of planning

Preproposal work is all-important. Your marketing people must always be working with potential customers to find out their plans and explain your company's capabilities. No RFP [request for proposal] should be a surprise; if it is, you've already lost.

on developing a win strategy

At the very start you have to identify what the customer wants, and then figure out what it is that you have that will give you a competitive advantage over other companies bidding on the project.

on learning how to write winning proposals

You start with a good course in how to control stress. Second, you have to know your business. You can't lag in any aspect of your business. And you have to be able to work with people; you have to be able to keep things on track. A good proposal writer has to maintain the lines of communication with management and try to anticipate problems before putting all those resources into a proposal that's not going to fly.

on analyzing your audiences

You're writing to different audiences, and this is one of the things your marketing people have to find out. They try to find out how the proposal is going to be graded and who's going to do it—no one person sits down and reads that whole proposal; they'll probably split it off into parts, maybe according to disciplines, and they'll give it to different people. They look for certain words in the RFP. They look to see if you responded to certain points and how thoroughly you seem to have responded, and they give you a grade depending on that. And when the scores are totaled up, the company that came in with the best score wins the contract, all other things being equal.

on meeting the deadline

You have to get the proposal in when it's due, and you can't allow for any possibility of its not happening. If they say it has to be delivered by twelve noon on a certain date, they mean that, and if it comes in at one second after then they can't take it; they have to reject it. So what we always did [when submitting proposals to federal government agencies in Washington, D.C.] was send three copies of the proposal. We would send one copy by overnight express; another copy would go by air; the marketing rep would take the third copy down in the trunk of his car. We had millions of dollars tied up in that proposal; we couldn't afford to miss the deadline.

Jim Hill, now retired, had a 30-year career as a proposal writer and manager for such companies as DuPont, Westinghouse, and E-Systems.

The Structure of the Proposal

Most proposals follow a basic structural pattern. If the authorizing agency provides an IFB, an RFP, or a set of guidelines, follow it to the letter. If guidelines have not been supplied, or you are writing an unsolicited proposal, use the following conventional structure, shown in Figure 15.4 on page 470, as a starting point.

FIGURE 15.4

Structure of a Proposal

Then modify it according to your subject, your purpose, and the needs of your audience.

Summary   For any proposal of more than a few pages, provide a summary. Many organizations impose a length limit—for example, 250 words—and ask the writer to present the summary, single spaced, on the title page. The summary is crucial, because in many cases it will be the only item the readers study in their initial review of the proposal.

The summary covers the major elements of the proposal but devotes only a few sentences to each. To write an effective summary, first define the problem in a sentence or two. Next describe the proposed program. Then provide a brief statement of your qualifications and experience. Some organizations wish to see the completion date and the final budget figure in the summary; others prefer that this information be displayed separately on the title page along with other identifying information about the supplier and the proposed project.

For more information on summaries, see Chapters 8 and 10.

Figure 15.5 shows an effective summary taken from a proposal (Farrington, 1993) submitted by a student working for BCE, Inc., a small company that fabricates computer boards. (An additional example of a proposal is presented at the end of this chapter.)

Note: Proposals are double spaced if they are presented as reports, not memos.

The background.

The research the author has already done.
The proposal and the deliverable.

The schedule, budget, and the writer's credentials.

Summary

   BCE, Inc., uses the hand-soldering method to attach components to its PC boards (approximately 10,000 connections per day). Because of a recent increase in orders, we will be unable to meet shipping deadlines without modifying our assembly process.

   Recent articles in the trade journals suggest that the easiest and fastest way to satisfy the increased demand would be to purchase a wave-solder machine.

   I am proposing that I be permitted to compare wave-flow technology with hiring additional workers or subcontracting as a means of improving our productivity. After completing my research, if wave flow still appears to be the best method for reducing costs and meeting our production deadlines, I will recommend the most appropriate machine and provide you costs and an implementation schedule.

   Performing the research and writing the report should require about 35 hours over four weeks, for a total cost of $303. As a marketing major with more than a year's fabrication experience at BCE, I feel qualified to perform this research effectively.

FIGURE 15.5

Summary of a Proposal

Introduction   The introduction of a proposal is like that of any other kind of document. Its purpose is to make sure the reader knows the answers to the following seven questions:

- *What is the problem or opportunity?* Be as specific as possible. Whenever you can, *quantify.* Describe the problem or opportunity in monetary terms because the proposal itself will include a budget of some sort and you want to convince your readers that spending money on what you propose is smart. Don't say that a design problem is slowing down production; say that it is costing $4,500 a day in lost productivity.

- *What is the purpose of the proposal?* Even though it might seem obvious to you, the purpose of the proposal is to describe a problem or opportunity and propose a course of action that will culminate in some deliverable. Be as specific as you can in explaining what you want to do.

- *What is the background of the problem or opportunity?* In answering this question, you probably will not be telling your readers anything they don't already know (except, perhaps, if your proposal is unsolicited). Your goal here is to show them that *you* understand the problem or opportunity: the circumstances that led to the discovery, the relationships or events that will affect the problem and its solution, and so on.

- *What are your sources of information?* Review the relevant literature, either internal reports and memos or external published articles or even books, so that your readers will understand the context of your work.
- *What is the scope of the proposal?* If appropriate, indicate what you are proposing to do, as well as what you are *not* proposing to do.
- *What is the organization of the proposal?* Indicate the organizational pattern you will use in the proposal.
- *What are the key terms that will be used in the proposal?* If you will use any new terms, the introduction is an appropriate place to define them.

Figure 15.6 is the introduction to the BCE proposal.

Proposed Program   Once you have defined the problem or opportunity, say what you want to do about it. As noted earlier, the proposed program demonstrates clearly how much work you have already done. Be specific. You won't persuade by saying that you plan to "gather the data and analyze it." How will you gather it? What techniques will you use to analyze it? Every word you say—or don't say—will give your readers evidence on which to base their decision. If you know your subject, the proposed program will show it. If you don't, you'll inevitably slip into meaningless generalities or include erroneous information that undermines the whole proposal.

Note: Proposals are double spaced if they are presented as reports, not memos.

The background and options.

The research the writer has done.

The proposal and the deliverables.

The logic and organization of the proposal.

Introduction

The current manufacturing process for PC boards involves hand soldering an average of 10,000 connections per day. Because of a recent increase in orders, we will be unable to meet shipping deadlines without modifying our assembly process. We have three alternatives:

- add at least one full-time employee
- subcontract some of the soldering
- purchase a single-wave solder machine

The two most recent articles on single-wave solder machines are by Lau (1996) and Woodgate (1993), who argue that the machine would greatly increase production speed while maintaining or improving product quality.

The purpose of the proposal is to request resources to investigate these three alternatives, and then present my findings in a report to you.

The report will compare the costs and benefits of the three options, focusing on cost and quality-control criteria. For the wave-solder machine, I will present the criteria necessary for selecting one, compare the available machines, and describe the steps we would need to take to purchase and set up a machine.

FIGURE 15.6

Introduction to a Proposal

If your project concerns a subject written about in the professional literature, show your familiarity with the scholarship by referring to the pertinent studies. However, don't just toss a bunch of references onto the page. For example, don't write, "Carruthers (1996), Harding (1997), and Vega (1996) have all researched the relationship between acid-rain levels and groundwater contamination." Rather, use the recent research to sketch the necessary background and provide the justification for your proposed program. For instance:

> Carruthers (1996), Harding (1997), and Vega (1996) have demonstrated the relationship between acid-rain levels and groundwater contamination. None of these studies, however, included an analysis of the long-term contamination of the aquifer. The current study will consist of . . .

You might include just one reference to recent research. However, if you have researched your topic thoroughly, you might devote several paragraphs or even several pages to recent scholarship.

Whether your project calls for primary research, secondary research, or both, the important point is that the proposal will be unpersuasive if you haven't already done a substantial amount of the research. For instance, say you are writing a proposal to do research on chain saws. You are not being persuasive if you write that you are going to visit Sears, JC Penney, and Coast to Coast Hardware to see what kinds of chain saws they carry. This statement is unpersuasive for two reasons:

- You need to justify why you are going to visit those three retailers rather than all the others. Anticipate your readers' questions: Why did you choose these three retailers? Why didn't you choose more specialized tool dealers?

- You need to have visited the appropriate stores already—and done any other preliminary research. If you haven't done the homework, readers have no assurance that you will in fact do it and that it will pay off. If your supervisor has already authorized the project but then learns that none of the chain saws on the market meets your organization's needs, you will have to go back and submit a different proposal—an embarrassing move.

Unless you can show in your proposed program that you have done the research—and that the research indicates that the project will likely succeed—the reader has no reason to authorize the project.

Figure 15.7 on page 474 shows the proposed program for the BCE project.

Qualifications and Experience

After you have described how you would carry out the project, demonstrate your ability to undertake it. Unless you convince your readers that you can turn an idea into action, your proposal will be unpersuasive.

Note: Proposals are double spaced if they are presented as reports, not memos.

Overview of this section of the proposal.

The writer shows that she has already done some research and thought about the problem.

Again the writer shows the research and thinking she has already done. Notice how the writer integrates her sources.

Proposed Procedure

To help you determine the best course of action for BCE, I will perform the following tasks:

1. determine BCE's soldering needs
2. evaluate the three options for meeting those needs
3. prepare a report that presents my methods and findings

Determining BCE's Soldering Needs

To determine our current production and predict our near-term soldering needs, I have studied the 1996 Production Logs and interviewed two key people: Larry Abrams, Production Supervisor, and Karen Kiramidjian, Assistant Production Supervisor. According to our Production Logs and the interviews, our needs can be described by the following criteria:

*   Volume: BCE is currently averaging production of approximately 100 new boards per week.
*   Design/type of boards: Most of BCE's boards are less than twelve inches in diameter and less than four layers thick, with a combination of through-hole and surface-mount components.
*   Physical requirements: We have an area approximately 9×5 feet. The production area is already outfitted with a suitable power supply, exhaust venting system, air compressor, and clean air source, should we decide to buy a machine.
*   Quality: BCE has a hard-earned reputation for high quality. Any solution we choose has to enable us to maintain and enhance that reputation.

Evaluating the Three Options for Meeting BCE's Needs

Three feasible options are available to BCE for increasing its production capacity.

*   Hiring an additional employee. I have contacted the Ada County Job Bank, the BSU College of Technology, and ITT Technical Institute. This preliminary research suggests that qualified workers are available, and that the current wage would be approximately $7.50/hour. For this option to be attractive, our workload would have to be relatively steady; if our workload varies greatly, we would have one of two problems: we would still have a problem meeting production needs or we would be paying a worker for not working. I am currently reviewing the Production Logs to determine whether our soldering needs are steady or whether they fluctuate according to industry conditions.
*   Subcontracting the work to an independent solder company. Woodgate (1993) lists the industry groups that I can contact to inquire about subcontracting. I showed this list to Larry and Karen, who recommended two other groups to contact to determine the availability and requirements of subcontractors. I have drafted a letter and faxed it to the groups and am now awaiting the replies. The letter solicits information about the costs, availability,

FIGURE 15.7

Proposed Program of a Proposal

(Figure 15.7 continued)

quality-control, and turnaround time of subcontractors. There are no sub-contractors in the Boise area, so we would have at least a three-day shipping time added to the fabrication time.

An additional factor needs to be determined: lot size. Much of BCE's work involves small, very specialized orders. Most of the large wave-solder machines are designed to handle runs of hundreds or thousands of boards, while many of our boards are manufactured in lots of 10 or fewer.

- Purchasing automated solder equipment. Studying this option involves answering two questions: (1) Would acquiring such equipment be the best solution to our production problem? and (2) If so, which brand and model should we acquire, and which supplier should we contact?

  Bahr (1994), Lau (1996), and Woodgate (1993) offer full descriptions of the different kinds of automated solder equipment. There are two main types of automated soldering systems: wave soldering and reflow. Bahr (1994) defines wave soldering as "a process in which printed boards are brought in contact with the surface of continuously flowing and circulating solder." The other type, reflow soldering, is defined as "a process for joining parts by tinning the mating surfaces, placing them together, heating until the solder fuses, and allowing to cool in the joined position." Reflow soldering is effective for surface-mounted components, but almost all of BCE's PC boards contain a mixture of through-hole and surface-mounted components. Therefore, the wave process would be indicated for BCE.

  On the basis of my conversations with Larry and Karen and my research, I will examine the available wave-solder machines according to the following criteria:
  - What kind of fluxer does it have? I will look for versatility, reliability, accuracy, failure-warning, easy cleaning, and corrosion resistance.
  - Does it have air knife wipe off? This optional feature aids in removing excessive flux from the boards before they enter the preheater.
  - What is the quality, reliability, etc., of the solder pot, the pumps and nozzles, conveyors, and automated process controls?
  - What is its provision for disposing of wastes and contaminants? Some fluxes are considered hazardous waste so we'll need to make sure we're in compliance with state, local, and EPA requirements.
  - What are the projected maintenance/service costs and frequency?
  - Are field reps and service technicians readily available?
  - How easy is it for us to convert to an automated process? How much training would our operators need?

Preparing a Report That Presents My Methods and Findings

After completing my research, I will prepare a report that summarizes my findings in comparing and contrasting the three alternative solutions to our problem. This report should enable you to make an informed decision.

The writer has already determined what criteria she will use in evaluating the machines.

The writer describes the deliverable.

The more elaborate the proposal, the more substantial the discussion of qualifications and experience has to be. For a small project, a few paragraphs describing your technical credentials and those of your coworkers will usually suffice. For larger projects, the résumés of the project leader—often called the *principal investigator*—and the other important participants should be included.

External proposals should also discuss the qualifications of the supplier's organization. Essentially similar to a discussion of personnel, this section outlines the pertinent projects the supplier has completed successfully. For example, a company bidding for a contract to build a large suspension bridge should describe other suspension bridges it has built. The discussion also focuses on the necessary equipment and facilities the company already possesses, as well as the management structure that will ensure successful completion of the project. Everyone knows that young, inexperienced persons and new firms can do excellent work. But when it comes to proposals, experience wins out almost every time.

Figure 15.8 shows the qualifications-and-experience section of the BCE proposal.

Budget Good ideas aren't good unless they're affordable. The budget section of a proposal specifies how much the proposed program will cost.

Budgets vary greatly in scope and format. For simple internal proposals, the writer adds the budget request to the statement of the proposed program: "This study will take me about two days, at a cost of about $400" or "The variable-speed recorder currently costs $225, with a 10 percent discount on orders of five or more." For more complicated internal proposals and for all external proposals, a more explicit and complete budget is usually required.

Most budgets are divided into two parts: direct costs and indirect costs.

- *Direct costs* include such expenses as salaries and fringe benefits of program personnel, travel costs, and necessary equipment, materials, and supplies.

- *Indirect costs* cover the intangible expenses that are sometimes called *overhead*. General secretarial and clerical expenses not devoted exclusively to any one project are part of the overhead, as are other operating expenses such as utilities and maintenance. Indirect costs are usually expressed as a percentage—ranging from less than 20 percent to more than 100 percent—of the direct expenses. In many external proposals, the client imposes a limit on the percentage of indirect costs.

Figure 15.9 shows the budget section from the BCE proposal.

Appendices Many types of appendices might accompany a proposal. Most organizations have *boilerplate* descriptions (standard descriptions that a writer can modify or place directly into a document) of other projects they have performed. Another

Notice how the writer links her practical work with her academic experience in doing the kind of research she is proposing, making a persuasive argument that she can fulfill the promise of her proposal.

FIGURE 15.8

Qualifications-and-Experience Section of a Proposal

Credentials

   I have more than one year of electronic assembly experience at BCE, including hand soldering, rework, operating a reflow solder machine, and prepping PC boards for through-hole and surface-mount soldering. I am also in my junior year as a BSU marketing major, with a concentration in entrepreneurship. In my academic work, I have written several feasibility studies comparing capital expenditures and hiring personnel. I am familiar with basic techniques of calculating return on investment and payback periods.

The writer presents and justifies the costs associated with the project.

FIGURE 15.9

Budget Section of a Proposal

Budget

   Table 1 is the budget itemization. The budget is based on my estimate of the time required for me to carry out the research, analyze the data, and prepare the report. I have included telephone and travel costs to accommodate the faxing, miscellaneous phone calls to out-of-state vendors and professional organizations, and travel to libraries.

Table 1. Budget Itemization

Category	Item	Cost
Wages	Laura Farrington, Equipment Intern, 35 hours labor @ $7.25	253.75
Miscellaneous	Long-distance telephone	30.00
	Travel	20.00
	Total:	$303.75

popular kind of appendix is the *supporting letter*—a testimonial to the supplier's skill and integrity, written by a reputable and well-known person in the field. Two other kinds of appendices deserve special mention: the task schedule and *the evaluation* description.

   The *task schedule* is almost always drawn in one of three graphical formats:

- table
- bar chart
- network diagram

Activity	Start Date	Finish Date
Design the security system	4 Oct. 97	19 Oct. 98
Research available systems	4 Oct. 97	3 Jan. 98
etc.		

FIGURE 15.10

Tabular Schedule

Tables

The simplest but least informative way to portray a schedule is to use a table, as shown in Figure 15.10. As with all graphics, you should provide a textual reference that introduces and, if necessary, explains it.

Although creating a table is better than writing out the information in sentences, readers cannot "see" the information; they have to read it and figure out how long each activity will last. In addition, readers cannot tell whether any of the activities are interdependent. That is, they cannot tell what would happen to the overall project schedule if one of the activities were delayed.

Bar Charts

Bar charts, also called *Gantt charts* after the early twentieth-century British civil engineer who first used them, are a distinct improvement over tables. The basic bar chart, as shown in Figure 15.11, allows readers to see how long each activity will take; in addition, bar charts let readers see different activities occurring simultaneously. Like tables, however, bar charts do not indicate the interdependency of tasks.

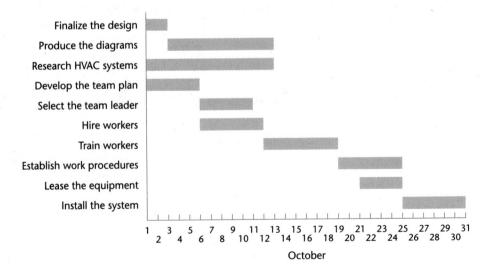

FIGURE 15.11

Bar Chart

Network Diagrams

There are many types of network diagrams, but they all show interdependence, clearly indicating which activities must be completed before other activities can begin.

Of the different kinds of network diagrams, the most popular are PERT (Program Evaluation and Review Technique) and CPM (Critical Path Method). Both techniques use nodes and arrows. As shown in Figure 15.12, PERT and CPM use slightly different symbols.

Figure 15.13 on page 480 shows a relatively simple network diagram. As you can see, understanding a network diagram requires experience. For this reason, a network diagram might be inappropriate for general readers. But even this brief look at scheduling graphics shows that a network diagram provides more useful information than either a table or a bar chart.

For more information on scheduling graphics, I recommend the following book:

Spinner, M. P. (1989). *Improving project management skills and techniques.* Englewood Cliffs, NJ: Prentice Hall.

Many companies produce project-management software.

Figure 15.14 on page 480 is the task schedule for the BCI proposal.

Other Appendices

Much less clear-cut than the task schedule is the description of *evaluation techniques*. In fact, the term *evaluation* means different things to different people, but in general an evaluation technique is any procedure for determining whether the proposed program is both effective and efficient. Evaluation techniques can range from simple progress reports to sophisticated statistical analyses. Some proposals provide for evaluation by an outside agent—a consultant, a testing laboratory, or a university. Other proposals describe evaluation techniques that the supplier itself will perform, such as cost/benefit analyses.

PERT (Events Labeled)

CPM (Activities Labeled)

FIGURE 15.12

Symbols Used in PERT and CPM Diagrams

FIGURE 15.13

Network Diagram

The subject of evaluation techniques is complicated by the fact that some people think in terms of *quantitative evaluations*—tests of measurable quantities, such as production increases—whereas others think in terms of *qualitative evaluations*—tests of whether a proposed program is improving, say, the workmanship of a product. And some people imply both qualitative and quantitative testing when they refer to evaluations. An additional complication is that projects can be tested both while they are being carried out (*formative evaluations*) and after they have been completed (*summative evaluations*). Formative evaluations are intended to help people improve the quality of the work they are doing; summative evaluations are intended to assess the quality of a completed project.

When an RFP calls for "evaluation," experienced writers of proposals know that it's a good idea to contact the sponsoring agency's representatives to determine precisely what they mean.

Figure 15.15 is a collection of appendices that describe the evaluation techniques for the BCE project.

Task Schedule

Determine BCE's soldering needs
Evaluate options for meeting those needs
Prepare completion report
Submit completion report

0 5 10 15 20 25 30
Days

FIGURE 15.14

Task Schedule of a
Proposal

Appendix A. Evaluation Techniques

I will submit a progress report at the end of week two of the project, at which time I should have completed almost all my research.

Appendix B. List of Solder Machine Vendors in North America

[Here the writer lists the names, addresses, and phone numbers of the 12 dealers she will contact.]

Appendix C. References

[Here the writer cites the sources referred to in the proposal.]

FIGURE 15.15

Appendices of a Proposal

The Format of the Proposal

Proposals are presented in many formats. Sometimes the organization receiving the proposal specifies the format. For large proposals—100 pages or longer—many organizations use a book format. As shown in Figure 15.16, a two-page unit has text on one page and graphics on the facing page. The two-page format is intended to force the writers to present material in small, easy-to-read units, with conveniently located graphics. All the information a reader needs to understand the topic is contained on the two pages.

For more information on page design, see Chapter 11.

FIGURE 15.16

Two-Page Format

Sample Internal Proposal

Following is an internal proposal (Wells, Tommack, & Tuck, 1997). The authors were students working as interns in their university's Office of Student Special Services.

The progress report written after the project was under way is included in Chapter 16. The completion report is in Chapter 17. Marginal notes have been added.

This proposal is presented as a memo and therefore is single spaced; if it were presented as a report, it would be double spaced.

DATE: April 7, 1997

TO: Margie Voren, Dean of Student Special Services, Central State University

FROM: Megan Wells, Danny Tommack, and Brandon Tuck

SUBJECT: Proposal to research whether Central State University should recognize American Sign Language as a foreign language

The writers identify the document as a proposal in the subject heading and the purpose statement.

Purpose

This proposal recommends that we be authorized to use our time and the Student Special Service's resources to research the possibility of classifying Central State University's American Sign Language classes as a foreign language in the Area I core group. We would then write a recommendation on whether American Sign Language should be considered a foreign language and how we could incorporate it.

The summary discusses the background, the proposal, the schedule, the deliverable, and the writers' credentials.

Summary

Despite the widespread use of sign language in America, Central State University does not give Amercian Sign Language (ASL) any special place among its offered courses. Currently, the ASL courses are only elective credits. Almost half of the states have recognized ASL as a foreign language and grant academic credit for completion of these courses. Six states have universities that independently recognize ASL as a foreign language, while 18 more have passed legislation on the state level recognizing ASL. Central State University is one of 15 states that grant only elective credit to ASL courses.

ASL itself is gaining prominence in mainstream culture and is also being recognized as a foreign language in the educational arena. The CSU English Department, for example, does grant ASL foreign-language status for linguistics majors; but in other departments, ASL is excluded. We propose to research the standards held by Central State University in deeming a language foreign. Our purpose is to determine whether ASL fulfills those requirements. In addition, we propose to research the legislative measures passed in the state of Texas that give ASL statewide recognition as a foreign language. If ASL fulfills CSU's requirements, we will submit a recommendation explaining how it can be implemented. As a safeguard and in the interest of Oklahoma education in general, we propose to investigate the legislative measures needed to attain formal recognition in the same manner as was achieved by the state of Texas.

We have several contacts in the deaf and educational communities in Oklahoma and Texas. With their cooperation we can begin on April 25 and have our research completed and our recommendations prepared by May 12. At that time, we will submit a completion report for your review.

The writers describe the problem that they wish to study.

Introduction

Sign language is the fourth most prominent language in the United States. In usage, it ranks below Spanish but above French, German, Russian, and Japanese, which together constitute all the foreign languages taught at CSU. But despite its prominence, Central State University offers few classes in ASL. With a student population of approximately 15,000, only 25 students per semester are given the opportunity to begin study in American Sign Language. These courses offered are not foreign-language credits. We want to research whether this course can or should be included.

Teresa O'Malley, Chair of the Modern Languages Department at CSU, said that her office receives several calls per semester regarding the inclusion of ASL in the core group of foreign languages. She is supportive of ASL's inclusion and cited other state-funded institutions that recognize ASL as a foreign language. The states have various approaches to ASL in the classroom, as shown in Figure 1. The University of Arizona and the University of Delaware, for example, are part of the 12% that have state-funded institutions that independently recognize ASL as a foreign language. An additional 36% of the states have passed legislative bills that formally recognize ASL as a foreign language, such as Texas. CSU is part of the 30% of states that recognize ASL as an elective credit but not a foreign language. As indicated in gray, roughly one-half of the state-funded institutions in the United States recognize ASL as a foreign language. CSU does not.

The problem has both a local and national dimension.

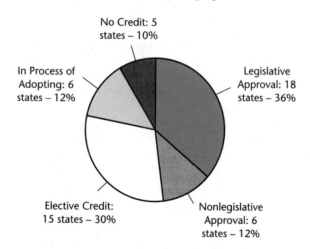

No Credit: 5 states – 10%

In Process of Adopting: 6 states – 12%

Legislative Approval: 18 states – 36%

Elective Credit: 15 states – 30%

Nonlegislative Approval: 6 states – 12%

Source: Rhonda Jacobs, *ASL as a Truly Foreign Language,* 1996.

Figure 1: Treatment of ASL in the 50 States

Current research strongly supports ASL's inclusion in the foreign-language category. Both Teresa O'Malley and Suzanne Christopher, the ASL instructor on campus, feel that ASL probably fits into the foreign-language category. Blaine Lee, the Student Special Services Coordinator, and June Yunker, the CSU

Some of the primary and secondary research the writers have already done.

(Continued)

(Continued)

Interpreter Coordinator, both feel that ASL is qualified to be given foreign-language status. The English Department at Central State does accept Suzanne Christopher's upper-level ASL courses for foreign-language credit in fulfillment of its linguistics-emphasis language requirement. However, this recognition doesn't apply elsewhere on campus. If the English Department feels ASL is foreign enough to be given formal academic credit, then ASL may be a good candidate for formal recognition. Suzanne provided us research by Jacobs (1993) that explores the characteristics of ASL, supporting its qualification as a foreign language, and the various methods of achieving this recognition. In addition, Armstrong (1988) presents strong legal arguments supporting this recognition.

The purpose of this proposal is to request resources to research whether ASL is a foreign language by CSU standards and whether we should petition the university and/or the Oklahoma state legislature to recognize ASL. However, we will not look into significantly expanding the existing ASL curriculum.

In the following sections, we will outline the proposed procedure for our research, our qualifications to do the research, and the budget needed to ensure an effective evaluation of our possibilities.

Proposed Program

We will perform the following tasks for your review in determining whether Central State University should recognize ASL as foreign language and how this recognition could be achieved at Central State University (by a university petition) and at all Oklahoma postsecondary institutions (by state legislation):

1. compile a summary of the criteria required by Central State University for a language to be considered foreign
2. research whether ASL fulfills the university-mandated requirements
3. investigate the broad approach of state legislative recognition as used in the state of Texas
4. provide a completion report detailing our findings and recommendations

Task 1. Compile a summary of the criteria required by Central State University for a language to be considered foreign

We interviewed Teresa O'Malley, Chair of the Modern Languages Department. She did not know the exact criteria for languages to be considered foreign at CSU, but she did know that only languages taught at the intermediate level counted as foreign-language credits. At present, CSU has two ASL courses that could qualify as intermediate courses: CM221: Intermediate American Sign Language, and CM321: Conversational American Sign Language. All beginner-level courses taught are counted as elective credits, including CM121: Beginning Sign Language. She can provide us the exact specifications for a class to count as a foreign-language credit at the end of this week. We request time on Friday morning, April 25, to visit with Teresa O'Malley to discuss the exact qualifications for a language to be considered foreign.

An advance organizer for the rest of the proposal.

An advance organizer for this section of the proposal.

Describing the proposed program in terms of clear tasks makes the proposal look credible.

(Continued)

A further discussion of the research the writers will complete.

Task 2. Research whether ASL fulfills the university-mandated requirements

We have also met with Suzanne Christopher, the current ASL instructor at Central State. She has done considerable research in American Sign Language and has prepared various proposals to expand the ASL program at the university. Currently, we have several works by Armstrong (1988), Wilcox (1977), and Jacobs (1996) that will assist us in understanding the current status of ASL in university curricula and in making our recommmendation. In addition, we have contacted the Oklahoma chapter of the Registry for Interpreters for the Deaf (ORID) and the Oklahoma Association for the Deaf (OAD). Kelly Eastwick, President of ORID, and Janis Seymour, committee member of OAD, are both interested and willing to assist us in our research and deliberations. After we learn the exact criteria for a langauge to be considered foreign, we request time to meet with Suzanne, Kelly, and Janis to compile our research concerning ASL's qualifications. We request a four-hour reservation of the Centennial Room in the Student Union Building on Monday morning, April 28, to meet with Suzanne Christopher, Kelly Eastwick, and Janis Seymour to prepare our recommendation. After our analysis of the university's requirements and ASL's qualifications, we will submit to you a progress report detailing whether an internal approach to approve ASL is viable.

Task 3. Investigate the broad approach of state legislative recognition as used in the state of Texas

Regardless of whether ASL as a foreign language is approved at Central State University, we would like to investigate the measures needed to have the state of Oklahoma extend legal recognition of the language. As employees of the state, we see CSU as only one of the several postsecondary institutions in Oklahoma. If ASL is deserving of this status, we wish to extend these benefits to other universities by investigating the feasibility of passing a bill recognizing ASL as a foreign language. If ASL as a foreign language is approved by CSU, it will strengthen our petition to the state. If internal approval is not feasible, petitioning the state will give us one more opportunity. We hope to work on the internal and external research simultaneously.

The alternative method of granting ASL foreign-language recognition is to have a bill passed in the state legislature. According to Figure 1, 18 states have legislative approval for ASL to be considered a foreign language in postsecondary schools. Texas is one of these states. We have contacted Sha H. Cowan from the Texas Education Agency, Services for the Deaf Department. He referred us to three educators involved with the legislative measures passed in Texas: Dr. Jean Andrews, Communication Disorders, Lamar University in Beaumont, Texas, and Carol Seeger and Lisa Bissin, deaf ASL instructors at the University of Texas at Austin. We request Friday afternoon, April 25, and Monday afternoon, April 28, to communicate with these individuals and gather research. In addition we request Tuesday, April 29, to follow up with the legislative proposal plan, analyze a feasible Oklahoma legislature approach, contact Oklahoma state representatives, compile our research, and prepare a completion report.

(Continued)

(Continued)

Task 4. Provide a completion report detailing our findings and recommendations

The writers remember to describe creating the report as one of their tasks.

Following the completion of our research, we will submit to you a report summarizing our findings and recommendations. We will explain the criteria required by CSU for a language to be considered foreign, whether ASL fulfills these requirements, and how we could implement a legislative aproach to obtain formal recognition for ASL. In this last area, we will report on the strategies used in the state of Texas and provide a list of contact personnel in the Oklahoma state government with whom we will work. This completion report will be on your desk May 12 at 9 A.M.

Figure 2 is our estimated task schedule, beginning on April 25 and ending with our recommendation to you on May 12.

A simple Gantt chart such as this can be made with the drawing tools on a word processor.

Figure 2: Task Schedule

Qualifications and Experience

We have the skills and experience to conduct this research. Each of us brings to the team talents and strengths that, when combined, help us to work effectively and efficiently.

The qualifications of each of the students are described.

Danny Tommack, our team leader, has the skills to direct the team's efforts and to help us keep our goals in focus. He has developed strong management and communication skills while working at McU Sports. His professional approach in meetings and planning sessions will help us work closely with members of our community and our long-distance contacts in Texas.

Brandon Tuck has been an American Sign Language interpreter for three years. He works for the Department of Student Special Services at Central State Univer-

(Continued)

sity and is a freelance interpreter. He brings to the group a strong knowledge of deafness and sign language, and contacts in the deaf community.

Megan Wells has highly developed research skills she obtained while working for three years in a research laboratory. She has excellent communication and interviewing skills. Her experience at several universities will help us to investigate Central State University and other postsecondary institutions.

Budget

Our proposed budget is itemized in Figure 3. We have estimated the time needed to carry out the research, funds to pay for telephone communication, and funds to provide refreshments at our conference with the representatives from ORID and OAD.

Item	Cost
30 hours labor ($9 per hour)	270.00
Long-Distance Telephone	20.00
Refreshments	20.00
Total	$310.00

Figure 3: Itemized Budget

References

Armstrong, D. F. (1988). Some notes on ASL as a "foreign" language. *Sign Language Studies 59:* 231-239.

Cowan, S. H. <shc@tenet.edu> "Re: ASL credit as a foreign language list." 1 Apr. 1997. (distribution list) (1 Apr. 1997).

Jacobs, R. (1996). ASL as a truly foreign language. In C. Lucas (Ed.), *Multicultural aspects of sociolinguistics in deaf communities* (pp. 183–226). Washington, D.C.: Gallaudet University Press.

Wilcox, S. (1997). "American Sign Language as a foreign language fact sheet." <www.unm.edu/~wilcox/ASLFL/asl_fl.html> (1 Apr. 1997).

CASES

1. Ethics and Proposals

You work for Devon Electronics, a company that does contract work for the federal government. Your department has submitted a proposal to build a friend-or-foe electronic device to be used in military aircraft. You have just learned that Martha Ruiz, the chief technical person for the project, has accepted a position with another company and will be leaving in two weeks. Your company has not yet decided who would replace her if the project is funded. Your Director of Operations favors delaying a decision until the company learns whether it has won the contract. The government agency is currently reviewing all the proposals and is not scheduled to announce the winner of the contract for three weeks. Should you notify them of Ruiz' impending resignation? If not, why not? If you think you should notify the government agency, draft the letter that you would send.

2. Selecting Project-Management Software

You work in the proposal department of Allied Marine Services, a small manufacturer of electronic equipment used in civilian and military boats and ships. Currently, your company does not use project-management software in writing its proposals; a graphic artist draws the Gantt charts. You think that project-management software might enable you to complete the charts more quickly and less expensively. You ask your supervisor if he will let you spend a few hours searching the Internet to learn about the software: what is available, what it does, how easy it is to use, and what it costs. Using a search engine, find the sites of three manufacturers of project-management software. Download a few pages from each site. In a memo to your supervisor, describe the different products and state whether you think it is a good idea to download demo copies and try them on your next proposal. See Chapter 16 for a discussion on memos.

WRITER'S CHECKLIST

The following checklist covers the basic elements of a proposal. Guidelines established by the recipient of the proposal should of course take precedence over these general suggestions.

1. Does the summary provide an overview of
   - ❑ the problem or the opportunity?
   - ❑ the proposed program?
   - ❑ your qualifications and experience?

2. Does the introduction indicate
   - ❑ the problem or opportunity?
   - ❑ the purpose of the proposal?
   - ❑ the background of the problem or opportunity?
   - ❑ your knowledge of the professional literature?
   - ❑ the scope of the proposal?
   - ❑ the organization of the proposal?
   - ❑ the key terms that will be used in the proposal?

3. Does the description of the proposed program provide a clear and specific plan of action?

4. Does the description of qualifications and experience clearly outline
   - ❑ your relevant skills and past work?
   - ❑ the skills and background of the other participants?
   - ❑ your department's (or organization's) relevant equipment, facilities, and experience?

5. Is the budget
   - ❑ complete?
   - ❑ correct?
   - ❑ accompanied by a textual reference?

6. Do the appendices include the relevant supporting materials, such as a task schedule, a description of evaluation techniques, and evidence of other successful projects?

EXERCISES

Exercises 2 and 3 call for you to write a memo. See Chapter 16 for a discussion of memos.

Collaborative Exercise

1. Write a proposal for a research project that will constitute a major assignment in this course. Your instructor will tell you whether the proposal is to be written individually or collaboratively. Start by defining a technical subject that interests you. (This subject could be one that you are involved with at work or in another course.) Using abstract services and other bibliographic tools, create a bibliography of articles and books on the subject. (See Chapter 7 for discussions of choosing a topic and finding information.) Create a reasonable real-world context. Here are three common scenarios from the business world:

   - Our company uses Technology X to perform Task A. Should we instead be using Technology Y to perform Task A? For instance, our company subcontracts the writing and production of our monthly employee newsletter. Should we be producing it ourselves using desktop-publishing? What kinds of personnel are needed? What skills do they require? How much time does it take? What kinds of hardware and software are required?

   - Our company has decided to purchase a particular kind of tool to perform Task A. Which make and model of the tool should we purchase, and from which supplier should we buy or lease it? For instance, our company has decided to purchase 10 multimedia computers. Which brand and model should we buy, and whom should we buy them from?

   - Our company does not currently perform Function X. Is it feasible to perform Function X? For instance, we do not currently offer daycare for our employees. Should we? What are the advantages and disadvantages of doing so? What forms can daycare take? How is it paid for?

   Following are some additional ideas for topics.
   - the need to provide Internet access to students

   - the value of using the Internet to form ties with another technical-communication class on another campus
   - the need for expanded opportunities for internships in your major
   - the need to create an advisory board of people from industry to provide expertise about your major
   - the need to raise money to keep the college's computer labs up-to-date
   - the need to evaluate your major to ensure it is responsive to students' needs
   - the advisability of starting a campus branch of a professional organization in your field
   - improving parking facilities on campus
   - creating or improving organizations for minorities or for women on campus

   Notice that these topics can be approached from different perspectives. For instance, the first one—on providing Internet access to students—could be approached in several ways:
   - Our college currently purchases journals but does not provide Internet access for students. Should we consider taking some money from the library's journal budget to subsidize Internet access for students?
   - Our college has decided to provide Internet access for its students. How should it do so? What vendors provide such services? What are the strengths and weaknesses of each vendor?
   - Our college does not offer Internet access to students. Should we make it a goal to do so? What are the advantages of doing so? The disadvantages?

2. Form groups according to major. Using CBDNET (http://cbdnet.gpo.gov), find a request for proposals (RFP) for a project related to your academic field. Study the RFP. What can you learn about the needs of the organization that issued it? How effectively does the RFP describe what the issuing organization expects to see in the proposal? Is the RFP relatively general or specific? What sorts of evaluation techniques does the RFP call for? Write a list of questions that you would ask the issuing organization if you were considering responding to the RFP. Present your results in a memo to your instructor.

 ## Web Exercise

3. Using the Web, locate three companies that provide consulting services about proposal writing. In a brief

memo to your instructor, list their names and URLs and briefly describe and evaluate their sites and services.

REFERENCES

CBDNET. (1997). *Commerce Business Daily Online.* <http://cbdnet.access.gpo.gov> (1997, Apr. 10).

Farrington, L. (1993). Proposal to research wave-solder machines for possible use at BCE. Unpublished document.

Wells, M., Tommack, D., & Tuck, B. (1997). Proposal to research whether Central State University should recognize American Sign Language as a foreign language. Unpublished document.

Writing Memos
and Informal Reports

Guidelines for Informal Reports 492

Formats for Informal Reports 493
  Memos *494*
  Other Formats *499*

Writing Directives 499

Writing Trip Reports 500

Writing Field and Lab Reports 502

Writing Progress and Status Reports 502
  Organizing Progress and Status Reports *505*
  Concluding Progress and Status Reports *506*

Writing Meeting Minutes 506
  Recording the Logistical Details of the Meeting *512*
  Providing an Accurate Record of the Meeting *512*
  Reflecting Positively on the Participants and the Organization *512*

Case: Amending a Proposal *514*

Writer's Checklist 514

Exercises 515
  Collaborative Exercise *517*
  Web Exercise *518*

A report is a statement—oral or written—that helps listeners or readers understand, analyze, or take action on some situation, idea, or action. This book classifies reports into two categories: formal and informal. This chapter discusses informal reports; Chapter 17 discusses formal reports.

Here are a few examples of informal reports:

- a memo that requests authorization to buy a piece of equipment for $1,200
- an e-mail describing a problem in the management of an office and recommending action to solve it
- a meeting at which an employee explains to her colleagues her analysis of a set of architectural plans for the company's new offices
- a memo describing the projects undertaken in the quality-assurance department of a company over the last quarter
- a memo recording what occurred at a recent meeting of a department
- a memo responding to a request for information from a supervisor

As you can see from this list, informal reports cover many subjects, have many purposes, and take many forms. What exactly is *informal* about an informal report? There is no single answer to this question. In one organization, a particular document might be considered a formal report; in another, an informal one. In general, however, an informal report might be said to be a routine communication of information that concerns an everyday matter.

This is not to say that informal reports are unimportant. An accident report written by police officers at the scene of an automobile accident might be only a page or two long, but it might have enormous implications for the people in the accident, and for many other people, too. By contrast, the report of the presidential commission investigating the explosion of the space shuttle *Challenger* was certainly not informal. It was a multivolume report that culminated many months of investigation, costing millions of dollars and involving dozens of people.

This chapter discusses the process of writing informal reports, the common formats, and typical kinds of informal reports, including trip reports, progress reports, and meeting minutes.

Guidelines for Informal Reports

The process of writing informal reports involves the same planning, drafting, and revising techniques that are used for most other kinds of technical communication. This section explains some of the major ways in which these techniques are applied in writing informal reports.

- *Analyze your audience and purpose.* In some cases, it is easy to identify your audience and purpose. For instance, a set of meeting minutes is addressed to all the members of the committee or department, and its purpose is to serve as the official record of a meeting. A quarterly status report about your department's activities goes to the department head; it describes the projects you worked on that quarter and might include other comments or recommendations that will help the reader understand your department's situation and needs.

In other cases, the audience and purpose are not routine. For instance, you might be reporting on an accident at your company's facility. To whom should you address it? Your direct supervisor? Your direct supervisor and others? What is your purpose: to describe what happened? To recommend steps that might reduce the chance that such accidents recur? In other words, sometimes the audience and purpose of an informal report are routine, deriving clearly from the occasion for writing. At other times, however, you need to analyze the audience and purpose just as carefully as you would in writing any other kind of document. For more information on analyzing audience and purpose, see Chapters 5 and 6.

- *Gather and create your information.* Sometimes, assembling your information is as simple as printing a database file. Other times it requires sophisticated information-gathering techniques, using primary and secondary research. See Chapter 7 for more information on research techniques.

- *Choose an appropriate format.* The most common format for informal reports is the memo, but in many cases there are preprinted forms or templates for you to follow. In some organizations, title pages, covers, and other elements usually associated with more formal reports are used for all reports. (See Chapter 10.) The choice of format is therefore often a local matter determined by your organization, your audience, and your purpose. The next section of this chapter discusses formats in more detail.

- *Draft the report.* For routine informal reports, writers sometimes can use sections of previous reports. In a status report, for instance, the description of your current project can be copied from the previous report. Some informal reports are drafted on site. For instance, a site study might be "drafted" as an engineer walks around a site, using a hand-held electronic device or a clipboard. For more information on drafting, see Chapters 8 through 10.

- *Revise the report.* Informal reports must be revised, just like any other kind of technical communication. *Informal* does not mean *careless*. (See Chapters 13 and 14.)

Formats for Informal Reports

Reports are communicated in two media: speech and writing. This chapter discusses the kinds of documents used for transmitting informal reports. Four basic formats are used commonly for informal reports:

- memos
- forms
- e-mail
- letters

Because memos are used so frequently for informal reports, this section concentrates on techniques for writing effective memos.

Memos   Although there are minor variations in the format of a memo from one organization to another, almost every organization uses memos. Your memos are most likely to be successful if you follow these seven guidelines:

Include the Identifying Information

Almost all memos have five elements at the top:

- the logo or a brief letterhead of the organization
- the "to" line
- the "from" line
- the "subject" line
- the "date" line

Some organizations have a "copies" or "c" (copy) heading as well. Some organizations still use the "cc" notation, even though carbon copies are a thing of the past. Figure 16.1 shows several styles for presenting the identifying information.

Organizations sometimes have preferences about the headings. Some prefer full names of the writer and reader; others want only the first initials and the last names. Some prefer job titles; some do not. If your organization does not object, include your job title and your reader's. The memo will then be informative for someone referring to it after you or your reader has moved on to a new position, as well as for readers elsewhere in the organization who might not know either of you. List the names of persons receiving photocopies of the memo, either alphabetically or in descending order of organizational rank. In listing the date, write out the month (March 4, 19XX or 4 March 19XX). Do not use the all-numeral format (3/4/XX); people from other countries use different notations and therefore could be confused.

Most writers put their initials or signature next to their typed names or at the end of the memo; the initials or signature shows that the writer has reviewed the memo and is accepting responsibility for it.

The subject heading—the title of the memo—deserves special mention. Don't be too concise. Avoid naming only the subject, such as *Tower Load Test*; rather, specify what aspect of the test you wish to address. For instance, *Tower Load Test Results* or *Results of Tower Load Test* would be much more informative

AMRO   MEMO

To: B. Pabst
From: J. Alonso
Subject: MIXER RECOMMENDATION FOR PHILLIPS
Date: 11 June 19XX

INTEROFFICE

To: C. Cleveland   c: B. Aaron
From: H. Rainbow   K. Lau
Subject: Shipment Date of Blueprints   J. Manuputra
           to Collier   W. Williams
Date: 2 October 19XX

NORTHERN PETROLEUM COMPANY
INTERNAL CORRESPONDENCE

Date: January 3, 19XX
To: William Weeks, Director of Operations
From: Helen Cho, Chemical Engineering Dept.
Subject: Trip Report—Conference on Improved Procedures
           for Chemical Analysis Laboratory

Identifying Information
in a Memo

than *Tower Load Test,* which does not tell the reader whether the memo is about the date, the location, the methods, the results, or any number of other factors related to the test. In short, each subject line should be informative, accurate, and unique. (See Chapter 6 on purpose and Chapter 13 on titles.)

Print the second and all subsequent pages of memos on plain paper rather than on letterhead. Include three items in the upper right-hand corner of each page:

- the name of the recipient
- the date of the memo
- the page number

You might even define the communication as a memo and repeat the primary names. A typical second page of such a memo begins like this:

Memo to: J. Alders   April 6, 19XX

From: R. Rossini   Page 2

Clearly State Your Purpose

The first sentence of the body of a memo should explain the purpose of the memo. Following are a few examples of purpose statements:

I want to tell you about a problem we're having with the pressure on the main pump, because I think you might be able to help us.

The purpose of this memo is to request authorization to travel to the Brownsville plant Monday to meet with the other quality inspectors.

This memo presents the results of the internal audit of the Phoenix branch, an audit that you authorized March 13, 19XX.

I want to congratulate you on your division's quarterly record.

Notice that the best purpose statements are concise and direct. Make sure your purpose statement has a verb that clearly communicates what you want the memo to accomplish, such as to request, to explain, or to authorize. (See Chapter 6 for a detailed discussion of purpose.) Some people object to a direct statement of purpose—especially when the writer is communicating bad news or asking for something, as in the example about requesting authorization to travel to the Brownsville plant. Rather than beginning with a direct statement of purpose, an argument for such a request would open with the reasons for the trip, the trip's potential benefits, and so on. Then the writer would conclude the memo with the actual request: "For these reasons, I am requesting authorization to. . . ." Although some readers would rather have the reasons presented first, far more would prefer to know immediately why you have written. There are two basic problems with stating the purpose at the end of the argument:

- Some readers will toss the memo aside if you seem to be rambling on about the Brownsville plant without getting to the point.
- Some readers will suspect that you are trying to manipulate them into doing something they don't want to do.

Although the purpose statement sacrifices subtlety for directness, it helps your readers understand why you are writing.

Use Headings to Help Your Readers

Headings are useful in all kinds of technical communication. Use them liberally in memos. Headings help your readers in two main ways:

- *Headings help your readers decide what to read.* If the section is labeled *What Is the Function of a Browser?*, those readers who already know will simply skip to the next section.
- *Headings help readers understand the information.* A simple heading such as *Summary* tells readers the function of the section, enabling them to concentrate on the information without having to worry about why it is included and how it relates to other information in the memo.

If Appropriate, Summarize Your Message

Brief memos, of one page or less, do not generally contain summaries. But longer ones, particularly those addressed to multiple readers, can benefit from a summary placed at the beginning. A summary has three main goals:

- to help all readers understand the body of the memo
- to enable executive readers to skip the body if they so desire
- to remind readers of the main points

Following are some examples of summaries:

> The conference was of great value. The lectures on new coolants suggested techniques that might be useful in our Omega line, and I met three potential customers who have since written inquiry letters.

> The analysis shows that lateral stress was the cause of the failure. We are now trying to determine why the beam did not sustain a lateral stress weaker than that it was rated for.

> In March, we completed Phase I (Design) on schedule. At this point, we anticipate no delays that will jeopardize our projected completion date.

The summary should reflect the length and complexity of the memo. It might range from one simple sentence to a long and technical paragraph. If possible, the memo's summary should reflect the sequence of the information in the body.

Provide Adequate Background for the Discussion

Writers sometimes mistakenly assume that their readers already know the background of the memo: the events that led to the current situation. If you have any doubt about whether your readers need a background statement, include one. (Readers who do not need to read the background will simply skip it.)

Each background discussion is, of course, unique; however, some basic guidelines are useful. If the memo defines a problem—for example, a flaw detected in a product line—the background might discuss how the problem was discovered or present the basic facts about the product line: what the product is, how long it has been produced, and in what quantities. If the memo reports the results of a field trip, the background might discuss why the trip was undertaken, what its objectives were, who participated, and so on.

Following is a background paragraph from a memo requesting authorization to have a piece of equipment retooled:

Background
The stamping machine, a Curtiss Model 6143, is used in the sheet-metal shop. We bought it in 1986 and it is in excellent condition. However, since we switched the size of the tin rolls last year, the stamping machine no longer performs to specifications.

Organize the Discussion

The discussion is the section in which you present your main arguments. You might divide the detailed discussion into the subsections of a more formal report: methods, results, conclusions, and recommendations. Or you might give it headings that pertain specifically to the subject you are discussing. You might include small tables or figures but should attach larger ones as appendices. (See Chapter 10.)

The discussion section of the memo can be developed according to any of the basic patterns for structuring arguments, including chronological, spatial, more important to less important, and cause and effect. These patterns are discussed in detail in Chapter 8.

Highlight Action

Some reports require follow-up action, by either the writer or the reader. For example, a memo addressing a group of supervisors might discuss a problem and end by stating what the writer is going to do about it. A supervisor might delegate tasks for other employees. In writing a memo, be sure to define clearly who is to do what and when. Following are two examples of ways to highlight action:

Action:
I would appreciate it if you would work on the following tasks and have your results ready for the meeting on Monday, June 9.

1. Henderson: recalculate the flow rate.

2. Smith: set up meeting with the regional EPA representative for sometime during the week of February 13.

3. Falvey: ask Armitra in Houston for his advice.

Action:
To follow up these leads, I will do the following this week:

1. Send the promotional package to the three companies.

2. Ask Customer Relations to work up a sample design to show the three companies.

3. Request interviews with the appropriate personnel at the three companies.

In the first example, notice that although the writer is the supervisor of his readers, he uses a polite tone in the lead-in to the list of tasks.

Other Formats   In addition to memos, informal reports are sometimes presented as forms, e-mail, and letters.

Forms

Routine informal reports are often completed on forms, either preprinted forms on which you write or type, or templates you use with a word processor. Forms prompt you to fill in the information that the readers need. For example, some companies use a template for trip reports; the template has a space in which to write the destination, another space for the purpose of the trip, and so forth. These spaces not only prompt you, they also indicate how much information is called for: a one-line space calls for a short answer, whereas a space two inches tall signals a longer response. Sometimes forms are used as cover sheets for informal reports: the first page is the form, but the following pages are created by the writer.

E-mail

E-mail is an increasingly popular format for transmitting informal reports because it is convenient and digital.

E-mail is convenient because almost everyone in an organizational setting has access to it. In addition, the ability to create address books and distribution lists makes it easy to automate the distribution of the report. If a set of minutes routinely goes to the members of a committee, the writer sets up a distribution list of those committee members.

The fact that e-mail is digital also makes it a convenient format for informal reports. That is, when the reader receives the e-mail, it is already in digital format, ready to be copied to another document. For example, if a supervisor receives four status reports every month, she can easily combine the files into a single report to transmit to her supervisor.

For more information on e-mail, see Chapter 4.

Letters

Letters are the preferred format when the reader and the writer work at different organizations. See Chapter 19 for a discussion of letters.

Writing Directives

In a *directive,* you report on a policy or procedure you want your readers to follow. If possible, explain the reason for the directive; otherwise, it might seem like an irrational order rather than a thoughtful request. For brief directives, the explanation should precede the directive itself, to prevent the appearance of bluntness.

Notice that the writer uses a polite, informal tone throughout, but that he clearly defines the problem and how he would like his readers to address it.

Quimby	Interoffice

Date: March 19, 19XX
To: All supervisors and sales personnel
From: D. Bartown, Engineering
Subject: Avoiding customer exposure to sensitive information
          outside Conference Room B

Recently I have learned that customers meeting in Conference Room B have been allowed to use the secretary's phone directly outside the room. This practice presents a problem: the proposals that the secretary is working on are in full view of the customers. Proprietary information such as pricing can be jeopardized unintentionally.

In the future, would you please escort any customers or non-Quimby personnel needing to use a phone to the one outside the Estimating Department? Thanks very much.

FIGURE 16.2

A Directive

For longer reports, the actual directive might precede the detailed explanation, to ensure that readers will not overlook the directive. Of course, the body of the directive should begin with a polite explanatory note, such as the following:

> I'd like to establish a uniform policy for dealing with customers who fall more than 60 days behind in their accounts. The policy is defined below under the heading "Policy Statement." Following the statement is our rationale.

Figure 16.2 shows an example of a directive.

Writing Trip Reports

A *trip report* is a record of a business trip written after the employee returns to the office. Most often, it takes the form of a memo, although some organizations use computerized templates. The key to effective trip reports is to remember that your reader is less interested in an hour-by-hour narrative of what happened than in a carefully structured discussion of what was important. If, for instance, you attended a professional conference, don't list all the presentations—simply attach the agenda or program if you think your reader will be interested. Communicate the important information you learned, or describe the important questions that didn't get answered. If you traveled to meet a client (or a potential client), focus on what your reader is interested in: how to follow up on the trip and maintain a good business relationship with the client.

Figure 16.3 shows a typical trip report.

This memo uses headings for the main sections. These headings make the memo easier to read—and easier to write, because they prompt the writer to provide the kind of information the readers need.

The writer and the reader appear to be relatively equal in rank. Still, the writer goes to the trouble of organizing the discussion clearly so it is easy to read and to refer to later.

Dynacol Corporation

INTEROFFICE COMMUNICATION

To:	G. Granby, R&D
From:	P. Rabin, Technical Services
Subject:	Trip Report—Computer Dynamics, Inc.
Date:	September 20, 19XX

This memo presents my impressions of the Computer Dynamics technical seminar of September 18. The purpose of the seminar was to introduce their new PQ-500 line of high-capacity storage drives.

Summary

In general, I was impressed with the technical capabilities and interface of the drives. Of the two models in the 500 series, I think we ought to consider the external drives, not the internal ones. I'd like to talk to you about this issue when you have a chance.

Discussion

Computer Dynamics offers two models in its 500 series: an internal drive and an external drive. Each model has the same capacity (100 M of storage), and they work the same way. They act just like any other kind of drive, preserving the user's directory structure.

Although the internal drive is convenient—it is already configured for the computer—I think we should consider only the external drive. So many of our employees do teleconferencing that the advantages of portability outweigh the disadvantage of inconvenience. The tech rep from Computer Dynamics walked me through the process of reconfiguring the drive for a second machine; the process will take most of our employees only a few minutes to learn. A second advantage of the external drive is that it can be salvaged easily when we take a computer out of service.

Recommendation

I'd like to talk to you, when you get a chance, about negotiating with Computer Dynamics for a quantity discount. I think we should ask McKinley and Rossiter to participate in the discussion. Give me a call (x3442) and we'll talk.

FIGURE 16.3

A Trip Report

Writing Field and Lab Reports

A common kind of informal report describes inspections, maintenance, and site studies. These reports, often known as *field reports* or *lab reports,* include the same type of information that high-school lab reports do—the problem, methods, results, and conclusions—but they deemphasize the methods and can include recommendations.

A typical field or lab report, therefore, answers the following questions:

- What is the purpose of the report?
- What are the main points covered in the report?
- What were the problems leading to the decision to perform the procedure?
- What methods were used?
- What were the results?
- What do the results mean?
- What should be done next?

The report in Figure 16.4 illustrates some of the possible variations on this standard report structure.

Writing Progress and Status Reports

A *progress report* communicates to a supervisor or sponsor the current status of an ongoing project. A *status report* is an update on the entire range of operations of a department or division of an organization. For example, the director of marketing for a manufacturing company might submit a monthly status report.

As its name suggests, a progress report is an intermediate communication between the proposal (the argument that a project be undertaken) and the completion report (the comprehensive record of a completed project). For more information on proposals, see Chapter 15; for completion reports, see Chapter 17.

Progress reports let you check in with your audience. Supervisors are vitally interested in the status of their projects, because they have to integrate them with other present and future commitments. Sponsors (or customers) have the same interest, plus an additional one: they want the projects to be done right—and on time—because they are paying for them.

Regardless of whether the project is proceeding smoothly or has encountered difficulties, you need to explain clearly and fully what happened and how it will affect the overall project. Your tone should be objective, neither defensive nor casual. Unless ineptitude or negligence caused the problem, you're not to blame. Regardless of what kind of news you are delivering—good, bad, or mixed—your

Because the writer and the reader work for different companies, the letter is the appropriate format for a brief informal report.

The word "visual" is all that needs to be said about the methods.

The writer states the purpose of the inspection.

Notice how the writer uses generic terms such as "summary" and "recommend" throughout the letter. It would be easy to add headings to this letter by using the generic terms. Some writers like to use headings in letters; others don't.

Lobate Construction
3311 Industrial Parkway
Speonk, NY 13508

Quality Construction Since 1957

April 1, 19XX

Ms. Christine Amalli, Head
Civil Engineering
New York Power
Smithtown, NY 13507

Dear Ms. Amalli:

We are pleased to report the results of our visual inspection of the Chemopump after Run #9, a 30-day trial on Kentucky #10 coal.

The inspection was designed to determine if the new Chemopump is compatible with Kentucky #10, the lowest-grade coal that you anticipate using. In preparation for the 30-day test run, the following three modifications were made by your technicians:

- New front-bearing housing buffer plates of tungsten carbide were installed.
- The pump-casting volute liner was coated with tungsten carbide.
- New bearings were installed.

Our summary is as follows. A number of small problems with the pump were observed, but nothing serious and nothing surprising. Normal break-in accounts for the wear. The pump accepted the Kentucky #10 well.

The following four minor problems were observed:

- The outer lip of the front-end bell was chipped along two-thirds of its circumference.
- Opposite the pump discharge, the volute liner received a slight wear groove along one-third of its circumference.
- The impeller was not free-rotating.
- The holes in the front-end bell were filled with insulating mud.

The following three components showed no wear:

- The 5½" impeller
- The suction neck liner
- The discharge neck liner

Our conclusion is that the problems can be attributed to normal break-in for a new Chemopump. The Kentucky #10 coal does not appear to have caused any extraordinary problems. In general, the new Chemopump seems to be operating well.

FIGURE 16.4

A Field Report

(Continued)

(Figure 16.4 continued)

We would recommend, however, that the pump be modified as follows:
1. Replace the front-end bell with a tungsten carbide-coated front-end bell.
2. Replace the bearings on the impeller.
3. Install insulation plugs in the holes in the front-end bell.

Further, we recommend that the pump be reinspected after another 30-day run on Kentucky #10.

If you have any questions, or would like to authorize these modifications, please call me at 555-1241. As always, we appreciate the trust you have placed in us.

Sincerely,

Marvin Littridge
Director of Testing and Evaluation

The writer concludes politely.

job is the same: to provide a clear and complete account of your activities and to forecast the next stage of the project.

Progress reports are crucial because they enable managers to make informed decisions when things go wrong. What can go wrong? Problems generally fall into three categories:

- *The deliverable won't be what you thought it would be.* The deliverable is what you plan to deliver to your manager or sponsor at the end of the project; it could be a report or a product or service. The progress report is your opportunity to explain that even though you thought you would be able to recommend which earth mover under $200,000 to purchase for your contracting company, you will not be able to do so. Then you explain why. Or you explain that the deliverable will not meet one of the specifications in the proposal.
- *You won't meet your schedule.* Explain why you are going to be late and specify when the project will be complete.
- *You won't meet the budget.* Explain why you need more money, and specify how much more you will need.

Seldom do projects exceed their initial projections. Rarely does the deliverable exceed the specifications agreed on in the proposal. Rarely is the project completed early, and rarely does it cost less than the initial budget specified.

A note about ethics: you might be tempted to cover up the problems and hope that you can solve them before the next progress report. This course of action is unwise and unethical. It is unwise because chances are the problems will multiply, and you will have a harder time explaining why you didn't alert your readers earlier. It is unethical because by withholding bad news you are violating your readers' rights to full and complete information. Because they are the sponsors or supervisors of the project, they have a right to know how it is going.

Organizing Progress and Status Reports

For a progress report, two basic organizational patterns are used frequently:

- *The time pattern.* This is the simplest way to organize the discussion: past work/future work. After describing the problem, you describe all the work that has been completed in the present reporting period and then sketch in the work that remains. This scheme is easy to read and easy to write. Some writers use a past/present/future pattern, which enables them to focus on a long or complex task that they are working on at the moment.
- *The task pattern.* If the project requires that you work on several tasks simultaneously, the task pattern is particularly effective; it enables the writer to describe, in order, what has been accomplished on each task. Often, the task-oriented structure incorporates the past work/future work structure.

Table 16.1 shows the difference between the two organizational patterns.

A status report, on the other hand, is usually organized according to task; by its nature a status report covers a specified period of time. Because the purpose of the status report is to describe a department's work in, say, July, the logical pattern is to focus on each task or project carried out by that department during July.

TABLE 16.1	Organizational Patterns in Progress and Status Reports

The Time Pattern	**The Task Pattern**

III. Discussion
    A. Past Work
    B. Future Work
	III. Discussion
	    A. Task I
	        1. Past work
	        2. Future work
	    B. Task II
	        1. Past work
	        2. Future work

Concluding Progress and Status Reports

In the conclusion of a progress or status report, you evaluate how the project is proceeding. In the broadest sense, you have one of two messages:

- Things are going well.
- Things are not going as well as anticipated.

If the news is good, convey your optimism, but avoid overstatement.

OVERSTATED: We are sure the device will do all that we ask of it, and more.

REALISTIC: We expect that the device will perform well and that, in addition, it might offer some unanticipated advantages.

Beware of promising early completion. Such optimistic forecasts rarely prove accurate, and it is always embarrassing to report a failure to meet the promise.

However, don't panic if the preliminary results are not as promising as you had planned, or if the project is behind schedule. Readers know that the most levelheaded and conservative proposal writers cannot anticipate all problems. As long as the original proposal was well planned and contained no wildly inaccurate computations, don't feel responsible. Just do your best to explain the unanticipated problems and the current status. If you suspect that the results will not match earlier predictions—or that the project will require more time, personnel, or equipment—say so, clearly. If your news is bad, at least give your reader as much time as possible to deal with it effectively.

If appropriate, append supporting materials to the progress report, such as computations, printouts, schematics, diagrams, charts, tables, or a revised task schedule. Be sure to cross-reference these appendices in the body of the report, so that the reader can consult them at the appropriate stage of the discussion.

Figure 16.5 is the progress report written as a follow-up to the proposal included in Chapter 15. (The completion report on this study is included in Chapter 17.) Marginal notes have been added to the progress report.

Writing Meeting Minutes

Minutes are an organization's official record of a meeting. Minutes are distributed to all people who belong to the committee or other unit that was represented at the meeting. In writing a set of minutes, your goal is to be clear, comprehensive, objective, and diplomatic. Your job is *not* to interpret what happened. You have three main goals in recording a set of minutes:

- to record the logistical details of the meeting
- to provide an accurate record of the meeting
- to reflect positively on the participants and the organization

DATE: April 28, 1997
TO: Margie Voren, Dean of Student Special Services, Central State University
FROM: Megan Wells, Danny Tommack, and Brandon Tuck
SUBJECT: Progress report on whether Central State University should recognize American Sign Language as a foreign language

Purpose

In this progress report, we review our research concerning the possibility of recognizing ASL as a foreign language and incorporating courses in ASL into the foreign-language curriculum at CSU. We also include the results of our initial research into legislation passed in Texas that recognizes ASL as a foreign language.

Summary

We have been researching whether ASL should be included in the foreign-language curriculum at CSU. Our research thus far indicates that ASL meets the criteria by which CSU defines a language as foreign and that ASL should be incorporated into the CSU foreign-language curriculum.

We have also been researching the methods by which ASL might be recognized as a foreign language, focusing on internal recognition (through revision of university policy) and external recognition (through state legislation). Our research indicates that internal recognition is most likely the preferable method of recognizing ASL. We will continue to study the option of external recognition, but our focus will now be on comparing the two approaches.

Our research is on schedule, and we anticipate that the project will be completed on schedule.

Introduction

On April 24, we received approval to research whether CSU should recognize ASL as a foreign language. In addition, we received approval to investigate how this recognition could be achieved, either through internal recognition (through revision of university policy) or by external recognition (through state legislation). Our proposed program included the following activities:

1. compile a summary of the criteria by which CSU defines a language as a foreign language
2. determine whether ASL meets the criteria by which CSU defines a language as a foreign language
3. investigate the broad approach in Texas, which uses state legislation to recognize ASL as a foreign language at all postsecondary institutions in the state
4. write a report that documents our research and recommends whether CSU should classify ASL as a foreign language. The report will also contain our

---

The writers summarize their progress and forecast that their completion report will be submitted on time.

The writers summarize the tasks they set out in their proposal.

FIGURE 16.5

Progress Report

(Continued)

(Figure 16.5 continued)

recommendations for using state legislation to recognize ASL at all postsecondary institutions in Oklahoma.

An advance organizer for the progress report.

This progress report presents the results of the first two activities and a portion of the third. Specifically, this report discusses CSU's criterion for defining a language as foreign, describes the extent to which ASL meets that criterion, and provides some comments on using state legislation to recognize ASL.

Results of Research

The following sections include our research findings for the completed tasks and comments on our research into obtaining recognition through state legislation.

1. *A brief description of the criterion by which CSU defines a language as foreign.* In an interview conducted on April 25, Dr. Teresa O'Malley, the Chair of the Modern Languages Department, explained that the sole criterion for a language to be considered foreign at CSU is that it be "non-English" in such matters as structure, grammar, syntax, and vocabulary.

2. *A summary of how ASL fulfills the university-mandated requirement.* Originally, we had planned to meet with the following people to discuss whether ASL meets the criterion by which CSU defines a foreign language.
   – Suzanne Christopher, ASL instructor
   – Kelly Eastwick, president, Oklahoma Chapter of the Registry for Interpreters for the Deaf (ORID)
   – Janis Seymour, committee member, Oklahoma Association for the Deaf (OAD)

The writers honestly explain their inability to complete one aspect of their task.

   Because of scheduling conflicts, however, we were unable to meet with the ORID and OAD representatives. We did meet with Suzanne Christopher and discussed the characteristics of ASL that make it "non-English." Ms. Christopher gave us current research concerning ASL's linguistic identity. After analyzing this research, we concluded that ASL has a distinct grammar, morphology, modality, structure, syntax, markers, and idioms. It therefore meets the criterion by which CSU defines a language as foreign. (Not all sign languages meet the criterion, though ASL clearly does. For example, Signed Exact English cannot be considered a "non-English" language; it assigns a unique sign to each English word and uses standard English grammar and syntax.)

   Scholarly research also supports the idea that ASL is a foreign language. Drawing upon interpreter/linguist Rhonda Jacobs' article "ASL as a Truly Foreign Language," we created a chart that classifies languages into one of four categories according to the degree to which they differ from English (see appendix.) According to the Department of State's Foreign Service Institute, ASL is classified with Arabic, Japanese, Chinese, and Korean in

(Figure 16.5 continued)

the highest level, Group Four; the higher the level, the more difficult it is to learn the language, owing largely to its being substantially different from English.

In *Learning to See: Teaching American Sign Language as a Second Language,* Sherman Wilcox and Phyllis Wilcox analyze the educational benefits of recognizing ASL as a foreign language. We are compiling this research so that we may present it to the Curriculum Board.

The writers explain why they are revising their initial description of their task.

3. *Comments on obtaining recognition through state legislation.* Initially, we planned to speak with the following people about the Texas legislation that grants statewide recognition to ASL.
   – Dr. Jean Andrews, Communication Disorders, Lamar University, Beaumont, Texas
   – Carol Seeger, deaf ASL instructor, University of Texas at Austin
   – Lisa Bissin, deaf ASL instructor, University of Texas at Austin

   We have been unable to conduct substantive discussions with any of these people, again because of scheduling conflicts. Dr. Andrews did refer us to two other educators familiar with the legislation: Dr. Marietta L. Yeates and Dr. Sherman Wilcox. We were able to talk at length with both Dr. Yeates and Dr. Sherman, who supplied us valuable information that helped us to refine this aspect of our research project. We will contact the individuals listed above if we need additional information, though at this point we do not feel we will need to do so.

   Marietta L. Yeates, professor in the Deaf/Hard of Hearing Program at Stephen F. Austin State University, explained that her school's recognition of ASL preceded the statewide legislation. The university passed measures internally to grant recognition and then revised its curriculum to reflect the fact that ASL was officially a foreign language.

   Sherman Wilcox, coauthor of *Learning to See: Teaching American Sign Language as a Second Language* and associate professor of linguistics at the University of New Mexico, cautioned us about attempting to recognize ASL through state legislation. He reminded us that "universities cherish their independence" and that politicians attempting to mandate changes in the curriculum often meet resistance from universities. He suggested seeking recognition internally (by revising university policy) as a logical first step and perhaps a more effective way of obtaining recognition. We found his arguments convincing.

   We intend to continue researching the legislative measures in the state of Texas. However, after speaking with Dr. Wilcox, we will now focus on the positive and negative aspects of pursuing recognition internally, as compared to pursuing recognition through state legislation.

(Continued)

(Figure 16.5 continued)

Figure 1 is our updated task schedule. Completed tasks are shown in black.

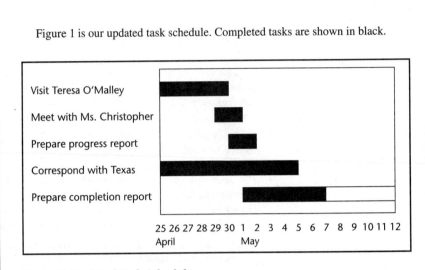

Figure 1: Updated Task Schedule

Conclusion

We were successful in completing our first two tasks. We researched the criteria by which CSU defines a language as foreign, and we have concluded that ASL fulfills these requirements. While researching Task 3, we were impressed by Sherman Wilcox's comments about academic independence being fiercely protected by university administration. Accordingly, we have shifted our research focus to the benefits and complications of internal recognition, as compared to recognition through state legislation. We anticipate that our project will be completed on schedule.

Appendix

The U.S. Foreign Service Institute, the division of the State Department that trains diplomats for foreign service, has classified languages into four categories, depending on how difficult they are for native English-speakers to learn. The level of difficulty is measured by the number of hours of study required for a native English-speaker to attain certain levels of proficiency, as determined by standardized tests.

Figure 2 shows the four categories (Series 1 through Series 4). Series 1 languages include Spanish, French, and German, all of which are very similar to English. Series 4 languages include Arabic, Japanese, Chinese, Korean, and American Sign Language, all of which are very different from English.

Also shown in Figure 2 are the levels of proficiency and the number of hours necessary for a native English-speaker to attain these levels. The levels are defined as follows:

- Level 1 speakers have "elementary proficiency." They can speak conversationally on familiar topics.

The writers add this information about the prestigious Foreign Service Institute to provide additional credibility to their claim that ASL deserves to be thought of as a foreign language.

(Figure 16.5 continued)

Figure 2: Expected Levels of Speaking Proficiency in Languages Taught at the Foreign Service Institute (1993)
Source: (Jacobs, 1996)

- Level 2 speakers have "limited working proficiency." They can confidently handle limited social interaction, including conversation on current events, autobiographical information, and nonspecialized discourse.
- Level 3 speakers have "general professional proficiency." They can converse in most formal and informal settings. They speak with structural accuracy and have a strong vocabulary.

To attain Level 1 proficiency in Spanish, French, or German, a native English-speaker would need to spend 240 hours studying. However, to attain the same level of proficiency in ASL, an English-speaker would need to spend 690 hours studying—almost three times as much.

This difference between the two groups of languages increases proportionately as the level of proficiency increases. Only 720 hours are required to achieve Level 3 speaking proficiency in Spanish, French, or German. To achieve the equivalent level of proficiency in ASL, one would have to study an estimated 3630 hours, more than five times as long. Clearly, ASL demonstrates a high level of foreignness, one that far exceeds that of Spanish, French, and German, each of which CSU recognizes as a foreign language.

References

Jacobs, R. (1996). ASL as a truly foreign language. In C. Lucas (Ed.), *Multicultural aspects of sociolinguistics in deaf communities* (pp. 183–226). Washington, DC: Gallaudet University Press.

Wilcox, S., & Wilcox, P. (1991). *Learning to see: American Sign Language as a second language.* Englewood Cliffs, NJ: Prentice Hall Regents.

Recording the Logistical Details of the Meeting

According to *Robert's Rules of Order* (Robert & Patnode, 1989, p. 115), the authoritative reference work on parliamentary procedure, you should record the following information:

- the name of the group or committee that met
- the location, date, and time of the meeting
- the type of meeting (regular or special)
- the presence of the chair and secretary, or their substitutes
- the time at which the meeting was adjourned

Another respected source, the *Gregg Reference Manual* (Sabin, 1992), suggests including the names of people who attended and of those who did not, as well as a separate list of guests.

Also record what action was taken about the minutes of the previous meeting. For example, the minutes were read (or distributed) and approved (or amended and approved). You should record any changes to the previous minutes.

Providing an Accurate Record of the Meeting

Because meetings rarely follow the agenda perfectly, you might find it challenging to provide an accurate record of the meeting. Be sure to write down the major topics discussed at the meeting, as well as any actions taken. For example, record the names of reports read or approved, motions made (and whether they were approved, defeated, or tabled), and resolutions adopted. Record the outcomes of votes. Record the names of the people who made motions, read reports, and so forth; for example, "Barry Young presented a report on the June activities of the Safety Department." If the conversation is going too fast for you to keep an accurate record, interrupt the discussion to request a clarification.

Reflecting Positively on the Participants and the Organization

Your job in recording the minutes includes separating the substance of the meeting from the emotions. Therefore, it is appropriate to write, "After considerable discussion, the motion to add a new position in the QA department was defeated by a vote of 7 to 6." It is inappropriate to write:

> The motion to add a new position in the QA Department was defeated 7 to 6 after a heated argument in which Bob Minor complained that "Alice states that she supports my department, but when it comes to action, she's all talk."

Sometimes the smartest thing a recording secretary can do is choose not to record what was said.

Figure 16.6 is an example of an effective set of minutes.

Robbins Junior High School
Weekly Planning Committee Minutes
Minutes of the February 14, 19XX Meeting

The meeting was called to order by Ms. De Grazia at 3:40 p.m. in the conference room. In attendance were Mr. Sipe, Ms. Leahy, Mr. Zaerr, and Principal Barson. Ms. Evett was absent.

The minutes of the February 7, 19XX meeting were read. The following correction was made: In paragraph 2, "800 hours" was replaced with "80 hours." The minutes were then unanimously approved.

The subject of the meeting was authorization for the antidrug presentations by motivational speaker Alan Winston.

Principal Barson reported on his discussion with Peggy Giles of the School District. She offered positive comments about Winston's presentations at other schools in the district last year. Principal Barson has also been in contact with the three other principals who invited Mr. Winston last year; they were all enthusiastic. He moved that the committee authorize Winston's visit, to be scheduled in late May. The motion was seconded.

Mr. Sipe asked whether the individual school or the district was to pay for the expenses (approximately $2,800) for the visit. Principal Barson replied that the school and the district would split the costs evenly, as had been done last year. The school has more than $4,000 surplus in operating expenses.

Mr. Zaerr expressed serious concerns about the effect of the visit on the teaching schedule. The visit would effectively disrupt one whole day for all three grades in the school. Principal Barson acknowledged this but replied that in his weekly meeting with department chairs earlier in the week, they gave their approval to the idea. Since student participation would be voluntary, teachers were to offer review sessions to those students who elected not to attend Winston's presentation.

There being no more discussion, Ms. De Grazia called for a vote on the motion. The motion passed 4-0, with one abstention.

Principal Barson asked the committee if they would assist him in planning and publicizing Winston's visit. The committee agreed.

Ms. De Grazia asked if there was any new business. There was none. She adjourned the meeting at 4:20 p.m.

Laura Weddle
Recording Secretary

Notice that the minutes say that Mr. Zaerr "expressed serious concerns . . ." not "Mr. Zaerr was very angry about . . ."

FIGURE 16.6

Meeting Minutes

CASE

Amending a Proposal

You are the chief technician for United Paper Company, an Oregon manufacturer of paper and paper products. The company has decided to replace two of its older rolling machines with newer models, each of which will cost more than $300,000. Last month, you formally proposed to your supervisor that you be permitted to visit two of the leading manufacturers of this equipment to study their products; your proposal was approved. You have carried out more research and are preparing your first progress report on this project. You have described the two competing machines and explained how each would fit in your company's manufacturing processes, and you have completed arrangements for visiting the two companies. With a budget of $6,000, next week you will visit American Equipment in Hawthorne, CA, and Consolidated Industrial Equipment, in Newark, NJ. Your supervisor had balked a little at the cost of the two trips, but he agreed that they were necessary, given the importance of the new machines to the company.

Studying your papers in preparing your progress report, your heart sinks. You discover that you made a serious omission in your proposal: there is a third manufacturer, Southern Printing Equipment, in Atlanta, Georgia. You look back through your materials and discover that Southern's equipment is fully competitive with that of the other two manufacturers. You remember that when you wrote the proposal you were having some problems at home. Your spouse had been laid off, and with your oldest child beginning college next year, you had spent some sleepless nights.

You realize now that adding Southern to your itinerary would delay the completion date, and you would have to request an increased budget. What would be the costs and benefits to you and to the company of not explaining the oversight to your supervisor? Of explaining it? What should you do? If you think it is best to explain the oversight, write a memo to your supervisor. If you think it is best not to mention it, write a memo to your instructor, explaining your thinking.

WRITER'S CHECKLIST

1. Have you chosen an appropriate format for the informal report?

2. Does the directive
   - ☐ clearly and politely explain your message?
   - ☐ explain your reasoning, if appropriate?

3. Does the trip report
   - ☐ clearly explain the purpose of the trip?
   - ☐ highlight the important information?
   - ☐ append appropriate supporting materials?

4. Does the field or lab report
   - ☐ clearly explain the important information?
   - ☐ use, if appropriate, a problem-methods-results-conclusion-recommendations organization?

5. Does the progress or status report
   - ☐ clearly announce that it is a progress report or status report?
   - ☐ use an appropriate organization?
   - ☐ clearly and honestly report on the subject and forecast the problems and possibilities of the future work?
   - ☐ append supporting materials that substantiate the discussion?

6. Do the minutes
   - ☐ provide the necessary housekeeping details about the meeting?
   - ☐ explain the events of the meeting accurately?
   - ☐ maintain a diplomatic tone?

EXERCISES

1. As the manager of Lewis Auto Parts Store, you have noticed that some of your salespeople smoke in the showroom. You have received several complaints from customers. Write a memo defining a new policy: salespeople may smoke in the employees' lounge but not in the showroom.

2. There are 20 secretaries in the six departments at your office. Although they are free to take their lunch hours whenever they wish, sometimes several departments have no secretarial coverage between 1:00 and 1:30 P.M. Write a memo to the secretaries, explaining why this lack of coverage is undesirable and asking for their cooperation in staggering their lunch hours.

3. If you have attended a lecture or presentation in your area of academic concentration, write a trip report to your instructor assessing its quality.

4. Write up a recent lab or field project for the instructor of the course.

5. The reports (a., b., c.) below and on pages 516–517 could be improved in tone, substance, and structure. Revise them to increase their effectiveness, adding any reasonable details.

6. Write a progress report about the project you are doing for the major assignment you proposed in Chapter 15.

a.

KLINE MEDICAL PRODUCTS

Date:	1 September 19XX
To:	Mike Framson
From:	Fran Sturdiven
Subject:	Device Master Records

The safety and efficiency of a medical device depends on the adequacy of its design and the entire manufacturing process. To ensure that safety and effectiveness are manufactured into a device, all design and manufacturing requirements must be properly defined and documented. This documentation package is called by the FDA a "Device Master Record."

The FDA's specific definition of a "Device Master Record" has already been distributed.

Paragraph 3.2 of the definition requires that a company define the "compilation of records" that makes up a "Device Master Record." But we have no such index or reference for our records.

Paragraph 6.4 says that any changes in the DMR must be authorized in writing by the signature of a designated individual. We have no such procedure.

These problems are to be solved by 15 September 19XX.

b.

Diversified Chemicals, Inc.
Memo

Date: August 27, 19XX
To: R. Martins
From: J. Speletz
Subject: Charles Research Conference on Corrosion

The subject of the conference was high-temperature dry corrosion. Some of the topics discussed were

1–thin film formation and growth on metal surfaces. The lectures focused on the study of oxidation and corrosion by spectroscopy.
2–the use of microscopy to study the microstructure of thick film formation on metals and alloys. The speakers were from the University of Colorado and MIT.
3–one of the most interesting topics was hot corrosion and erosion. The speakers were from Penn State and Westinghouse.
4–future research directions for high-temperature dry corrosion were discussed from five viewpoints.

    1–university research
    2–government research
    3–industry research
    4–European industry research
    5–European government research

5–corrosion of ceramics, especially the oxidation of Si_3N_4. One paper dealt with the formation of Si ALON, which could be an inexpensive substitute for Si_3N_4. This topic should be pursued.

c.

FREEMAN, INC.   INTEROFFICE

To: C. F. Ortiz
From: R. C. Nedden
Subject: Testing of Continuous Solder Strip Alternative for Large-Scale Integrated Terminals
c: J. A. Jones
    M. H. Miller

(c. Continued)

We ordered samples of continuous solder strips in three thicknesses for our testing: 1.5 mil, 2.5 mil, and 4.0 mil. Then we manufactured each thickness into terminals to test for pull strength.

The 1.5 mil material had an average pull strength of 1.62 pounds, which is above our goal of 1.5 pounds. But 30 percent of these terminals did not meet the goal. The 2.5 mil material had an average pull strength of 2.4 pounds, with a minimum force of 1.65 pounds. The 4.0 mil material had an average pull strength of 2.6 pounds per terminal, with a minimum of 1.9. Even though there was 60 percent more solder available than with the 2.5 mil material, the average pull strength increased by only 8 percent.

We concluded from this that the limit to the pull strength of the terminal is dependent on the geometry of the terminal, not on the amount of solder.

Collaborative Exercise

7. You and another student are cochairs of the Associated Student Organization (ASO) at your college. Student organizations that receive university funding, including the ASO, are required to submit formal minutes of each of their meetings. You have been approached by the dean of students, who oversees student organizations. The dean is concerned that ASO minutes are unprofessional and could reflect poorly on the organization. The dean has asked you to revise the most recent set of minutes (the one you see printed here) and attach a memo describing the changes you have made and explaining your decisions. Submit to your instructor the revised minutes, along with the memo.

Minutes of the Associated Student Organization
May 13, 19XX

The meeting began at 3:10, even though it was scheduled to begin at 3:00. We had to wait for enough people to come to constitute a quorum.

The first issue we discussed was the proposed raise in fees for the General Parking Permit. This year it is $15, which it has been for the last six years. The university's plan to raise it to $30 is unreasonable, especially given the 8% increase in tuition and the 12% increase in dormitory fees. A resolution was passed unanimously to protest this increase.

(Continued)

(Continued)

The discussion turned next to the lineup for next semester's events. Bob commented that there were too many country acts, and not enough alternative acts. He said "a little country goes a long way." Marty got insulted at this and replied, "What do you want, some idiot in leather who spits up fake blood?" He added that country acts are popular with the whole community, whereas a lot of the alternative acts draw very poorly. With the current policy of up-front deposits required by Elite Productions, we have to be sure we don't book any weak acts, which could deplete our whole budget for the year. Bob said he didn't mean to be insulting about country acts, and the two agreed to keep talking about the issue. The meeting adjourned at 4:15.

Web Exercise

8. You are one of three members of the Administrative Council of your college's student association. Recently, the three of you have concluded that your weekly meetings have become chaotic, largely because you do not use rules for parliamentary procedure and because controversial issues have arisen recently that have attracted numerous students (the meetings are open to all students). You have concluded that it is time to consider adopting some parliamentary procedures to make the meetings more orderly and more effective. Using a search engine, look on the Web for models of parliamentary procedure. Is there one that you can adopt? Are there elements of several that can be combined to create an effective model? Find or create a brief set of procedures, being sure to cite your sources. Explain in a memo to your instructor the advantages and disadvantages of the model you propose.

REFERENCES

Robert, H. M., & Patnode, D. (1989). *Robert's rules of order.* New York: Berkley.

Sabin, W. (1992). *Gregg reference manual*, 7th ed. Lake Forest, IL: Glencoe.

CHAPTER
17

Writing Formal Reports

Understanding the Types of Formal Reports 520
  Informational Reports *520*
  Analytical Reports *521*
  Recommendation Reports *522*

A Problem-Solving Model for Preparing Formal Reports 524
  Analyze Your Audience *524*
  Analyze Your Purpose *525*
  Identify the Questions That Need to Be Answered *526*
  Carry Out Appropriate Research *527*
  Draw Conclusions from the Results *527*
  Formulate Recommendations Based on the Conclusions *527*

A Problem-Solving Model for Preparing Feasibility Reports 528
  Identify the Problem or Opportunity *529*
  Establish Criteria for Responding to the Problem or Opportunity *530*
  Determine the Options *530*
  Study Each Option according to the Criteria *531*
  Draw Conclusions about Each Option *532*
  Formulate Recommendations Based on the Conclusions *533*

Organizing the Body of the Formal Report 533
  Introduction *534*
  Methods *535*
  Results *536*
  Conclusions *537*
  Recommendations *537*

Sample Formal Report 542

Case: Writing a Recommendation Report *553*

Writer's Checklist 554

Exercises 554
  Collaborative Exercise *554*
  Web Exercise *555*

A formal report can be the final link in a chain of documents that began with a proposal (Chapter 15) and one or more progress reports (Chapter 16). In these cases, the report is often called a *final report, project report,* or *completion report.* An example of such a report is presented later in this chapter—the report about American Sign Language.

A formal report can also be a freestanding document, one that was not preceded by a proposal or progress reports. For instance, your supervisor could ask you to find out employee attitudes toward comp pay: letting employees who work overtime choose to be compensated with time off rather than with overtime pay. In this case, you do the research and write the report.

This chapter begins with a discussion of the types of formal reports—informational, analytical, and recommendation—and a problem-solving model for writing them. Next, the chapter discusses an important type of recommendation report: the feasibility report. Then the chapter discusses the typical components of formal reports. Finally, the chapter presents a sample report written by a group of students.

Understanding the Types of Formal Reports

One of the challenges in talking about reports is that there is no standard terminology. There are *lab reports* and *site studies, progress reports, fact-finding reports, environmental-impact statements, analytical reports, completion reports, annual reports, informational reports,* and *interim reports.* The list goes on and on. Some of these terms refer to the topic of the report (*environmental-impact statements, lab reports*), some to the phase of the investigation (*progress reports, annual reports*), and some to the function or purpose of the report (*informational reports, analytical reports*). Some of these terms have special meanings in particular fields and in particular organizations. One piece of advice: when you are asked to write a report, talk to more-experienced people at your organization to determine what is meant by the word *report.*

This section presents three kinds of reports, classified according to their main purpose, as shown in Figure 17.1.

Informational Reports An informational report presents facts so that readers can understand a situation. These facts are often called *results.* For instance, the circulation department at a library might present an informational report to the library planning board,

FIGURE 17.1

Formal Reports, Classified According to Purpose

Informational Report: | Presents results |

Analytical Report: | Presents results + draws conclusions |

Recommendation Report: | Presents results + draws conclusions + makes recommendations |

providing statistics about the number of patrons who visit the library, the kinds of items they borrow, the number and kinds of items the library purchases, the number and kinds of items stolen, and so forth.

This report might be a routine *status report,* covering a particular period, such as the last quarter (see Chapter 16). Or it might be a nonroutine report requested by the planning board because it believes the library is not being run well or because the government body that funds the library requested it.

Following are some examples of questions addressed in informational reports:

- *What is the status of Project X?* What is the current status of the work on the project to reinforce the levees on the river? How many open-heart surgeries did the hospital perform last year, and what were the outcomes?

- *How do we do Function X?* What procedures do we currently use to assess the performance of our clerical personnel? What are our quality-control procedures for the laptops we manufacture?

- *What are the most popular ways of doing Function X?* What are our major competitors' policies on returns? What is the ratio of surgical to laser operations for treating gallstones in the United States? What are the three top-selling brands and models of ink-jet printers?

- *What do our people think of Situation X?* How satisfied are our employees with our medical benefits? What do our production personnel think we could do to improve the quality of our products?

An informational report usually involves the primary- and secondary-research techniques discussed in Chapter 7: using the library and the Internet, conducting interviews, and distributing questionnaires.

Analytical Reports   Like informational reports, analytical reports provide information, but they go one step further: they provide an analysis of it and present *conclusions.* A conclusion is an interpretation of the results, an answer to the question, "What do these results mean?" For instance, in writing the report to the planning board of the library, the staff in the circulation department might present all the statistics, highlighting the fact that thefts of compact discs have increased substantially. With further analysis, the writers can conclude that theft is high for one type of music disc but not for other types. An analytical report, therefore, tries to describe *why* something happens or *how* it happens or what it *means.*

As with an informational report, an analytical report can be routine (for example, a status report) or special.

Here are some examples of the questions addressed in analytical reports:

- *What is the best way to do Function X?* What is the most effective method for tracking manuscripts being edited by different people? Which therapy for people with spinal-cord injuries has the best outcome?

- *What causes Situation X?* Why is there a high turnover in the Information Systems Department? Why is our competitor's new product doing so much better than ours?
- *What are the results of Situation X?* If we create a Web site for our company, how much increase in business can we expect? If our accounting functions were performed by an outside firm, would we save money?
- *Could we do Function X?* Do we currently have the infrastructure to compete with Company A in bidding on these government contracts? Could we afford to offer tuition reimbursement for *all* college classes our employees wish to take?

Analytical reports usually contain an informational element as well as an analytical one. For instance, if you work for a software manufacturer that provides customer support, you might commission an analytical report to learn why your customer-support costs are high. In the informational section—the results— you describe exactly your customer-support costs, focusing on which products, and which aspects of the products, are causing your customers the most problems. For example, most of the problems occur when your customers try to perform a specific type of calculation using your spreadsheet program. In the analytical section—the conclusion—you use these data to explain why this calculation is difficult to perform.

Analytical reports involve analyzing the data to determine cause-and-effect relationships (see Chapter 8). Until you understand the problem, you can't fix it. An analytical report helps you—and your readers—understand the problem.

Recommendation Reports

A recommendation report goes one step further than presenting information and analyzing it; a recommendation report advocates a certain course of action. (Sometimes, however, that course of action is to do nothing.) What is the difference between drawing conclusions and presenting recommendations? If your conclusion is that Product A meets our needs better than Product B or Product C, the recommendation would seem obvious: purchase Product A. Maybe, but maybe not. A report might recommend that, even though a high number of compact discs is being stolen, the library should take no action: we can't afford any of the available antitheft devices, none of the devices does a good enough job, all the devices create unacceptable problems, or next year there will be a better device on the market. A report can even recommend that, even though Product A is better for us than B or C, we should buy Product B, because B does 90 percent of what A does but costs only 50 percent as much as A.

Following are some examples of questions that a recommendation report addresses:

- *What should we do about Problem X?* What should we do about the fact that the number of calls to our Technical Support lines is very high?

How should we celebrate our company's one hundredth anniversary next year?

- *Should we do Function X?* Although we currently have the infrastructure to compete with Company A in bidding on these government contracts, could we increase our business more effectively in some other way? Although we cannot afford to offer tuition reimbursement for *all* college classes our employees wish to take, can we reimburse them for classes highly relevant to their work?

- *Should we use Technology A or Technology B to do Function X?* Should we buy copiers or should we lease them? Should we buy several high-output copiers, or a larger number of low-output copiers?

- *We currently use Technology A to do Function X. Should we be using Technology B?* Presently, we have a Technical Publications Department; should we subcontract that department's work instead? Currently, we sort our barcoded mail by hand; should we buy an automatic sorter instead?

This brief look at informational, analytical, and recommendation reports might suggest that there is a clear demarcation among them and that the process of creating them is neat and orderly. The working world, however, is messy. Often, a routine informational report reveals—or confirms—a problem or an opportunity in the organization. Then a report is commissioned to analyze the situation or present recommendations. As problems and opportunities arise and events occur, people make the best decisions they can with the information they have at the moment. Sometimes they think that a recommendation report outlines a wise course of action, only to realize later that it won't work; they then have to start over again by gathering more information and analyzing it again. Even if the recommendations are implemented and the situation appears to improve, in time a new problem arises and the whole process begins again. As with the writing process itself, writing reports is a process that is more cyclical than linear.

Two notes about how organizational culture affects the writing of reports:

- *In some organizations, only certain people are authorized to draw conclusions and make recommendations.* For instance, bench chemists at a pharmaceuticals manufacturer might be entitled to present results, but not to go on to the next two steps: drawing conclusions and making recommendations. Those steps are left to more senior people.

- *Some organizations like to segment the report-writing process so that different collaborative groups work on results, conclusions, and recommendations.* In other words, one group writes the informational report, then hands over the project to a different group, which writes the analytical report. Then a third group writes the recommendation report. Each group in effect checks the work of the group or groups that preceded it. The rationale for this method is that it can help reduce the chances that inaccurate or incomplete results are used to derive conclusions, and that invalid conclusions are used to derive unwise recommendations.

A Problem-Solving Model for Preparing Formal Reports

Although the process of researching and writing every report is unique, many writers find that a problem-solving approach helps them to create an effective report. A problem-solving approach has six basic steps. Figure 17.2 shows a basic problem-solving model, as well as the relationship of the problem-solving model to various types of reports.

Analyze Your Audience The writing process used for most documents—prewriting, drafting, and revising—is as effective for reports as it is for other documents. As usual, analyzing your audience is critical because it helps you determine how much and what kinds of information you need to provide. In addition, analyzing your audience helps you determine where to put the information. For instance, if your readers are knowledgeable and do not need much detail to understand the project, you might decide to put much of the information in appendices rather than in the body of the document. This way, the information is available for reference, but you are not forcing anyone to read it.

Before you start to write, think about the following five questions related to audience:

- *How well do your readers know your field or the subject of your report?* The better they know the subject, the less explanation you need to provide in the body of the report. If some of your readers are managers who do not know the subject well, consider adding detailed explanations as

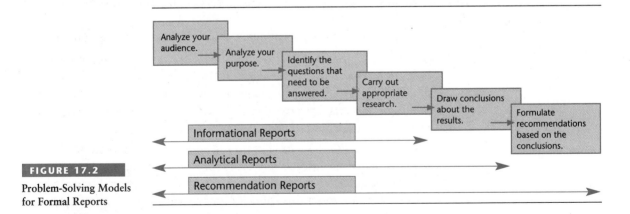

FIGURE 17.2

Problem-Solving Models for Formal Reports

appendices. Less-knowledgeable readers can also benefit from glossaries (see Chapter 10).

- *Why are the readers reading the report?* If they merely need to understand the subject, you can provide less information than if they will use your report as the basis for some further action. For instance, if some of your readers will duplicate your methods, you need to provide complete details about those methods. Again, it might be appropriate to put the complete description of the methods in an appendix.

- *Are your readers negative, neutral, or positive about the project?* The attitude of the reader might affect the organizational pattern, the amount of detail, and the vocabulary you choose, as described in Chapter 8.

- *What is your standing within the organization?* Writers with strong technical credentials and experience do not need to justify their assertions, conclusions, or recommendations as fully as writers without an extensive track record.

- *How routine is the project?* Some fields have accepted methods for carrying out certain projects. For example, if in assessing hurricane damage to a physical structure you always perform a visual inspection first, there is no need to explain how or why you performed the visual inspection. But if the project is unusual or unique, your readers will want to understand why you chose the approach you used.

Analyze Your Purpose

Key to the success of any technical communication is a sensitive analysis of your purpose. When you plan a formal report, however, it is especially important to analyze your purpose well, if only because a formal report is likely to be a long and important document. You want to make sure you get the purpose right.

In the broadest sense, you have one of three basic purposes:

- to present information
- to present information and analyze it
- to present and analyze information and make recommendations

However, you also have a purpose or set of purposes that relates more directly to your subject. Think about why you are doing the research, why your readers have requested it, and what they are going to do with it once you present it to them. Is the report merely a routine kind of communication—a status report, for instance—or do your readers believe there is a specific problem or opportunity that they need to examine in great detail? If, for instance, you work for a library and know that your research will help your readers understand the broad problem of theft of library materials, your purpose might be to help them understand how antitheft systems work and what the antitheft devices could be expected to do, or it might be to help them determine which of the devices (if any) would best solve the library's problems.

Before you proceed too far in your planning, make sure your principal reader agrees with your understanding of your purpose. Writing the memo described in Chapter 6—in which you explain to your supervisor the basic audience, purpose, and scope of the report—can prevent you from going off in the wrong direction.

Identify the Questions That Need to Be Answered

After you have analyzed your audience and purpose, you can start thinking in detail about the subject of the investigation. Obviously, you have already thought about your subject a great deal. But now it is time to think specifically about the critical questions you will have to answer. Although sometimes it is your responsibility to determine these questions, often they are established before you begin.

Study your purpose statement carefully. If you are writing an informational report on library antitheft devices, here are some questions you might consider:

- What are antitheft devices?
- What technologies do they use?
- How effective are they in reducing thefts?
- What kinds of library materials do they protect?
- What have been the experiences of libraries like ours that have used them?
- Are there reliable benchmark tests of the different devices?
- How easy are they to install and maintain?
- How often do they break down?
- Do they pose any health risks to library workers or patrons?
- Is there some sort of financial support—grants, for instance—to help finance the purchase of the antitheft devices?

State these questions as precisely as you can.

VAGUE:	How do library antitheft devices work?
SPECIFIC:	What different technologies are used in currently available library antitheft devices?

The problem with the vague question above is that it could refer to the different technologies, or it could refer to the steps that library personnel have to follow to operate these devices.

VAGUE:	What are some of the disadvantages of library antitheft devices?
SPECIFIC:	Do any of the technologies pose any health risks to our workers or patrons?

This specific question doesn't address all the concerns suggested by "some of the disadvantages," of course, but it is a start. The writer could easily pose a half dozen more questions.

Carry Out Appropriate Research

The questions you need to answer will determine the appropriate kinds of research to carry out. Often, you will consult company records, interview people, distribute questionnaires, perform experiments and observations, and consult books, journals, and Internet sources. As discussed in Chapter 7, you should try to do as much research as time and other resources permit, and you must analyze the information to be sure it is valid and timely.

Keep in mind, as well, that the needs and preferences of your audience will help you determine the kinds of research to perform. For example, if you know your readers are very concerned about the attitudes of your organization's workers toward changes in policies, you will naturally want to conduct interviews and distribute questionnaires.

Draw Conclusions from the Results

As discussed in the section on causal reasoning in Chapter 8, there is no foolproof way to draw valid conclusions from data. You have to think about the data carefully, looking for a body of evidence that suggests a causal relationship between two factors. Some kinds of evidence are more valuable than others; benchmark tests from disinterested groups are obviously more persuasive than claims from manufacturers about their own products. And look for a number of different kinds of evidence that all point in the same direction; one or two pieces of evidence is generally insufficient.

As you draw your conclusions, pay particular attention to the problems in logic discussed in Chapter 8: inadequate sampling, *post-hoc* reasoning, and oversimplifying. Just as washing your car does not cause rain, new cost-cutting measures implemented in your company are not necessarily responsible for decreased costs. Those decreased costs might have nothing to do with the new measures; in fact, costs might have decreased *despite* the cost-cutting measures.

Formulate Recommendations Based on the Conclusions

In most cases, if you have carefully formulated the questions you need to answer, carried out appropriate research, and drawn valid conclusions, the recommendation will flow directly and inevitably from those conclusions. For instance, if you have identified the reasons that customers are having problems with one of your software products and have concluded that the best course of action would be to send all customers new disks, your recommendation will probably be to go ahead with this plan as soon as possible.

But this is not *necessarily* the recommendation you will present. For example, you might know that your company does not have the funds to go ahead with this plan right away. Or maybe your company is planning a new upgrade next quarter, and you feel your plan would be more trouble than it is worth to your customers.

Sometimes managers decide not to implement recommendations, even though the recommendations seem perfectly valid and reasonable. This fact can surprise and dismay the report writers, who have spent many hours on the project. In some cases, managers simply change their minds about priorities, making the recommendations irrelevant. In other cases, the situation changes, making the recommendations irrelevant. For example, your company is being acquired by another company, and management wants to get authorization to carry out the recommendations from its new supervisors next month. You might be unaware of some of the factors that management is considering.

This gap between the technical information about the subject and management's priorities can lead to an ethical dilemma for writers. As a writer, you might know that the best course of action is to implement Plan A, but that management favors Plan B and has no intention of implementing Plan A. Should you just tell them what they want to hear by recommending Plan B, adjusting your results and conclusions to lead to this recommendation? In most instances, the answer is simple: No. Recommending Plan B would be unethical, because it would entail lying or misleading. On the other hand, an appropriate course of action might be to recommend each plan equally, provided that the following conditions are present:

- There is nothing unsafe or unethical about Plan B.
- Plan B is not clearly inferior to Plan A.
- You do not have to lie or mislead to recommend Plan B.

This course of action would give you the opportunity to make the case for Plan A, while still making it easy for management to choose Plan B if they remain unconvinced of the superiority of Plan A.

A Problem-Solving Model for Preparing Feasibility Reports

One kind of recommendation report is written so often that it deserves special discussion. A *feasibility report* documents a study that evaluates at least two alternative courses of action. For example, should we have our Web site developed and maintained by our employees or by an outside company? Should we expand our product line to include a new item, or should we make changes in an existing product?

A feasibility report is an argument that answers three kinds of questions:

- *Questions of possibility.* We would like to build a new rail line to link our warehouse and our retail outlet, but if we cannot raise the money the project is not possible at this time. Even if we have the money, do we have the necessary permission from government authorities? If we do, are the soil conditions adequate for the rail link?
- *Questions of economic wisdom.* Even if we can raise the money to build the rail link, is it wise to do so? If we use all our credit on this pro-

ject, what other projects will we have to postpone or cancel? Is there a less expensive—or a less financially risky—way to achieve the same goals?

- *Questions of perception.* If your company's workers have recently accepted a temporary wage freeze, they might view the rail link as an unnecessary expenditure. The truckers' union might see it as a threat to truckers' job security. Some members of the general public also might be interested parties because any sort of large-scale construction might affect the environment. Even though your plan might be perfectly acceptable according to its environmental-impact statement—the study required by the government—some citizens might disagree with the statement or might oppose the project on aesthetic grounds. Whether or not you agree with the objections, going ahead with the project could create adverse publicity.

Figure 17.3 shows the eight steps in a basic problem-solving model for feasibility reports.

The following discussion explains the six steps that are particular to feasibility reports. Of course, you must begin by analyzing your audience and purpose, as you would for any technical document.

Identify the Problem or Opportunity

What is not working, or not working as well as it might? What situation presents an opportunity for our organization to decrease its costs or improve the quality of its product or service? Without a clear statement of your problem or opportunity—that is, without answering these kinds of questions—you cannot plan your research.

For example, you know that a number of employees in your organization smoke, and that these employees are absent and ill more often than are

FIGURE 17.3

A Problem-Solving
Model for Preparing
Feasibility Reports

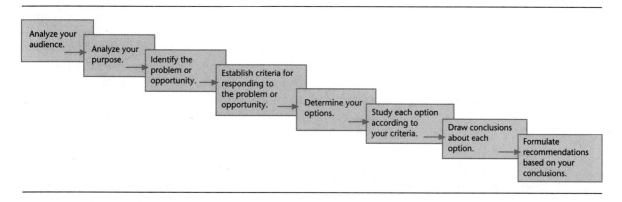

nonsmoking employees. Your supervisor has asked you to investigate whether your company should offer employees a free smoking-cessation program. In talking with your supervisor, you learn that the only way the company can offer the program is if the company's insurance carrier will pay for the program or reduce premiums for those people who successfully complete the program. This fact tells you that the first thing you need to do is talk with the insurance agent; if the agent says the carrier will pay for the program or reduce the premiums, then you can proceed with your investigation. If the agent says no, the problem might shift; now the problem is to determine whether another insurance company offers your organization better coverage.

Establish Criteria for Responding to the Problem or Opportunity

As explained in Chapter 8 in the discussion about comparison and contrast, *criteria* are standards against which you measure your options. Criteria can take two forms: *minimum specifications* and *evaluative criteria*. For example, if you need to find an apartment to live in when you return to school in the fall, your minimum specifications might include that the apartment cost less than $500 per month and be within one mile of campus. On the other hand, your evaluative criteria might include that the kitchen be separated from the living room by a wall and that laundry facilities be available. In other words, minimum specifications define what you *need*, while evaluative criteria define what you *want*. Note that these examples of evaluative criteria can be refined: by *laundry facilities,* do you mean a washer and dryer in the apartment or a laundromat located on the premises? Which would be preferable?

When you solve problems in the working world, you use the same process of establishing what you *need* and what you *want*. For example, in selecting photocopiers for your department, a minimum specification might be that each copy cost less than two cents to produce and that the photocopier be able to handle oversized documents. Evaluative criteria might include that the photocopier do double-sided copying and stapling. You could get by without these features, but they are desirable; therefore, you might rate a photocopier offering these features more highly than one without the features, all other things being equal. That is the function of evaluative criteria: to enable you to make distinctions among a variety of similar objects, objectives, actions, or effects.

Until you can establish your criteria, you have no way of knowing what your options are. Sometimes you inherit your criteria; your supervisor tells you how much money you can spend, for instance, and that figure becomes part of your minimum specifications. Sometimes you derive your criteria from your research; for instance, you study your photocopying needs and determine the standards to evaluate the different copiers.

Determine the Options

After you establish your criteria, you can determine your options. *Options* are potential solutions. For example, your options in choosing an apartment are the apartments that have vacancies and that meet your minimum specifications and, if possible, many of your evaluative criteria. Likewise, your options for the photocopying project are the different photocopiers that meet your minimum specifications and evaluative criteria.

Determining your options is a critically important phase in the problem-solving process because if you fail to identify all the appropriate options, the rest of the analysis will be flawed. For instance, if you do not consider the photocopiers made by Xerox, the results, conclusions, and recommendations you present later in your report might well be invalid, because Xerox might manufacture the best copier for your needs.

For this reason, you have to be especially careful that you carry out your research thoughtfully and thoroughly. You must use the appropriate methods of primary and secondary research (see Chapter 7) to discover what companies manufacture the kind of copier you are looking for.

Study Each Option according to the Criteria   Once you have established your criteria and determined your options, you study each option according to the criteria. The process of studying might include both primary and secondary research. In the case of the photocopier project, secondary research would include studying articles about photocopiers in technical journals, as well as specification sheets from the different manufacturers. Primary research might include observing product demonstrations, as well as interviewing representatives from the different manufacturers and people who have purchased different brands of machines.

Studying the options usually involves collecting objective and subjective information. An objective piece of information would be the number of copies per minute that a photocopier produces. A subjective piece of information would be an assessment of ease of use or quality.

To try to make the analysis of the options as objective as possible, investigators sometimes create a *matrix,* a method for systematically entering an evaluation for each option according to each criterion. A matrix is really just a table (or a spreadsheet), as shown in Figure 17.4 on page 532.

To use a matrix, you assign a value (0–10 is a popular range) for each criterion for each option. Then you add up the values for each option and compare the totals. In Figure 17.4, option 3 scores the highest, with 35 points; option 2 scores the lowest, with 26 points.

Does using a matrix ensure an objective analysis? Not at all. Subjectivity can creep in at three stages:

- *Determining which criteria to examine.* Option 2—Xerox Model 4500—did very poorly on criterion 4; if criterion 4 were removed from the analysis, or if many other criteria were added, option 2 might score much higher.
- *Deciding the range of values for each criterion.* If one of your criteria is whether the copier can staple, how do you decide what value to assign to a machine that does stapling? To a machine that does not? If you give 10

Option	Criterion 1: Pages per Minute	Criterion 2: Ability to Duplex	Criterion 3: Extra Paper Bins	Criterion 4: Color Printing	Total Points
Option 1: Ricoh Model 311	9	8	7	9	**33**
Option 2: Xerox Model 4500	8	9	7	2	**26**
Option 3: Savin Model 12X	10	8	8	9	**35**
Option 4: Sharp Model C112	7	8	8	9	**32**

FIGURE 17.4

A Matrix

points for that criterion to a machine that does stapling and 0 points to a machine that doesn't, you have probably eliminated the nonstapling machine from contention. However, if the ability to staple is not very important, it might be more reasonable to assign a value of 8 points to nonstapling machines.

- *Assigning values to criteria.* If, for example, one of your criteria is ease of operation, you might give one machine a 9, whereas someone else might give that same machine a 3 on that same criterion. Other criteria are equally difficult to assess objectively. For example, what value do you give to the cost criterion if one machine costs $12,500 per year to operate and another costs $14,300?

As you can see, evaluating options according to your criteria is ultimately a subjective process. Still, for you as the writer, the principal advantage of a matrix is that it makes it easier to do a methodical analysis. For your readers, a matrix makes it easier to see what you did. You are helping your readers by clarifying the presentation of your methods and results.

Draw Conclusions about Each Option Whether you use a matrix or some less formal means of recording your evaluations, your next step is to draw conclusions—that is, write evaluative statements—about the options you have studied.

For the study of photocopiers, the conclusion might be that a particular copier made by Ricoh is the best copier: it meets all of your minimum specifications and the greatest number of your evaluative criteria. Or the Ricoh scores highest on your matrix. Sometimes your readers want you to rank all of the options: option 3 is the best, option 1 is second best, and so forth. Sometimes your readers want you to classify all the options in two categories: acceptable and un-acceptable. Sometimes your readers want you to present a kind of compound conclusion: the copier that offers the most technical capabilities is the Ricoh, but the copier that offers the best value for us is the Savin.

Formulate Recommendations Based on the Conclusions
Everything said in the discussion of recommendations earlier in this chapter applies to feasibility reports. But one more note about recommendations in feasibility reports: don't hesitate to recommend that your organization take no action. Often, research projects yield mixed or bad news; that is, none of the options would be an unqualified success, or none would work at all. Don't feel that a negative recommendation reflects negatively on you. If the problem being studied were easy to solve, it probably would have been solved before you came along. Give the best advice you can, even if that advice is to do nothing. The last thing you want to do is recommend a course of action that will not live up to the organization's expectations.

Organizing the Body of the Formal Report

Every report should have a unique structure that accommodates the unique characteristics of the audience, purpose, and subject. Still, over the decades a basic structure for reports has developed that can serve as a good starting point to think about how to structure your report. The typical formal report contains many of the following standard elements:

- front matter
  - title page
  - abstract
  - table of contents
  - list of illustrations
  - executive summary
- body
- back matter
  - glossary
  - list of symbols
  - references
  - appendices

This chapter presents guidelines for writing the *body* of a completion report; the other elements common to many formal reports are discussed in Chapter 10.

In the *introduction,* you explain the context of the report: the problem you were investigating, as well as your purpose; you also forecast the organization and content of the report. In the *methods* section, you describe what you did in carrying out the project. In the *results* section, you present the information you found or created. Both the methods and results sections are simple statements of facts; you do not interpret the information, but merely document what you did and what resulted. In the *conclusions* section, you interpret the results by explaining what they mean. In the *recommendations* section, you suggest to your readers what they should do next. This pattern of organization, which is used throughout engineering and the sciences, enables different readers to find the information they need quickly and easily.

More often than most other kinds of technical communication, reports are read by multiple audiences. Some of your readers will be managers who are not technically competent in your field and who need only an overview of the project; others will be technical personnel who are competent in your field and who need detailed information. Other readers might include technical personnel in related fields, attorneys, or representatives of regulatory agencies.

To accommodate these multiple audiences, a formal report today generally consists of modular elements: components that remain independent even as they work together to form the whole report. For instance, a typical formal report contains an executive summary, which precedes the body. Some readers will skip the executive summary; others will read nothing but the executive summary.

Some writers like to draft these elements in the order in which they will be presented. These writers like to compose the introduction first because they want to be sure that they have a clear sense of direction before they draft the discussion and findings. Other writers prefer to put off the introduction until they have completed the other elements of the body. Their reasoning is that in writing the discussion and the findings, they will inevitably have to make some substantive changes; therefore, they would have to revise the introduction if they wrote it first. In either case, analyzing your audience and purpose, brainstorming, and careful outlining are necessary before you begin to draft.

Introduction As discussed in Chapter 8, the introduction enables readers to understand the technical discussion that follows. Start drafting an introduction by analyzing the audience and then considering the standard questions to which all introductions respond:

- *What is the subject of the report?* If the report is a follow-up document to a proposal and progress report, you can probably copy this information from one or the other of those documents, modifying it as necessary.
- *What is the purpose of the report?* The purpose of the report is not the purpose of the project. The purpose of the report is to present information to enable readers to understand a subject, to affect readers' attitudes toward some subject, or to enable readers to carry out a task. In stating the purpose of the report, be as specific as you can. See Chapter 6 for a discussion of purpose.

- *What is the background of the report?* You might be able to copy this information from a previous document. Remember, when describing the background, you are addressing readers who might not have read your previous documents and who therefore might be ignorant of the project; do not assume that they have been following along. Even those readers who have read the previous documents might need a refresher.

- *What are your sources of information?* Briefly describe the kinds of primary and secondary research you have carried out, to prepare your readers for a more detailed discussion of your sources in subsequent sections of the report.

- *What is the scope of the report?* Indicate what topics you are including, as well as what you are not including.

- *What are the most significant findings of the project?* Briefly summarize the most significant findings—what you learned from the project.

- *What are your recommendations?* In a short report containing a few simple recommendations, you might include those recommendations in the introduction. In a lengthy report containing numerous, complex recommendations, you might briefly summarize those recommendations in the introduction, and then refer your reader to the more detailed discussion in "Recommendations."

- *What is the organization of the report?* Indicate what organizational pattern you will use in the report, so readers can understand where you are going and why.

- *What key terms are used in the report?* If you will introduce new terms, the introduction is an appropriate place to define them. If many terms require definition, consider placing them in a glossary and using the introduction to refer your readers to the glossary.

For more information on introductions, see Chapter 8.

Figure 17.5 on page 536 is an introduction to the body of a formal report—a feasibility report, in this case. The subject of the report (Corder, 1993) is an investigation to determine whether the Quality Assurance (QA) lab at a microchip manufacturer should purchase a scanning electron microscope (SEM). The report is titled "QA Lab SEM Evaluation: Project Completion Report." Marginal notes have been added.

Methods The methods section answers the question "What did you do?" In drafting the methods section, analyze your audience, and then provide the appropriate kind and amount of information.

To analyze your audience, think about the questions mentioned earlier. Consider your readers' knowledge of the field, their perception of you, and the uniqueness of the project, as well as your readers' purposes in reading and their

Note: Reports are often
double spaced.

The background of
the project.

An overview of
the project.

The organization of
the report.

Introduction

In his memo of January 12, 1993, the Director of Quality Assurance authorized the purchase of a modern scanning electron microscope (SEM). However, the criteria for choosing a SEM were not established. In the current project I have established the criteria for evaluating the different SEMs on the market and evaluated them. This report describes the project and includes my recommendation.

Early in March I undertook a project to recommend which SEM to add to the Quality Assurance (QA) lab. This project was undertaken in three phases.

- Phase I involved evaluating SEM requests and interviewing QA SEM lab customers. From this information I developed a set of criteria.
- Phase II involved contacting the SEM manufacturers and evaluating their systems based on our criteria.
- Phase III involved compiling the information in this report, which presents my conclusions and recommendation.

First I present my methods, followed by the results, conclusions, and recommendation.

FIGURE 17.5

Introduction to the Body of a Formal Report

attitudes toward the project. Provide enough information to enable your readers to understand what you did—and why you did it that way.

Figure 17.6 shows the methods section from the report on the SEM.

Results The results are the data you discovered or created. Present the results objectively, without comments; save the interpretation of the results—your conclusions—for later. If you combine results and conclusions, your readers might be unable to follow your reasoning. Consequently, they will not be able to tell whether your conclusions are justified by the evidence—the results.

Just as the methods section answers the question "What did you do?" the results section answers the question "What did you see?"

The needs of your audience will help you decide how to structure the results. How much they know about the subject, what they plan to do with the report, what they anticipate your recommendation will be—these and many other factors will affect how you structure the discussion of the results.

For instance, suppose that your company is considering installing an intranet, a companywide computer network. In the introduction, you have discussed the company's current system for internal communication, explaining its disadvantages and limitations. In the methods section, you have described how you established the minimum specifications and evaluative criteria to apply to

Methods

To carry out this project, I performed four tasks:

1. I studied all the internal QA records of the requests that the SEM lab has received over the last year. For each request I noted the department of the requester and the kind of information requested. For example, some requesters needed to determine the location of a fault. Others needed to determine the nature of the fault or needed an image of the fault.

2. I interviewed the 12 requesters to find out how well the lab is meeting their needs and what further information would help them do their jobs. I used a standard script (see Appendix B) that combined closed-ended and open-ended questions.

3. I wrote to the six North American SEM manufacturers (see Appendix C) to get specifications, warranties, and prices on the available systems that would fulfill the requirements I established on the basis of the requests and the interviews.

4. I analyzed the information from the SEM manufacturers and the independent reviews of the equipment in the professional journals (see References).

FIGURE 17.6

Methods Section of a
Formal Report.

the available intranet systems, as well as the research procedures you carried out. In the results section, you provide the details of each intranet system you are considering, as well as the results of your evaluation of those systems.

Figure 17.7 on page 538 is an excerpt from the results section of the SEM report.

Conclusions Conclusions are the implications—the meaning—of the results. As discussed earlier in the chapter, you want to present a body of evidence, from current, reputable sources, that points to your conclusions.

Just as the results section answers the question "What did you see?" the conclusions section answers the question "What does it mean?"

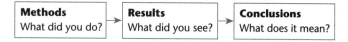

For more information on causal reasoning and on drawing conclusions, see Chapter 8.

The conclusions section of the SEM report is shown in Figure 17.8 on page 541.

Recommendations Recommendations are suggestions for taking particular actions. Just as the conclusions section answers the question "What does it mean?" the recommendations section answers the question "What should we do now?"

Job Requests and Interviews

I began by collecting all SEM job requests submitted to the QA SEM lab over a period of one year. In analyzing these job requests, I found that the largest proportion (some 75%) of the job requests came from either the failure-analysis engineers in QA and product engineering, or from the R&D design engineers. Most of the requests were similar: find a specific bit or location on an individual die, and image the area to find defects or failure mechanisms.

I simultaneously began interviewing the 12 requesters. I asked them if our lab provided them adequate images to define failure mechanisms, and if there were other functions that would make their identification of defects and locations simpler, quicker, or better in some other way.

The answers to these questions were that the images and data provided by the QA SEM lab are of sufficient resolution and contain adequate information. However, many requesters expressed a desire to have the images on the network and available for manipulation with an image processor, such as Semicaps. According to the requesters, this would accomplish three things for the engineers:

- It would allow them to include images in their reports through Harvard Graphics or other similar software.
- It would allow the engineers to make adjustments to the contrast, brightness, gamma, and other aesthetic characteristics, at their local workstation.
- It would allow the images to be archived electronically, rather than saving file cabinets full of Polaroid photos, as is currently done.

Determining Criteria

Using the information I obtained through the analysis of requests and the interviews of requesters, I developed a set of criteria for evaluating the various SEMs available on the market. The features I specified as criteria for the QA lab are the following:

- Turbomolecular vacuum pump (TMP)
- Lanthanum hexaboride electron gun (LaB6 Cathode)
- Digital imaging
- Solid-state backscattered electron detector (BSE)
- 100 mm or larger specimen stage
- Eucentric stage controls
- Minimal integral image processing

Turbomolecular Pump

Of the two major types of vacuum pumps on SEMs—diffusion and turbomolecular—turbomolecular will better meet our needs.

Diffusion pumps use a heated oil vapor to remove air molecules. These pumps are quite efficient and relatively inexpensive to operate, but simple errors can cause the oil to backstream into the specimen chamber and even into the column. Backstreaming oil causes many problems once inside the SEM column.

FIGURE 17.7

Excerpt from the Results Section of a Formal Report

(Figure 17.7 continued)

Turbomolecular pumps use a series of very high speed, rotating turbines to remove air molecules. These pumps are very dependable, very fast, and very efficient. Most important, however, they cannot backstream oil into the column; therefore, they are a much safer type of vacuum pump for a SEM.

Lanthanum Hexaboride Electron Gun

The lanthanum hexaboride electron gun (LaB6) is the best of the three common types of electron guns used in SEMs.

- The most common is a tungsten gun. Tungsten is the material used for filaments in light bulbs. Although the value of tungsten guns is well documented, they have problems. A tungsten filament (the consumable portion of the gun) is very inexpensive, at about $25 each, but typical lifetimes are only 20–25 hours of use.
- The LaB6 gun uses an LaB6 crystal instead of a tungsten filament. There are several advantages to this. First, LaB6 puts out more electrons at lower accelerating voltages than does tungsten. Second, LaB6 filaments last several hundreds of hours. The trade-off is that LaB6 filaments cost close to $600 each. The extended use and the enhanced signal, however, more than make up for the extra cost of replacement.
- The field-emission (FE) gun uses a slightly different technique. FE guns put out even more electrons at lower accelerating voltages than LaB6, but they are much more delicate to operate. FE guns require a higher vacuum; therefore, more vacuum pumps. FE guns are temperamental and difficult to use. They are also typically very expensive to operate, up to 50 percent more expensive. The difference in low accelerating voltage does not justify the use of an FE gun in the QA lab.

Digital Imaging

Digital imaging capability, a requirement for a new SEM, is a standard feature on most new SEMs.

Other Criteria

All modern SEMs meet our other four criteria:

- Solid-state backscattered electron detector (BSE)
- 100 mm or larger specimen stage
- Eucentric stage controls
- Minimal integral image processing

Contacting SEM Manufacturers

The writer explains the next phase of his project.

I contacted the SEM manufacturers and asked them to provide brochures and information on the SEMs they have available.

(Continued)

(Figure 17.7 continued)

> We had assumed that because 10 years have passed since we purchased our JEOL 840, many advances had been made in SEM technology. That assumption proved false. The technology of producing images with a scanning electron beam has not changed significantly. No dramatically different columns or detectors have been developed. The major changes include only two additions:
>
> - Nearly all SEMs now produce digital video images, as opposed to the analog imaging of our 840.
> - Nearly all SEMs now include much higher degrees of computerization and integral image processing.
>
> The computerization of SEMs is very much like the Semicaps software already mentioned. The difference is that the image processors that are integral to many SEMs are not compatible with systems we already have in use in QA. Nearly any image processor can manipulate a digital image, and most SEMs built today produce digital images. Therefore, "new" SEM technology is fairly old.

As you draft your recommendations, consider the following four factors:

- *Content.* When you tell readers what you think they should do next, be clear and specific. If the project you are describing has been unsuccessful, don't simply recommend that your readers "try some other alternatives." What alternatives do you recommend, and why?

- *Tone.* When you recommend a new course of action, you run the risk of offending whoever formulated the earlier course. Do not write that acting on your recommendations will "correct the mistakes" that have been made. Instead, write that acting on your recommendations "offers great promise for success." If the previous actions were not proving successful, your readers will probably already know that. A restrained, understated tone is not only more polite but also more persuasive: it indicates that you are interested only in the good of your company, not personal rivalries.

- *Form.* If the report leads to only one recommendation, use traditional paragraphs. If the conclusion of the report leads to more than one recommendation, consider a numbered list.

- *Location.* The recommendations section is always at the end of the body; however, because of its importance, it is often also summarized—or inserted verbatim—after the executive summary or in the introduction.

Figure 17.9 shows the recommendations section of the SEM report.

Note: Reports are often double spaced.

Conclusions

Based on its technical capabilities, price, and service record, the JEOL 5100 SEM is the best option for us.

In addition to Amray and JEOL, I interviewed representatives from Hitachi, Olympus, ISI, Cambridge, and Electroscan. Each of these companies offer SEMs that meet the basic criteria I established. I have evaluated the differences that each of these companies highlights. I found none of these features to be significant for our application. In addition, our experience with most of these companies is quite limited.

On the other hand, we have considerable experience with both Amray and JEOL. We have a good idea of their reliability; we have established service contracts with both companies. Therefore, we have established relationships with their field-service engineers. In addition, we have some stock of spares from each company, many of which are applicable to their new models.

With all these factors considered, Amray and JEOL were the only manufacturers whom I invited to visit our site for more thorough discussions and to provide system quotes.

Amray quoted a model 1810 SEM. The total price from Amray is $100,000. JEOL quoted a model 5100 SEM. The total price of the JEOL system is $115,500. While the JEOL appears 15% higher in price, we have learned from experience that the Amray will require a retrofitted Eddie Fjeld stage before it is operational. The Eddie Fjeld stage suitable for the Amray model 1810 is available for $28,000, installed. Therefore, the actual price for the Amray 1810 is $128,000.

Beyond the price difference, however, the SEM technicians, the equipment engineers, and I all have greater confidence in the JEOL service organization on the basis of our excellent relationship with them over the years.

The writer explains his reasoning carefully. Notice that he describes, in the last paragraph, the unanimous opinion of his colleagues.

FIGURE 17.8

Conclusions Section from a Formal Report

Note: Reports are often double spaced.

Recommendation

I recommend that we purchase a JEOL 5100 SEM configured with the features listed in the report and on the quotation. The total system price will be $115,000. Installation will be another $1,000.

FIGURE 17.9

Recommendations Section from a Formal Report

Sample Formal Report

The following report (Wells, Tommack, & Tuck, 1997) is a follow-up to the writers' proposal (Chapter 15) and progress report (Chapter 16).

May 12, 1997

Margie Voren, Dean
Student Special Services
Central State University
1910 University Avenue
Central, OK 63509

Dear Dean Voren:

The purpose of the study.

Attached is the completion report for our research project. As you will recall, we investigated whether Central State University (CSU) should recognize American Sign Language as a foreign language and incorporate it into the foreign-language curriculum. In addition, we researched two possible methods to attain formal recognition for ASL: a revision of university policy and the passage of a bill in the state legislature.

The methods.

First, we researched the criteria by which CSU defines a language as foreign; the sole criterion is that the language be "non-English" in such matters as syntax, grammar, and vocabulary. Next, we investigated the linguistic characteristics of ASL to determine if it meets this criterion. Our analysis of ASL yielded sufficient evidence that ASL is non-English in many major linguistic categories, including grammatical structure, topicalization, spatial indexing, use of multidirectional verbs, and rules governing questions and negations.

The results.

The main conclusion.

CSU could recognize ASL in one of two ways: through revision of university policy or through state legislation mandating such recognition at all state colleges and universities. We researched the legislative approach by compiling information about legislation passed recently in Texas that recognizes ASL as a foreign language. We concluded that revising university policy is the best method by which CSU might recognize ASL, primarily because such internal recognition avoids potential conflict between the legislature and the university.

A polite offer to provide more information.

We thank you for the opportunity to perform this research. If you have any questions about the project or this report, we would be glad to answer them at your convenience.

Sincerely,

Danny Tommack   Brandon M. Tuck   Megan Wells

The title indicates the subject and purpose of the report.

**RECOGNIZING AMERICAN SIGN LANGUAGE AS A FOREIGN
LANGUAGE AT CENTRAL STATE UNIVERSITY:
A RECOMMENDATION REPORT**

Prepared for: Margie Voren
Dean of Student Special Services
Central State University

Prepared by: Danny Tommack
Brandon M. Tuck
Megan Wells

May 12, 1997

Note: Reports are often double spaced.

Abstract

"Recognizing American Sign Language as a Foreign Language at Central State University: A Recommendation Report"

by Danny Tommack, Brandon M. Tuck, and Megan Wells

Currently, Central State University (CSU) does not recognize American Sign Language (ASL) as a foreign language, yet several other states and universities have recognized ASL as a foreign language and have incorporated ASL courses into foreign-language curricula. Our research project investigated the criterion by which CSU defines a language as foreign, the extent to which ASL meets this criterion, and the two methods by which CSU might recognize ASL as a foreign language (either by revising university policy or through state legislation). CSU's sole requirement for a language to be considered foreign is that it be "non-English" in such matters as grammar, syntax, and vocabulary. Our analysis shows that ASL is decidedly non-English in many major linguistic categories. This conclusion is in accord with policy at the Foreign Service Institute, a division of the U.S. State Department that trains foreign diplomats: the Foreign Service Institute classifies ASL as similar to Chinese and Japanese in its difference from English. In our research, we found that ASL differs from English in many ways, including grammatical structure, topicalization, spatial indexing, use of multidirectional verbs, and rules governing questions and negations. This report summarizes the foreign-language characteristics of ASL, compares those characteristics to the criterion by which CSU defines a language as foreign, and recommends CSU revise university policy to incorporate ASL into the foreign-language curriculum.

Keywords: American Sign Language, foreign language, linguistics, university curriculum

The abstract focuses on the technical aspects of the subject.

A keywords list ensures that if the report is searched electronically, it will register "hits" for each of the terms listed.

i

Contents

Abstract i

Executive Summary 1

Introduction 2

Methods 3

Results 3

   CSU's Requirement for a Language to Be Considered Foreign 4

   ASL's Fulfillment of CSU's Requirement 4

   Analysis of Two Methods by Which ASL Could Be Formally Recognized 5

Conclusions 6

Recommendations 7

References 7

Appendices 7

   Appendix A: Universities That Accept ASL in Fulfillment of
      Foreign-Language Requirements 7

   Appendix B: Classification of American Sign Language
      According to the Foreign Service Institute 8

   Appendix C: The Non-English Linguistic Characteristics of
      American Sign Language 8

An executive summary tells the story of the project with a focus on the managerial aspects, particularly the recommendations.

Executive Summary

Currently, Central State University (CSU) does not recognize American Sign Language (ASL) as a foreign language, meaning that students in CSU's three ASL courses may earn elective credits but not foreign-language credits. Dr. Teresa O'Malley, Chair of the Modern Languages Department, and Suzanne Christopher, CSU's current ASL instructor, have noticed a growing demand for ASL courses at CSU, in spite of the fact that the students in the courses do not earn foreign-language credits. This growing interest in ASL is reflected throughout the United States, as more and more universities formally recognize ASL as a foreign language. Formally recognizing ASL as a foreign language, as this report recommends, would help CSU remain current with modern linguistic and educational trends.

This report documents our research project investigating the possibility of CSU's recognizing ASL and incorporating ASL into the foreign-language curriculum. Specifically, we investigated the criterion by which CSU defines a language as foreign, the extent to which ASL meets this criterion, and the two methods by which CSU might recognize ASL as a foreign language (either by revising university policy or through state legislation).

For a language to be considered foreign at CSU, it must be linguistically "non-English" in such matters as grammar, syntax, and vocabulary. ASL is sufficiently non-English to be grouped by some language authorities with Japanese and Chinese, both of which are significantly different from English. Given that ASL meets the criterion by which CSU defines a language as foreign, we recommend that CSU formally recognize ASL as a foreign language and that CSU incorporate ASL into the foreign-language curriculum. Doing so would add the fourth-used language in the United States to the other foreign languages taught at CSU and would thus provide students with a rigorous and challenging addition to the Modern Languages Department. We further recommend revising university policy to recognize ASL, rather than seek state legislation granting recognition at all Oklahoma colleges and universities, primarily because this approach would prevent political conflict between the state legislature and administrators at Oklahoma colleges and universities.

In some organizations, every first-level heading begins on a new page.

The background and purpose of the report.

The problem.

The methods.

The scope.

Introduction

On April 24, we received approval to investigate whether Central State University (CSU) should formally recognize American Sign Language (ASL) as a foreign language and incorporate it into the foreign-language curriculum. We also received approval to investigate how this recognition could be achieved at CSU (through revision of university policy) and at all Oklahoma colleges (by state legislation). Our goals in this project were to identify the criteria by which CSU defines a language as foreign, determine the extent to which ASL meets those criteria, and determine which of the two methods of granting recognition would be more appropriate for CSU.

Currently, CSU does not recognize ASL as a foreign language, meaning that students in CSU's three ASL courses may earn elective credits but not foreign-language credits. Dr. Teresa O'Malley, Chair of the Modern Languages Department, and Suzanne Christopher, CSU's current ASL instructor, have noticed a growing demand for ASL courses at CSU, in spite of the fact that the students in the courses do not earn foreign-language credits. This growing interest in ASL is reflected throughout the United States as more and more universities formally recognize ASL as a foreign language. In fact, almost half of the states recognize ASL as a foreign language and grant academic credit for these courses; these states have recognized ASL either through state legislation or revision of university policy. Oklahoma is one of 15 states in which colleges and universities grant only elective credit to students who complete ASL courses. Formally recognizing ASL as a foreign language, as this report recommends, would help CSU remain current with modern linguistic and educational trends.

In conducting our investigation, we drew upon the expertise of several individuals knowledgeable about ASL and its role in the curricula of colleges and universities. For instance, Ceil Lucas's *Sociolinguistics of the Deaf Community* (1989) helped us analyze the non-English characteristics of ASL. We consulted interpreter/linguist Rhonda Jacobs's 1996 article "ASL as a Truly Foreign Language" to determine how the Foreign Service Institute of the U.S. State Department classifies languages based on their differences from English. Finally, Sherman Wilcox, coauthor of *Learning to See: Teaching American Sign Language as a Second Language* (1991) and associate professor of linguistics at the University of New Mexico, helped us evaluate the two methods by which CSU might recognize ASL as a foreign language.

We have focused our research on determining the linguistic characteristics that qualify ASL as a foreign language and on the methods by which CSU might recognize it as such. We have not looked into significantly expanding the current ASL program at CSU.

The results.

Our most significant findings are as follows:

- ASL meets the criterion by which CSU defines a language as foreign.
- Recognizing ASL through revision of university policy is preferable to recognizing it through state legislation.

The recommendation.

We therefore recommend that CSU revise university policy to formally recognize ASL as a foreign language. We further recommend that CSU incorporate ASL into the foreign-language curriculum.

An advance organizer.

The following sections provide additional detail about the research methods we used, the results we obtained, the conclusions we drew from those results, and our recommendations based on those conclusions.

Methods

In conducting our research, we completed the following tasks:

- We interviewed Dr. Teresa O'Malley, Chair of the Modern Languages Department at CSU, to determine the criterion by which CSU defines a language as foreign.
- We studied the linguistic characteristics of ASL to determine if it met the criterion used at CSU. Suzanne Christopher, the ASL instructor at CSU, assisted with this portion of our project, primarily by directing us to sources of information and helping us to interpret that information. In addition, we consulted several books and periodicals to determine whether ASL meets the criterion by which CSU defines a language as foreign. These sources are listed in "References."
- We spoke with educators and specialists familiar with Texas legislation that recognizes ASL as a foreign language, to determine if CSU should grant recognition internally (through revision of university policy) or if recognition should be granted externally (through state legislation). Sha H. Cowan from the Texas Education Agency, Services for the Deaf Department, referred us to deaf educators in his area. We interviewed Marietta L. Yeates, professor in the Deaf/Hard of Hearing Program at Stephen F. Austin State University, and Dr. Sherman Wilcox, associate professor of linguistics at the University of New Mexico, to learn their views on the legislation passed in Texas. Using the information supplied by Dr. Yeates and Dr. Wilcox, we analyzed the positive and negative aspects of internal recognition of ASL as a foreign language as compared to external recognition.

Results

An advance organizer for the results section.

Our research project investigated whether CSU should recognize ASL and incorporate it into the foreign-language curriculum. Specifically, we investigated

3

the criterion by which CSU defines a language as foreign, the extent to which ASL meets this criterion, and the two methods by which CSU might recognize ASL as a foreign language (either internally, by revising university policy or externally, through state legislation). We obtained the following results:

CSU's Requirement for a Language to Be Considered Foreign

According to Dr. Teresa O'Malley, Chair of the CSU Department of Modern Languages, the sole criterion for a language to be considered foreign at Central State University is that it be "non-English" in such matters as structure, grammar, syntax, and vocabulary. Dr. O'Malley also noted that any language that can be shown to be "non-English" could be incorporated as a foreign language.

After reviewing the ASL courses offered at CSU, Dr. O'Malley said that two of them—CM221, Intermediate American Sign Language, and CM321, Conversational American Sign Language—would be eligible for foreign-language credit if CSU recognized ASL and incorporated it into the foreign-language curriculum. Courses at the beginner level would be ineligible; such courses are considered equivalent to high school courses, so students who complete them receive no credit.

ASL's Fulfillment of CSU's Requirement

Initially, we had planned to meet with Suzanne Christopher, the ASL instructor at CSU; Kelley Eastwick, the president of the Oklahoma Chapter of the Registry of Interpreters for the Deaf; and Janis Seymour, committee member of the Oklahoma Association of the Deaf. However, because of scheduling conflicts we were able to meet only with Suzanne Christopher. She assisted us in researching ASL's linguistic identity and determining the extent to which ASL meets the criterion by which CSU defines a language as foreign.

We found that ASL is a "highly elaborate and fully expressive language whose structure is determined in part by the articulating dynamics of the body" (Isenhath, 1990, p. 126); in this respect, ASL differs significantly from English. In conducting this portion of our research, we drew upon several linguistic manuals and other sociolinguistic resources to identify the "non-English" characteristics of ASL. This investigation of ASL's linguistic identity included research into grammatical structure, topicalization, spatial indexing, the use of multidirectional verbs, and rules governing questions and negations. (Appendix C contains summaries and examples of ASL's "non-English" characteristics in these five areas.)

We studied ASL's non-English characteristics in these five areas because they are linguistically simple yet illustrate profound differences between ASL and English. Three of these areas—topicalization, spatial indexing, and multidirectional verb usage—are key to ASL but are not found in English. Grammatical structure

Note that the three main parts of the results section match the tasks defined in the methods section.

4

and rules governing questions and negation are key components of nearly all languages; however, ASL's unique grammatical structure, as well as rules governing questions and negation, are quite different from those of English.

Analysis of the Two Methods by Which ASL Could Be Formally Recognized

Two methods exist by which ASL could be formally recognized as a foreign language: internal recognition (through revision of university policy) and external recognition (through state legislation).

Our research into these methods included communication with several members of the deaf-education community in Texas, a state that has recently passed legislation recognizing ASL as a foreign language at all Texas colleges and universities. We could have looked to Maryland, New York, or Washington, DC, where ASL recognition has been strongly advocated since the establishment of schools such as Gallaudet University and the National Technical Institute for the Deaf. But the per capita deaf population there far exceeds Oklahoma's.

Sha H. Cowan from the Texas Education Agency, Services for the Deaf Department, referred us to deaf educators in his area. Consequently, we interviewed Marietta L. Yeates, professor in the Deaf/Hard of Hearing Program at Stephen F. Austin State University, and Dr. Sherman Wilcox, associate professor of linguistics at the University of New Mexico.

Dr. Wilcox maintains a list of universities that recognize ASL as a foreign language. Dr. Wilcox's list (Appendix A) shows that ASL is being recognized as a foreign language internally by colleges and universities more often than it is being recognized through state legislation. Moreover, his research was most recently updated on January 30, 1997, making it both more current and more inclusive than the original research we outlined in our proposal.

In our proposal, we referred to Rhonda Jacobs's research into the percentage of states that have passed legislation recognizing ASL as a foreign language. Though published in 1996, her article referred to raw data collected in 1993, making us question the relevance of her article. Also, she defined "legislative approval" (in our terms, "legislative recognition") as state-mandated recognition that was applicable to all state colleges and universities. However, we are interested only in approval legislatively granted to postsecondary institutions, though other relevant legislation would naturally give the petitioner greater weight. Dr. Wilcox's list, found in Appendix A, shows that more colleges are independently recognizing ASL as a foreign language.

Using the data from Dr. Yeates and Dr. Wilcox, we analyzed the positive and negative aspects of recognizing ASL internally (by revising university policy) and externally (through state legislation). Dr. Yeates explained that Stephen F. Austin State University recognized ASL as a foreign language *before* Texas state legislation

5

recognized it for all Texas colleges and universities. Dr. Yeates encouraged us to seek state legislation that would recognize ASL at all Oklahoma colleges and universities.

However, Dr. Wilcox cautioned us about using this approach. Specifically, he said that conflict might arise if the legislature tried to mandate the university's curriculum; in his experience, the faculty and administration of colleges and universities often respond with distrust and resistance when legislation mandates changes to the curriculum. With this idea in mind, he suggested that the best course of action would be to recognize ASL as a foreign language internally, by revising university policy. In Dr. Wilcox's view, proponents of ASL as a foreign language could then use CSU's recognition to garner support for state legislation designating ASL as a foreign language at all Oklahoma colleges and universities.

As a result of our communication with Dr. Wilcox, he has offered to assist CSU in formally recognizing ASL as a foreign language and incorporating it into the foreign-language curriculum. He is a strong advocate for the study of ASL and has experience in organizing presentations and petitions in support of ASL.

Conclusions

After carefully analyzing the non-English characteristics of American Sign Language (ASL) and the criterion by which Central State University (CSU) defines a language as foreign, we have concluded that ASL qualifies as a foreign language. CSU's criterion stipulates that a language must be "non-English" in such matters as grammar, syntax, and vocabulary; ASL clearly differs from English, particularly in such areas as grammatical structure, topicalization, spatial indexing, the use of multidirectional verbs, and rules governing questions and negations.

ASL's unique linguistic identity has been established in current literature and has been deemed sufficiently distinct for dozens of state universities to recognize ASL as a foreign language. Moreover, the complexity of ASL would make the intermediate ASL courses a challenging and rewarding addition to the foreign-language curriculum at CSU.

After reviewing the process by which the Texas state legislature officially recognized ASL as a foreign language, we have concluded that internal recognition would be in the best interests of CSU and would be the most effective method of officially recognizing ASL. Recognizing ASL internally would require revision of university policy, whereas external recognition would require passage of state legislation. State legislation would bring with it the benefit of statewide recognition: each college or university in Oklahoma would be required to recognize ASL as a foreign language. However, state legislation might also create political friction between the members of the legislature and the faculty and administration of Oklahoma colleges and universities; in short, representatives of Oklahoma colleges and universities may resent the state legislature's mandating a change in curriculum and, therefore, work to defeat state legislation recognizing ASL as a foreign language, regardless of their feelings about the merits of ASL as a foreign language.

Recommendations

On the basis of the conclusions we have drawn from the results of our research, we recommend the following:

- Central State University (CSU) should recognize American Sign Language (ASL) as a foreign language and should incorporate ASL into the foreign-language curriculum.
- CSU should internally recognize ASL, through revision of university policy, rather than seek external recognition through state legislation.
- CSU should seek the advice of Dr. Sherman Wilcox, of the University of New Mexico, in implementing the first two recommendations.

References

Isenhath, J. O. (1990). *The linguistics of American Sign Language.* Jefferson, NC: McFarland.

Jacobs, R. (1996). ASL as a truly foreign language. In C. Lucas (Ed.), *Multicultural aspects of sociolinguistics in deaf communities* (pp. 183–226). Washington, DC: Gallaudet University Press.

Lucas, C. (1989). *Sociolinguistics of the deaf community.* San Diego, CA: Academic Press.

Valli, C., & Lucas, C. (1992). *Linguistics of American Sign Language: A resource text for ASL users.* Washington, DC: Gallaudet University Press.

Wilcox, S., & Wilcox, P. (1991). *Learning to see: American Sign Language as a second language.* Englewood Cliffs, NJ: Prentice Hall Regents.

Wilcox, S. (1997). Universities that accept ASL in fulfillment of foreign language requirements. <www.unm.edu/~wilcox/ASLFL/univlist.html> (1997, May 10).

Appendix A
Universities That Accept ASL in Fulfillment
of Foreign-Language Requirements

[Here the writers list the 72 universities; they cite Sherman (1997) as the source of this information.]

7

Appendix B
Classification of American Sign Language
According to the Foreign Service Institute

[Here the writers provide the information from the Foreign Service Institute that they provided in their progress report (see Chapter 16).]

Appendix C
The Non-English Linguistic Characteristics
of American Sign Language

Introduction

John O. Isenhath, in *The linguistics of American Sign Language,* and Clayton Valli and Ceil Lucas, in *Linguistics of American Sign Language: A resource text for ASL users,* explore the linguistic characteristics that make American Sign Language (ASL) different from English. This document briefly summarizes those characteristics and explains how they contribute to the unique linguistic identity of ASL. Included are descriptions and examples of differences evident in grammatical structure, topicalization, spatial indexing, multidirectional verbs, and rules governing questions and negations.

[In this six-page appendix, excerpted here, the writers summarize the major research that shows the significant differences between ASL and English in terms of the five areas mentioned at the end of the previous paragraph.]

8

CASE

Writing a Recommendation Report

Go to the Web site for the U.S. Government Printing Office (http://www.access.gpo.gov), where you will find links to the sites of dozens of government departments and agencies. Select a department or agency that relates to your group's interests. Imagine that you work for this organization and that you have been asked by a supervisor to review all the information on the site and to write a recommendation report about the site. Your supervisor wants to know "what we have on the site, how well it communicates to our viewers, and what changes (if any) we should make to it." Write the recommendation report.

WRITER'S CHECKLIST

1. In planning your informational, analytical, or recommendation report, have you
   - ❏ analyzed your audience?
   - ❏ analyzed your purpose?
   - ❏ identified the questions that need to be answered?
   - ❏ carried out appropriate research?
   - ❏ drawn valid conclusions about the results (if appropriate)?
   - ❏ formulated recommendations based on the conclusions (if appropriate)?

2. In planning to write your feasibility report, have you
   - ❏ analyzed your audience?
   - ❏ analyzed your purpose?
   - ❏ identified the problem or opportunity?
   - ❏ established criteria for responding to the problem or opportunity?
   - ❏ determined your options?
   - ❏ studied each option according to your criteria?
   - ❏ drawn valid conclusions about each option?
   - ❏ formulated recommendations based on your conclusions?

3. Does the introduction
   - ❏ explain the subject of the report?
   - ❏ explain the purpose of the report?
   - ❏ explain the background of the report?
   - ❏ describe your sources of information?
   - ❏ indicate the scope of the report?
   - ❏ briefly summarize the most significant findings of the project?
   - ❏ briefly summarize your recommendations?
   - ❏ explain the organization of the report?
   - ❏ define key terms used in the report?

4. Does the methods section describe your methods in sufficient detail?

5. Are the results presented
   - ❏ clearly?
   - ❏ objectively?
   - ❏ without interpretation?

6. Are the conclusions
   - ❏ presented clearly?
   - ❏ drawn logically from the results?

7. Are the recommendations
   - ❏ clear?
   - ❏ objective?
   - ❏ polite?
   - ❏ in an appropriate form (list or paragraph)?
   - ❏ in an appropriate location?

EXERCISES

Exercises 3 and 4 call for you to write a memo. See Chapter 16 for a discussion of memos.

1. An important element in carrying out a feasibility study is determining the criteria by which to judge each option. For each of the topics listed here, list five important criteria you might apply in evaluating the options.
   a. buying a computer printer
   b. selecting a major
   c. choosing a company to work for
   d. buying a car
   e. choosing a place to live while you attend college

2. Interview a local professional who writes completion reports. (See Chapter 7 for a discussion of interviewing.) What process does he or she use in creating the report, and what does he or she see as the most important aspects of writing a completion report? To what extent does the information provided by the professional match the information in this chapter? Respond in a memo to your instructor.

 Collaborative Exercise

3. Write the completion report for the research project you proposed in Chapter 15. For a list of possible topics and approaches, see the exercises in Chapter 15. Your instructor will tell you whether the report is to be written individually or collaboratively. Regardless, work with a partner in reviewing and revising your report. (See Chapter 3 for a discussion of collaboration.) You and your partner will of course work together closely at the end of the project, as you revise your reports, but keep in mind that a partner can be very helpful during the planning phase, too, as you choose a topic, refine it, and plan your research.

Web Exercise

4. Secure a completion report for a project subsidized by a city or federal agency, a private organization, or a university committee or task force. (Be sure to check your university's Web site; universities routinely publish strategic planning documents and other sorts of self-study reports. Also check http://www.nas.edu, which is the site for the National Academy of Sciences, the National Academy of Engineering, the Institute of Medicine, and the National Research Council, all of which publish reports on the Web.) In a memo to your instructor, describe how the completion report differs from the models in this chapter. What challenges have the writers faced in describing their project, and how have they met those challenges? Overall, how effective is the report? If possible, submit a copy of the report along with your memo.

REFERENCES

Corder, Z. (1993). QA Lab SEM evaluation: Project completion report. Unpublished document.

Wells, M., Tommack, D., & Tuck, B. (1997). Recognizing American Sign Language as a foreign language at Central State University: Recommendation report. Unpublished document.

CHAPTER 18

Preparing Job-Application Materials

Planning the Job Search 557

Understanding Seven Ways to Look for a Position 558

Writing Résumés 561
  Appearance of the Résumé 562
  Content of the Résumé 563
  Elements of the Chronological Résumé 564
  Elements of the Analytical Résumé 571

Writing Electronic Résumés 576
  Content of an Electronic Résumé 577
  Format of an Electronic Résumé 577

Writing Job-Application Letters 580
  Content of the Job-Application Letter 580
  Elements of the Job-Application Letter 580

Writing Follow-Up Letters 585

Case: Updating Career-Center Materials 590

Writer's Checklist 590

Exercises 591
  Collaborative Exercise 592
  Web Exercises 593

For most of you, the first nonacademic test of your technical writing skills will come when you seek a job. Job-application materials will inform employers about your academic and employment experience, personal characteristics, and reasons for applying. But they will provide much more information, too. Employers have learned that one of the most important skills an employee can bring to a job is the ability to communicate effectively. Therefore, potential employers look carefully for evidence of writing skills. Job-application materials pose a double hurdle: the tasks of showing employers both what you can do and how well you can communicate.

Some students think that once they get a satisfactory job, they will never again have to worry about résumés and application letters. Statistics suggest otherwise. The typical professional changes jobs more than five times. Although the discussion in this chapter pays special attention to the student's first job hunt, the skills and materials discussed here also apply to established professionals who wish to change jobs.

Planning the Job Search

Planning the job search is a lot of work, stretching over weeks and months, not days. Three main tasks are involved in planning the search, as shown in Figure 18.1.

- *Do a self-inventory.* Before you can start thinking of where you want to work, you need to answer some questions about yourself:

    - *What are your strengths and weaknesses?* Are your skills primarily technical? Do you work well with people? Do you work best with supervision or without?
    - *What are your preferences?* Think about what you have liked or disliked about your previous positions and college courses. In addition to a salary, what do you want to get out of a job?
    - *What kind of organization would you like to work for?* Profit or nonprofit? Government or private industry? Small or large?
    - *What are your geographical preferences?* If you are free to move, where would you like to live? How do you feel about commuting?

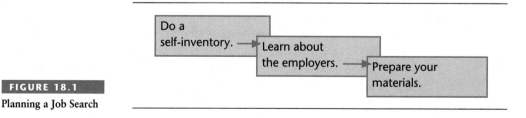

FIGURE 18.1

Planning a Job Search

- *Learn about the employers.* Don't base your job search exclusively on the information in a job advertisement. Learn as much as you can about the organizations through other means as well:

  - *Attend job fairs.* Your college and perhaps your community hold job fairs, where different employers provide information about their organizations.
  - *Find out about trends in your field.* Read the *Occupational Handbook,* published by the U.S. Department of Labor, for information about your field and related fields. Talk with professors in your field and with the people at your job-placement office.
  - *Research the companies you are interested in.* Visit their Web sites. Scan the index of the *Wall Street Journal* for articles. Study their annual reports, many of which are collected in your college library.

- *Prepare your materials.* You know you will need to write application letters and résumés and that you will go on interviews. Start planning early by obtaining materials from the career-placement office. Talk with friends who have gone through the process successfully; study their application materials. Read some of the books on different aspects of job searching.

  One more very important part of preparing your materials: make a *portfolio,* a collection of your best work. You'll want to give a prospective employer a copy of the portfolio as an example of the kind of work you can do. For a technical communicator, the portfolio will include a variety of documents you made in courses and in previous positions. For technical professionals, the portfolio will include documents such as project reports, as well as other materials, including computer simulations or output. A portfolio is often presented in a loose-leaf notebook, with each item preceded by a descriptive and evaluative statement. Frequently, a portfolio contains a table of contents and an introductory statement.

Understanding Seven Ways to Look for a Position

Once you have done your planning, you can start to look for a position. Seven major ways to find a job are shown in Figure 18.2.

- *Through a college or university placement office.* Almost all colleges and universities have placement offices, which bring companies and students together. Generally, students submit a résumé—a brief list of credentials—to the placement office. The résumés are then made available to representatives of business, government, and industry, who use the placement office to arrange on-campus interviews with selected students. Those who do best in the campus interviews are then invited by the rep-

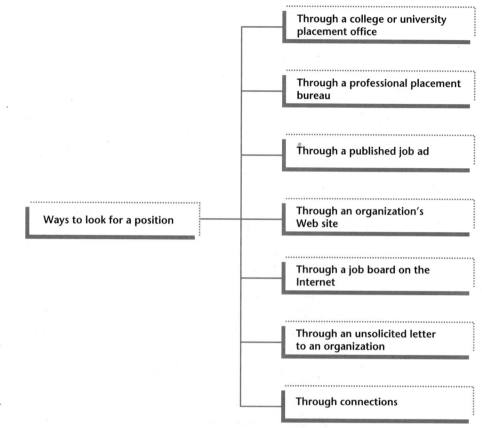

FIGURE 18.2

Seven Ways to Look for
a Position

resentatives to visit the organization for a tour and another interview. Sometimes a third interview is scheduled; sometimes an offer is made immediately or shortly thereafter. The advantage of this system is twofold: it is free and it is easy. You merely deliver a résumé to the placement office and wait to be contacted.

- *Through a professional placement bureau.* A professional placement bureau offers essentially the same service as a college placement office, but it charges either the employer (or, rarely, the new employee) a fee— often 10 percent of the first year's salary—when the client accepts a position. Placement bureaus cater primarily to more advanced professionals who are looking to change jobs.

- *Through a published job ad.* Published job ads generally offer good opportunities for both students and professionals. Organizations advertise in three basic kinds of print publications: public-relations catalogs (such as *College Placement Annual*), technical journals, and newspapers. Check the major technical journals in your field and the large metropolitan newspapers, especially the Sunday editions. In responding to an ad, you must include with the résumé a job-application letter that highlights the crucial information on the résumé.

- *Through an organization's Web site.* Many organizations list their job offerings on their Web sites. A job ad on a Web site usually tells candidates how to apply: by filling out an electronic form, by e-mailing an application, by sending a scannable résumé, by sending a traditional letter and résumé, or by coming to the organization in person. (See the discussion later in this chapter about electronic résumés.)

- *Through a job board on the Internet.* Job boards are sites, sponsored by federal agencies, Internet service providers, and private organizations, that exist to help introduce employers to prospective employees. Some of these sites offer their services for free; others charge. Some sites merely list positions, to which you respond by regular mail or by e-mail; others let you submit your résumés electronically, so that employers can get in touch with you. Use the reference function in your search engine to search for "employment," "careers," and "jobs." Or combine one of these terms with the name of your field, as in "careers and forestry." Among the biggest job boards are the following:

America's Job Bank (sponsored by the U.S. Department of Labor):
    http://www.ajb.dni.us
Monster board: http://www.monster.com
CareerPath: http://careerpath.com
Career Mosaic: http://www.careermosaic.com
Online Career Center: http://www.occ.com

Many of these and other sites have links to articles about searching for jobs electronically, including how to write electronic résumés and how to prepare for interviews.

One caution about using job boards: once you post something to an Internet site, you probably have lost control of it. Here are four questions to ask before you post to a job board:

- Who has access to your résumé? You might wish to remove your home address and phone number from it if anyone can view it.
- How will you know if an employer requests your résumé? Will you be notified by the job board?
- Can your current employer see your résumé? If your employer knows you are searching for a new job, your current position could be jeopardized.
- Can you update your résumé at no cost? Some job boards let you post for free, then charge you each time you update your résumé.

- *Through an unsolicited letter to an organization.* Instead of waiting for an ad or a notice on a Web site, consider writing unsolicited letters of application to organizations you would like to work for. The disadvantage of this technique is obvious: there might not be an opening. Yet many professionals favor this technique, because there are fewer competitors for those jobs that do exist, and organizations do not advertise all available positions. And sometimes an impressive unsolicited application can prompt an organization to create a position.

  Before you write an unsolicited application, make sure you learn as much as you can about the organization (see Chapter 7): current and anticipated major projects, hiring plans, and so forth. You should be knowledgeable about any organization you are applying to, of course, but when you are submitting an unsolicited application you have no other source of information on which to base your résumé and letter. The business librarian at your college or university will be able to point out sources of information, such as the Dun and Bradstreet guides, the *F&S Index of Corporations,* and the indexed newspapers such as the *New York Times,* the *Washington Post,* the *Los Angeles Times,* and the *Wall Street Journal.* And, of course, you should study the organization's Web site.

- *Through connections.* A relative or an acquaintance who can exert influence or at least point out a new position can help you get a job. Other good contacts include employers from your past summer jobs and faculty members in your field. Become active in the student chapter of your field's professional organization, through which you can meet professionals in your area.

Writing Résumés

This section concentrates on techniques of preparing printed résumés: paper-based résumés that you will mail or carry to a potential employer. The following section discusses electronic résumés. Even if you know you will be sending electronic résumés, however, you should read this section; it provides useful advice about both printed and electronic résumés.

Many students wonder whether to write the résumé themselves or use a résumé-preparation agency. It is best to make your own résumé, for three reasons:

- *You know yourself better than anyone else does.* No matter how thorough and professional the work of a résumé-preparation agency, you can do a better job communicating the important information about yourself.

- *Employment officers know the style of the local agencies.* Readers who realize that you did not write your own résumé might wonder what kinds of deficiencies you are trying to hide.

- *If you write your own résumé, you will more likely vary it to meet the needs of different situations.* You are not likely to go back to a résumé-preparation agency and pay them an additional fee to make a minor revision.

Whether you write the résumé by yourself or get help, it communicates in two ways: by its appearance and by its content.

Appearance of the Résumé

Because potential employers normally see your résumé before they see you, it has to make a good first impression. Employers believe—often correctly—that the appearance of the résumé reflects the writer's professionalism. A sloppy résumé implies that you would do sloppy work. A neat résumé implies that you would do professional work. When employers look at a résumé, they see the documents they will be reading if they hire you.

Some colleges and universities advise students to have their résumé professionally printed. A printed résumé is attractive, and that's good—provided, of course, that the information on it is consistent with its professional appearance. Most employers agree, however, that a neatly word-processed résumé, printed on a laser printer and photocopied on good-quality paper, is just as effective.

People who photocopy a word-processed résumé are more likely to tailor different versions to the needs of the organizations to which they apply—a good strategy. People who go to the trouble and expense of a professional printing job are far less likely to make different résumés. The résumé looks so good that they don't want to tinker with it. This strategy is dangerous because it encourages the writer to underestimate the importance of tailoring the résumé to a specific audience.

However they are reproduced, résumés should appear neat and professional. They should have

- *Generous margins.* Leave a 1-inch margin on all four sides.
- *Clear type.* Use a good-quality laser printer.
- *Balance.* Arrange the information so that the page has a balanced appearance.
- *Clear organization.* Use adequate white space. Make sure the line spacing *between* items is greater than the line spacing *within* an item. That is, the line spacing between your education section and your employment section should be greater than that between items within one of those sections. You should be able to see the different sections clearly if you stand while the résumé is on the floor in front of your feet.

Use indentation clearly. When you arrange items in a vertical list, indent turnovers a few spaces. (*Turnovers* are the second and subsequent lines of any item.) Notice, for example, that the following list from the computer-skills section of a résumé can be confusing:

Computer Experience
Systems: PC, Macintosh, UNIX, Andover AC–256, Prime 360
Software: Lotus 1–2–3, DBase V, PlanPerfect, Micrografx Designer, Adobe PageMaker, Microsoft Word
Languages: Pascal, C++, HTML

With the second line of an entry indented, the arrangement is much easier to understand:

Computer Experience
Systems: PC, Macintosh, UNIX, Andover AC–256, Prime 360
Software: Lotus 1–2–3, DBase V, PlanPerfect, Micrografx Designer, Adobe PageMaker,
    Microsoft Word
Languages: Pascal, C++, HTML

See Chapter 11 for more information on page design.

Content of the Résumé

Although different experts advocate different approaches to résumé writing, everyone agrees on three things:

- *The résumé must be honest.* There are no accurate statistics on how many résumés contain lies, but the figure is probably considerable. You have probably read about employees who have been caught including not only exaggerations but outright lies on their résumés. Naturally, they are fired immediately. Many employers today routinely check candidates' credentials. So, for practical as well as ethical reasons, tell the truth.

- *The résumé must be completely free of errors.* Grammar, punctuation, usage, and spelling errors cast doubt on the accuracy of the information in the résumé. Ask for assistance after you have written the draft, and proofread the finished product at least twice. Then have someone else proofread it, too.

- *The résumé must provide clear, specific information, without generalizations or self-congratulation.* Your résumé is a sales document, but you are both the salesperson and the product. You cannot say "I am a terrific job candidate" as if the product were a toaster or an automobile. Instead, you have to provide the details that will lead the reader to the conclusion that you are a terrific job candidate. You must *show* the reader. Telling the reader is graceless and, worse, unconvincing.

A résumé should be long enough to include all the pertinent information but not so long that it will bore or irritate the reader. Generally, you should keep it to one page. If, however, you have special accomplishments—such as journal articles, patents, or substantial service in student government—a two-page résumé is appropriate. If your information comes to just over one page, either eliminate or condense some material to make it fit onto one page, or modify the layout so that it fills a substantial part of the second page. According to a study by Harcourt,

TABLE 18.1	Preferred Length for a Résumé
Preferred Length	**Hiring Officials (Percent)**
No longer than one page	23.6
No longer than two pages	41.8
Depends on applicant's information	32.7
Other	01.8

Krizan, & Merrier (1991), summarized in Table 18.1, most hiring officials prefer that a new college graduate's résumé be brief.

Two major styles of résumés are *chronological* and *analytical.* In the chronological résumé, you organize your information chronologically; that is, you use time as the organizing factor for each section of the résumé, including education and experience. In the analytical résumé, you present a section called *skills,* which is organized according to your skills, not according to time.

Recent graduates commonly use the chronological résumé because in most cases they haven't built the record of skills and accomplishments needed to fill out an analytical résumé. However, if you have professional work experience, you might consider the analytical style.

Elements of the Chronological Résumé Almost every chronological résumé has the six basic elements shown in Figure 18.3. However, your résumé should reflect one particular person: you. Many people have special skills or background that could be conveyed in additional sections.

Identifying Information

Include your full name, address, and phone number at the top of the page. Generally, you should present your name in boldface letters, centered at the top. Use your complete address, with the zip code. Use the two-letter state abbreviations used by the U.S. Postal Service. Also give your complete phone number, with the area code.

If your address during the academic year differs from your home address, list both and identify them clearly. An employer might call during an academic holiday to arrange an interview.

Job Objectives

After the identifying information, add a statement of objectives, in the form of a brief phrase or sentence—for example, "Objective: Entry-level position as a dietitian in a hospital." According to one study, 88.5 percent of managers making personnel decisions consider a statement of objectives important, because it gives

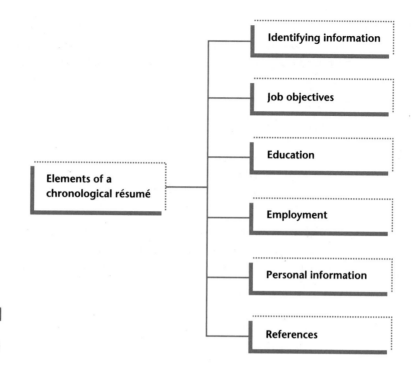

FIGURE 18.3

Elements of a
Chronological Résumé

the appearance that the writer has a clear sense of direction and goals (Harcourt, Krizan, & Merrier, 1991). In drafting the statement, follow two guidelines:

- *State only goals or duties explicitly mentioned in the job advertisement, or at least clearly implied.* If you unintentionally suggest that your goals are substantially different from the job responsibilities, the organization might infer that you would not be happy working there and not consider you further.

- *Avoid meaningless generalities.* You accomplish nothing by writing, "Position offering opportunities for advancement in the field of health science, where I can use my communication and analytical skills."

Education

If you are a student or a recent graduate, add the education section after the identifying information. If you have substantial professional experience, place the employment experience section before the education section.

Include at least the following information in the education section:

- *The degree.* After the degree abbreviation (such as B.S., B.A., A.A., or M.S.), list your academic major (and, if you have one, your minor)—for example, "B.S. in Materials Engineering, minor in General Business."

- *The institution.* Identify the institution by its full name: "Louisiana State University," not "LSU."

- *The location of the institution.* Include the city and state.
- *The date of graduation.* If your degree has not yet been granted, write "Anticipated date of graduation" or "Degree expected in" before the month and year.
- *Information about other schools you attended.* List any other institutions you attended beyond high school—even those at which you did not earn a degree. Employers are generally impressed to learn that a student began at a community or junior college and was able to transfer to a four-year college or university. The listing for other institutions attended should include the same information as the main listing. Arrange the entries in reverse chronological order: that is, list first the school you attended most recently.

The basic information about the educational experience is necessary, but it is not enough. It implies that you merely endured an institution and received a degree. The following four strategies can be useful in elaborating on your education:

- *List your grade-point average.* If your grade-point average is significantly above the median for the graduating class, list it. Or list your average in the courses in your major, if that is more impressive.
- *Create a list of courses.* Include courses that will be of particular interest to your reader. Advanced courses in an area of your major concentration might be appropriate, especially if the prospective employer has mentioned that area in the job ad. Also useful would be a list of communications courses: technical communication, public speaking, organizational communications, and the like. For example, a list of business courses on an engineer's résumé shows special knowledge and skills. The only kind of course list that is *not* particularly helpful is one that merely names the traditional required courses. Include the substantive titles of listed courses; employers won't know what "Chemistry 450" is. Call it "Chemistry 450. Organic Chemistry."
- *Describe a special accomplishment.* If you did a special senior design or research project, for example, a two- or three-line description would be informative. Include in this description the title and objective of the project, any special or advanced techniques or equipment you used, and—if you know them—the major results. Such a description might be phrased as follows: "A Study of Composite Substitutes for Steel—a senior design project intended to formulate a composite material that can be used to replace steel in car axles." Even a traditional academic course in which you conducted a sophisticated project and wrote a major report can be described profitably. A project discussion makes you look less like a student—someone who takes courses—and more like a professional—someone who designs and carries out projects.

- *List honors and awards you received.* Scholarships, internships, and academic awards offer evidence of exceptional ability. If you have received a number of such honors, or some that were not exclusively academic, you might list them separately (in a section called *Honors* or *Awards*) rather than in a subsection of the education section. Often, some information could logically be placed in two or even three locations, and you must decide where it will make the best impression.

The education section is the easiest part of the résumé to adapt for different positions. For example, a student majoring in electrical engineering who is applying for a position that calls for strong communications skills can list communications courses in one version of the résumé. The same student can list advanced electrical engineering courses in a résumé directed to another prospective employer. As you compose the education section of your résumé, carefully emphasize those aspects of your background that meet the requirements for the particular job.

Employment

The employment section, like the education section, conveys at least the basic information about each job you've held:

- dates of employment
- the organization's name and location
- your position or title

However, such a skeletal list would not be very informative or impressive. As in the education section, you should provide carefully selected details.

What readers want to know, after they have learned where and when you were employed, is what you actually did. Therefore, you should provide at least a two- to three-line description for each position. For particularly important or relevant jobs, give a larger description. Focus the description on one or more of the following factors:

- *Documents.* What kinds of documents did you write or assist in writing? List, especially, various governmental forms and any long reports, manuals, or proposals.
- *Clients.* What kinds of, and how many, clients did you transact business with as a representative of your organization?
- *Skills.* What kinds of technical skills did you learn or practice on the job?
- *Equipment.* What kinds of technical equipment did you operate or oversee? Mention, in particular, any computer skills you demonstrated; they can be very useful in almost every kind of position.
- *Money.* How much money were you responsible for? Even if you considered your bookkeeping position fairly simple, the fact that the

organization grossed, say, $2 million a year shows that the position involved real responsibility.

- *Personnel.* How many people did you supervise? Students sometimes supervise small groups of other students or technicians. Naturally, supervision shows maturity and responsibility.

Whenever possible, emphasize *results.* If you reorganized the shifts of the weekend employees you supervised, state the results:

Reorganized the weekend shift, resulting in a cost savings of more than $3,000 per year.

Wrote and produced (with desktop publishing) a parts catalog that is still used by the company and that increased our phone inquiries by more than 25 percent.

When you describe your positions, use the active voice—"supervised three workers"—rather than the passive voice—"three workers were supervised by me." The active voice emphasizes the verb, which shows the action. In thinking about your functions and responsibilities in various positions, keep in mind the strong action verbs that clearly communicate your activities. (See Chapter 14 for more information on strong verbs.) Also note that résumés often omit the *I* at the start of sentences. Rather than writing, "I prepared bids. . . ." many people write, "Prepared bids. . . ." Whichever style you use, be consistent.

Figure 18.4 lists some strong verbs.

Here is a sample listing:

June–September 19XX: Millersville General Hospital, Millersville, TX. Student Dietitian. Gathered dietary histories and assisted in preparing menus for a 300-bed hospital. Received "excellent" on all items in evaluation by head dietitian.

In just a few lines, you can show that you sought and accepted responsibility and that you acted professionally. Do not write, "I accepted responsibility"; rather, present facts that lead the reader to that impression.

Naturally, not all jobs entail such professional skills and responsibilities. Many students find summer work as laborers, sales clerks, short-order cooks, and so forth. If you have not held a professional position, list the jobs you have held, even if they were completely unrelated to your career plans. If the job title

FIGURE 18.4

Strong Action Verbs for Use in Résumés

administered	coordinated	evaluated	maintained	provided
advised	corresponded	examined	managed	purchased
analyzed	created	expanded	monitored	recorded
assembled	delivered	hired	obtained	reported
built	developed	identified	operated	researched
collected	devised	implemented	organized	solved
completed	directed	improved	performed	supervised
conducted	discovered	increased	prepared	trained
constructed	edited	instituted	produced	wrote

is self-explanatory—such as waitperson or service-station attendant—don't elaborate. Every job is valuable. You learn that you are expected to be someplace at a specific time, wear appropriate clothes, and perform specific duties. Also, every job helps pay college expenses. If you can write that you earned, say, 50 percent of your annual expenses through a job, employers will be impressed by your self-reliance. Most of them probably started out with nonprofessional positions. And any job you have held can yield a valuable reference.

One further word: if you have held a number of nonprofessional positions as well as several professional positions, you can group the nonprofessional ones:

> Other Employment: Cashier (summer, 1993), salesperson (part-time, 1993), clerk (summer, 1992).

This technique prevents the nonprofessional positions from drawing the reader's attention away from the more important positions.

List your jobs in reverse chronological order on the résumé, to highlight your most recent positions.

Personal Information

Most résumés do not include such information as the writer's height, weight, date of birth, and marital status; federal legislation prohibits organizations from requiring this information. In addition, most people now feel that such personal information is irrelevant to a person's ability.

However, the personal information section of the résumé *is* the appropriate place for a few items about your outside interests:

- participation in community-service organizations—such as Big Brothers/Big Sisters—or volunteer work in a hospital

- hobbies related to your career—for example, amateur electronics for an engineer

- sports, especially those that might be socially useful in your professional career, such as tennis, racquetball, and golf

- university-sanctioned activities, such as membership on a team, participation on the college newspaper, or election to a responsible position in an academic organization or a residence hall

Do not include activities that might create a negative impression: hunting, gambling, performing in a rock band. And always omit such activities as meeting people and reading—everybody meets people and reads.

References

Potential employers will want to learn more about you from your professors and previous employers. In providing these references, follow three steps:

- *Decide how you want to present the references.* You may list the names of three or four referees—people who have written letters of recommendation or who have agreed to speak on your behalf—on your résumé. Or you may simply say that you will furnish the names of the referees upon request. The length of your résumé sometimes dictates which approach to use. If the résumé is already long, the abbreviated form might be preferable. If it does not fill out the page, the longer form might be the better one. However, each style has advantages and disadvantages that you should consider carefully.

    Furnishing the referees' names appears open and forthright. It shows that you have already secured your referees and have nothing to hide. If one or several of the referees are prominent people in their fields, the reader is likely to be impressed. And, perhaps most important, the reader can easily phone the referees or write them a letter. Listing the referees makes it easy for the prospective employer to proceed with the hiring process. The only disadvantage is that it takes space on the résumé that might be needed for other information.

    On the other hand, writing "References will be furnished upon request" takes only one line. In addition, it leaves you in a flexible position. You can still secure referees after you have sent the résumé. Also, you can send selected letters of reference to prospective employers according to your analysis of what they want. Using different references for different positions is sometimes just as valuable as sending different résumés. However, some readers will interpret the lack of names and addresses as evasive or secretive or assume that you have not yet asked prospective referees. A greater disadvantage is that if the readers are impressed by the résumé and want to learn more about you, they cannot do so quickly and directly.

    What do personnel officers prefer regarding references? According to the study referred to earlier (Harcourt, Krizan, & Merrier, 1991), 58.4 percent of the hiring officials surveyed want to see the full references, including the name, title, organization, mailing address, and phone number for each referee. For example:

Dr. Robert Ariel
Assistant Professor of Biology
Central University
1910 Westerly Parkway
Portland, OR 97202
(503) 636–5746

- *Choose your referees carefully.* Solicit references only from persons who know your work best and for whom you have done the best work—for instance, a previous employer with whom you worked closely, or a professor from whom you have received A's. It is unwise to ask prominent professors who do not know your work well; the advantage of having a famous name on the résumé will be offset by the referee's brief and uninformative letter. Often, a less prominent professor who knows you can write the most informative letter or provide the best recommendation.

- *Give the potential referee an opportunity to decline gracefully.* Sometimes the person has not been as impressed with your work as you think. Also, if you simply ask, "Would you please write a reference letter for me?" the potential referee might accept and then write a lukewarm letter. It is better to follow up the first question with "Would you be able to write an enthusiastic letter for me?" or "Do you feel you know me well enough to write a strong recommendation?" If the potential referee shows any signs of hesitation or reluctance, you can withdraw the request at that point. The scene is a little embarrassing, but it is better than receiving a half-hearted recommendation.

Other Elements

So far, the discussion has concentrated on the sections that appear on virtually everyone's résumé. Other sections are discretionary or are appropriate only for certain writers.

- *Military experience.* If you are a veteran of the armed forces, include a military service section on the résumé. Define your military service as if it were any other job, citing the dates, locations, positions, ranks, and tasks. Often a serviceperson receives regular evaluations from a superior; these evaluations can work in your favor.
- *Language ability.* If you have a working knowledge of a foreign language, your résumé should include a *Language Skills* section. Language skills are particularly relevant if the potential employer has international interests and you could be useful in translation or foreign service.
- *Willingness to relocate.* If you are willing to relocate, state that fact explicitly. Many organizations today will find you a more attractive candidate if they know you are willing to move around as you learn the business.

Elements of the Analytical Résumé

The analytical résumé differs from the chronological one in that instead of using the experience section to convey the job skills and experience, it adds a separate section, usually called *skills* or *skills and abilities,* to convey this information. Figure 18.5 on page 572 shows the seven basic elements of an analytical résumé. The employment section of an analytical résumé becomes a briefer list of identifying information about the candidate's employment history: name of the company, dates of employment, and name of the position. The skills section usually is placed prominently near the top of the résumé.

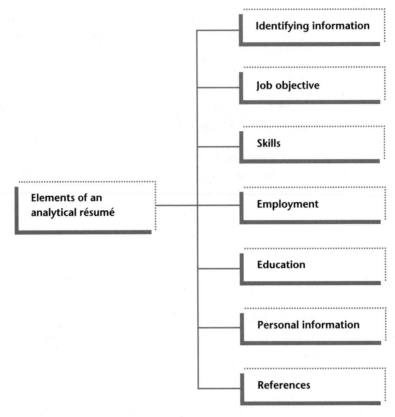

Elements of an Analytical
Résumé

Following is an example of a skills section.

Skills and Abilities

Management
   Served as weekend manager of six employees in the retail clothing business. Also
   trained three summer interns at a health-maintenance organization.

Writing and Editing
   Wrote status reports, edited performance appraisals, participated in assembling and
   producing an environmental impact statement by desktop publishing.

Teaching and Tutoring
   Tutored in the University Writing Center. Taught a two-week course in electronics for
   teenagers. Coach youth basketball.

In a skills section, you choose the headings, the arrangement, and the level of detail. Your goal, of course, is to highlight those skills that the employer is seeking.

Figures 18.6, 18.7, and 18.8 show effective résumés. Figure 18.6 is a chronological résumé for Carlos Rodriguez, a student seeking a full-time job. He is what is known as a traditional student: he entered college right after high school and has proceeded on schedule. He has no dependents and is willing to relocate, so he makes that point. Because some of his former employers are high-powered referees, he lists their names at the end of the résumé.

Figure 18.7 on page 574 is a chronological résumé written by Alice Linder, a nontraditional student: a single mother who returned to school after having raised three children. She is applying for an internship. Figure 18.8 on page 575 is an analytical résumé by Alice Linder.

CARLOS RODRIGUEZ

3109 Vista Street Philadelphia, PA 19136 (215) 555-3880 crodrig@dragon.du.edu

Objective   Entry-level position in signal processing

Education   Bachelor of Science in Electrical Engineering
Drexel University, Philadelphia, PA
Anticipated June 1999
Grade-Point Average: 3.67 (of 4.0)
Senior Design Project: "Enhanced Path-Planning Software for Robotics"
Advanced Engineering Courses

Digital Signal Processing	Computer Hardware
Introduction to Operating Systems I, II	Systems Design
Digital Filters	Computer Logic Circuits I, II

Employment

6/96–1/97   RCA Advanced Technology Laboratory, Moorestown, NJ
Developed and tested new LISP software to implement the automated production of VLSI standard cell family databooks. Assisted senior engineer in designing and testing two Gate Array (2700 and 5000 gates) VLSI integrated chips.

6/94–1/95   RCA Advanced Technology Laboratory, Moorestown, NJ
Verified and documented several integrated circuit designs. Used CAD software and hardware to simulate, check, and evaluate these designs. Gained experience on the VAX and Applicon.

Honors, Awards, and Organizations
Eta Kappa Nu (Electrical Engineering Honor Society)
Tau Beta Pi (General Engineering Honor Society)
Institute of Electrical and Electronics Engineers, Inc.

Willing to relocate.

References

Ms. Anita Feller	Mr. Fred Borelli	Mr. Sam Shamir
Engineering Consultant	Unit Manager	Comptroller
700 Church Road	RCA Corporation	RCA Corporation
Cherry Hill, NJ 08002	Route 38	Route 38
(609) 555–7836	Moorestown, NJ 08057	Moorestown, NJ 08057
	(609) 555–2435	(609) 555–7849

FIGURE 18.6

Chronological Résumé of a Traditional Student

Alice P. Linder   1781 Weber Road
                          Warminster, PA 18974
                          (215) 555-3999

Objective: A position in molecular research that uses my computer skills

Education: Harmon College, West Yardley, PA
                  Major: Bioscience and Biotechnology
                  Expected Graduation Date: June 2000

                  Related Course Work
                  General Chemistry I, II, III   Biology I, II, III
                  Organic Chemistry I, II   Statistical Methods for Research
                  Physics I, II   Technical Communication
                  Calculus I, II

Employment Experience:

June 1997–present   *Smith Kline & French Laboratory,* Upper Merion, PA
(20 hours per week) Analyze molecular data on E & S PS300, Macintosh, and
                     IBM PCs. Write programs in C+, and wrote a user's guide
                     for an instructional computing package. Train and consult
                     with scientists and deliver in-house briefings.

June 1986–Jan. 1990 *Anchor Products, Inc.,* Ambler, PA
                     Managed 12-person office in a $1.2 million company. Also
                     performed general bookkeeping and payroll.

Oct. 1994–present   *Children's Hospital of Philadelphia,* Philadelphia, PA
                     Volunteer in the physical therapy unit. Assist therapists and
                     guide patients with their therapy. Use play therapy to en-
                     hance strengthening progress.

Honors: Awarded three $5,000 tuition scholarships (1995–1998)
                  from the Gould Foundation.

Additional Information:
                  Member, Harmon Biology Club, Yearbook Staff
                  Raised three school-age children
                  Tuition 100% self-financed

References: Available upon request.

FIGURE 18.7

Chronological Résumé
of a Nontraditional
Student

Alice P. Linder	1781 Weber Road Warminster, PA 18974 (215) 555-3999
Objective:	A position in molecular research that uses my computer skills
Skills and Abilities:	*Laboratory Skills* • Analyze molecular data on E & S PS300, Macintosh, and IBM PCs. Write programs in C+. • Have taken 12 credits in biology and chemistry labs. *Communication Skills* • Wrote a user's guide for an instructional computing package. • Train and consult with scientists and deliver in-house briefings. *Management Skills* • Managed 12-person office in a $1.2 million company.
Education:	Harmon College, West Yardley, PA Major: Bioscience and Biotechnology Expected Graduation Date: June 2000

Related Course Work

General Chemistry I, II, III	Biology I, II, III
Organic Chemistry I, II	Statistical Methods for
Physics I, II	Research
Calculus I, II	Technical Communication

Employment Experience:

June 1997–present (20 hours per week)	*Smith Kline & French Laboratory,* Upper Merion, PA Laboratory Assistant Grade 3
June 1983–Jan. 1987	*Anchor Products, Inc.,* Ambler, PA Office Manager
Oct. 1994–present	*Children's Hospital of Philadelphia,* Philadelphia, PA Volunteer in the physical therapy unit. Assist therapists and guide patients with their therapy. Use play therapy to enhance strengthening progress.

Honors:	Awarded three $5,000 tuition scholarships (1995–1998) from the Gould Foundation.
Additional Information:	Member, Harmon Biology Club, Yearbook Staff Raised three school-age children Tuition 100% self-financed
References:	Available upon request.

FIGURE 18.8

Analytical Résumé of a
Nontraditional Student

A note about Alice Linder's résumés: there are two strategies for presenting the information that a candidate is a nontraditional student. Alice Linder feels that her nontraditional status is an asset: her maturity and experience will make her a more effective employee than the traditional student. For this reason, she mentions this information on her résumé. Other people would deemphasize this information because they feel it is irrelevant and draws attention away from important credentials. As a compromise between these two strategies, a student could mention the nontraditional status briefly in the letter but omit it from the résumé.

Writing Electronic Résumés

More than a half million jobs are posted on the Internet today, and within the next year or two electronic job searches will become the major way people are hired (Robinson, 1997). The digital revolution is changing the way people look for work.

The key change is that electronic résumés are becoming as popular as traditional paper résumés. An electronic résumé is a résumé that you send in a digital format. There are two main reasons for the popularity of electronic résumés:

- *Electronic communication is fast.* You can get your materials to the prospective employer in a matter of seconds.

- *Electronic résumés can be searched electronically.* Many organizations cannot afford to read every application; Hewlett-Packard, for instance, receives more than 1,000 résumés every day (Robinson, 1997). Therefore, large organizations often create electronic databases, so that when they need a person with a particular set of skills, they search for keywords electronically.

An electronic résumé can take different forms:

- *A résumé sent as an e-mail message.* You can format your résumé as an e-mail message. The recipient can print it, scan it, or add the information to a database.

- *A résumé attached to an e-mail message.* You can attach the word-processing file to an e-mail message, and the recipient can print it or treat it like an e-mail message. Keep in mind, however, that some electronic-mail software cannot read attached files. If the job notice requests "a plain text document sent in the body of the message," do not send an attached file.

- *A printed résumé that will be scanned into an organization's database.* You can send a printed résumé that the recipient will scan and enter into a database. There are several popular database programs for this purpose, such as ResTrac or Resumix.

- *A Web-based résumé.* You can put your résumé on your own Web site and hope that employers will come to you, or you can post it to a Web board.

Ways of creating and sending résumés will doubtless change as the technology changes. For now, you need to know that the traditional printed résumé is just one of the several ways to present your credentials, and you need to keep abreast of the new techniques for applying for positions. Which form should your résumé take? Whichever form the organization prefers. If you learn of a position from an ad on the organization's own site, it will tell you how to apply.

Content of an Electronic Résumé

Most of what was said earlier about the content of a printed résumé applies to an electronic résumé. The résumé must be honest, it must be free of errors, and it must provide clear, specific information, without generalizations.

But if the résumé is to be scanned instead of read by a person, you need to include industry-specific jargon. That means you need to include all the keywords that the employment officer might use to search for qualified candidates. If the employment officer is looking for someone with experience writing Web pages, be sure you include the terms "Web page," "Internet," "HTML," and any other relevant keywords. If your position requires an understanding of programming languages, name the languages you know. In short, whereas a traditional printed résumé focuses on verbs—tasks you have done—an electronic résumé focuses on nouns. Electronic résumés often contain a section called *keywords,* which is just a list of industry-specific jargon.

Two other points:

- Be sure to use the phrase "willing to relocate" if you are applying for a position that would require you to move. Some organizations use willingness to relocate as a criterion as they search their database.

- List your e-mail address in your résumé.

Format of an Electronic Résumé

Here are five tips for preparing an electronic résumé that is easy to read and scan:

- *Use ASCII text.* ASCII text includes the letters, numbers, and basic punctuation marks. Avoid boldface, italics, and underlining, as well as special characters such as "smart quotation marks" or math symbols. Also avoid using horizontal or vertical lines or graphics. To be sure you are using only ASCII characters, save your file as "text only." Open it up using your software's text editor, such as Notepad, and check to be sure it contains only ASCII characters.

- *Use a simple sans-serif typeface, such as Ariel.* Scanners can easily interpret large, open typefaces such as Ariel.

- *Use a single-column format.* A double-column text will not scan accurately. Align everything on a uniform left margin.

- *Use wide margins.* Instead of an 80-character width, set your software for 60 or 65; this way, regardless of the equipment your reader is using, the lines will break as you intend them.
- *Use the space bar instead of the tab key.* Tabs will be displayed according to the settings on your reader's equipment, not according to the settings on yours. Therefore, use the space bar to move text horizontally.

If you are mailing a paper résumé that will be scanned, follow these three additional guidelines:

- *Use a good-quality laser printer.* The better the resolution, the better the scanner will work.
- *Use white paper.* Even a slight tint to the paper can increase the chances that the scanner will misinterpret a character.
- *Do not fold the résumé.* The fold line can cause problems for the scanner.

Figure 18.9 is an example of an electronic résumé.

This is an electronic version of the résumé in Figure 18.8. Notice that the writer uses ASCII text and left justification.

Alice P. Linder
1781 Weber Road
Warminster, PA 18974
(215) 555-3999

Objective: A position in molecular research that uses my computer skills

Skills and Abilities:
Laboratory Skills. Analyze molecular data on E & S PS300, Macintosh, and IBM PCs. Write programs in C+. Have taken 12 credits in biology and chemistry labs.

Communication Skills. Wrote a user's guide for an instructional computing package. Train and consult with scientists and deliver in-house briefings.

Management Skills. Managed 12-person office in a $1.2 million company.

Education:
Harmon College, West Yardley, PA
Major: Bioscience and Biotechnology
Expected Graduation Date: June 2000

Related Course Work:
General Chemistry I, II, III

FIGURE 18.9

Electronic Résumé

(Figure 18.9 continued)

Organic Chemistry I, II
Physics I, II
Calculus I, II
Biology I, II, III
Statistical Methods for Research
Technical Communication

Employment Experience:
June 1997-present (20 hours per week)
Smith Kline & French Laboratory, Upper Merion, PA
Laboratory Assistant Grade 3

June 1983-Jan. 1987
Anchor Products, Inc., Ambler, PA
Office Manager

Oct. 1994-present
Children's Hospital of Philadelphia, Philadelphia, PA
Volunteer in the physical therapy unit. Assist therapists and guide patients with their therapy. Use play therapy to enhance strengthening progress.

Honors:
Awarded three $5,000 tuition scholarships (1995–1998) from the Gould Foundation.

Additional Information:
Member, Harmon Biology Club, Yearbook Staff
Raised three school-age children
Tuition 100% self-financed

Willing to relocate.

Keywords:
molecular research, programming, presentations, management, laboratory assistant, volunteer, physical therapy

References:
Available upon request.

Electronic résumés are not as attractive as paper-based ones, because they lack the formatting. In addition, they are longer, because they use only a single, narrow column of text.

The writer has added "willing to relocate" and a keywords list.

For more information on electronic résumés, use a search engine and look for "careers" or "jobs" or "employment." The search engine will direct you to job boards, as well as to Usenet newsgroups.

Writing Job-Application Letters

If you send a traditional paper résumé, you will also include a job-application letter. You might also send a letter with an electronic résumé.

The job-application letter is crucial because it is the first thing your reader sees. If the letter is ineffective, the reader probably will not bother to read the résumé.

If students had infinite time and patience, they would make a different résumé for each prospective employer, highlighting their appropriateness for that one job. But because they don't have unlimited time and patience, they usually make only one or two versions of their résumés. As a result, the typical résumé makes a candidate look only relatively close to the ideal the employer has in mind. Thus, the job-application letter must appeal directly and specifically to the needs and desires of the particular employer.

See Chapter 19 for a discussion of letter formats.

Content of the Job-Application Letter

Like the résumé, the job-application letter is a sales document. Its purpose is to convince the reader that you are an outstanding candidate who should be called in for an interview. Of course, you accomplish this purpose through evidence, not empty self-praise.

The job-application letter is *not* an expanded version of everything in the résumé. The key to a good application letter is selectivity. In the letter, you choose from the résumé two or three points of greatest interest to the potential employer and develop them into paragraphs. Emphasize results, such as improved productivity or quality or decreased costs. If one of your previous part-time positions called for specific skills that the employer is looking for, that position might be the subject of a substantial paragraph in the letter, even though the résumé devotes only a few lines to it. If you try to cover every point on your résumé, the letter will be fragmented. The reader then will have a hard time forming a clear impression of you, and the purpose of the letter will be thwarted.

In most cases, a job-application letter should fill the better part of a page. Like all business correspondence, it should be single spaced, with double spaces between paragraphs. For more experienced candidates, the letter may be longer, but most students find that they can adequately describe their credentials in one page. If you write at length on a minor point, you become boring. Worse still, you appear to have poor judgment. Employers are always seeking candidates who can say much in a short space.

Elements of the Job-Application Letter

Among the mechanical elements of the job-application letter, the inside address—the name, title, organization, and address of the recipient—is most important. If you know the correct form of this information from an ad, there is no problem. However, if you are uncertain about any of the information—the recipient's name, for example, might have an unusual spelling—verify it by phoning the organization. Many people are very sensitive about such matters, so you should not risk beginning the letter with a misspelling or an incorrect title.

When you do not know who should receive the letter, do not address it to a department of the company—unless the job ad specifically says to do so—because nobody in that department might feel responsible for dealing with it. Instead, phone the company to find out who manages the department. If you are unsure of the appropriate department or division to write to, address the letter to a high-level executive, such as the president. The letter will be directed to the right person. Also, because the application includes both a letter and a résumé, include an enclosure notation.

The four-paragraph letter that will be discussed here is only a basic model. Because every substantial job-application letter has an introductory paragraph, two body paragraphs, and a concluding paragraph, the minimum number of paragraphs for the job-application letter has to be four. But there is no reason that an application letter cannot have five or six paragraphs. The four basic paragraphs are shown in Figure 18.10.

Because this is such an important letter, you should plan it carefully. Choose information from your background that best responds to the needs of the potential employer. Draft the letter and then revise it. Let it sit for a while, then revise it again. Spend as much time on it as you can. Make each paragraph a unified, functional part of the whole letter. Supply clear transitions from one paragraph to the next. (See Chapter 13 for more information on paragraphs.)

The Introductory Paragraph

The introductory paragraph establishes the tone of the letter and captures the reader's attention. It has four specific functions.

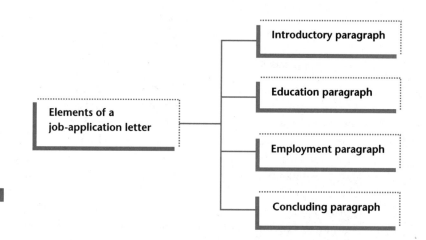

FIGURE 18.10

Elements of a Job-Application Letter

- *It identifies your source of information.* In an unsolicited application, all you can do is ask if a position is available. For most applications, however, your source of information is a published advertisement or an employee already working for the organization. If your source is an ad, identify specifically the publication and its date of issue. If an employee told you about the position, identify that person by name and title.

- *It identifies the position you are interested in.* Often, the organization you are applying to has a number of positions advertised; without the title, your reader might not know which position you seek.

- *It states that you wish to be considered for the position.* Although the context makes your wish obvious, you should mention it because the letter would be awkward without it.

- *It forecasts the rest of the letter.* Choose a few phrases that forecast the body of the letter, so that the letter flows smoothly.

These four aspects of the introductory paragraph need not appear in any particular order, nor need each be covered in a single sentence. The following examples of introductory paragraphs demonstrate different ways to provide the necessary information.

RESPONSE TO A JOB AD:
I am writing in response to your notice in the May 13 *New York Times.* I would like to be considered for the position in system programming. I hope you find that my studies at Eastern University in computer science, along with my programming experience at Airborne Instruments, would qualify me for the position.

UNSOLICITED:
My academic training in hotel management and my experience with Sheraton International have given me a solid background in the hotel industry. Would you please consider me for any management trainee position that might be available?

UNSOLICITED PERSONAL CONTACT:
Mr. Howard Alcott of your Research and Development Department suggested that I write. He thinks that my organic chemistry degree and my practical experience with Brown Laboratories might be of value to you. Do you have an entry-level position in organic chemistry for which I might be considered?

As these sample paragraphs indicate, the important information can be conveyed in any number of ways. The difficult part of the introductory paragraph—and of the whole letter—is to achieve the proper tone: quiet self-confidence. Because the letter will be read by someone who is professionally superior to you, the tone must be modest, but it should not be self-effacing or negative. Never say, for example, "I do not have a very good background in computers but I'm willing to learn." The reader will take this kind of statement at face value and probably stop reading right there. You should show pride in your education and experience, while at the same time suggesting by your tone that you have much to learn.

The Education Paragraph

For most students, the education paragraph should come before the employment paragraph because the content of the education paragraph will be stronger. If, however, your work experience is more pertinent than your education, discuss your work first.

In devising your education paragraph, take your cue from the job ad (if you are responding to one). What aspect of your education most directly fits the job requirements? If the ad stresses versatility, you might structure your paragraph around the range and diversity of your courses. Also, you might discuss course work in a field related to your major, such as business or communication skills. Extracurricular activities are often very valuable; if you were an officer in a student organization in your field, you could discuss the activities and programs that you coordinated. Perhaps the most popular strategy for developing the education paragraph is to discuss skills and knowledge gained from advanced course work in the major field.

Whatever information you provide, the key to the education paragraph is to develop one unified idea, rather than to toss a series of unrelated facts onto the page. Notice how each of the following education paragraphs develops a unified idea.

> At Eastern University, I have taken a wide range of courses in the sciences, but my most advanced work has been in chemistry. In one laboratory course, I developed a new aseptic brewing technique that lowered the risk of infection by more than 40 percent. This new technique was the subject of an article in the *Eastern Science Digest.* Representatives from three national breweries have visited our laboratory to discuss the technique with me.

> To broaden my education at Southern, I took eight business courses in addition to my requirements for the Civil Engineering degree. Because your ad mentions that the position will require substantial client contact, I believe that my work in marketing, in particular, would be of special value. In an advanced marketing seminar, I used PageMaker to produce a 20-page sales brochure describing Oppenheimer Properties' different kinds of building structures that were for sale to industrial customers in our section of the city. That brochure is now being used at Oppenheimer Properties, where I am serving as an intern.

> The most rewarding part of my education at Western University took place outside the classroom. My entry in a fashion-design competition sponsored by the university won second place. More important, through the competition I met the chief psychologist at Eastern Regional Hospital, who invited me to design clothing for people with physical handicaps. I have since completed six different outfits. These designs are now being tested at the hospital. I hope to be able to pursue this interest once I start work.

Notice that each of these paragraphs begins with a topic sentence—a forecast of the rest of the paragraph—and uses considerable detail and elaboration to develop the main idea. Notice, too, a small point: if you haven't already specified your major and college or university in the introductory paragraph, be sure to do so in the education paragraph.

The Employment Paragraph

Like the education paragraph, the employment paragraph should begin with a topic sentence and then elaborate a single idea. That idea might be that you have a broad background or that a single job has given you special skills that make you particularly suitable for the available job. The following examples show effective experience paragraphs.

> For the past three summers and part-time during the academic year, I have worked for Redego, Inc., a firm that specializes in designing and planning industrial complexes. I began as an assistant in the drafting room. By the second summer, I was accompanying a civil engineer on field inspections. Most recently, I have used CAD to assist an engineer in designing and drafting the main structural supports for a 15-acre, $30 million chemical facility.

> Although I have worked every summer since I was 15, my most recent position, as a technical editor, has been the most rewarding. I was chosen by Digital Systems, Inc., from among 30 candidates because of my dual background in computer science and writing. My job was to coordinate the editing of computer manuals. Our copy editors, who are generally not trained in computer science, need someone to help verify the technical accuracy of their revisions. When I was unable to answer their questions, I was responsible for interviewing our systems analysts to find the correct answer and to make sure the computer novice could follow it. This position gave me a good understanding of the process by which operating manuals are created.

> I have worked in merchandising for three years as a part-time and summer salesperson in men's fashions and accessories. I have had experience running inventory control software and helped one company switch from a manual to an online system. Most recently, I assisted in clearing $200,000 in out-of-date men's fashions: I coordinated a campaign to sell half of the merchandise at cost and was able to convince the manufacturer's representative to accept the other half for full credit. For this project, I received a certificate of appreciation from the company president.

Notice how these writers carefully define their duties to give their readers a clear idea of the nature and extent of their responsibilities.

Although you will discuss your education and experience in separate paragraphs, try to link these two halves of your background. If an academic course led to an interest that you were able to pursue in a job, make that point clear in the transition from one paragraph to the other. Similarly, if a job experience helped shape your academic career, tell the reader about it.

The Concluding Paragraph

The concluding paragraph of the job-application letter, like that of any sales letter, is intended to stimulate action. In this case, you want the reader to invite you for an interview. In the preceding paragraphs of the letter, you provided the information that should have convinced your reader to give you another look. In the last paragraph, you want to make it easy for him or her to do so. The concluding paragraph contains three main elements:

- a reference to your résumé
- a request for an interview
- your phone number and e-mail address

If you have not yet referred to the enclosed résumé, do so at this point. Then, politely but confidently request an interview—making sure to use the phrase *at your convenience*. Don't make the request sound as if you're asking a personal favor. And be sure to include your phone number and the time of day you can be reached—even though the phone number is also on your résumé. Many employers will pick up a phone and call promising candidates personally, so make it as easy as possible for them to do so. Adding an e-mail address gives the employer one more way to get in touch with you.

Following are examples of effective concluding paragraphs.

> The enclosed résumé provides more information about my education and experience. Could we meet at your convenience to discuss further the skills and experience I could bring to Pentamax? You can leave a message for me anytime at (303) 555-5957 or cfilli@claus.cmu.edu.

> More information about my education and experience is included on the enclosed résumé, but I would appreciate the opportunity to meet with you at your convenience to discuss my application. You can reach me after noon on Tuesdays and Thursdays at (212) 555-4527.

Figures 18.11 and 18.12 on pages 586 and 587 show examples of effective job-application letters corresponding to the résumés presented in Figures 18.6 and 18.7.

The letter in Figure 18.12 needs some comment. An earlier section discussed strategies for discussing a student's nontraditional status. The writer of the letter in Figure 18.12 decided to discuss her nontraditional background—she is a single mother who has been out of school for more than 10 years—without overemphasizing it. Accordingly, the first two paragraphs of the body discuss her qualifications for the internship position. She wants to make sure that the reader evaluates her as he would any candidate. Then she adds a paragraph that attempts to explain how her additional experience—as an office manager and a single mother—have enabled her to develop skills that would be of value to anyone in any field. She exploits her situation gracefully, always appealing to the reader's needs, not to his sympathy.

Writing Follow-Up Letters

If you do get an interview, you should write two of the four types of follow-up letters: the first type plus one of the other three:

- *The letter of appreciation after an interview.* Your purpose in writing a follow-up letter is to thank the representative for taking the time to see you and to emphasize your particular qualifications. You can also take

3109 Vista Street
Philadelphia, PA 19136

January 19, 1996

Mr. Stephen Spencer, Director of Personnel
Department 411
Boeing Naval Systems
103 Industrial Drive
Wilmington, DE 20093

Dear Mr. Spencer:

I am writing in response to your advertisement in the January 16 *Philadelphia Inquirer.* Would you please consider me for the position in Signal Processing? I believe that my academic training at Drexel University in Electrical Engineering, along with my experience with RCA Advanced Technology Laboratory, would qualify me for the position.

My education at Drexel University has given me a strong background in computer hardware and system design. I have concentrated on digital and computer applications, developing and designing computer and signal-processing hardware in two graduate-level engineering courses I was permitted to take. For my senior-design project, I am working with four other undergraduates in using OO programming techniques to enhance the path-planning software for an infrared night-vision robotics application.

While working at RCA I was able to apply my computer experience to the field of the VLSI design. In one project I used my background in LISP to develop and test new LISP software used in the automated production of VLSI standard cell family databooks. In another project, I used CAD software on a VAX to evaluate IC designs.

The enclosed résumé provides an overview of my education and experience. Could I meet with you at your convenience to discuss my qualifications for this position? Please write to me at the above address or leave a message anytime at (215) 555-2660. My e-mail address is crodrig@dragon.du.edu.

Yours truly,

Carlos Rodriguez

Carlos Rodriguez

Enclosure (1)

FIGURE 18.11

Job-Application Letter

1781 Weber Road
Warminster, PA 18974

January 17, 1995

Mr. Harry Gail
Fox Run Medical Center
399 N. Abbey Road
Warminster, PA 18974

Dear Mr. Gail:

Last April I contacted you regarding the possibility of an internship as a laboratory assistant at your center. Your assistant, Mary McGuire, told me then that you might consider such a position this year. With the experience I have gained since last year, I believe that I would be a valuable addition to your center in many ways.

At Harmon College, I have earned a 3.7 GPA in 36 credits in chemistry and biology; all but two of these courses had laboratory components. One skill stressed here is the ability to communicate effectively in writing and orally. Our science courses have extensive writing and speaking requirements; my portfolio includes seven research papers and lab reports of more than 20 pages each, and I have delivered four oral presentations, one of 45 minutes, to classes.

At Smith Kline & French, where I currently work part time, I analyze molecular data on an E&S PS300, a Macintosh, and an IBM PC. I have tried to remain current with the latest advances; my manager at Smith Kline has allowed me to attend two different two-day in-house seminars on computerized data analysis using SAS.

Having been out of school for more than a decade, I am well aware of how much the technology has changed. However, as the manager of a 12-person office for four years, I believe that I have acquired skills that would benefit Fox Run. In addition, as a single mother of three I know something about time management.

More information about my education and experience is included on the enclosed résumé, but I would appreciate the opportunity to meet with you at your convenience to discuss my application. If you would like any additional information about me or Harmon's internship program, please write to me at the above address, call me at (215) 555-4006, or e-mail me at alinder@harmon.edu.

Very truly yours,

Alice P. Linder

Alice P. Linder

Enclosure

FIGURE 18.12

Job-Application Letter

this opportunity to restate your interest in the position. The follow-up letter can do more good with less effort than any other step in the job-application procedure, because so few candidates take the time to write it. Figure 18.13 is an example of a letter of appreciation.

- *The letter of acceptance of a job offer.* This one is easy: express appreciation, show enthusiasm, and repeat the major terms of your employment. Figure 18.14 is an example of a letter of acceptance of a job offer.

- *The letter of rejection in response to a job offer.* This one is also easy. If you decide not to accept a job offer, write a polite letter, expressing your appreciation and, if appropriate, explaining why you decline the offer. Remember, you might want to work for this company some time. Figure 18.15 shows this kind of letter.

- *The letter acknowledging a job rejection.* This one is hard to write. Why should you write back after you have been rejected for a job? To maintain good relations with the company. You just might get a phone call the next week, explaining that the person who accepted the job has had to change her plans and offering you the position. Figure 18.16 on page 590 is an example of this kind of letter.

Dear Mr. Weaver:

Thank you for taking the time yesterday to show me your facilities and to introduce me to your colleagues.

Your advances in piping design were particularly impressive. As a person with hands-on experience in piping design, I can appreciate the advantages your design will have.

The vitality of your projects and the good fellowship among your employees further confirm my initial belief that Cynergo would be a fine place to work. I would look forward to joining your staff.

Sincerely yours,

Harriet Bommarito

Harriet Bommarito

FIGURE 18.13

Letter of Appreciation after an Interview

Dear Mr. Weaver:

Thank you very much for the offer to join your staff. I accept.

I look forward to joining your design team on Monday, July 19. The salary, as you indicate in your letter, is $31,250.

As you have recommended, I will get in touch with Mr. Matthews in Personnel to get a start on the paperwork.

I appreciate the trust you have placed in me, and I assure you that I will do what I can to be a productive team member at Cynergo.

Sincerely yours,

Mark Greenberg

Mark Greenberg

FIGURE 18.14

Letter of Acceptance
of a Job Offer

Dear Mr. Weaver:

I appreciate very much the offer to join your staff.

Although I am certain that I would benefit greatly from working at Cynergo, I have decided to take a job with a firm in Pittsburgh, where I have been accepted at Carnegie-Mellon to pursue my master's degree at night.

Again, thank you your generous offer.

Sincerely yours,

Cynthia O'Malley

Cynthia O'Malley

FIGURE 18.15

Letter of Rejection in
Response to a Job Offer

Dear Mr. Weaver:

I was disappointed to learn that I will not have a chance to join your staff, because I feel that I could make a substantial contribution. However, I appreciate that job decisions are complex, involving many candidates and many factors.

Thank you very much for the courtesy you have shown me. I have long believed—and I still believe—that Cynergo is a first-class organization.

Sincerely yours,

Paul Goicochea

Paul Goicochea

FIGURE 18.16

Letter of
Acknowledgment
of a Job Rejection

CASE

Updating Career-Center Materials

The members of your group are student interns in the Career Center at your school. The director of the Center, William Karey, would like your group to assist him in updating the materials he distributes to students looking for work. "What I'd like you to do," he tells you, "is take a look at what is out there on the Web, and integrate it with what we already have on résumés and job letters." If possible, secure an electronic copy of available materials. Then search the Web for information on job boards and electronic résumés. Is the information consistent, or are there different ideas on how to do an electronic search? Which sources seem to be of most use for students at your school? Decide whether you should integrate your ideas into the printed material the Career Center distributes, put part or all of the new and existing information on the school's Web site, or both. Regardless of your decision, submit a copy of the existing materials as well as a copy of the new materials.

WRITER'S CHECKLIST

Printed Résumé

1. Does the résumé respond to the needs of its readers?

2. Is the résumé honest?

3. Does the résumé have a professional appearance, with generous margins, a symmetrical layout, adequate white space, and effective indentation?

4. Is the résumé free of errors?

5. Does the identifying information section contain your name, address(es), and phone number(s)?

6. Have you included a clear statement of your job objectives?

7. Does the education section include your degree, your institution and its location, and your anticipated date of graduation,

as well as any other information that will help your reader appreciate your qualifications?

8. Does the employment section include, for each job, the dates of employment, the organization's name and location, and (if you are writing a chronological résumé) your position or title, as well as a description of your duties and accomplishments?

9. Does the personal information section include relevant hobbies or activities, including extracurricular interests?

10. Does the references section include the names, job titles, organizations, mailing addresses, and phone numbers of three or four referees? If you are not listing this information, does the strength of the rest of the résumé offset the omission?

11. Does the résumé include any other appropriate sections, such as military service, language skills, or honors?

Electronic Résumé

In addition to the items mentioned in the checklist for the printed résumé, have you
- used ASCII text?
- used a simple sans-serif typeface?
- used a single-column format?
- used wide margins?
- used the space bar instead of the tab key?

Job-Application Letter

1. Does the letter respond to the needs of its readers?

2. Is the letter honest?

3. Does the letter look professional?

4. Does the introductory paragraph identify your source of information and the position you are applying for, state that you wish to be considered, and forecast the rest of the letter?

5. Does the education paragraph respond to your reader's needs with a unified idea introduced by a topic sentence?

6. Does the employment paragraph respond to your reader's needs with a unified idea introduced by a topic sentence?

7. Does the concluding paragraph include a reference to your résumé, a request for an interview, your phone number, and your e-mail address?

8. Does the letter include an enclosure notation?

Letter of Appreciation

1. Does the letter of appreciation for a job interview thank the interviewer and briefly restate your qualifications?

EXERCISES

Several of the following exercises call for you to write memos. See Chapter 16 for a discussion of memos.

1. In a newspaper or journal or on the Internet, find a job advertisement for a position in your field for which you might be qualified. Write a résumé and a job-application letter in response to the ad. Include the job ad or a photocopy of it. You will be evaluated not only on the content and appearance of the materials, but also on how well you have targeted the materials to the job ad.

2. You are a human-resources specialist for XYZ Corporation. Your primary responsibility is to screen job applications, rejecting those of applicants who seem inappropriate for a particular position and forwarding those of applicants who seem qualified. Your supervisor is interested in knowing more about how you do this work; she has given you a résumé (shown on page 592) and asked you to comment on it. Write a 500-word memo to your supervisor evaluating the strengths and weaknesses of the résumé responding to an ad describing the following duties: "Research and develop key technology and system concepts for spectrally efficient digital radio frequency data networks such as digital cellular mobile radio telephones, public safety trunked digital radio systems, and satellite communications."

3. As a human-resources specialist for XYZ Corporation, write a 500-word memo to your supervisor evaluating the strengths and weaknesses of the letter on page 593 responding to an ad describing the following duties: "Buyer for a high-fashion ladies' dress shop. Experience required."

4. Write a brief memo to your instructor evaluating the letter of appreciation on page 594.

Rajiv Siharath
2319 Fifth Avenue
Waverly, CT 01603
Phone: 611-3356

Personal Data:	22 Years old Height 5'11" Weight 176 lbs.
Education:	B.S. in Electrical Engineering University of Connecticut, June, 19XX
Experience:	6/XX–9/XX Falcon Electronics Examined panels for good wiring. Also, I revised several schematics. 6/XX–9/XX MacDonalds Electrical Supply Co. Worked parts counter. 6/XX–9/XX Happy Burger Made hamburgers, fries, shakes, fish sandwiches, and fried chicken. 6/XX–9/XX Town of Waverly Outdoor maintenance. In charge of cleaning up McHenry Park and Municipal Pool picnic grounds. Did repairs on some electrical equipment.
Backround:	Born and raised in Waverly. Third baseman, Fisherman's Rest softball team. Hobbies: jogging, salvaging and repairing appliances, reading magazines, politics.
References:	Will be furnished upon request.

Collaborative Exercise

5. In a newspaper or journal or on the Internet, find a job advertisement for a position in your field for which you might be qualified. Write a résumé and a job-application letter in response to the ad. Include the job ad or a photocopy of it. Exchange your materials with another group member. Meet with that person to discuss your responses to his or her materials. Then revise your materials and exchange them again. Meet again to discuss the other person's response to them. Together, draft a memo to your instructor on the success of the exercise. Which aspects of the critique were useful? Which were less successful? When you apply for your first position, what kind of peer review will you use? Why?

April 13, 19XX

Marilyn Grissert
Best Department Store
113 Hawthorn
Atlanta, Georgia

Dear Ms. Grissert:

As I was reading the Sunday *Examiner,* I came upon your ad for a buyer. I have always been interested in learning about the South, so would you consider my application?

I will receive my degree in fashion design in one month. I have taken many courses in fashion design, so I feel I have a strong background in the field.

Also, I have had extensive experience in retail work. For two summers I sold women's accessories at a local clothing store. In addition, I was a temporary department head for two weeks.

I have enclosed a résumé and would like to interview you at your convenience. I hope to see you in the near future. My phone number is 555-6103.

Sincerely,

Brenda Sisneros

Web
Exercises

6. Using a job bank on the Web, list and briefly describe five positions in your field in your state. What skills, experience, and background does each position require? What is the salary range for each position?

7. Locate and provide the URLs of three job boards that provide interactive forms to create a résumé automatically.

REFERENCES

Harcourt, J., Krizan, A.C., & Merrier, P. (1991). Teaching résumé content: Hiring officials' preferences versus college recruiters' preferences. *Business Education Forum 45* (7), 13–17.

Robinson, K. (1997, Mar. 24). Job search engine. *Idaho Statesman,* p. D1.

914 Imperial Boulevard
Durham, NC 27708

November 13, 19XX

Mr. Ronald O'Shea
Division Engineering
Safeway Electronics, Inc.
Holland, MI 49423

Dear Mr. O'Shea:

Thanks very much for showing me around your plant. I hope I was able to convince you that I'm the best person for the job.

Sincerely yours,

Robert Harad

Robert Harad

CHAPTER 19

Writing Letters

Projecting the "You Attitude" 596

Avoiding Letter Clichés 598

Understanding the Elements of the Letter 599
 Heading *600*
 Inside Address *600*
 Attention Line *601*
 Subject Line *601*
 Salutation *601*
 Body *602*
 Complimentary Close *602*
 Signature *602*
 Enclosure Line *602*
 Copy Line *603*

Learning the Format of the Letter 603

Understanding Common Types of Letters 603
 Order Letters *603*
 Inquiry Letters *605*
 Responses to Inquiries *607*
 Sales Letters *607*
 Claim Letters *610*
 Adjustment Letters *612*

Case: Dangerous Wrenches 616

Writer's Checklist 617

Exercises 617
 Collaborative Exercise *620*
 Web Exercise *622*

Whether you are a technical professional or a technical communicator, you will find yourself writing letters frequently. Letters are the most traditional and formal of the basic formats you will write on the job. Although letters are sometimes sent from one person to another within the same organization, letters are used most often when people in different organizations communicate. In addition to showing the basic formats, this chapter discusses tone and strategy in letter writing, as well as several common kinds of letters.

The letter is still a basic means of communication between two organizations: close to 100 million business letters are written each workday. With the popularity of e-mail, why are people still writing letters? One reason is that letters provide a documentary record; the reader can see the writer's signature. A second reason is that letters convey a more formal appearance than e-mail. For communication between two organizations, and sometimes for communication within an organization, this formality can be important. (One problem with letters, compared to e-mail, is the time it takes for a letter to be delivered; however, fax machines solve this problem in many situations.)

Even as a new employee, you can expect to write letters regularly. And as you advance to positions of greater responsibility, you will write even more letters, because you will be representing your organization more often. Writing a letter is much like writing any other technical document. First you have to analyze your audience and determine your purpose. Then you have to gather your information, create an outline, write a draft, and revise it.

Projecting the "You Attitude"

Like any other type of technical communication, the letter should be ethical, clear, concise, comprehensive, accessible, correct, professional, and accurate. It must convey information in a logical order. It should not contain small talk; the first paragraph should get directly to the point without wasting words. And to enable the reader to locate information quickly and easily, a topic sentence should appear at the start of each paragraph in the body. Often, letters use headings and indentation just as reports do.

Moreover, because it is a communication from one person to another, a letter must also convey a courteous, positive tone. The key is the *you attitude*. This term means looking at the situation from your reader's point of view and adjusting the content, structure, and tone to meet the person's needs. The you attitude is largely common sense. If, for example, you are writing to a supplier who has failed to deliver some merchandise on the agreed-upon date, the you attitude dictates that you not discuss problems you are having with other suppliers—those problems don't concern your reader. Rather, you should concentrate on explaining clearly and politely that your reader has violated your agreement and that not having the merchandise is costing you money. Then you should propose ways to expedite the shipment.

INTERVIEW: ENGINEER JOHN MOLLER

on the kinds of short documents he writes routinely

My responsibility is business development, so I routinely write letters to prospective clients outlining my company's capabilities, proposals, meeting minutes that record what the people assigned to a project are supposed to do (and what they're not supposed to do), and brochures.

on planning the kind of information that goes into a document

There are certain items that go into each kind of document. For instance, minutes include information about where and when the meeting took place, who attended, what was discussed, and what was resolved. But the most important thing I think about when I sit down to write any kind of document is the needs of the people who will read it.

on using existing documents as a template

I start with an existing example of the document I'm going to write, then make the necessary changes. All our documents are in electronic form, so this is the easiest way. In addition, using an existing document as a template ensures that I don't leave out any necessary kinds of information.

on the characteristics he tries to achieve in his short documents

The first sentence or paragraph is usually a summary. The key is to get the summary into one or two sentences. Also, I try to keep everything to one page; it increases the probability that someone will read it. Big contracts and proposals, of course, have to be longer, but for routine documents I strive for clarity and conciseness.

on his education

I never took a technical writing course. Except for a few lab courses, writing wasn't really built into the engineering curriculum. At one company I worked for, I think the owner got fed up with the quality of the documents we produced, and he hired a writing consultant; the entire company took his course, discussing writing and doing a lot of exercises. And it worked. We got a lot better.

on the amount of writing he does

I write almost 100 percent of the time.

on the importance of writing to his career

It's absolutely critical to my career and to the success of this company. If we can't communicate clearly with our customers, we're in big trouble.

John Moller is vice president of Laramore, Douglass and Popham, a consulting engineering firm in New York City.

Looking at things from the other person's point of view would be simple if both parties always saw things the same way. They don't, of course. Sometimes the context of the letter is a dispute. Nevertheless, good letter writers always maintain a polite tone. Civilized behavior is good business, as well as a good way to live.

Following are examples of thoughtless sentences, each followed by an improved version that exhibits the you attitude.

EGOTISTICAL: Only our award-winning research and development department could have devised this revolutionary new sump pump.

BETTER: Our new sump pump features significant innovations that you may appreciate.

BLUNT: You wrote to the wrong department. We don't handle complaints.

BETTER: Your letter has been forwarded to the Customer Service Division.

ACCUSING:	You must have dropped the engine. The housing is badly cracked.
BETTER:	The badly cracked housing suggests that your engine must have fallen onto a hard surface from some height.
SARCASTIC:	You'll need two months to deliver these parts? Who do you think you are, the Post Office?
BETTER:	Surely you would find a two-month delay for the delivery of parts unacceptable in your business. That's how I feel too.
BELLIGERENT:	I'm sure you have a boss, and I doubt if he'd like to hear about how you've mishandled our account.
BETTER:	I'm sure you would prefer to settle the account between us rather than having it brought to your supervisor's attention.
CONDESCENDING:	Haven't you ever dealt with a major corporation before? A 60-day payment period happens to be standard.
BETTER:	Perhaps you were not aware of the standard 60-day payment period.
OVERSTATED:	Your air-filter bags are awful. They're all torn. We want our money back.
BETTER:	You will doubtless be surprised to learn that 19 of the 100 air-filter bags we purchased are torn. We hope you agree that refunding the purchase price of the 19 bags—$190.00—is the fair thing to do.

After you have drafted a letter, review it. Put yourself in your reader's place. How would you react if you received it? A calm, respectful, polite tone always makes the best impression and therefore increases the chances of achieving your goal.

Avoiding Letter Clichés

Related to the you attitude is the issue of letter clichés. Over the decades, a set of words and phrases has come to be associated with letters; one common example is *as per your request*. For some reason, many people think that these phrases are required. They're not. They make the letter sound stilted and insincere. If you would feel awkward or uncomfortable saying these clichés to a friend, avoid them in your letters.

Figure 19.1 is a list of some of the common clichés and their more natural equivalents.

Figure 19.2 on page 600 shows two versions of the same letter: one written in letter clichés, the other in plain language.

The letter on the right side in Figure 19.2 not only avoids clichés but also shows an understanding of the you attitude. Rather than building the letter around the violation of the warranty, as the writer on the left does, the writer on the right presents the message as good news: the snowmobile is not ruined, and it can be returned in less than a week for a low cost.

Letter Clichés	Natural Equivalents
attached please find	attached is
cognizant of	aware that
enclosed please find	enclosed is
endeavor (verb)	try
herewith ("We herewith submit . . . ")	(None. "Herewith" doesn't say anything. Skip it.)
hereinabove	previously, already
in receipt of ("We are in receipt of . . . ")	"We have received . . . "
permit me to say	(None. Permission granted. Just say it.)
pursuant to our agreement	as we agreed
referring to your ("referring to your letter of March 19, the shipment of pianos . . . ")	"As you wrote in your letter of March 19, the . . . " or subordinate the reference at the end of your sentence.
same (as a pronoun: "Payment for same is requested . . . ")	(Use the noun instead: "Payment for the merchandise is requested . . . ")
wish to advise ("We wish to advise that . . . ")	(The phrase doesn't say anything. Just say what you want to say.)
the writer ("The writer believes that . . . ")	"I believe . . . "

FIGURE 19.1

Letter Clichés and
Natural Equivalents

Understanding the Elements of the Letter

Almost every letter has a heading, inside address, salutation, body, complimentary close, signature, and reference initials. In addition, some letters contain one or more of the following notations: attention, subject, enclosure, and copy.

For brief, simple letters, you can compose the elements in the sequence in which they will appear. For longer, more complex letters, however, you might find it easier to start with the body and continue to the end, then add the preliminary elements.

In the following paragraphs, the elements of the letter are discussed in the order they would ordinarily appear. Six common types of letters will be discussed in detail later in this chapter.

Letter Containing Clichés

Dear Mr. Smith:

Referring to your letter regarding the problem encountered with your new Eskimo Snowmobile. Our Customer Service Department has just tendered its report.

It is their conclusion that the malfunction is caused by water being present in the fuel line. It is our unalterable conclusion that you must have purchased some bad gasoline. We trust you are cognizant of the fact that while we guarantee our snowmobiles for a period of not less than one year against defects in workmanship and materials, responsibility cannot be assumed for inadequate care. We wish to advise, for the reason mentioned hereinabove, that we cannot grant your request to repair the snowmobile free of charge.

Permit me to say, however, that the writer would be pleased to see that the fuel line is flushed at cost, $30. Your Eskimo would then give you many years of trouble-free service.

Enclosed please find an authorization card. Should we receive it, we shall endeavor to perform the above-mentioned repair and deliver your snowmobile forthwith.

Sincerely yours,

Letter in Natural Language

Dear Mr. Smith:

Thank you for writing to us about the problem with your new Eskimo Snowmobile.

Our Customer Service Department has found water in the fuel line. Apparently some of the gasoline was bad. While we guarantee our snowmobiles for one year against defects in workmanship and materials, we cannot assume responsibility for problems caused by bad gasoline. We cannot, therefore, grant your request to repair the snowmobile free of charge.

However, no serious harm was done to the snowmobile. We would be happy to flush the fuel line at cost, $30. Your Eskimo would then give you many years of trouble-free service.

If you will authorize us to do this work, we will have your snowmobile back to you within four working days. Just fill out the enclosed authorization card and drop it in the mail.

Sincerely yours,

FIGURE 19.2

Letter Clichés and
Natural Language

Heading The typical organization has stationery with its name, address, phone number, and perhaps a logo—the letterhead—printed at the top. The letterhead and the date the letter is sent (printed two lines below the letterhead) make up the heading. When printing on blank paper, use your address (without your name) and date as the heading. Print only the first page of any letter on letterhead stationery. Print the second and all subsequent pages on blank paper, with the name of the recipient, the page number, and the date in the upper left-hand corner. For example:

> Mr. John Cummings
> Page 2
> July 3, 19XX

Do not number the first page of a letter.

Inside Address The inside address is your reader's name, position, organization, and business address. If your reader has a courtesy title, such as *Dr., Professor,* or—for public officials—*Honorable,* use it. If not, use *Mr.* or *Ms.* (unless you know the reader prefers *Mrs.* or *Miss*). If your reader's position can fit conveniently on the same line as his or her name, add it after a comma; otherwise, place it on the line below. Spell the name of the organization the way the organization itself does: for example, International Business Machines calls itself IBM. Include the complete mailing address: the street, city, state, and zip code.

Attention Line

Sometimes you will be unable to address the letter to a particular person. If you don't know (and cannot easily find out) the person's first name or don't know the person's name at all, use the attention line:

Attention: Technical Director

Use the attention line if you want to make sure that the organization you are writing to responds even if the person you write to is unavailable. In this case, put the name of the organization or one of its divisions on the first line of the inside address:

Operations Department
Haverford Electronics
117 County Line Road
Haverford, MA 01765

Attention: Charles Fulbright, Director

Subject Line

On the subject line, put either a project number (for example, "Subject: Project 31402") or a brief phrase defining the subject of the letter (for example, "Subject: Price Quotation for the R13 Submersible Pump"):

Operations Department
Haverford Electronics
117 County Line Road
Haverford, MA 01765

Attention: Charles Fulbright, Director

Subject: Purchase Order #41763

Salutation

If you have no attention line or subject line, put the salutation two lines below the inside address. The traditional salutation is *Dear* followed by the reader's courtesy title and last name. Use a colon after the name, not a comma. If you are fairly well acquainted with your reader, use *Dear* followed by the first name. When you do not know the reader's name, use a general salutation:

Dear Technical Director:

Dear Sir or Madam:

When you are addressing a group of people, use one of the following salutations:

Ladies and Gentlemen:

Gentlemen: (if all the readers are male)

Ladies: (if all the readers are female)

Or you can tailor the salutation to your readers:

Dear Members of the Restoration Committee:

Dear Members of Theta Chi Fraternity:

This same strategy is useful for sales letters without individual inside addresses:

Dear Homeowner:

Dear Customer:

Body The body is the substance of the letter. In most cases, you will have at least three paragraphs: an introductory paragraph, a concluding paragraph, and one or more body paragraphs. For information on how to develop the argument of the body, see Chapter 8.

Complimentary Close After the body of the letter, include one of the traditional closing expressions: *Sincerely, Sincerely yours, Yours sincerely, Yours very truly, Very truly yours.* Capitalize only the first word in the complimentary close, and follow all such expressions by a comma. Today, all these phrases have lost whatever particular meanings and connotations they once possessed. They are interchangeable.

Signature Type your full name on the fourth line below the complimentary close. Sign the letter, in ink, above the typewritten name. Most organizations prefer that you add your position, beneath your typed name. For example:

Very truly yours,

José Santos

José Santos
Personnel Manager

Enclosure Line If the envelope contains documents other than the letter, identify the number of enclosures:

FOR ONE ENCLOSURE: Enclosure
or
Enclosure (1)

FOR MORE THAN ONE ENCLOSURE: Enclosures (2)
Enclosures (3)

In determining the number of enclosures, count only the separate items, not the number of pages. A 3-page memo and a 10-page report constitute only two enclosures. Some writers like to identify the enclosures:

Enclosure: 1997 Placement Bulletin

Enclosures (2): "This Year at Ammex"
1997 Annual Report

Copy Line If you want the reader to know that other people are receiving a copy, use the symbol *c* (for "copy") or *pc* (for "photocopy") followed by the names of the other recipients (listed either alphabetically or according to organizational rank). If you do not want your reader to know about other copies, use *bc* ("blind copy") on the copies only—not on the original.

Learning the Format of the Letter

Three popular formats are used for letters: modified block, modified block with paragraph indentations, and full block. Figure 19.3 on page 604 shows diagrams of letters written in these three formats.

Understanding Common Types of Letters

Dozens of kinds of letters exist for specific occasions. This section focuses on six types written most frequently in the technical world, as shown in Figure 19.4 on page 605. Two other types of letters are discussed in this book: the transmittal letter in Chapter 10 and the job-application letter in Chapter 18. For a more detailed discussion of business letters, consult one of the several full-length books on the subject.

Order Letters Perhaps the most basic form of business correspondence is the order letter, written to a manufacturer, wholesaler, or retailer. When writing an order letter, include all the information your reader will need to identify the merchandise: the quantity, model number, dimensions, capacity, material, price, and any other pertinent details. Also specify the method of delivery. A typical order letter is shown in Figure 19.5 on page 606. (Notice that the writer uses an informal table to describe the parts she orders.)

Many organizations have preprinted forms, called purchase orders, for ordering products or services. A purchase order calls for the same information that appears in an order letter.

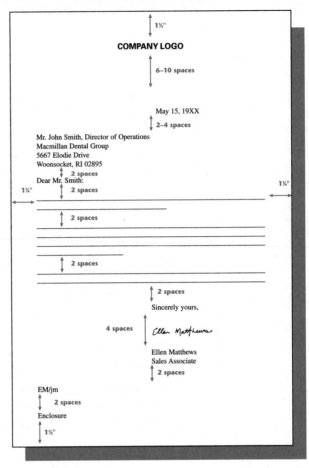

a. Modified block format

Paragraphs are generally indented five characters.

FIGURE 19.3

Popular Formats for Letters

b. Modified block with paragraph indentations

c. Full block style—everything aligned along the left margin

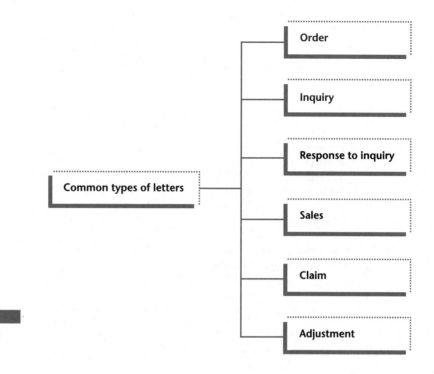

FIGURE 19.4

Common Types
of Letters

Inquiry Letters   Your purpose in writing an inquiry letter is to obtain information. Writing an inquiry letter is more or less difficult, depending on whether the reader is expecting your letter.

- If the reader is expecting the letter, your task is easy. For example, if a company that makes institutional furniture has advertised that it will send its 48-page, full-color brochure to prospective clients, you need write merely a one-sentence letter: "Would you please send me the brochure advertised in *Higher Education Today,* May 13, 19XX?" The manufacturer knows why you're writing and, naturally, wants to receive letters such as yours, so no explanation is necessary. You can also ask a technical question, or set of questions, about any product or service for sale. Your inquiry letter might begin "We are considering purchasing your new X-15 work stations for an office staff of 35 and would like some further information. Would you please answer the following questions?" The detail about the size of the potential order is not necessary, but it does make the inquiry seem serious and the potential sale substantial. An inquiry letter of this kind will get a prompt and gracious reply.

- If your reader is not expecting your letter, your task is more difficult. If the reader will not directly benefit from supplying information, you must ask a favor. Only careful, persuasive writing will make your reader *want* to respond despite the absence of the profit motive.

WAGNER AIRCRAFT
116 North Miller Road
Akron, OH 44313

September 4, 19XX

Mr. Frank DiFazio
Franklin Aerospace Parts
623 Manufacturer's Blvd.
Bethpage, NY 11741

Dear Mr. DiFazio:

Would you please send us the following parts by parcel post? All page numbers
refer to your 19XX catalog.

Quantity	Model No.	Page	Catalog Description	Price
2	36113-NP	42	Seal fins	$ 34.95
1	03112-Bx	12	Turbine-bearing support	19.75
5	90135-QN	102	Turbine disc	47.50
1	63152-Bx	75	Turbine-bearing housing	16.15
			Total Price:	$118.35

Yours very truly,

Christine O'Hanlon

Christine O'Hanlon
Purchasing Agent

FIGURE 19.5

Order Letter

Follow these four guidelines in writing an inquiry letter:

- *In the first paragraph, state why you are writing to* this *person or organization.* You might use subtle flattery at this point—for example, "I was
  hoping that, as the leader in solid-state electronics, your company might
  be able to furnish some information about. . . ." Then you might explain
  why you want the information. Obviously, a company will not furnish

information to a competitor. You have to show that your interests are not commercial—for instance, "I will be using this information in a senior project in agronomy at Illinois State University. I am trying to devise a . . . ." If you need the information by a certain date, this might be a good place to mention it: "The project is to be completed by April 15, 19XX."

- *List your questions.* Readers are understandably annoyed by thoughtless requests to send "everything you have" on a topic. They much prefer a set of carefully thought-out technical questions showing that the writer has already done substantial research. "Is your Model 311 compatible with Norwood's Model B12?" is obviously much easier to respond to than "Would you please tell me about your Model 311?" If your questions can be answered in a small space, simply leave room for your reader's reply after each question or in the margin.

- *Offer something in return, because you are asking someone to do something for you.* In many cases, all you can offer are the results of your research. If possible, state that you would be happy to send a copy of your report. Express your appreciation. Don't say, "Thank you for sending me this information." Such a statement is presumptuous, because it assumes that the reader is both willing and able to meet your request. A statement such as the following is more effective: "I would greatly appreciate any help you could give me in answering these questions." Finally, if the answers will be brief, enclose a stamped self-addressed envelope for your reader's reply.

- *Write a thank-you note to someone who has responded to your inquiry letter.*

Figure 19.6 on page 608 shows an example of a letter of inquiry.

Responses to Inquiries

If you ever receive an inquiry letter, keep the following suggestions in mind. If you wish to provide the information the writer asks for, do so graciously. If the questions are numbered, number your responses correspondingly. If you cannot answer the questions, either because you don't know the answers or because you cannot divulge proprietary information, explain the reasons and offer to assist with other requests. Figure 19.7 on page 609 shows a response to the inquiry letter in Figure 19.6.

Sales Letters

A large, sophisticated sales campaign costs millions of dollars—for marketing surveys and consulting fees, printing, postage, and promotions. This kind of campaign is beyond the scope of this book. However, you may well have to draft a sales letter for a product or service.

The you attitude is crucial in a sales letter. Your readers don't care why you want to sell your product or service. They want to know why they should buy it.

14 Hawthorne Ave.
Belleview, TX 75234

November 2, 19XX

Dr. Andrew Shakir
Director of Technical Services
Orion Corporation
721 West Douglas Avenue
Maryville, TN 31409

Dear Dr. Shakir:

I am writing to you because of Orion's reputation as a leader in the manufacture of adjustable X-ray tables. I am a graduate student in biomedical engineering at the University of Texas, and I am working on an analysis of diagnostic equipment for a seminar paper. Would you be able to answer a few questions about your Microspot 311?

1. Can the Microspot 311 be used with lead oxide cassettes, or does it accept only lead-free cassettes?
2. Are standard generators compatible with the Microspot 311?
3. What would you say is the greatest advantage, for the operator, of using the Microspot 311? For the patient?

My project is due on January 15. I would greatly appreciate your assistance in answering these questions. Of course, I would be happy to send you a copy of my report when it is completed.

Yours very truly,

Albert K. Stern

Albert K. Stern

FIGURE 19.6

Inquiry Letter

Because you are asking them to spend valuable time studying the letter, you must provide clear, specific information to help them understand what you are selling and how it will help them. Be upbeat and positive, but never forget that readers want facts.

Most writers of sales letters use a four-stage strategy:

ORION

721 WEST DOUGLAS AVE.   (615) 619-8132
MARYVILLE, TN 31409   TECHNICAL SERVICES

November 7, 19XX

Mr. Albert K. Stern
14 Hawthorne Ave.
Belleview, TX 75234

Dear Mr. Stern:

I would be pleased to answer your questions about the Microspot 311. We think it is the best unit of its type on the market today.

1. The 311 can handle lead oxide or lead-free cassettes.
2. At the moment, the 311 is fully compatible only with our Duramatic generator. However, special wiring kits are available to make the 311 compatible with our earlier generator models—the Olympus and the Saturn. We are currently working on other wiring kits.
3. For the operator, the 311 increases the effectiveness of the radiological procedure while at the same time cutting down the amount of film used. For the patient, it cuts down the number of repeat exposures and therefore reduces the total dose.

I am enclosing a copy of our brochure on the Microspot 311. If you would like additional copies, please let me know. I would be happy to receive a copy of your analysis when it is complete. Good luck!

Sincerely yours,

Andrew Shakir, M.D.

Andrew Shakir, M.D.
Director of Technical Services

AS/le

Enclosure
c: Robert Anderson, Executive Vice-President

FIGURE 19.7

Response to an Inquiry

1. *Gain the reader's attention.* Unless the opening sentence seems either interesting or important, the reader will put the letter aside. To attract the reader, use interesting facts, quotations, or questions. In particular, try to identify a problem that will interest your reader. A few examples of effective openings follow.

   How much have construction costs risen since your plant was built? Do you know how much it would cost to rebuild at today's prices?

   The Datafix copier is better than the Xerox—and it costs less, too. We'll repeat: it's better and it costs less!

   If you're like most training directors, we bet you've seen your share of empty promises. We've heard all the stories, too. And that's why we think you'll be interested in what *Fortune* said about us last month.

2. *Describe the product or service you are trying to sell.* What does it do? How does it work? What problems does it solve?

   The Datafix copier features automatic loading, so your people don't waste time watching the copies come out. Datafix copies from a two-sided original—automatically! And Datafix can turn out 90 copies a minute—which is 25 percent faster than our fastest competitor. . . .

3. *Convince your reader that your claims are accurate.* Refer to users' experience, testimonials, or evaluations by reputable experts or testing laboratories.

   In a recent evaluation conducted by *Office Management Today,* more than 85 percent of our customers said they would buy another Datafix. The next best competitor: 71 percent. And Datafix earned a "Highly Reliable" rating, the highest recommendation in the reliability category. All in all, Datafix scored *higher* than any other copier in the desktop class. . . .

4. *Tell your reader how to find out more about your product or service.* If possible, provide a postcard that the reader can use to request more information or arrange for a visit from one of your sales representatives. Make it easy to proceed to the next step in the sales process.

Figure 19.8 shows an example of a sales letter.

Claim Letters   A claim letter is a polite and reasonable complaint. If as a private individual or a representative of an organization you purchase a defective or falsely advertised product or receive inadequate service, your first recourse is a claim letter.

The purpose of the claim letter is to convince your reader that you are a fair and honest customer who is justifiably dissatisfied. If the letter does this, your chances of receiving an equitable settlement are good. Most organizations today pay attention to reasonable claims, because they realize that unhappy customers are bad for business. In addition, claim letters indicate the weak points in their product or service.

Writing a claim letter often calls for a four-stage strategy:

DAVIS TREE CARE
1300 Lancaster Avenue
Berwyn, PA 19092

May 13, 19XX

Dear Homeowner:

Do you know how much your trees are worth? That's right—your trees. As a recent purchaser of a home, you know how much of an investment your house is. Your property is a big part of your total investment.

Most people don't know that even the hardiest trees need periodic care. Like shrubs, trees should be fertilized and pruned. And they should be protected against the many kinds of diseases and pests that are common in this area.

At Davis Tree Care, we have the skills and experience to keep your trees healthy and beautiful. Our diagnostic staff is made up of graduates of major agricultural and forestry universities, and all of our crews attend special workshops to keep current with the latest information on tree maintenance. Add to this our proven record of 43 years of continuous service in the Berwyn area, and you have a company you can trust.

May we stop by to give you an analysis of your trees—absolutely without cost or obligation? A few minutes with one of our diagnosticians could prove to be one of the wisest moves you've ever made. Just give us a call at 555-9187 and we'll be happy to arrange an appointment at your convenience.

Sincerely yours,

Jasmine Brown

Jasmine Brown
President

FIGURE 19.8

Sales Letter

1. *Identify the product or service.* List the model numbers, serial numbers, sizes, and any other pertinent data.
2. *Explain the problem.* State the symptoms clearly and specifically. What function does not work? What exactly is wrong with the service?
3. *Propose an adjustment.* Define what you want the reader to do: for example, refund the purchase price, replace or repair the item, improve the service.

4. *Conclude courteously.* Suggest that you trust the reader, in the interest of fairness, to abide by your proposed adjustment.

The you attitude is just as important as content in a claim letter. You must project a calm and rational tone. A complaint such as "I'm sick and tired of being ripped off by companies like yours" will hurt your chances of an easy settlement. If, however, you write "I am very disappointed in the performance of my new Eversharp Electric Razor," you sound like a responsible adult. There is no reason to show anger in a claim letter, even if the other party has made an unsatisfactory response to an earlier one. Calmly explain what you plan to do, and why. Your reader will then more likely see the situation from your perspective. Figure 19.9 shows an example of a claim letter that the writer faxed to the reader.

Adjustment Letters   In an adjustment letter, you respond to a claim letter and tell the customer how you plan to handle the situation. Whether you are granting the customer everything proposed in the claim letter, part of it, or none of it, your purpose remains the same: to show that your organization is fair and reasonable, and that you value the customer's business.

If you can grant the request, the letter is easy to write. Express your regret about the situation, state the adjustment you are going to make, and end on a positive note by encouraging the customer to continue doing business with you.

If you cannot grant the request, try to salvage as much goodwill as you can. Obviously, your reader will not be happy. If your letter is carefully written, however, it can show that you have acted reasonably. While denying a request, you are attempting to explain your side of the matter, thus educating your reader about how the problem occurred and how to prevent it in the future.

This more difficult kind of adjustment letter generally has a four-stage structure:

1. *Attempt to meet the customer on some neutral ground.* Consider an expression of regret but not an apology. You might even thank the customer for bringing the matter to the attention of the company. If you admit that the customer is right in this kind of adjustment letter—such as by writing "We are sorry that the engine you purchased from us is defective"—the customer would have a good case against you if the dispute ended up in court.

2. *Explain why your company is not at fault.* Most often, you explain to the customer the steps that led to the failure of the product or service. Do not say "You caused this." Instead, use the less blunt passive voice: "The air pressure apparently was not monitored."

3. *Clearly state that your company is denying the request for the reasons you have noted in the letter.* This statement must come late in the letter. If you begin with it, most readers will not finish reading the letter, and therefore you will not be able to achieve your twin goals of education and goodwill.

ROBBINS CONSTRUCTION, INC.
255 Robbins Place Centerville, MO 65101 (417) 555-1850

August 19, 19XX

Mr. David Larsen
Larsen Supply Company
311 Elmerine Avenue
Anderson, MO 63501

Dear Mr. Larsen:

As steady customers of yours for over 15 years, we came to you first when we
needed a quiet pile driver for a job near a residential area. On your recommendation, we bought your Vista 500 Quiet Driver, at $14,900. We have since found,
much to our embarrassment, that it is not substantially quieter than a regular pile
driver.

We received the contract to do the bridge repair here in Centerville after promising
to keep the noise to under 90 db during the day. The Vista 500 (see enclosed copy
of bill of sale for particulars) is rated at 85 db, maximum. We began our work and,
although one of our workers said the driver didn't seem sufficiently quiet to him,
assured the people living near the job site that we were well within the agreed
sound limit. One of them, an acoustical engineer, marched out the next day and
demonstrated that we were putting out 104 db. Obviously, something is wrong with
the pile driver.

I think you will agree that we have a problem. We were able to secure other equipment, at considerable inconvenience, to finish the job on schedule. When I
telephoned your company that humiliating day, however, a Mr. Meredith informed
me that I should have done an acoustical reading on the driver before I accepted
delivery.

I would like you to send out a technician—as soon as possible—either to repair the
driver so that it performs according to specifications or to take it back for a full
refund.

Your truly,

Jack Robbins

Jack Robbins, President

Enclosure

FIGURE 19.9
Claim Letter

4. *Try to create goodwill.* You might, for instance, offer a special discount on another, similar product. A company's profit margin on any one item is almost always large enough to permit attractive discounts as an inducement to continue doing business.

Larsen Supply Company
311 Elmerine Avenue
Anderson, MO 63501

August 21, 19XX

Mr. Jack Robbins, President
Robbins Construction, Inc.
255 Robbins Place
Centerville, MO 65101

Dear Mr. Robbins:

I was very unhappy to read your letter of August 19 telling me about the failure of the Vista 500. I regretted most the treatment you received from one of my employees when you called us.

Harry Rivers, our best technician, has already been in touch with you to arrange a convenient time to come out to Centerville to talk with you about the driver. We will of course repair it, replace it, or refund the price. Just let us know your wish.

I realize that I cannot undo the damage that was done on the day that a piece of our equipment failed. To make up for some of the extra trouble and expense you incurred, let me offer you a 10 percent discount on your next purchase or service order with us, up to $1,000 total discount.

You have indeed been a good customer for many years, and I would hate to have this unfortunate incident spoil that relationship. Won't you give us another chance? Just bring in this letter when you visit us next, and we'll give you that 10 percent discount.

Sincerely,

Dave Larsen, President

FIGURE 19.10

"Good News"
Adjustment Letter

Figures 19.10 and 19.11 show examples of "good news" and "bad news" adjustment letters. The first letter is a reply to the claim letter shown in Figure 19.9.

QUALITY VIDEO PRODUCTS

February 3, 19XX

Mr. Dale Devlin
1903 Highland Avenue
Glenn Mills, NE 69032

Dear Mr. Devlin:

Thank you for writing us about the videotape you purchased on January 11, 19XX.

You used the videotape to record your daughter's wedding. While you were playing it back last week, the tape jammed and broke as you were trying to remove it from your VCR. You are asking us to reimburse you $500 because of the sentimental value of that recording.

As you know, our videotapes carry a lifetime guarantee covering parts and workmanship. We will gladly replace the broken videotape. However, the guarantee states that the manufacturer and the retailer will not assume any incidental liability. Thus we are responsible only for the retail value of the blank tape.

However, your wedding tape can probably be fixed. A reputable dealer can splice tape so skillfully that you will hardly notice the break. It's a good idea to make backup copies of your valuable tapes.

Attached to this letter is a list of our authorized dealers in your area, any of whom would be glad to do the repairs for you. We have already sent out your new videotape. It should arrive within the next two days.

Please contact us if we can be of any further assistance.

Sincerely yours,

Paul R. Blackwood

Paul R. Blackwood, Manager
Customer Relations

FIGURE 19.11

"Bad News"
Adjustment Letter

CASE

Dangerous Wrenches

Your group works for the Customer Service Department at United Tools, a manufacturer of hand tools. Recently, you received the following letter from a hardware store that carries your products.

Handyman Hardware, Inc.
Millersville, AL 61304

December 4, 19XX

United Tools
19 Central Avenue
Dover, TX 76104

Gentlemen:

I have a problem I'd like to discuss with you. I've been carrying your line of hand tools for many years.

Your 9" pipe wrench has always been a big seller. But there seems to be something wrong with its design. I have had two complaints in the last few months about the handle snapping off when pressure is exerted on it. In one case, the user cut his hand seriously enough to require stitches.

Frankly, I'm hesitant to sell any more of the 9" pipe wrenches, but I still have more than two dozen in inventory.

Have you had any other complaints about this product?

Sincerely yours,

You decide to investigate the problem. You learn that in the last two quarters you have sold more than 300 of these wrenches to retailers, and you have received four complaints from retailers and retail customers about the handle snapping. No one reported injuries. You go to the engineers, who two months ago discovered a design flaw that accounts for the problem. The engineers redesigned the wrench, manufactured samples, and tested them thoroughly. The engineers found no problems. Like the old design, the new design exceeds ANSI standards for this type of tool. The old design is no longer sold; all new orders are being filled with the new design.

What additional steps should you take? Are there other people in the company with whom you should meet? Draft a letter, to be sent to the retailers and the retail customers who wrote claim letters to your company, offering an appropriate adjustment. Does the situation you have uncovered within your company merit further action? If so, write an appropriate memo or letter to an appropriate officer in the company.

WRITER'S CHECKLIST

Letter Format

1. Is the first page typed on letterhead stationery?

2. Is the date included?

3. Is the inside address complete and correct? Is the appropriate courtesy title used?

4. If appropriate, is an attention line included?

5. If appropriate, is a subject line included?

6. Is the salutation appropriate?

7. Is the complimentary close typed with only the first word capitalized? Is the complimentary close followed by a comma?

8. Is the signature legible and the writer's name typed beneath the signature?

9. If appropriate, are the reference initials included?

10. If appropriate, is an enclosure line included?

11. If appropriate, is a copy line included?

12. Is the letter typed in one of the standard formats?

Types of Letters

1. Does the order letter
   ❑ include identifying information, such as quantities and model numbers?
   ❑ specify, if appropriate, the terms of payment?
   ❑ specify the method of delivery?

2. Does the inquiry letter
   ❑ explain why you chose the reader to receive the inquiry?
   ❑ explain why you are requesting the information and to what use you will put it?
   ❑ specify by what date you need the information?
   ❑ list the questions clearly and, if appropriate, provide room for the reader's responses?
   ❑ offer, if appropriate, the product of your research?

3. Does the response to an inquiry letter answer the reader's questions or explain why they cannot be answered?

4. Does the sales letter
   ❑ gain the reader's attention?
   ❑ describe the product or service?
   ❑ convince the reader that the claims are accurate?
   ❑ encourage the reader to find out more about the product or service?

5. Does the claim letter
   ❑ identify specifically the unsatisfactory product or service?
   ❑ explain the problem(s) clearly?
   ❑ propose an adjustment?
   ❑ conclude courteously?

6. Does the "good-news" adjustment letter
   ❑ express your regret?
   ❑ explain the adjustment you will make?
   ❑ conclude on a positive note?

7. Does the "bad-news" adjustment letter
   ❑ meet the reader on neutral ground, expressing regret but not apologizing?
   ❑ explain why the company is not at fault?
   ❑ clearly deny the reader's request?
   ❑ attempt to create goodwill?

EXERCISES

1. Write an order letter to John Saville, general manager of White's Electrical Supply House (13 Avondale Circle, Los Angeles, CA 90014) to order these items: one SB11 40-ampere battery backup kit, at $73.50; twelve SW402 red wire kits, at $2.50 each; ten SW400 white wire kits, at $2.00 each; and one SB201 mounting hardware kit, at $7.85. Invent any reasonable details about methods of payment and delivery.

2. You are the marketing director of the company that makes your bicycle (or calculator, stereo set, running shoes, etc.). Write a sales letter that might be sent to retailers to encourage them to sell the product.

3. You are an intern in the Admissions Office at your college or university. Write a letter that can be sent to high-school guidance counselors in your state. The letter, which will be included with a copy of your catalog, is intended to motivate the counselors to study the catalog and interest appropriate students in applying to your school.

4. A thermos you just purchased for $8.95 has a serious leak. The grape drink you put in it ruined a $15.00 white tablecloth. Inventing any reasonable details, write a claim letter to the manufacturer of the thermos.

5. The gasoline you purchased from New Jersey Petroleum contained water and particulate matter; after

using it, you had to have your automobile tank flushed at a cost of $50. You have a letter signed by the mechanic explaining what he found in the bottom of your tank. As a credit-card customer of New Jersey Petroleum, write a claim letter. Invent any reasonable details.

6. As the recipient of one of the claim letters described in exercises 4 and 5, write an adjustment letter granting the customer's request.

7. You are the manager of a private swimming club. A member has written saying that she lost a contact lens (value $55) in your pool and she wants you to pay for a replacement. The contract that all members sign explicitly states that the management is not responsible for loss of personal possessions. Write an adjustment letter denying the request. Invent any reasonable details.

8. As manager of a stereo equipment retail store, you guarantee that you will not be undersold. If a customer who buys something from you can prove within one month that another retailer sells the same equipment for less money, you will refund the difference in price. A customer has written to you, enclosing an ad from another store showing that it is selling the equipment he purchased for $26.50 less than he paid at your store. The advertised price at the other store was a one-week sale that began five weeks after the date of his purchase from you. He wants his $26.50 back. Inventing any reasonable details, write an adjustment letter denying his request. You are willing, however, to offer him a blank cassette tape worth $4.95 for his equipment if he would like to come pick it up.

9. Revise the letters below and on pages 619–620 to increase their effectiveness, adding any reasonable details.

a.

MODERN LABORATORIES, INC.
DEAUVILLE, IN 43504

July 2, 19XX

Adams Supply Company
778 North Henson Street
Caspar, IN 43601

Gentlemen:

Would you please send us the following items:

one dozen petri dishes
one gross pyrex test tubes
three bunsen burners

Please bill us.

Sincerely,

b.

1967 Sunset Avenue
Rochester, N. Y. 06803

November 13, 19XX

Admissions Department
University of Pennsylvania Law School
Philadelphia, Pa. 19106

Gentlemen:

I am a senior considering going to law school. Would you please answer the following questions about your law school:

1. How well do your graduates do?
2. Is the LSAT required?
3. Do you have any electives, or are all the courses required?
4. Are computers skills required for admission?

A swift reply would be appreciated. Thank you.

Sincerely yours,

c.

GUARDSMAN PROTECTIVE EQUIPMENT, INC.
3751 PORTER STREET
NEWARK, DE 19304

April 11, 19XX

Dear Smith Family:

A rose is a rose is a rose, the poet said. But not all home protection alarms are alike. In a time when burglaries are skyrocketing, can you afford the second-best alarm system?

(Continued)

(c. *continued*)

Your home is your most valuable possession. It is worth far more than your car. And if you haven't checked your house insurance policy recently, you'll probably be shocked to see how inadequate your coverage really is.

The best kind of insurance you can buy is the Watchdog Alarm System. What makes the Watchdog unique is that it can detect intruders before they enter your home and scare them away. Scaring them away while they're still outside is certainly better than scaring them once they're inside, where your loved ones are.

At less than two hundred dollars, you can purchase real peace of mind. Isn't your family's safety worth that much?

If you answered yes to that question, just mail in the enclosed postage-paid card, and we'll send you a 12-page, fact-filled brochure that tells you why the Watchdog is the best on the market.

Very truly yours,

 Collaborative Exercise

10. Form small groups for this exercise on claim and adjustment letters. Have each member of your group study the two letters on pages 621–622. Meet and discuss your reactions to the two letters. How effectively does the writer of the claim letter present his case? How effective is the adjustment letter? Does its writer succeed in showing that the company's procedures for ensuring hygiene are effective? Does its writer succeed in projecting a professional tone? Write a memo to your instructor discussing the two letters. Attach to the memo a revision of the adjustment letter. (See Chapter 16 for a discussion of memos.)

19 Lowry's Lane
Morgan, TN 30610

April 13, 19XX

Sea-Tasty Tuna
Route 113
Lynchburg, TN 30563

Gentlemen:

I've been buying your tuna fish for years, and up to now it's been OK.

But this morning I opened a can to make myself a sandwich. What do you think was staring me in the face? A fly. That's right, a housefly. That's him you see taped to the bottom of this letter.

What are you going to do about this?

Yours very truly,

SEA-TASTY TUNA
Route 113
Lynchburg, TN 30563

April 21, 19XX

Mr. Seth Reeves
19 Lowry's Lane
Morgan, TN 30610

Dear Mr. Reeves:

We were very sorry to learn that you found a fly in your tuna fish.

Here at Sea-Tasty we are very careful about the hygiene of our plant. The tuna are scrubbed thoroughly as soon as we receive them. After they are processed, they are inspected visually at three different points. Before we pack them, we rinse and sterilize the cans to ensure that no foreign material is sealed in them.

Because of these stringent controls, we really don't see how you could have found a fly in the can. Nevertheless, we are enclosing coupons good for two cans of Sea-Tasty tuna.

We hope this letter restores your confidence in us.

Truly yours,

 Web Exercise

11. You work in the marketing department for one of the following manufacturers of high-quality bicycles: Cannondale (http://www.cannondale.com), High Velocity (http://www.highvelocity.on.ca/), or Kestrel (http://www.kestrel-usa.com/). You have been asked to draft a one-page sales letter that can be sent to bicycle retailers in your state to interest them in carrying your products. Use information from your company's Web site in your letter.

Writing Instructions and Manuals

Understanding the Role of Instructions and Manuals 624

Analyzing Your Audience 625

Planning for Safety 625
  Writing Safety Information 626
  Designing Safety Information 627
  Placing Safety Information in the Appropriate Locations 628

Drafting Effective Instructions 628
  Drafting General Introductions 628
  Drafting Step-by-Step Instructions 630
  Drafting Conclusions 631

Drafting Effective Manuals 633
  Drafting Front Matter 637
  Drafting the Body 638
  Drafting Back Matter 644
  Drafting Revisions of Manuals 644

Writing Instructions and Manuals for Nonnative Speakers of English 646

Hypertext 647
  Advantages of Hypertext 648
  Disadvantages of Hypertext 649

Usability Testing of Instructions and Manuals 650
  The Goals of Usability Testing 651
  Basic Concepts of Usability Testing 651
  Preparing for a Usability Test 653
  Conducting a Usability Test 654
  Interpreting and Reporting the Data from a Usability Test 655

Case: Writing a Set of Instructions 656

Writer's Checklist 656

Exercises 657
  Collaborative Exercise 657
  Web Exercise 657

The customer-support staff at Dell Computers can tell you that no matter how hard their technical communicators try, their manuals will never please everyone. One customer called, asking how to install batteries in her new laptop. When told the instructions were on page one of the manual, she replied, "I just paid $2,000 for this [deleted] thing, and I'm not going to read a book" ("Befuddled," 1994, p. B1). Instructions and manuals have acquired a bad reputation over the years, and so many people don't even bother trying to read them. Perhaps that is why many technical communicators use the phrase RTFM, which (roughly translated) stands for Read the Manual. But many people apparently don't. This situation is regrettable, because instructions and manuals are fundamentally important to carrying out procedures and using products safely and effectively.

Chapter 9 discussed process descriptions, which explain how a process occurs. This chapter discusses instructions, which are process descriptions written to help the reader perform a specific task. For instance, a set of instructions shows how to install a water heater in your home.

This chapter also discusses manuals, which are larger documents consisting primarily of instructions. Often manuals are printed and bound, like books. Manuals can be classified according to function. One common type is a user's manual. Like a set of instructions, its function is to instruct by explaining, for example, how to use a software program, maintain inventory, or operate a piece of machinery. Other types include installation manuals, maintenance manuals, and repair manuals.

Three other aspects of instructional writing are also discussed in this chapter: writing instructions and manuals for nonnative speakers of English, hypertext, and usability testing.

Effective instructions and manuals are challenging to write. You must make sure that your audience will want to read the document and be able to understand it easily. You also must make sure that in performing the tasks they won't damage any equipment or, more important, injure themselves or other people.

Understanding the Role of Instructions and Manuals

Instructions and manuals are central to technical communication. If you are a technical professional such as an engineer, you will probably be asked to write or contribute to instructions and manuals often. If you are a technical communicator, you will write them more often than any other kind of major technical document.

Whereas just a decade or two ago little attention was paid to the quality of instructions and manuals, today the goal is to make systems safe and "user-friendly."

- Safety is an important issue largely because of liability suits. As discussed in Chapter 2, a company can lose millions if a court finds that its instructions and manuals failed to explain how to use the product properly or to alert the user about dangers related to using its product.
- User-friendliness is critical, because if the product is hard to use, it could fail in the marketplace. Major advertising campaigns for all kinds of products stress clear, simple, easy-to-use manuals.

Instructions and manuals are no longer an afterthought but rather an integral part of the planning process. At the most progressive companies, the people who create the documentation—the documents that come with the product—are part of the research and development team.

Analyzing Your Audience

When instructions and manuals are ineffective, chances are that the writer has inaccurately assessed the audience. Performing a function, such as assembling a backyard swing or maintaining a conveyor belt, is easy for the expert but not necessarily for the person reading the documentation. A reader who doesn't know what a self-locking washer is or what the calibrations on a timing control mean might not be able to complete the process.

Before you start to write a set of instructions or a manual, think carefully about the background and skill level of your audience. If you are writing to people who are experienced in the field, use technical vocabulary and concepts. But if you are addressing general readers, define technical terms and provide detailed directions. Don't be content to write "Make sure the tires are rotated properly." Instead, define proper rotation and describe how to achieve it.

The best way to make sure you have assessed your audience effectively is to find people whose backgrounds resemble those of your intended readers and then test the effectiveness of the documentation. This process, called *usability testing*, is discussed later in this chapter.

Planning for Safety

Your most important responsibility in writing documentation is to make sure you do everything you can to ensure the safety of your readers. Even though some kinds of tasks do not involve safety risks, many do. Planning for safety includes three steps:

- writing the safety information
- creating a design for the safety information
- placing the safety information in appropriate locations

Writing Safety Information

Be clear and concise. Avoid complicated sentences.

COMPLICATED: It is required that safety glasses be worn when inside this laboratory.

SIMPLE: You must wear safety glasses in this laboratory.

SIMPLE: Wear safety glasses in this laboratory.

Sometimes a phrase works better than a sentence: "Safety Glasses Area."

Because a typical set of instructions or manual can contain dozens of comments—both safety and nonsafety comments—experts have created terms to indicate the seriousness of the advice. Unfortunately, this terminology is not consistent.

For instance, the American National Standards Institute (ANSI) and the U.S. military's MILSPEC publish definitions that differ significantly, and many private companies have their own definitions that don't conform to either ANSI or MIL-SPEC. The following explanation of four common terms, presented here from most to least serious, points out the significant differences between ANSI and MILSPEC.

- *Danger.* MILSPEC does not use this term, but for ANSI and many companies *danger* alerts the reader to an immediate and serious hazard that will likely be fatal.

  DANGER. EXTREMELY HIGH VOLTAGE. STAND BACK.

  Often, safety warnings are written in all-uppercase letters.

- *Warning.* For MILSPEC, *warning* is the most serious level, indicating an action that could result in serious injury or death. For ANSI, it also suggests the potential for serious injury or death. Among different companies, *warning* ranges from serious injury or death to serious damage to equipment.

  WARNING: TO PREVENT SERIOUS INJURY TO YOUR ARMS AND HANDS, MAKE SURE THE ARM RESTRAINTS ARE IN PLACE BEFORE YOU OPERATE THIS MACHINE.

- *Caution.* For MILSPEC, *caution* warns of the potential for both equipment damage and long-term health hazards. For ANSI, it indicates the potential for minor or moderate injury. Among companies, *caution* signals the potential for anything from moderate injury to serious equipment damage or destruction.

  Caution: Do not use nonrechargeable batteries in this charging unit; they could damage the charging unit.

- *Note.* A *note* is a tip or suggestion to help the readers carry out the procedure successfully.

  Note: Two kinds of washers are provided: regular and locking. Be sure to use the locking washers here.

If your organization does not have guidelines for safety labeling, you might recommend the following definitions:

- *Danger.* Likelihood of serious injury, including death.
- *Warning.* Potential for minor, moderate, or serious injury.
- *Caution.* Potential for damage to equipment.
- *Note.* A suggestion on how to carry out a task.

Designing Safety Information

Whether in a document or on machinery or equipment, safety information should be prominent and easy to read. Many organizations use visual symbols to accompany their levels of safety comments, but these symbols are not yet standardized. If your organization doesn't have a set of symbols that you can use in your document, create different designs for each kind of comment. For instance, you could present warnings in 18-point type, boldfaced, within a box. Of course, the more critical the safety comment, the larger and more emphatic it should be. Safety information is often printed in color. Text is presented against a yellow, orange, or red background. Symbols are printed in color, too. Flames are presented in red, for instance. See the discussion of color in Chapter 12 for more information.

The clip art in Figure 20.1 (*Arts and Letters,* 1990) is typical of symbols for safety information. The three symbols represent fire danger, electrical danger, and the need to wear safety glasses. Note that the triangle indicates that the danger is severe, whereas the image inside the triangle conveys the specific danger.

FIGURE 20.1

Common Symbols in Safety Information

Placing Safety Information in the Appropriate Locations

What are appropriate locations? This question has no easy answer because you cannot control how your audience reads your document. Be conservative: put safety information wherever you think the reader is likely to see it, and don't be afraid to repeat yourself. Naturally, you don't want to repeat the same piece of advice in front of each of 20 steps, because that will merely teach your readers to stop paying attention to you. But a reasonable amount of repetition—such as including the same safety comment at the top of each page—might be effective. If your company's procedure format calls for a safety section near the beginning of the document, put that information there and right before the appropriate step in the step-by-step section.

The Occupational Safety and Health Administration Guidelines (Chapter XVII, Sections 1910.145 and 1926.155) describe proper standards for placing safety messages on products and manuals. These standards address the following questions:

- Is the message prominently displayed so that the user sees it?
- Is the message large enough and clearly legible, under operating conditions?
- Are the graphics and the words of the message clear and informative?

Figure 20.2, from an operator's manual for a John Deere lawnmower, shows one company's approach to placing safety warnings on machinery.

Drafting Effective Instructions

Instructions can be brief—a small sheet of paper—or extensive, up to 20 pages or more. Brief instructions might be produced by one or two people: a writer, or a writer and an artist. Sometimes a subject-matter expert—an expert in the technical subject being written about—is added to the team. For more extensive instructions, other people—marketing and legal personnel, for example—might be added. The team could consist of 10 or even 20 professionals working with a budget of many thousands of dollars.

As is the case with process descriptions, you will probably find it easiest to write instructions sequentially; that is, in the order in which they will appear. As they write instructions, many writers perform the task they are explaining, a process that enables them to detect errors and omissions.

Regardless of the size of the project, most instructions are structured like process descriptions. (See Chapter 9.) The main difference is that the conclusion of a set of instructions is less a summary than an explanation of how to make sure the reader has followed the instructions correctly. Most sets of instructions contain the three components shown in Figure 20.3 on page 630.

Drafting General Introductions

The general introduction provides the preliminary information that readers will need to follow the instructions safely and easily. In writing the introduction, answer these three questions about the process:

- *Why should the reader carry out this task?* Often, you do not need to answer this question because the reason is obvious. For example, you don't need to explain why a backyard barbecue grill should be assembled. Sometimes, however, you do need to explain, as in the case of many preventive-maintenance chores, such as changing radiator antifreeze every two years.

  If appropriate, answer two more questions:

  - *When should the reader carry out the task?* Some tasks, such as rotating tires or planting crops, need to be performed at particular times or at particular intervals.
  - *Who should carry out the task?* Sometimes you need to describe or identify the person or persons who are to carry out a task. Some kinds of aircraft maintenance, for example, may be carried out only by persons certified to do that maintenance.

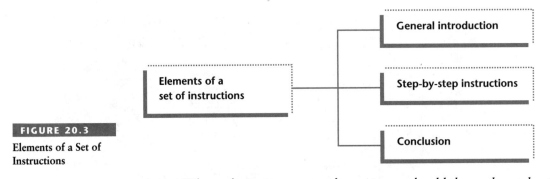

FIGURE 20.3
Elements of a Set of
Instructions

- *What safety measures or other concerns should the reader understand?* In addition to stating the safety measures that apply to the whole task, state any tips that will make your reader's job easier. For example:

  NOTE: For ease of assembly, leave all nuts loose. Give only 3 or 4 complete turns on bolt threads.

- *What tools and materials will the reader need?* Include the list of necessary tools and equipment in the introduction so that readers will not have to interrupt their work to hunt for something. If you think your readers might not be able to identify the tools or materials easily, include drawings next to the names (see Chapter 12).

Following is a list of tools and materials from a set of instructions on replacing broken window glass:

You will need the following tools and materials:

Tools	Materials
glass cutter	putty
putty knife	glass of proper size
window scraper	paint
chisel	hand cleaner
electric soldering iron	work gloves
razor blade	linseed oil
pliers	glazier's points
paint brush	

Drafting Step-by-Step Instructions

Follow these five guidelines when writing step-by-step instructions:

- *Number the instructions.* For long, complex instructions, use two-level numbering, such as a decimal system.

  1
    1.1
    1.2
  2
    2.1
    2.2
  etc.

- *Put the right amount of information in each step.* Each step should define a single task that the reader can carry out easily, without having to refer to the instructions.

TOO MUCH INFORMATION:

1. Mix one part of the cement with one part water, using the trowel. When the mixture is a thick consistency without any lumps bigger than a marble, place a strip of the mixture about 1" high and 1" wide along the face of the brick.

TOO LITTLE INFORMATION:

1. Pick up the trowel.

RIGHT AMOUNT OF INFORMATION:

1. Mix one part of the cement with one part water, using the trowel, until the mixture is a thick consistency without any lumps bigger than a marble.

2. Place a strip of the mixture about 1" high and 1" wide along the face of the brick.

- *Use the imperative mood.* For example, "Attach the red wire. . . ." The imperative is more direct and economical than the indicative mood ("You should attach the red wire . . ." or "The operator should attach the red wire . . ."). Make sure your sentences are grammatically parallel. Avoid the passive voice ("The red wire is attached . . ."), because it can be ambiguous: Is the red wire already attached?
- *Include graphics.* When appropriate, accompany each step with a photograph or drawing that shows the reader what to do. Some kinds of activities—such as adding two drops of a reagent to a mixture—do not need illustration, but they might be clarified by charts or tables.

  Figure 20.4 on page 632 shows the extent to which a set of instructions can integrate words and graphics. This excerpt is from the operating-instructions booklet for a videocassette recorder. See Chapter 12 for a discussion of graphics.

- *Do not omit the articles (*a, an, the*) to save space.* Omitting the articles makes the instructions hard to read and, sometimes, unclear. In the sentence "Locate midpoint and draw line," for example, the reader cannot tell if "draw line" is a noun ("the draw line") or a verb and its object ("draw the line").

Drafting Conclusions

Instructions often conclude with *maintenance tips.* Another popular conclusion for a set of instructions is a *troubleshooter's guide,* usually in the form of a table, that identifies common problems and explains how to solve them.

A portion of the troubleshooter's guide included in the operating instructions for a lawnmower is shown on page 633.

EASY OPERATION GUIDE

First   Refer to BASIC OPERATION.

To watch TV normally

Recording

To watch and record the same TV program

To watch one TV program while recording another

Instructions That
Integrate Words
and Graphics

Problem	Cause	Correction
THE MOWER DOES NOT START.	1. The mower is out of gas. 2. The gas is stale. 3. The spark plug wire is disconnected from the spark plug.	1. Fill the gas tank. 2. Drain the tank and refill it with fresh gas. 3. Connect the wire to the plug.
THE MOWER LOSES POWER.	1. The grass is too high. 2. The air cleaner is dirty. 3. There is a buildup of grass, leaves, or trash.	1. Set the mower to the "higher cut" position. 2. Replace the air cleaner. 3. Disconnect the spark plug wire, attach it to the retainer post, and clean the underside of the mower housing.

Some troubleshooter's checklists refer the reader to the page where the recommended action is described. For example, the Correction column in the lawn-mower troubleshooting checklist might say, "1. Fill the gas tank. See page 4."

Figure 20.5 (Iomega, 1997) on pages 634–636 is an excerpt from instructions for a Zip™ drive, a storage device for a computer. One note about these instructions. Some readers like the cartoon figure of the little fellow; others don't. The purpose of the cartoon is to make setting up the equipment seem less intimidating; many people dread the process of configuring new hardware and software.

Drafting Effective Manuals

Most of what has been said in this chapter about instructions applies to manuals. For example, you have to analyze your audience, explain procedures clearly, and complement your descriptions with graphics.

However, because manuals are usually much more ambitious projects than instructions, they require more planning. A bigger investment is at stake for the organization, not only in the costs of writing and producing the manual but also in the potential effects of the published manual. In addition to reducing the possibility of injuries and liability, a good manual attracts customers and reduces costs, because the organization needs fewer customer-support people. A poor manual is expensive to produce because it must be revised more often and alienates customers.

Drafting a manual, like drafting a book, is so complex that the following discussion can provide only an introduction. If you are involved in creating a manual, consult the sources in the bibliography (Appendix D).

Writing a manual is almost always a collaborative project. A full-size manual is simply too big and requires too many skills for one person to write: technical skills in the subject area, writing skills, graphics skills, production skills, as well

Installation Guide

This guide shows how to install the parallel port model of the Zip drive using your PC's printer connection. If you encounter any problems while installing your Zip drive, refer to "Troubleshooting" in the Zip User's Guide.

CAUTION *This model of the Zip drive cannot be used on Macintosh computers. To connect to a Macintosh (including the PowerBook), use the SCSI model of the Zip drive.* ■

Using Your Printer with Zip

Your Zip drive lets you use a printer and your Zip drive at the same time, even if your computer has only one parallel port. Step 5 of this guide shows you how to connect a printer to the Zip drive. Your Zip **User's Guide** and the README.TXT file on the **Install** floppy disk have additional information on using a printer with your Zip drive.

Enhancing Zip Drive Transfer Speed

The parallel port model of the Zip drive supports three data transfer modes which have different transfer speeds. To ensure the broadest base of compatibility for the Zip drive, the Iomega installation software automatically uses the most compatible, but slowest, mode. After installing your Zip drive, you can run the Parallel Port Accelerator included in your Tools software package to automatically set up your system to use the fastest transfer mode for the Zip drive that is supported by your computer's parallel port.

Parallel Port
Accelerator

Windows® 95 Users: Double click on the Parallel Port Accelerator icon in the Iomega Tools folder on your Start Programs menu.

Windows 3.1 Users: Double click on the Parallel Port Accelerator icon in your Iomega Zip Tools group.

DOS Users: Run OPTPPM1.EXE from the \IOMEGA directory on drive C. Refer to README.TXT for additional information.

Windows NT or OS/2 Support

You can purchase Iomega Driver and Tools software packages that provide full support for Zip drives under Windows NT or OS/2. Call 1-800-MY-STUFF to order software

FIGURE 20.5

Excerpt from Instructions

(Figure 20.5 continued)

The steps are numbered prominently.

1 Unpack your Zip drive and accessories.

Zip Manuals and the
Windows/DOS Install Diskette

Data Cable

External Power
Supply

Zip Drive

100MB Zip Tools Disk

The components are drawn clearly and labeled.

Because it is important to turn off the computer before proceeding, the writers devote a step to the task, even though there is not much information to "show."

2 Turn off your computer's power switch and pick up the Zip data cable.

WARNING *Always turn off power to the computer and all devices connected to it before connecting the Zip drive.* ■

(Open up for Step 3)

(Continued)

(Figure 20.5 continued)

5 OPTIONAL: You can connect a printer to the Zip drive using a standard printer cable.

The Zip drive must have power in order for the printer pass-through feature to work.

A note refers readers to
another place in the in-
structions.

■ **NOTE:** *If you use other peripherals or software keys (dongles) that hook to the parallel port, refer to the information under #8 in the "Troubleshooting" section of the Zip User's Guide.* ■

6 Turn your computer ON, and connect the Zip drive's power supply. *(The power cord fits in the groove on the side of the drive.)*

Always turn on power to your computer before or at the same time as you supply power to your parallel port Zip drive. Iomega recommends using a power strip to turn on all your devices at the same time. Power must be applied to the Zip drive before DOS begins to load.

WARNING *Use only an approved Iomega Zip power supply. Other power supplies may damage your Zip drive.* ■

This blow-up helps
readers see the details.

Computer Power Cord
Printer Power Cord

🌀 *To make it easier to move your Zip drive between home and work, you may want to purchase an extra power supply, cable, and carrying case. Call 1-800-MY-STUFF to order Zip accessories.*

as the knowledge of contracts and law to prevent lawsuits. In addition, obtaining other people's perspectives can prevent little problems from becoming big problems. See Chapter 3 for more information on collaboration.

The following discussion comments on the three stages of writing a manual.

- *Prewriting.* The most important stage of manual writing is prewriting. First, you must analyze your audience and purpose.

    The documentation for a sophisticated procedure or system almost always accommodates a diverse audience. Often you will decide to produce a set of manuals—one for the user, one for the manager, one for the installer, one for the maintenance technicians, and so on. Or you might decide on a main manual and some other, related documents, such as brochures, flyers, and workbooks.

    While your primary purpose varies with the kind of manual, you also have a secondary purpose: to motivate your readers. The mere fact that the manual might be 200 pages or even longer means that you have to persuade people to read it. In many cases, readers are uncomfortable with the new system. Your job is to make the task of learning it less overwhelming. A well-designed manual, with plenty of white space and graphics, will help considerably. In planning a manual, follow the guidelines discussed in Chapters 5 through 7 of this text.

- *Drafting.* Drafting a manual is much the same as drafting any other kind of technical document. Sometimes one person collects information from all the people on the documentation team and creates a draft from it. Sometimes different subject-matter experts write their own drafts, which a writer or a writing team then revises or rewrites. In drafting a manual, you also need to create and integrate graphics, and to design the whole manual and each page.

- *Revising.* Because manuals are bigger and more complex than most other kinds of technical documents, revision is a more complicated process. In fact, revision is a series of different kinds of checks: from checks of the technical information down to checks of structure, organization, emphasis, and style. See the discussion of usability testing later in this chapter.

Drafting Front Matter

The front matter, which helps readers understand the content of the manual and the best ways to use it, consists of everything before the body. Most manuals have a cover or title page and a table of contents. In addition, most manuals contain a preface and a how-to-use-this-manual section. Sometimes, however, these last two items are simply combined in an introduction.

To decide whether to use a cover or just a title page, consider the manual's size and intended use. Manuals that will receive some wear and tear—such as maintenance manuals for oil-rigging equipment—need a hard cover, usually of a water-resistant material. Manuals used around an office usually don't need hard covers, unless they are so big that they require extra strength. See Chapter 11 for more information on covers.

The title page, usually designed by a graphic artist, contains at least the title of the manual, plus the company's name, address, and logo. An extensive table of contents is especially important in a manual because people refer to manuals for specific information; they don't read them straight through. The headings in an effective table of contents are clear and focus on the task the reader wants to accomplish. See Chapters 13 and 14 for more information on phrasing headings.

The other introductory information, which sometimes precedes and sometimes follows the table of contents, takes a number of forms. The first page of text in the manual, for instance, can be called "Introduction" or "Preface" or have no heading at all. Another popular strategy is to use several "about" phrases: "About Product X," "About the Product X Documentation Set," and so on.

Regardless of what you call this introductory information, it should answer five basic questions for your reader:

- Who should use this manual?
- What product, procedure, or system does the manual describe?
- What is the manual's purpose?
- What are the manual's major components?
- How should the manual be used?

These questions need not be answered in separate paragraphs or even in separate sentences.

Some manuals need to answer one other question: what does the typography signify? If you use type creatively in the manual, and you want your readers to understand your conventions, define the conventions in the front matter.

Figure 20.6 (Group 1 Software, 1994, pp. vii–ix, xi–xii) shows selected pages from the front matter of a manual. For more information on front matter, see Chapter 10.

Drafting the Body The body of a manual might look like the body of a traditional report, or it might look radically different. Its structure, style, and graphics will depend on its purpose and audience. For instance, the body of a manual might include summaries and diagnostic tests to help readers determine whether they have understood the discussion. A long manual might have more than one "body"; that is, each chapter might be a self-contained unit with its own introduction, body, and conclusion.

Follow these four guidelines when drafting the body of a manual:

- *Structure the body according to the way the reader will use it.* For example, if the reader is supposed to carry out a process, structure the body chronologically. If the reader is supposed to understand a concept, structure the body from more important elements to less important elements. The various argument patterns discussed in Chapter 8 are appropriate, but, as always, the writing situation might call for another pattern.

An overview of the manual.

The audience for the manual.

FIGURE 20.6

Pages from the Front Matter of a Manual

Preface

About This Manual

The Purpose of This Guide

This User's Guide provides you with the information you need to effectively use the AccuMail System. With this guide, you will learn about fundamental name and address matching concepts, AccuMail's many features and functions, how to define and run jobs with AccuMail, and more.

This guide is divided into three sections:

Part I: **Before You Begin**	Provides basic mailing and functional concepts of the AccuMail System, and important instructions for installing and configuring your AccuMail software. This section also provides definitions of the terms used in this guide, and an overview of basic mouse and keyboard functions.
Part II: **Using AccuMail**	Provides instructions for performing the various tasks needed to work with jobs. This section also covers step-by-step instructions for working with filters, databases, mailing lists, and other AccuMail features.
Part III: **Appendices**	Provides a glossary, index, and appendices regarding special issues such as error messages, optimizing for speed, troubleshooting, and calling AccuMail from your own application.

Who Should Use This Guide?

This guide was developed for three types of users:

- Novice AccuMail users
- Experienced AccuMail users
- Programmers wanting to call AccuMail from their own applications.

> **NOTE:** This book assumes some familiarity with the Windows and DOS environments.

User's Guide   vii

(Continued)

(Figure 20.6 continued)

A conventions section helps readers understand the typographical styles that will be used in the manual.

Conventions Used in This Guide

This guide uses the following documentation conventions:

NOTE:	Indicates important information or warnings.
①	Indicates a numbered step in a process.
Italics	Indicates words or phrases that require emphasis or that appear in the glossary. The glossary words only appear in italics the first time they appear in the manual. Italics are also used for all dialog box and window titles referenced in this manual.
Bold	Indicates any of the following: menu name or menu option; icon; name of a list box, text box, check box, or radio button; group label and fields.
"Quoted Text"	Indicates one of the following (1) text that appears on a screen, window, or dialog box, such as a message or list box option or (2) text that you are to type (you do not type the quotation marks).
<Key>	Indicates a keyboard key such as <Return> or <Shift>. When you are to hold two or more keys down at the same time, they are shown in a single set of brackets, such as <Ctrl-C>.
[OK]	Indicates the name of a button on a dialog box.
ALL CAPITALS	Indicates file names, directory names, and job names. This typeface is also used to reference drives and DOS prompts.
[text]	Indicates user-specific names, directories, drives, and other information.

(Figure 20.6 continued)

Summary of Chapters

The following table summarizes the chapters in this guide. Novice users will find Chapter 4, "Example Job" and all of Part I most useful, while experienced users will find the more advanced chapters of Parts II and III valuable.

Many manuals begin with a chapter-by-chapter summary.

Chapter	Topics Discussed
Part I: Before You Begin	
Chapter 1. AccuMail Basics	This chapter provides an overview of AccuMail's features and functions, including sections on address standardization, CASS, and other issues to consider before using AccuMail.
Chapter 2. Setting Up AccuMail	This chapter describes how to install AccuMail. It includes software and hardware requirements and separate installation and log-on instructions for the DOS, and Windows LAN versions.
Chapter 3. Getting Started	This chapter begins with an overview of the AccuMail interface, identifying all components of windows and dialog boxes used in the AccuMail system. It then describes navigation basics such as using the arrow and tab keys, entering and highlighting text, and using the mouse.
Part II: Using AccuMail	
Chapter 4. Example Job	This chapter functions as a tutorial for the novice user. It describes certain key concepts such as a job and the basic features of the **File** menu. It demonstrates the necessary steps to create and run a job, and to open, close, and copy a job. It includes sections on selecting lists, defining and matching fields, selecting databases, setting job defaults, and printing the CASS Form 3553.
Chapter 5. Looking for an Address	This chapter explains how to perform single-address lookups, browse the USPS National Database, and print envelopes and labels.
Chapter 6. Using Filters	This chapter explains how to select certain records based on criteria by using filters that you specify.

(Continued)

(Figure 20.6 continued)

Related References

The front matter often directs readers to other documents.

The following pamphlets and handbooks can be obtained from the United States Postal Service:

- *Domestic Mail Manual (DMM)*: USPS Document ID DMM
- *Addressing Conventions*: USPS Document ID #MI DM-940-89-3
- *Postal Addressing Standards:* USPS Document ID #28
- *A Guide to Business Mail Preparation:* USPS Document ID #25
- *Addressing for Success:* USPS Notice #221
- *Automation Plan For Business Mailers:* USPS Document ID #67.

For further information, please contact:

National Address Information Center
United States Postal Service
6060 Primary Pkwy Suite 101
Memphis, TN 38188-0001

NOTE: The USPS may charge a fee for some or all of these publications.

(Figure 20.6 continued)

About This Manual

If You Need More Help

A manual helps readers understand how to get technical support.

Group 1 Software provides free telephone hotline support on all Group 1 microcomputer products for 90 days from the date of purchase. When calling Technical Support:

- Have your phone near the computer.
- Have the AccuMail program up on the screen.
- Know the name of the person and the company to whom AccuMail is registered.
- Know the AccuMail version number that you are using.
- Know any specific system configuration information (it is best if you have a printout of your CONFIG.SYS and AUTOEXEC.BAT files readily available).
- Be prepared to describe where you are in the program and what you are trying to accomplish.
- Have all product documentation within reach.

The **Technical Support Hotline** number for customers in the United States and Canada is **(301) 731-2316** between 9:00 a.m. and 7:00 p.m. Eastern Standard Time. The **facsimile number** for Technical Support is **(301) 306-4373**. Please address your facsimile cover sheet to Microproducts Customer Support.

> **NOTE:** To be eligible for this service, you MUST complete and return the Product Registration Card.

Your Comments Are Welcome

The company solicits comments and suggestions.

We appreciate and welcome your comments concerning this guide! If you have suggestions, please let us know. For your convenience, a Documentation Comment Form is provided with your new system. If the Documentation Comment Form is missing, however, you may address your comments to:

Group 1 Software, Inc.
Documentation Department
4200 Parliament Place, Suite 600
Lanham, MD 20706-1844

---

xii   AccuMail

- *Write clearly.* Simple, short sentences work best. Use the imperative when giving instructions.
- *Be informal, if appropriate.* For some kinds of manuals, especially those used by readers untrained in the subject, an informal style is appropriate. Contractions and everyday vocabulary are common. A caution about informality: safety warnings and information about serious subjects such as disease or war are usually presented in a formal style.
- *Use graphics.* They break up the text and improve comprehension. Whenever you want your readers to perform an action with their hands, include a drawing or photograph showing the action. Where appropriate, use tables and figures. Deere & Co., the farm-equipment manufacturer, uses the term *illustration* to describe the writing style of its owner/operator manuals: each step is an illustration with a brief explanatory comment. See Chapter 12 for more information on graphics.

Figure 20.7 (Microsoft, 1995) shows a page from the body of a user's manual for a software program.

Drafting Back Matter

The audience and purpose of the manual will determine what items to include in the back matter. However, two items are common: a glossary and an index.

A *glossary* is an alphabetized list of definitions of important terms in the document. An *index* is common for most manuals of 20 to 30 pages or more. Glossaries and indexes are discussed in Chapter 10.

Appendices to manuals contain many kinds of information. Procedures manuals often have flowcharts or other graphics that picture the processes described in the body of the manual. Sometimes, these graphics are produced so that they can be removed from the manual and taped to the office wall for frequent reference. User's guides often have diagnostic tests and reference materials—error messages and sample data lists for the computer system, troubleshooting guides, and the like. For a further discussion of appendices, see Chapter 10.

Drafting Revisions of Manuals

In high-tech industries, a new generation of a product might come along as often as every 18 months. As products evolve, technical communicators need to revise the manuals that go with them. When a new version of the product is released, you can take one of two approaches to revising.

- *Publish a "new" manual.* Although a new manual will likely contain elements from the old manual, it makes no explicit reference to the old manual. Someone who has not used the previous version of the product or seen its manual would not be aware that they even existed.

  The advantage of a new manual is that it can reflect the look of the new product. It encourages people to take out their checkbooks. The disadvantage of a new manual is that people who are switching from

An advance organizer for this section.

Notes about how to prepare to begin the task.

The use of an icon in the manual.

Step-by-step instructions, complemented by screen shots.

Directing the reader to the next page.

For more information.

FIGURE 20.7

A Page from the Body of a Manual

---

32   Introducing Microsoft Windows 95

Printing

Windows 95 offers significant improvements to printing – in particular, to the way you set up a printer to print from Windows. This section covers the basics of printer setup and printing. Additional printing topics are covered in Help.

Setting Up a Printer

Setting up your printer is easy and fast, thanks to the Add Printer *wizard*, a step-by-step printer setup program.

Note Before you begin, make sure your printer is correctly connected to your computer and that you know the make and model of the printer. If you're planning to use a shared printer over the network, you may also need to know the *path* to it (for example, \\Accounting\Printer1 might be the path to a network printer). Or you can browse for the printer in Network Neighborhood and then double-click its icon to set it up.

To set up a printer for use with Windows

◄ **1** Click the Start button, point to Settings, and then click Printers.

TIP
If any printers are already set up, their icons appear in the Printers folder. These printers are ready for you to use.

◄ **2** Double-click Add Printer.

■ *See Also*
For information about Network Neighborhood, see Chapter 3, "Introducing Networks."

For more information about paths, see "Starting a Program by Using the Run Command," in Chapter 1, "The Basics."

Continued on next page ↗

the old product to the new will have to start all over again; they cannot distinguish the new information from the old.

- *Publish a "revised" manual.* Revised manuals, like new ones, contain a lot of information from the old version, but the new information is flagged so that users can find it easily. The new information is usually marked with an icon in the margin or with a change bar: a vertical bar in the margin. Sometimes a section in the front matter describes the changes to this version of the product.

  The advantage of a revised manual is that it is easier for previous users to work with. The disadvantage is that it can look like a patch job.

  Some software companies make a point of reminding the reader to study the descriptions of the new material: the text that appears on the screen during installation contains a reference to the new material; if the reader does not acknowledge the reference by hitting a particular key, the installation stops.

Writing Instructions and Manuals for Nonnative Speakers of English

International trade is fundamental to the financial health of many U.S. companies. When a company wishes to provide a set of instructions or a manual to a non–English-speaking audience or to an audience whose first language is not English, it can either translate the document into the reader's native language or try to make the English easy to understand.

Translation is sometimes the best or only alternative, but companies often use a simplified form of English, particularly when readers need to acquire a basic competence for long-term use of English. For instance, thousands of non–English-speaking aircraft technicians train for months in the United States to learn how to maintain U.S.-made civilian and military aircraft purchased by their countries. The manufacturers would find it prohibitively expensive to translate the many volumes of documentation into all the languages used by their customers; therefore, they use a simple form of English.

The different forms of simple English include Simplified English, Controlled English, and Fundamental English. Some of these forms have been devised by scholars, others by companies such as Caterpillar Tractor and Kodak. They share three basic characteristics:

- *A limited vocabulary.* Some forms use vocabularies of as few as 500 words. Every word has only one meaning: *right* is the opposite of *left*, not of *wrong*; and *correct* is the opposite of *incorrect* (Sanderlin, 1988).
- *A limited grammar and sentence structure.* Sentences are short and simple. The imperative mood dominates. Following is an example of a sentence that has been revised from standard English into a simple English.

ORIGINAL: After visually inspecting the gap to ensure the gap is no more than
              0.025 inches, replace the housing.

REVISED: Use the gauge to make sure the gap is no more than 0.025 inches.
              Then replace the housing.

- *A reliance on graphics.* As discussed in Chapter 12, graphics are used extensively in communicating with speakers whose first language is not English. If you can use a drawing, diagram, or photograph instead of words, do so. But be careful—graphics can project a cultural bias that interferes with the communication or offends your readers.

The length of instruction to learn different forms of Simplified English ranges from 30 hours to 3 months.

For more information, see Sanderlin (1988). Also see the *Proceedings of the 1st International Workshop on Controlled Language Applications* (http://wwwots.let.ruu.nl/cgi-bin/gen.pl?claw96). For further discussion of Simplified English, see Chapter 14.

Hypertext

The word *hypertext* was coined in 1965 by Ted Nelson, whose vision was to create a "docuverse," an electronically linked document universe in which every piece of information is linked to every other piece of information. As discussed in Chapter 4, hypertext refers to a multilinear technique for presenting information.

A book is linear not only in the obvious way—the pages are numbered—but also in the most essential way: the author has laid a clear path for the reader to follow. Even though you can skip around in the book by reading chapters out of order (as you may be doing with this textbook in your class), the idea behind paper-based documents is that the author has created a linear structure for the information within which the reader is intended to work. In contrast, hypertext is multilinear. Viewers are encouraged to jump from one piece of information to another, according to their interests and needs. Viewers create their own paths through the information.

Although hypertext is still a relatively new technology, we are beginning to understand its potential and its limitations. Obviously, it cannot be used without a computer. And it should not be used for brief documents intended to be read from cover to cover. It is most effective when the writer needs to present a large, complex body of information to numerous people, each of whom has different needs and goals.

For this reason, hypertext is used most to present three basic kinds of instructional material:

- *Instructional material for computer users.* The most obvious role for hypertext is in helping people use computers. Software tutorial manuals appear in it, as do online help functions, error messages, and reference manuals.

- *Instructional material for technicians.* Hypertext is used commonly in training manuals, particularly for repair and maintenance personnel. Flight mechanics, auto mechanics, and other technicians use hypertext manuals to progress from the novice to the expert level.

- *Advertising and consumer information.* Hypertext is popular in the following applications:

  – point-of-sale demonstrations (the computerized presentations you see set up in department stores) for consumer and industrial products
  – trade show presentations such as those intended for automobile dealers introducing the new year's product line
  – electronic kiosks in museums and shopping malls to help users find their way, do educational learning exercises, look up product information, and order goods in self-service stores

Figure 20.8 is a paragraph (Sematech, 1997) from a hypertext manual about gas delivery systems (GDS), in which each underlined phrase is a link to another file of text and graphics.

Advantages of Hypertext

Although we don't fully understand the promise and limitations of hypertext, scholars are fairly confident that it offers the following seven advantages:

- *It accommodates different kinds of viewers.* The chief advantage of hypertext is that viewers create their own documents to match their aptitudes, interests, and needs. Information can be layered, so that viewers who need only a superficial understanding of the topic can skim the top layer of nodes, whereas those who need more detailed information can delve into deeper layers.

The components of the GDS are interconnected using rigid, metal lines called tubing. The choice of tubing material, diameter, and wall thickness depends upon the type of gas and the <u>working pressure</u>. These tubings typically have an outer diameter of 1/4" and are made of stainless steel (SS-316L). The tubing is neatly routed, and secured using mounting brackets to a panel in such a manner that minimizes excessive stress on the tubing and the rest of the GDS. <u>Dead ends</u> in tubing are eliminated and minimized because they accumulate contaminants and are difficult to purge. The tubing which leads from the cylinder to the gas panel is typically shaped into two or more coils called the cylinder lead or <u>"pigtail."</u> The pigtail allows the cylinder changes to be performed without unduly stressing and damaging the tubing. It also lessens the shock associated with sudden surges in the high-pressure line. <u>Double-walled tubing</u> is used for safety reasons in hazardous gas service. If a gas leak develops on the inner gas line, the leak is confined by the outer tubing. Often, the vacuum in the outer tubing is constantly sampled by a hazardous gas detector.

FIGURE 20.8

Excerpt from a Hypertext Manual

- *It enables viewers to customize information.* Some hypertext systems keep track of the path the viewer has chosen, enabling the viewer to reproduce that path later. And because some hypertext systems allow viewers to create their own nodes and links, they can write notes for their own future reference. Authoring software—software that lets you create hypertext documents—is widely available now; much of it is suitable even for nonspecialists.

- *It saves space.* The documentation for an F-18 fighter is 300,000 pages and consumes 68 cubic feet of space. In hypertext on a CD-ROM, it requires 0.04 cubic feet (Nielsen, 1990, pp. 13–14).

- *It is relatively inexpensive to produce.* The cost of producing a CD-ROM, the most common delivery medium for hypertext, is less than a dollar, using a machine that costs about $6,000. If your business involves printing and sending large quantities of information to many people, hypertext documents can be much cheaper to produce than paper documents. The savings increase if the documents have to be updated frequently: you just produce and send out new disks.

- *It can be used like other digital information.* For example, documents can be put on networks so that they can be used by many people simultaneously. And they can be revised to make new documents.

- *It can reduce training costs.* Hypertext manuals can reduce the need for some kinds of training that require human instructors, because viewers study the hypertext document at their own pace.

- *It can reduce training hazards.* Some kinds of training—involving electricity, explosives, and some chemicals, for instance—are inherently dangerous. The more knowledgeable the viewers become through hypertext training, the less danger they face.

Disadvantages of Hypertext

Hypertext, however, has the following six disadvantages:

- *Viewers can become "lost in hyperspace."* Some viewers find themselves unsure of where they are or have been, how to find the information they want, or how to get out of the system. Hypertext authors are trying to solve this problem by providing maps and hierarchical diagrams.

- *Viewers can be swamped by information.* Using hypertext media, viewers learn by association, jumping from one idea to another according to their needs and desires. The writer therefore does not have the opportunity to create a frame of reference; the viewer is in control. Consequently, viewers can be overwhelmed by new information and not learn as effectively as they would if the writer exerted more control over the pace and sequence.

- *Viewers can skip important information.* Because viewers are on their own, writers cannot ensure that they read important information.

- *Hypertext is expensive to write.* Writing hypertext documents is quite different from writing traditional documents. Few people are practiced at it. For this reason, good hypertext documents can be very expensive to create.
- *Hypertext is a new medium.* In most cases, companies subcontract the creation of hypertext documents. But there is no assurance that the sub-contractor you hire today will exist tomorrow. Investments can be lost.
- *Hypertext software and hardware are not yet standardized.* One piece of hypertext software might not be compatible with another. This can create logistical complications when you try to use a hypertext document on a system other than the one on which it was created.

Software for creating hypertext documents is becoming easier to use and less expensive every year. And colleges and universities are offering courses about how to create hypertext. Departments of technical communication, instructional technology, media studies, communication, and art are most likely to be involved in educating students about hypertext.

For more information on hypertext, consult the following books:

Dillon, P. M., & Leonard, D. C. (1995). *Multimedia technology from A to Z.* Phoenix, AZ: Oryx.

Horton, W. (1994). *Designing and writing online documentation: Hyperme-dia for self-supporting products* (2nd ed.). New York: Wiley.

Nielsen, J. (1990). *Hypertext and hypermedia.* Boston: Academic Press.

Stevens, G. H., & Stevens, E. F. (1995). *Designing electronic performance support tools: Improving workplace performance with hypertext, hyper-media, and multimedia.* Englewood Cliffs, NJ: Educational Technology Publications.

Usability Testing of Instructions and Manuals

Writers of instructions and manuals have always tried to write clearly. They have used checklists, style programs, and spell checkers. And they have had the documents reviewed by subject-matter experts and other kinds of information specialists, such as legal counsel and marketing specialists. These reviews are helpful in pointing out potential problems.

In recent years, however, many organizations have also begun to do usability testing on their instructions and manuals. *Usability testing* is the process of performing experiments with people who represent real users to see how easily they can use a product or carry out a process. Usability testing has been performed for decades on all kinds of machinery and consumer products. However, it has become popular as a way to test documents only in the last decade. Usability testing is different from other forms of review in that it allows you to examine the document from the perspective of the reader, the person who will be using the document.

This section covers five topics:

- the goals of usability testing
- the basic concepts of usability testing
- preparing for a usability test
- conducting a usability test
- interpreting and reporting the data from a usability test

The information in this section is based on two excellent books:

Dumas, J. S., & Redish, J. C. (1993). *A practical guide to usability testing.* Norwood, NJ: Ablex.

Rubin, J. (1994). *Handbook of usability testing: How to plan, design, and conduct effective tests.* New York: Wiley.

The Goals of Usability Testing

Usability testing has six main goals:

- *To help a company design safe products.* Usability testing helps a company see whether users can carry out tasks safely, effectively, and quickly.
- *To help a company market its products.* Usability testing helps the company understand and exploit the product's competitive advantages.
- *To help reduce the expense of service calls and customer-support phone lines.* Sophisticated products are often sold with maintenance and technical support included for a certain period, such as 90 days. The less often a technician is sent out to the customer, or the less often the customer calls for help, the lower the costs.
- *To help reduce the number of updates to the product.* Computer and software companies do usability testing to kill the bugs before the product is released, so that it doesn't have to be updated often.
- *To help a company learn how to improve other products.* The insights gained from usability testing improve not only the product being tested but also the process of making all similar products.
- *To help increase customer satisfaction with the product.* Unhappy customers share two characteristics: they try very hard not to buy another product from the same company, and they tell everyone they can think of why they are unhappy.

Basic Concepts of Usability Testing

There are five basic principles of usability testing:

- *Usability testing is an idea that permeates product development, not a single task that must be accomplished.* Usability testing involves a

INTERVIEW: USABILITY TESTER STEPHANIE ROSENBAUM

on the kinds of companies that do usability testing

All kinds do it. But the companies that do it most are the ones that realize that their customers are having problems—that's why you see so much usability testing in the computer industries—and those companies with mission-critical products and systems—for example, the banking and insurance industry. When the systems in place must be used correctly, and if they are so complicated that people make a lot of mistakes, then it's really important to do usability testing.

on the methodology of usability testing

There are quite a few accepted methods, but many occasions arise for creating new methods. Among experienced usability engineers, there is quite a bit of thinking and research, but when you are just starting, you begin with accepted methods.

on whether to leave it to professionals

It's comparable to technical communication: an untrained person could buy a few books about writing a manual, and if you studied them and took it seriously you could write a better manual than if you hadn't read the books, but you wouldn't be as skilled as if you had gone through a good university program or a good apprenticeship program. When you're just getting started, it's best to work with experienced usability professionals. You don't need to make every mistake that has been made for the last 15 years.

on teaching companies to create their own usability programs

Maybe a third to a half of our projects are skill-transfer projects: the purpose is to teach the client's own staff how to do that kind of project, so that over the course of several projects and over time they will build their own skills. It's the same process we do internally: the people we hire often require training. They read the research literature, go to seminars, observe projects in progress, and assist experienced practitioners.

on the biggest challenge of usability programs

Many people do not realize that study design and administration must be rigorous if you want useful results. It's awfully easy to lead the witness to the answer you are hoping to get rather than an objective answer—all without realizing it. Also, when planning a study if you don't recognize, for example, that people who have used Product X will perform differently on Product Y from people who haven't used Product X, you're going to get biased results. Designing studies that collect valid and reliable information about relevant issues is the biggest challenge.

on doing quick-and-dirty tests

People always talk about doing a quick-and-dirty test. It doesn't have to be quick and dirty; it can be quick and clean. You just have to narrow the scope of the study. It doesn't do you any good to get information that you cannot trust.

return on investment

If you establish a usability testing program early in a product-development project, it will reduce your development and support costs. But no single test is going to do everything; you need an ongoing program. One test can identify obvious usability problems, but there's always going to be another product or another version. A lot of what you learn in usability testing can't be applied to the current product but can be applied only to future products. You're never done with usability testing, just as you're never done with development or support.

on the future of usability testing

There's going to be a lot more usability work in the future. As more of this work moves earlier in the development cycle, as it should, there might be less "testing" but more usability projects. Testing will become more a validation of the user-centered design you did early on. All usability work pays for itself many times over. What we have to do is collect enough data and communicate the benefits to management. But it would be hard for anyone who has observed user studies—and seen the difference they make in the success of the product or system—not to be a believer.

Stephanie Rosenbaum is president of Tec Ed, a company specializing in usability testing and other technical-communication services.

systematic program of testing the product rigorously and often to make sure it is easy to use. The organization devotes substantial resources to training and equipping the testers and to enabling the designers to use the information generated by the tests.

- *Usability testing involves teamwork by different kinds of specialists.* Most high-technology products are so complicated that no one person could understand every aspect of how they work. A usability test involves the collaborative efforts of a number of people with different areas of expertise, including product experts, usability experts, technical communicators, and videocamera operators.

- *Usability testing involves studying real users as they use the product.* The people who designed the product are not the best people to test it because they know it too well. Usability testing requires real users, or people who are as close to real users as can be found. A company will always learn important information from real users that it would not have learned from people in the organization.

- *Usability testing involves setting measurable goals and determining whether the product meets them.* Usability testing involves determining in advance what the user is supposed to be able to do. For instance, in testing a user's guide for a word-processing program, the testers might decide that the user should be able to find the section on saving a file and carry out that task successfully in less than one minute. With a clear, measurable goal, the testers can study a number of participants and record meaningful data.

- *Usability testing involves repeatedly revising the product.* If the test shows that a large percentage of participants are having trouble with a task, testers examine all the information—including questionnaires the participants filled out, oral comments they made while performing the task, and videotapes of the test—to determine what went wrong. Then the testers try to fix the problem; test more participants, and fix the problem; and so on. They keep at it, until they run out of time or money—or problems!

Preparing for a Usability Test

Usability testing requires careful planning over a period of weeks or months. According to Kantner (1994), planning accounts for one-half to three-quarters of the time devoted to testing. Eight main tasks must be accomplished in planning a usability test:

- *Understand your users' needs.* Companies conducting usability tests frequently conduct *focus groups*. A focus group is a group of people brought together for a few hours to talk about a product or an issue. In addition, companies test existing products (either their own or a competitor's), have experts review the product, and conduct on-site interviews and observations of real users in the workplace, to observe them doing their work and to interview them.

- *Determine the purpose of the test.* Because *usability* is such a broad term, it is not sufficient to say that you want to test an item's usability. You can test an idea, before the product is even designed, to see if people understand it and like it. Or you can test a prototype to see if it is easy to use, or a finished product to see if there are any last-minute changes you can make to improve it.

- *Staff the test team.* Extensive programs in usability testing involve a number of specialists, each of whom does one, and only one, job. In many companies, however, only two or three people do all the tasks involved in the tests.

- *Set up the test environment.* A basic test environment includes a room for the test participant, equipped with VCRs to record the test, and another room for the test observers.

- *Develop a test plan.* A *test plan* is a document that describes and justifies what the testers plan to do. The test plan is a kind of proposal, a statement requesting approval and resources (see Chapter 15 on proposals).

- *Select participants.* Testers need to recruit participants who match the profile of the intended users of the product. Generally, it is best not to use company employees, who might know more about the product than the real user and thereby bias the test.

- *Prepare the test materials.* The most time-consuming task in preparing for a usability test is creating the test materials. Most tests require legal forms, an orientation script to help the participant understand the purpose of the test, background questionnaires, instructions for the participant to follow in the test, and a log for the testers to record data during the test.

- *Conduct a pilot test.* A pilot test is, in effect, a usability test for the usability test. A pilot test can uncover problems with the equipment, the instructions or manual, the test materials, and the test design.

Conducting a Usability Test

Although every tester makes mistakes every time, the goal is to minimize the number and severity of the mistakes so that as much as possible can be learned from each test. There are three important aspects of conducting the test:

- *Staying organized.* Experienced testers recommend making a checklist and a schedule for the test day. The schedule should cover everything from before the first participant arrives until after the last one is escorted out—including packing the equipment in preparation for the next day. In addition, experts recommend keeping each participant's materials in a separate folder.

- *Interacting with the participant.* A popular way to elicit useful information from the participant is called a *think-aloud protocol*. In this technique, the participant speaks what he or she is thinking—for instance, "I guess I'm supposed to press ENTER here, but I'm not sure because the manual didn't say to do it." If the participant feels comfortable thinking aloud, the information can prove invaluable. However, many people become very self-conscious when thinking aloud. In addition, some testers mistrust the technique because they feel it makes the participants slow down and act more carefully than they normally would.

  A second important aspect of interacting with the participant is to make sure not to bias the participant. When testers ask a participant a question, they try not to reveal the answer they want. They do not say "Well, that part of the test was pretty simple, wasn't it?" Regardless of whether the participant thought it was simple or difficult, his or her impulse will be to answer yes. Dumas and Redish (1993, p. 298) recommend neutral phrasing, such as "How was it performing that procedure?" or "Did you find that procedure easy or difficult?" In responding to questions, testers are indirect. If the participant asks "Should I press ENTER now?" they respond, "Do you think you should?" or "I'd like to see you decide."

- *Debriefing the participant.* After the test, testers usually have questions about the participant's actions. For this reason, they debrief—that is, interview—the participant. The debriefing is critically important; once the participant walks out the door, it is difficult and expensive to ask any questions, and the participant likely will have forgotten the details. Consequently, the debriefing can take as long as the test itself did.

Interpreting and Reporting the Data from a Usability Test

After completing a usability test, testers have a great deal of data, including notes, questionnaires, and videotapes. The three steps in turning that data into useful information are the following:

- *Tabulate the data.* This step involves gathering all the information, both quantitative and qualitative, and, when appropriate, performing statistical procedures on it. The quantitative data include performance measures, such as how long it took a participant to complete a task, and some attitude measures, such as data on how easy the participant found the task to perform. The qualitative data include nonnumerical information, such as notes taken by the test team and quotations from participants during or after the test.

- *Analyze the data.* Once testers have compiled all the data, they analyze it to understand and fix the usability problems. Naturally, they want to concentrate on the most important ones. Therefore, they want to determine the severity and the frequency of each problem.

- *Report the data.* Although writing a report can require weeks and add 20 to 25 percent to the cost of the test (Kantner, 1994), the rigor required in compiling a report often leads the testers to insights they might not have achieved otherwise.

Usability testing might seem like an extremely expensive and difficult undertaking, and certainly it is frustrating. But testers who are methodical, open-minded, and curious about how people use the product find that usability testing is the least expensive and most effective way to improve its quality.

CASE

Writing a Set of Instructions

The instructor in your technical-communication class would like your help in familiarizing students with the process of downloading software from the Internet. She has asked you to write a set of instructions that students in the class can use to download an Internet browser from Microsoft (http://www.microsoft.com) or Netscape (http://home.netscape.com). Visit one of these sites, download the software, and configure it either for a home computer or a computer in one of your labs. Then, write a set of instructions that are clear to a person who is basically familiar with using a computer but who has not downloaded and configured software.

WRITER'S CHECKLIST

Instructions

1. Does the introduction to the set of instructions
   □ state the purpose of the task?
   □ describe safety measures or other concerns that the readers should understand?
   □ list necessary tools and materials?

2. Are the step-by-step instructions
   □ numbered?
   □ expressed in the imperative mood?
   □ simple and direct?

3. Are appropriate graphics included?

4. Does the conclusion
   □ include any necessary follow-up advice?
   □ include, if appropriate, a troubleshooter's guide?

Manuals

1. Does the manual include, if appropriate, a cover?

2. Does the title page provide all the necessary information to help readers understand whether they are reading the appropriate manual?

3. Is the table of contents clear and explicit? Are the items phrased to indicate clearly the task that readers are to carry out?

4. Does the other front matter clearly indicate
   □ the product, procedure, or system the manual describes?
   □ the purpose of the manual?
   □ the major components of the manual?
   □ the best way to use the manual?

5. Is the body of the manual organized clearly?

6. Are appropriate graphics included?

7. Is a glossary included, if appropriate?

8. Is an index included, if appropriate?

9. Are all other appropriate appendix items included?

10. Is the writing style clear and simple throughout the manual?

EXERCISES

Several of the following exercises call for you to write a memo. See Chapter 16 for a discussion of memos.

1. You work in the customer-relations department of a company that makes plumbing supplies. The head of product development has just handed you the draft of installation instructions for a sliding tub door that you see on page 658. He has asked you to comment on their effectiveness. Write a memo to him, evaluating the instructions and suggesting improvements.

2. Write a brief manual for some process you are familiar with. Consider writing a procedures manual for a school activity or part-time job, such as your work as the business manager of the school newspaper or as a tutor in the Writing Center.

3. Write a memo to your instructor evaluating the front matter of a manual. Does it explain the purpose of the manual and how to use it? Is its tone welcoming and professional? How might it be improved? Attach to your memo copies of the pages you are evaluating.

4. If a local company conducts usability testing, see whether you can sign up to be a test participant. After the test, write a 1,000-word memo to your instructor describing the experience, focusing on what you learned about usability testing.

Collaborative Exercise

5. Write instructions for one of the following activities or for a process used in your field. Include appropriate graphics. In a brief note preceding the instructions, indicate your audience and purpose. Exchange these materials with a partner. Observe your partner and take notes as he or she attempts to carry out the instructions. Then, revise your instructions and share them with your partner. Discuss with your partner whether and how the revised instructions are easier to understand and apply. Submit your instructions to your instructor.
   a. how to change a bicycle tire
   b. how to parallel-park a car
   c. how to light a fire in a fireplace
   d. how to make a cassette-tape copy of a compact disc
   e. how to tune up a car

 Web Exercise

6. Go to the Web site for Hewlett Packard (http://www.hp.com) and look at the technical support for home or office printers. Follow the instructions for finding and downloading a driver for a popular piece of software, such as a word processor or spreadsheet. Then check out the site of one of the company's major competitors and carry out the same process for the same kind of driver. Compare the ease of doing so from the two sites. Write a memo comparing and contrasting the two sites. Attach printouts of any relevant pages from the two sites.

REFERENCES

Arts and letters clip art handbook. (1990). 15th ed. 1990. Dallas: Computer Support Corporation.

Befuddled PC users flood help lines, and no question seems to be too basic. (1994, March 1). *Wall Street Journal,* sec. B, p. 1.

Dumas, J. S., & Redish, J. C. (1993). *A practical guide to usability testing.* Norwood, NJ: Ablex.

Group 1 Software, Inc. (1994). *AccuMail User's Guide.* Lanham, MD: Group 1 Software, Inc.

Iomega Corporation. (1997). Zip™ Drive 100 Parallel Port user's guide. Roy, UT: Iomega Corporation.

Kantner, L. (1994). The art of managing usability tests. *IEEE Transactions on Professional Communication 37,* 143–148.

Microsoft Corporation. (1995). *Introducing Microsoft® Windows® 95.* Redmond, WA: Microsoft Corporation.

Nielsen, J. (1990). *Hypertext and hypermedia.* Boston: Academic Press.

Rubin, J. (1994). *Handbook of usability testing: How to plan, design, and conduct effective tests.* New York: Wiley.

Sanderlin, S. (1988). Preparing instruction manuals for non-English readers. *Technical Communication 35,* 96–100.

Sematech (1997). Electronic Documentation—Demonstration. <http://www.sematech.org/public/elecdoc/html/level3a.htm> (1997, May 14).

INSTALLATION INSTRUCTIONS

<u>CAUTION</u>: SEE BOX NO. 1 BEFORE CUTTING ALUMINUM HEADER OR SILL

1 Measure the wall to wall opening at the tub rim.

<u>CAUTION</u>: Do not forget to add 2" to inside tape measurement when required.

USE YOUR TAPE CORRECTLY.

2 Cut the <u>bottom sill</u> track 1/4" less than opening.

3 If desired, use a good all purpose caulk on the under side of sill. Press sill down on tub rim. Be sure drain holes face into tub.

4 Set wall jambs against the wall. Align vertically, mark wall with pencil or crayon.

5 Peel backing from installation tape on jambs, install by setting each jamb firmly over and down upon the sill. Press firmly to the wall for a good bond.

6 Measure the width inside the installed jambs, cut header bar 1/8" less.

7 *Wall Jamb* *Header Bar* *Bottom Sill* *Door Panel*

Mount nylon rollers on top of each door panel (see sketch) using the center hole. Other holes will raise or lower the doors for wall alignment. Thread door panels onto header bar with smooth side of panels facing inside the tub.

8 Push doors to the center of header bar. Lift and lower into place, easing bottom nylon door guides into the proper channel of sill.

TRIDOR MODEL ONLY:

To reverse direction of panels, raise panels out of bottom track and slide catches past each other thereby reversing direction so that shower head does not throw water between the panels.

HARDWARE KIT CONTENTS
TUDOR MODEL
4 nylon bearings
4 ball bearing screws # 8-32 × 3/8"
TRIDOR MODEL
6 nylon bearings
6 ball bearing screws # 8-32 × 3/8"

Making Oral Presentations

Understanding the Role of Oral Presentations 660

Preparing an Oral Presentation 662

Assessing the Speaking Situation 663
Choosing Effective Language 664
Preparing an Outline or Note Cards 666
Preparing Presentation Graphics 668
Rehearsing the Presentation 672

Giving the Oral Presentation 675

Dealing with Nerves 675
Using Your Voice Effectively 676
Using Your Body Effectively 677

Answering Questions after the Presentation 678

Sample Evaluation Form 679

Case: Publicizing the School Newspaper 680

Speaker's Checklist 680

Exercises 681

Collaborative Exercises 681
Web Exercise 681

So far, this book has focused on writing. In this chapter, the focus shifts to speaking. Even though documents and presentations are different, the techniques for preparing them are quite similar; you analyze your audience, gather information, organize it, and create graphics. The big difference between writing and speaking is, of course, the form of delivery.

There are four basic types of presentations:

- *Impromptu presentations.* You deliver an impromptu presentation without advance notice. For instance, at a meeting, your supervisor calls on you to speak for a few minutes about a project you are working on. You did not know you were going to be asked; you did not prepare any materials or rehearse.

- *Extemporaneous presentations.* In an extemporaneous presentation, you might refer to notes or an outline, but you actually make up the sentences as you go along. Regardless of how much you have planned and rehearsed the presentation, you create it as you speak.

    The extemporaneous presentation is preferable for all but the most formal occasions. At its best, it is clear and seemingly spontaneous. If you have planned and rehearsed your presentation sufficiently, the information will be accurate, complete, organized, and easy to follow. And if you can think well on your feet, the presentation will have a naturalness that will help your audience concentrate on what you are saying.

- *Scripted presentations.* In a scripted presentation, you read a text that you (or someone else) wrote out completely in advance. You sacrifice naturalness for increased clarity and precision.

- *Memorized presentations.* In a memorized presentation, you speak without notes or a script. Memorized presentations are not appropriate for most technical subjects because of the difficulty of remembering technical data. In addition, few people other than trained actors can memorize presentations of more than a few minutes.

This chapter discusses extemporaneous and scripted presentations. It explains how to prepare an outline or speaking script, how to prepare graphics, and how to rehearse. Finally, the chapter addresses how to give the presentation and respond to questions afterward.

Understanding the Role of Oral Presentations

In certain respects, an oral presentation is inefficient. For the speaker, preparing and rehearsing it generally take more time than writing the same information in a document would take. For the audience, physical conditions during the presentation—such as noises, poor lighting or acoustics, or an uncomfortable temperature—can make concentration difficult.

INTERVIEW: HOSPITAL ADMINISTRATOR VINCENT KURAITIS

on the importance of oral communications skills to his job

They're really important. I have to build consensus, I need to persuade people, I need to get my ideas across, and much of that is done orally.

on where he learned to give presentations

It's an acquired skill over time. I debated in college and high school, so I had some very specific skill building from back then. And I took a speech course in college. In my first job there were ongoing training sessions, and public speaking was part of that, and then I guess on-the-job training over time is another way you build your skills.

on the need to practice giving presentations

Practice is extremely important. Through practice almost everybody can become more proficient at public speaking. I think Toastmasters is one of the most economical, practical ways to practice, and I've been involved in that and recommended it to other people. You get together and practice a wide variety of different types of speeches with other people who will give you some candid feedback.

on how to prepare to give an important presentation

It's really important to reflect on who the audience is and to tailor each presentation with the audience in mind. I address a wide diversity of audiences. On the one hand, I make presentations to hospital employees, and some of them look at their work as a job, not a career. They may have very little knowledge of what's going on in health care or in the industry, and I've got to be able to speak to that level. At the other extreme, I make presentations to our hospital board of directors and the people in our corporate office, and they are very knowledgeable about business and social issues. I think it's important to plan the presentation according to what your audience knows, to the way they like information presented; some people like numbers, some people like pictures, some people like words.

on the need to consider your purpose in planning a presentation

You also need to think about how you can most effectively achieve your goal. Sometimes the goal is to inform, sometimes it's to persuade, sometimes it's a combination. You have to be conscious of what you're trying to achieve when you're talking to them.

on planning resources for making graphics

We use transparencies more often than slides. It varies, depending on the audience. If I'm going to make something for a presentation and use it once, I'm probably not going to bother to go out and spend $50 per slide for a presentation, but there are some presentations that we deliver to different groups of our employees over a period of time. So it makes a little more sense to invest some time and energy so that the presentations look a little more professional.

on dealing with nervousness

I look at nervousness as an opportunity to try and channel that adrenaline, to use it as something that will help me do my best rather than detract from my ability to be persuasive. So I can get a little bit more alert, a little bit more in touch with my audience, and I look at the adrenaline rush that comes with speaking in front of an audience as something to look forward to because it makes me do my best.

Vincent Kuraitis, J.D./M.B.A., is vice president of corporate development and specialty services at Saint Alphonsus Regional Medical Center in Boise, Idaho.

Yet an oral presentation has one big advantage over a written one: it permits a dialogue between the speaker and the audience. Listeners can offer alternative explanations and viewpoints, or simply ask questions that help the speaker clarify the information. And the speaker and listeners can converse before and after the presentation.

Oral presentations are therefore a popular means of technical communication. You can expect to give oral presentations to four types of audiences:

- *Clients and customers.* An oral presentation can be a valuable sales technique. Whether you are trying to interest clients in a silicon chip or a bulldozer, you will present its features and its advantages over the competition. Then, after the sale, you will likely provide detailed oral operating instructions and maintenance procedures to the users.

- *Colleagues in your organization.* If you are the resident expert on a mechanism, procedure, or technical subject, you will instruct fellow workers, both technical and nontechnical. After you return from an important conference or an out-of-town project, your supervisors will want a briefing—an oral report. If you have an idea for improving operations at your organization, you probably will write a brief informal proposal and then present the idea orally to a small group of managers. Your presentation will help them determine whether it is prudent to devote resources to studying the idea.

- *Fellow professionals at technical conferences.* You might speak on your own research project or on a team project carried out at your organization. Or you may be invited to speak to professionals in other fields. If you are an economist, for example, you might be invited to speak to real-estate agents about interest rates.

- *The general public.* As you assume greater prominence in your field, you will receive more invitations to speak to civic organizations and governmental bodies. Your organization will encourage you to give these presentations, because they reflect positively on the organization.

You might not have had much experience in public speaking, and perhaps your few attempts have been difficult. Natural speakers who can talk off the cuff are rare; for most of us, an oral presentation requires deliberate and careful preparation.

Preparing an Oral Presentation

Professional speakers make presentations look easy. But when you see an excellent 20-minute presentation, you are seeing only the last 20 minutes of a process that took many hours. How much time should you devote to preparing an oral presentation? Experts recommend devoting 20 to 60 minutes for each minute of the presentation (Smith, 1991, p. 6). At an average of 40 minutes, you would need more than 13 hours to prepare a 20-minute presentation. Obviously, there are many variables, including your knowledge of the subject, experience in creating presentation graphics, and experience in giving presentations on the same subject. But the point is clear: good presentations don't just happen.

Preparing an oral presentation requires the five steps shown in Figure 21.1.

Assessing the
Speaking Situation

The first step in assessing the speaking situation is to analyze your audience and purpose. Then you determine how much information you can deliver in the allotted time.

Analyzing Your Audience and Purpose

Consider audience and purpose, just as you would in writing a document.

- *Audience.* Who are the people in the audience? How much do they know about your subject? Answers to these questions determine the level of technical vocabulary and concepts appropriate to the audience. Speaking over an audience's head puzzles them; oversimplifying makes you appear condescending and insulting.

  Ask yourself the same kinds of questions you would ask about readers: Why is the audience there? What do they want to accomplish in listening to your presentation? Are they likely to be hostile, enthusiastic, or neutral? A presentation on the virtues of free trade, for instance, will be received one way by conservative economists and another way by U.S. steelworkers.

- *Purpose.* Are you attempting to inform your audience, or to both inform and persuade them? If you are explaining how windmills can be used to generate power, you have one type of argument. If you are explaining

FIGURE 21.1

Preparing an Oral
Presentation

why *your* windmills are an economical means of generating power, you have another type.

Your analysis of your audience and purpose will affect the strategy—the content and the form—of your presentation. You might have to emphasize some aspects of your subject or ignore some altogether. You might have to arrange topics to accommodate the audience's needs.

Budgeting Your Time

At most professional conferences, the organizers clearly state a maximum time, such as 20 minutes, for each speaker. If the question-and-answer period is part of your allotted time, plan accordingly. Even at an informal presentation, you probably will have to work within an unstated time limit that you must determine from the speaking situation. Claiming more than your share of an audience's time is rude and egotistical, and eventually your listeners will start to resent you or simply stop paying attention.

For a 20-minute presentation, the time allotment in Table 21.1 is typical.

For scripted presentations, most speakers need more than a minute to deliver a double-spaced page of text effectively.

Choosing Effective Language

Delivering an oral presentation is more challenging than writing a document, for two reasons:

- Listeners can't flip backward in the presentation to listen again to something they didn't understand.
- Because you are speaking live, you must maintain your listeners' attention, even if they are hungry or tired or the room is too hot. (Readers can be distracted, too, but they have more freedom to deal with the distractions.)

Using language effectively helps you meet these two challenges.

TABLE 21.1	Time Allotment for a 20-Minute Presentation
Task	**Time (minutes)**
• Introduction	2
• Body	
– First Major Point	4
– Second Major Point	4
– Third Major Point	4
• Conclusion	2
• Questions	4

Using Language to Signal Advance Organizers, Summaries, and Transitions

Even if you use graphics effectively, listeners cannot "see" the organization of a presentation as well as readers can. For this reason, use language to alert your listeners to advance organizers, summaries, and transitions.

- *Advance organizers.* An advance organizer is a statement that tells the listener what you are about to say. You will use an advance organizer in the introduction, of course, when you tell your audience your purpose, scope, main points, and organization. In addition, you will use advance organizers when you introduce main ideas in the body of the presentation.

  Advance organizers have to be explicit:

  In the next 20 minutes, I'd like to discuss the implications of the new RCRA regulations on the long-range waste-management strategy for Radnor Township. I want to make three major points. First, that. . . . Second, that. . . . And third, that. . . . After the presentation, I'll be happy to field your questions.

  Notice that the speaker numbers his points. He can use this numbering system throughout the presentation to help listeners follow him.

- *Summaries.* The major summary is in the conclusion, but you might also summarize at strategic points in the body of the presentation. For instance, after a three- to four-minute discussion of a major point, you might summarize it in one sentence before going on to the next major point. Here is a sample summary from a conclusion:

  Let me conclude by summarizing my three main points about the implications of the new RCRA regulations on the long-range waste-management strategy for Radnor Township. The first point: . . . . The second point: . . . . The third point: . . . . I hope this presentation will give you some ideas as you think about the RCRA. If you have any questions, I'd be happy to try to answer them at this time.

- *Transitions.* As you move from one point to the next, signal the transition clearly. Summarize the old point, and then announce that you are moving to the next point:

  It is clear, then, that the federal government has issued regulations without indicating how it wants county governments to comply with them. I'd like to turn now to my second main point:. . . .

Using Memorable Language

If people doze off while reading a document you have written, you probably won't know it, at least until they complain or you realize that the document failed to accomplish your purpose. But if they doze off while you are speaking to them, you'll know it right away. Effective presentations require memorable language.

Three techniques can help make your language more memorable.

- *Involve the audience.* People are more interested in their own concerns than in yours. Talk to the audience about their problems and their solutions. In the introduction, establish a link between your topic and the audience's interests. For instance, the presentation to the Radnor Township Council about waste management might begin like this:

  Picture yourself on the Radnor Township Council two years from now. After exhaustive hearings, proposals, and feasibility studies, you still don't have a waste-management plan that meets federal regulations. What you do have is a mounting debt: the township is being assessed $1,000 per day until you implement an acceptable plan.

- *Refer to people, not to abstractions.* People remember specifics; they forget abstractions. To make a point memorable, describe it in human terms:

  What could you do with that $365,000 every year? You could buy almost 200 personal computers; that's a computer for almost every classroom in every elementary school in Radnor Township. You could expand your school-lunch program to feed every needy child in the township. You could extend your after-school programs to cover an additional 3,000 students.

- *Use interesting facts, figures, and quotations.* Do your research and find interesting information about your subject. For instance, you might find a brief quotation from an authoritative figure in the field or a famous person not generally associated with the field (for example, Abraham Lincoln on waste management).

A note about humor: There are about 500 people in the United States who make a good living being funny. In comparison, there are millions of engineers, accountants, scientists, and technical communicators. Don't plan to tell a joke. If something happens in the context of the presentation that provides an opening for a witty remark, and you are good at it, fine. But don't *prepare* to be funny.

Preparing an Outline or Note Cards

After assessing your audience, purpose, and strategy, prepare an outline or a set of note cards. Keep in mind that an oral presentation should, in general, be simpler than a written version of the same material. For example, keep statistics and equations to a minimum.

Structure the presentation logically. The argument patterns presented in Chapter 8 are as useful for an oral presentation as for a document.

Write an outline or note cards just as you would for a document. Your own command of the facts—and your ability to remember them while you are under stress—will determine the degree of specificity. Many people prefer a sentence outline (see Chapter 8) because of its specificity; others feel more comfortable with a topic outline, especially when using note cards. Figure 21.2 shows a

Purpose: to describe, to a group of civil engineers, a new method of treatment and disposal of industrial waste.

1. Introduction
   1.1 The recent Resource Conservation Recovery Act places stringent restrictions on plant engineers.
   1.2 With neutralization, precipitation, and filtration no longer available, plant engineers will have to turn to more sophisticated treatment and disposal techniques.

2. The Principle Behind the New Techniques
   2.1 Waste has to be converted into a cementitious load-supporting material with a low permeability coefficient.
   2.2 Conversion Dynamics, Inc., has devised a new technique to accomplish this.
   2.3 The technique is to combine pozzolan stabilization technology with the traditional treatment and disposal techniques.

3. The Applications of the New Technique
   3.1 For new low-volume generators, there are two options.
       3.1.1 Discussion of the San Diego plant.
       3.1.2 Discussion of the Boston plant.
   3.2 For existing low-volume generators, Conversion Dynamics offers a range of portable disposal facilities.
       3.2.1 Discussion of the Montreal plant.
       3.2.2 Discussion of the Albany plant.
   3.3 For new high-volume generators, Conversion Dynamics designs, constructs, and operates complete waste-disposal management facilities.
       3.3.1 The Chicago plant now processes up to 1.5 million tons per year.
       3.3.2 The Atlanta plant now processes up to 1.75 million tons per year.
   3.4 For existing high-volume generators, Conversion Dynamics offers add-on facilities.
       3.4.1 The Roanoke plant already complies with the new RCRA requirements.
       3.4.2 The Houston plant will be in compliance within six months.

4. Conclusion
   The Resource Conservation Recovery Act will necessitate substantial capital expenditures over the next decade.

FIGURE 21.2

Sentence/Topic Outline for an Oral Presentation

combined sentence/topic outline, and Figure 21.3 on page 668 shows a topic outline. (Of course, the outline could be written on note cards.) The speaker is the specialist in waste-treatment facilities. The audience is a group of civil engineers interested in gaining a general understanding of new developments in industrial-waste disposal. The speaker's purpose is to provide this information and to suggest that his company is a leader in the field.

Notice the differences in content and form between these two versions of the same outline. Whereas some speakers prefer to have full sentences before them, others find that full sentences interfere with spontaneity. (For presentations read from manuscript, of course, the outline is only preliminary to writing the text.)

Purpose: to describe, to a group of civil engineers, a new method of treatment and disposal of industrial waste.

Introduction
– implications of the RCRA

Principle behind New Technique
– reduce permeability of waste
– use pozzolan stabilization technology

Applications of New Technique
– for new low-volume generators (San Diego, Boston)
– for existing low-volume generators (Montreal, Albany)
– for new high-volume generators (Chicago, Atlanta)
– for existing high-volume generators (Roanoke, Houston)

Conclusion

FIGURE 21.3

Topic Outline for an
Oral Presentation

**Preparing
Presentation
Graphics**

Graphics fulfill the same purpose in an oral presentation that they do in a written one: they clarify or highlight important ideas or facts. Statistical data, in particular, lend themselves to graphical presentation, as do descriptions of equipment or processes. Research reported by Smith (1991) indicates that presentations that include transparencies are judged more professional, persuasive, and credible than those without them, and that audiences remember information better if it is accompanied by graphics. Smith (1991, p. 58) offers these figures:

	Retention after	
	3 hours	3 days
Without graphics	70%	10%
With graphics	85%	65%

If you have not had a lot of experience creating graphics for presentations, review Chapter 12; everything said there about graphics applies to graphics for presentations.

Characteristics of an Effective Graphic

Effective graphics have five characteristics:

- *Visibility.* The most common problem with presentation graphics is that they are too small. Many speakers mistakenly try to transfer information from an 8.5 × 11-inch page to a slide or transparency. As a general rule, text has to be in 24-point type or larger to be visible on the screen.

To save space, reduce traditional sentences to brief phrases:

TEXT IN A DOCUMENT:	The current system has three problems: • It is expensive to maintain. • It requires nonstandard components. • It is not compliant with the new MILSPEC.
SAME TEXT ON A SCREEN:	Three Problems: • Expensive Maintenance • Nonstandard Components • Noncompliant with MILSPEC

- *Legibility.* Use clear, legible lines for drawings and diagrams: black on white works best for transparencies. Use legible typefaces for text; a boldfaced sans serif typeface such as Helvetica is effective because it reproduces clearly on a screen. Avoid shadowed and outlined letters. See Chapter 11 for more information on type. And see Chapter 12 for information on color in presentation graphics and for examples of a black-and-white transparency and a color slide.

- *Simplicity.* Both text and drawings must be simple. Each graphic should present only one idea. Remember that your listeners have not seen the graphic before and will not be able to linger over it.

- *Clarity.* Of course, the information has to make sense to your audience. In cutting verbiage and simplifying concepts and visual representations, make sure the point of the graphic remains clear.

- *Correctness.* Rare is the presentation that does not contain at least one graphic with a typo or some other error. Everyone makes mistakes, but mistakes are particularly embarrassing when they are 10 inches tall on a screen.

See Chapter 11 on design and Chapter 12 on graphics. These two chapters provide a detailed look at the principles of using visual elements to enhance communication. One point made in Chapter 12 is worth repeating: be careful when you use graphics templates in your software. Some of the templates violate basic principles of design. And be careful when you use clip art. If it helps you communicate, go ahead. But don't use clip art just to fill blank space on a transparency or slide.

Graphics and the Speaking Situation

To plan your graphics, analyze four aspects of the speaking situation:

- *Length of the presentation.* How many graphics should you have? A guideline is to have a different graphic for every 30 seconds of the presentation. It is far better to have a series of simple graphics than to have one complicated one that stays on the screen for 10 minutes.

- *Audience aptitude and experience.* What kinds of graphics can your audience understand easily? You don't want to present scatter graphs if your listeners have had no experience interpreting them.

- *Size and layout of the room.* The graphics for a presentation in a small meeting room differ from those for a presentation in a 500-seat auditorium. Think first about the size of the images, then about the layout of the room. For instance, will a window create glare that you will have to consider as you plan the type or placement of the graphics?

- *Equipment.* Find out ahead of time what kind of equipment is available in the presentation room. Inquire about backups in case of equipment failure. If possible, bring your own equipment. That way, you know it works and you know how to use it. Some speakers bring graphics in two media just in case; that is, they have slides but they also have transparencies of the same graphics.

When experienced speakers make presentations away from the office, they often bring a set of supplies with them just in case. The following list, based on Smith (1991, pp. 148–149), specifies some of these items:

electrical plug adapter

extension cord

spare bulbs for overhead projector and slide projector

chalk and eraser

transparency pens

blank transparency sheets

transparent tape

scissors

Using Graphics to Signal the Organization of the Presentation

Used effectively, graphics can help you communicate the organization of the presentation. You can use the transition from one graphic to the next to indicate the transition from one point to the next.

As you create your graphics, try to reflect the basic organization of the presentation. Start with a title graphic that lists the title of the presentation, as well as your name, affiliation, and job title. Sometimes speakers also indicate the occasion for the presentation and the date. Figure 21.4 is a typical title graphic.

Next, display a graphic showing the first-level headings from your outline. For the presentation outlined in Figure 21.2, the first-level headings would appear as in Figure 21.5.

RCRA: Implications for the Future

Charles E. Harim

Department of Civil Engineering
The Johns Hopkins University

Radnor Township Council
Resource Committee

October 16, 1997

Overview

1. **Introduction: Implications of the RCRA**

2. **Principles behind the New Technique**

3. **Applications of the New Technique**

4. **Conclusion**

To move from one topic to the next, display a graphic for the new topic, complete with its subheadings. For instance, Figure 21.6 on page 672 shows the graphic that introduces the topic "Principles behind the New Technique."

Notice that the outline numbering system helps listeners understand the organization of the talk. This system of layering the graphics can be extended down to a third level if needed.

For your last graphic, consider a summary of your main points or a brief set of questions. The questions give you an occasion to restate your main points as you prompt the audience to synthesize the information.

2. Principles behind the New Technique

2.1   Reduce permeability of waste

2.2   Use pozzolan stabilization technology

Different Media Used for Graphics

Table 21.2 describes the basic media for graphics.

If you are using presentation graphics software, such as PowerPoint, keep in mind that many of the templates provided with the software are unnecessarily ornate, full of fancy shading and designs and colors. Choose a simple template, then modify it to meet the needs of your audience, purpose, and subject. You want to focus on your delivery of information, not on the complex design of the graphic.

One other note about presentation graphics software: set the software so that you use the mouse to control the rate of advance from one graphic to the next. Do not set it so that it advances automatically at a specified interval, such as 60 seconds. You don't want to be speeding up or slowing down your presentation so that you are synchronized with the graphics.

Rehearsing the Presentation Even the most gifted speakers need to rehearse. It is a good idea to set aside enough time to rehearse your speech thoroughly.

Rehearsing an Extemporaneous Presentation

Rehearse your extemporaneous presentation at least three times.

- *First rehearsal.* Don't worry about posture or voice projection. Just try to compose your presentation out loud with your outline before you. Your goal is to see if the speech makes sense—if you can explain all the points you have listed and can forge effective transitions from point to point. If you have any trouble, stop and try to figure out the problem. If you need more information, get it. If you need a better transition, create one.

  You might have to revise your outline or notes. This is very common and no cause for alarm. Pick up where you left off and continue through the presentation, stopping again where necessary to revise the outline. When you have finished your first rehearsal, put the outline away and do something else.

- *Second rehearsal.* Once you are rested, try the presentation again. This time, it should flow more easily. Make any necessary revisions to the

TABLE 21.2	Basic Media for Oral Presentations		

Medium	Advantages	Disadvantages
COMPUTER PRESENTATIONS: IMAGES APPEAR ON A COMPUTER SCREEN OR ON A PROJECTOR ATTACHED TO THE COMPUTER, OR CAN BE TRANSFERRED TO SLIDES, VIDEOTAPE, OR FILM.	• Very professional appearance. • You can produce any combination of static or dynamic images, from simple graphs to sophisticated, three-dimensional images. • You can control the rate at which the images change.	• The equipment is not available everywhere. • Preparing the graphics is very time consuming. • Presentations prepared on one system might not run on an incompatible system.
SLIDE PROJECTOR: PROJECTS PREPARED SLIDES ONTO A SCREEN.	• Very professional appearance. • Versatile—can handle photographs or artwork, color or black-and-white. • With a second projector, you can eliminate the pause between slides. • During the presentation, you can easily advance and reverse the slides. • Graphics software lets you create small paper copies of your slides to distribute to the audience after the presentation.	• Slides can be expensive to produce. • Room has to be kept relatively dark during the slide presentation.
OVERHEAD PROJECTOR: PROJECTS TRANSPARENCIES ONTO A SCREEN.	• Transparencies are inexpensive and easy to create. • You can draw transparencies "live." • You can create overlays by placing one transparency over another. • Lights can remain on during the presentation. • You can face the audience. • Graphics software lets you create small paper copies of your transparencies to distribute to the audience after the presentation.	• Not as professional-looking as slides. • Each transparency must be loaded separately by hand.
OPAQUE PROJECTOR: PROJECTS IMAGES ON PAPER ONTO A SCREEN.	• You can project single sheets or pages in a bound volume. • Requires no expense or advance preparation.	• Room must be kept dark during the presentation. • Cannot magnify sufficiently for a large auditorium. • Each page must be loaded separately by hand. • The projector is noisy.
POSTER: A GRAPHIC DRAWN ON OAK TAG OR OTHER PAPER PRODUCT.	• Inexpensive. • Posters can be drawn or modified "live."	• Too small for large rooms.
FLIP CHART: A SERIES OF POSTERS, BOUND AT THE TOP LIKE A LOOSE-LEAF BINDER; GENERALLY PLACED ON AN EASEL.	• Relatively inexpensive. • You can easily flip back or forward. • Posters can be drawn or modified "live."	• Too small for large rooms.

(Continued)

TABLE 21.2 *(continued)*		
Medium	**Advantages**	**Disadvantages**
FELT BOARD: A HARD, FLAT SURFACE COVERED WITH FELT, ONTO WHICH PAPER CAN BE ATTACHED USING DOUBLED-OVER ADHESIVE TAPE.	• Relatively inexpensive. • Effective if you wish to rearrange the items on the board during the presentation. • Versatile—can handle paper, photographs, cutouts.	• Informal appearance.
CHALKBOARD OR OTHER HARD WRITING SURFACE.	• Almost universally available. • You have complete control—can add, delete, or modify the graphic easily.	• Complicated or extensive graphics are difficult to create. • Ineffective in large rooms. • Very informal appearance.
OBJECTS: MODELS OR SAMPLES OF MATERIAL THAT CAN BE HELD UP OR PASSED AROUND THROUGH THE AUDIENCE.	• Interesting for the audience. • They provide a good look at the object.	• Audience members might not listen while they are looking at the object. • It can take a long while to pass an object around a large room. • The object might not survive intact.
HANDOUTS: PHOTOCOPIES OF WRITTEN MATERIAL GIVEN TO EACH AUDIENCE MEMBER.	• Much material can be fit on the page. • Audience members can write on their copies and keep them.	• Audience members might read the handout rather than listen to the speaker.

outline or notes. When you have complete control over the organization and flow, check to see if you are within the time limits.

- *Third rehearsal.* After a satisfactory second rehearsal, try the presentation again, under more realistic circumstances—if possible, in front of people. The listeners might offer constructive advice about parts they don't understand or about your speaking style. If people aren't available, tape-record the presentation and then evaluate your own delivery. If you can visit the site of the presentation to get the feel of the room and rehearse there, you will find giving the actual speech a little easier.

Rehearse again until you are satisfied with your presentation. Then stop. Don't attempt to memorize it; if you do, you will surely panic the first time you forget a phrase. During the presentation, you must be thinking of your subject, not about the words you used during the rehearsals.

Rehearsing a Scripted Presentation

Rehearsing a scripted presentation is a combination of revising the text and rehearsing it. As you revise, read the script out loud to hear how it sounds. Once you think the presentation says what you want to say, try reading it into a tape recorder. Revise it until you are satisfied, and then rehearse in front of real people. Again, do not memorize. There is no need to: you will have your script there on the podium.

Giving the Oral Presentation

After all the preparation, the time to give the presentation finally arrives. In giving the presentation, you will concentrate on what you have to say. You will have three additional concerns, however, as shown in Figure 21.7.

Dealing with Nerves

Most professional actors freely admit to being nervous before a performance, so it is no wonder that most technical speakers are nervous. You might well fear that you will forget everything or that no one will be able to hear you. These fears are common. But keep in mind three facts about nervousness:

- *You are much more aware of your nervousness than the audience is.* They are farther away from your trembling hands.
- *Nervousness gives you energy and enthusiasm.* Without energy and enthusiasm, your presentation will almost surely fail. If you *seem* bored and listless, your audience will *be* bored and listless.
- *After a few minutes, your nervousness will pass.* You will be able to relax and concentrate on the subject.

This advice, however, is unlikely to make you feel much better if you are distracted by nerves as you wait to give your presentation. Experienced speakers suggest different points to keep in mind when you find yourself getting nervous before a presentation:

- *Realize that you are prepared.* If you have done your homework and rehearsed the presentation, you'll be fine. You are in control of the presentation.

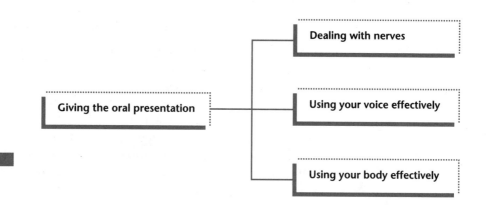

FIGURE 21.7

Giving the Oral Presentation

- *Realize that the audience is there to hear you, not to judge you.* Your listeners want to hear what you have to say. They are much less interested in your nervousness than you are.
- *Realize that your audience is made up of individual people who happen to be sitting in the same room.* If you tell yourself that the audience members are individuals like yourself who also get nervous before making presentations, you'll feel better.

And experienced speakers suggest actions you can take to lessen the nerves:

- *Walk around.* A brisk walk of a minute or two can calm you by dissipating some of your nervous energy.
- *Go off by yourself for a few minutes.* Some people find that getting away for a moment helps them compose their thoughts and realize that they can handle the nervousness.
- *Talk with someone for a few minutes.* For some speakers, distraction works best. Talk with someone who has come to the presentation a few minutes early.
- *Take several deep breaths, exhaling slowly.* Doing so will help control your nerves.

When it is time to begin, don't jump up to the lectern and start speaking quickly. Walk up slowly and arrange your text, outline, or note cards before you. If water is available, take a sip. Look out at the audience for a few seconds before you begin. It is polite to begin formal presentations with "Good morning" (or "Good afternoon," or "Good evening") and to refer to the officers and dignitaries present. If you have not been introduced, identify yourself. In less formal contexts, just begin your presentation.

So that the audience will listen to you and have confidence in what you say, try to project the same attitude that you would in a job interview: restrained self-confidence. Show interest in your topic and knowledge about your subject. You can convey this sense of control through your voice and your body.

Using Your Voice Effectively

Inexperienced speakers encounter problems with five aspects of vocalizing.

- *Volume.* Acoustics vary greatly from room to room, so you won't know how well your voice will carry in a room until you have heard someone speaking there. In some well-constructed auditoriums, speakers can use a conversational volume. Other rooms require greater voice projection. Some rooms have an annoying echo.

    These circumstances aside, more people speak too softly than too loudly. After your first few sentences, ask if the people in the back of the room can hear you. When people speak into microphones, they tend to bend down toward the microphone and end up speaking too loudly. Glance at your audience to see if you are having volume problems. The body language of the audience members will be clear.

- *Speed.* Nervousness makes people speak more quickly. Even if you think you are speaking at the right rate, you might be going a little too fast for some of your audience. Remember, you know what you will say; your listeners, however, are trying to understand new information. For particularly difficult points, slow down for emphasis. After finishing a major point, pause before beginning the next point.

- *Pitch.* In an effort to control their voices, many speakers end up flattening their pitch. The resulting monotone is boring—and, for some listeners, actually distracting. Try to let the pitch of your voice go up or down as it would in a normal conversation. In fact, experienced speakers often exaggerate pitch variations slightly.

- *Articulation.* The nervousness that goes along with an oral presentation tends to accentuate sloppy pronunciation. If you want to say *environment,* don't say *envirament.* Don't drop final *g*s. Say *trying,* not *tryin'.* A related pronunciation problem concerns technical words and phrases, especially the important ones. When a speaker uses a phrase over and over, it tends to get clipped and becomes difficult to understand. Unless you articulate carefully, *Scanlon Plan* will end up as *Scanluhplah.*

- *Nonfluencies.* Avoid such meaningless fillers as *you know, like, okay, right, uh,* and *um.* These phrases do not disguise the fact that you aren't saying anything; they call attention to it. A thoughtful pause is better than an annoying verbal tic.

Using Your Body Effectively

Besides listening to what you say, the audience will be looking at you. Effective speakers use their bodies to help the listeners follow the presentation. Keep in mind four guidelines about physical movement:

- *Maintain eye contact.* It is only polite to look at your audience. This is called *eye contact.* For small groups, look at each listener randomly; for larger groups, be sure to look at each segment of the audience frequently during your speech. Do not stare at your notes, at the floor, at the ceiling, or out the window. Eye contact is more than politeness; it is a way for you to see how the audience is receiving the presentation. You will see, for instance, if listeners in the back are having trouble hearing you.

- *Use natural gestures.* When people talk, they tend to gesture with their hands. Most of the time, these gestures make the presentation look natural and improve listeners' comprehension. You can supplement your natural gestures by using your arms and hands to signal pauses and to emphasize important points. When referring to graphics, walk to the screen and point to direct the audience's attention.

Avoid mannerisms—those physical gestures that serve no useful purpose. Don't play with your jewelry or the coins in your pocket. Don't tug at your beard or fix your hair. These nervous gestures can quickly distract an audience from what you are saying. Don't pace back and forth. Like verbal mannerisms, physical mannerisms are often unconscious. Constructive criticism from friends can help you pinpoint them.

- *Don't block the audience's view of the screen.* Don't stand at the overhead projector if doing so blocks some people's view of the screen. After you put on a transparency, step back to the side of the screen. Use a pointer to indicate key words or images on the screen.

- *Control the audience's attention.* People will listen to and look at anything that is interesting. Don't lose the audience's attention. If you hand out photocopies at the start of the presentation, some people will start to read them and stop listening to you. If you leave an image on the screen after you finish talking about it, some people will keep looking at it instead of listening to you. When you want the audience to look at you and listen to you, remove the graphics or turn off the projector.

Answering Questions after the Presentation

In some presentations, particularly informal ones, audience members ask questions throughout the presentation. This method of questioning helps you clarify points as you go along, but it can make it difficult for you to stay on track.

More often, an oral presentation is followed by a question-and-answer period. When you invite questions, don't abruptly say "Any questions?" This phrasing suggests that you don't really want any questions. Instead, say something like this: "If you have any questions, I'll be happy to try to answer them now." If invited politely, people will be much more likely to ask; in that way, you will more likely communicate your information effectively.

When you respond to questions, five situations occur frequently:

- *You're not sure everyone heard the question.* If there is no moderator to ask if the question was audible, ask if people have heard the question. If they haven't, repeat or paraphrase it, perhaps as an introduction to your response: "Your question is about the efficiency of these three techniques. . . ." Some speakers always repeat the question; that way, they are sure that everyone hears it, and they get an extra moment to think about their answer.

- *You don't understand the question.* Ask for a clarification. After responding, ask if you have answered the question adequately.

- *You don't know the answer to the question.* Tell the truth. Only novices believe that they ought to know all the answers. If you have some ideas about how to find out the answer—by checking a certain reference source, for example—share them. If the question is obviously important to the person who asked it, you might offer to meet with him or her after the question-and-answer period to discuss ways for you to give a more complete response, such as by e-mail.

- *You get a question that you have already answered in the presentation.* Restate the answer politely. Begin your answer with a phrase such as the following: "I'm sorry I didn't make that point clear in my talk. I wanted to explain how. . . ." Never insult the person by pointing out that you already covered that: "I already answered that question in my talk, but let me repeat it for you. . . ."

- *A belligerent member of the audience rejects your response and insists on restating his or her original point.* Offer politely to discuss the matter further after the session. This strategy will prevent the person from boring or annoying the rest of the audience.

If it is appropriate to stay after the session to talk individually with members of the audience, offer to do so. Remember to thank them for their courtesy in listening to you.

Sample Evaluation Form

Figure 21.8 is a list of questions that can help you focus your thoughts as you watch and listen to a presentation.

Evaluation Form

To the left of each of the following statements, write a number from 1 to 5, with 5 signifying strong agreement and 1 signifying strong disagreement.

( ) 1. In the introduction, the speaker made an attempt to relate the topic to my concerns.

( ) 2. In the introduction, the speaker explained the main points he or she wanted to make in the presentation.

( ) 3. In the introduction, the speaker explained the organization of the presentation.

( ) 4. The speaker used interesting, clear language to get the points across.

( ) 5. The speaker made the transitions from one point to the next clearly.

( ) 6. The speaker used clear and distinct enunciation.

( ) 7. The speaker exhibited no distracting vocal mannerisms.

( ) 8. The speaker exhibited no distracting physical mannerisms.

( ) 9. The speaker used graphics effectively to reinforce and explain the main points.

(Continued)

FIGURE 21.8

Sample Evaluation Form

(Figure 21.8 continued)

( ) 10. The speaker summarized the main points effectively in the conclusion.

( ) 11. The graphics helped me understand the organization of the presentation.

( ) 12. Throughout the presentation, the speaker paid attention to the audience.

( ) 13. The speaker seemed to be enthusiastic throughout the presentation.

( ) 14. The speaker used the allotted time effectively.

( ) 15. The speaker invited questions politely.

( ) 16. The speaker answered questions effectively.

Answer the following two questions on the other side of this sheet.

17. What did you particularly like about this presentation?

18. What would you have done differently if you had been the speaker?

CASE

Publicizing the School Newspaper

As the editor of your school's student newspaper, you want to publicize the newspaper on campus. You have contacted a number of teachers, who have agreed to give you 10 minutes of class time to make a presentation to their classes about the newspaper. Prepare and deliver a presentation—complete with graphics—on one of the following topics:

- the role of the newspaper in campus life
- five features you didn't know your newspaper provided
- proposed changes to the contents or design of the newspaper
- writing for the newspaper as a means of professional development
- how the newspaper is produced

SPEAKER'S CHECKLIST

This checklist covers the steps involved in preparing an oral presentation.

1. Have you assessed the speaking situation—the audience and purpose of the presentation?

2. Have you determined how much information you can communicate in your allotted time?

3. Have you chosen language to signal advance organizers, summaries, and transitions?

4. Have you chosen language that is vivid and memorable?

5. Have you outlined your information on note cards?

6. Have you prepared graphics that are
   □ visible?
   □ legible?
   □ simple?
   □ clear?
   □ correct?

7. In planning your graphics, have you considered your audience's aptitude and experience, the size and layout of the room, and the equipment?

8. Have you planned your graphics to help the audience understand the organization of your presentation?

9. Have you chosen appropriate media for your graphics?

10. Have you made sure that the presentation room will have the necessary equipment for the graphics?

11. Do you have appropriate supplies with you in case you need them?

12. Have you rehearsed your presentation several times with a tape recorder or a live audience?

EXERCISES

Exercise 2 calls for you to write a memo. See Chapter 16 for a discussion of memos.

1. Prepare a five-minute presentation, including graphics, on one of the topics listed here. For each presentation, your audience consists of the other students in your class, and your purpose is to introduce them to an aspect of your academic field.

    a. Define a key term or concept in your field.

    b. Describe how a particular piece of equipment is used in your field.

    c. Describe how to carry out a procedure common in your field.

    The instructor and the other students will evaluate the presentation by filling out the form in Figure 21.8.

2. Write a 1,000-word memo to your instructor in which you describe and evaluate a recent oral presentation of a guest speaker at your college or a politician on television.

Collaborative Exercises

3. Prepare a five-minute presentation based on your proposal for a research-report topic. Your audience consists of the other students in your class, and your purpose is to introduce them to your topic. The instructor and the other students will evaluate the presentation by filling out the form in Figure 21.8. If your instructor wishes, this assignment can be done collaboratively.

4. After the report is written, prepare a five-minute presentation based on your findings and conclusions. Your audience consists of the other students in your class, and your purpose is to introduce them to your topic. The instructor and the other students will evaluate the presentation by filling out the form in Figure 21.8. If your instructor wishes, this assignment can be done collaboratively.

Web Exercise

5. Many speeches of different kinds can be found on the Internet. Some of them are audio files, some are just text. These speeches are from the public sector (see http://www.access.gpo.gov to link to federal government sites) as well as the private sector (find the site of a large organization, such as Exxon—http://www.exxon.com—or AT&T—http://www.att. com—or Boeing—http://www.boeing.com). Find a speech on a topic that interests you. Download the text, then do the following tasks:

    • Create a set of presentation graphics in the form of bulleted text that might accompany the speech.

    • Describe and evaluate the speech. Consider such questions as the following: What is the audience of the presentation? In what ways has the speaker attempted to meet that audience's needs and goals? What is the organization of the speech? What aspects of language has the speaker used? On the basis of the text, how effective do you think the presentation was?

Present your findings in a memo to your instructor. Attach the transcript of the presentation.

REFERENCE

Smith, T. C. (1991). *Making successful presentations: A self-teaching guide*. New York: Wiley.

Appendix A
Handbook

Sentence Style, 684

Avoid Sentence Fragments, 684

Avoid Comma Splices, 685

Avoid Run-On Sentences, 686

Avoid Ambiguous Pronoun References, 686

Compare Items Clearly, 687

Use Adjectives Clearly, 688

Maintain Number Agreement, 689
Subject-Verb Disagreement, 689
Pronoun-Antecedent Disagreement, 689

Use Tenses Correctly, 689
Present Tense, 689
Past Perfect Tense, 690

Punctuation, 690

The Period, 690

The Exclamation Point, 690

The Question Mark, 691

The Comma, 691
Common Errors in Using Commas, 694

The Semicolon, 695
Common Error in Using Semicolons, 696

The Colon, 696
Common Error in Using Colons, 696

The Dash, 697
Common Error in Using Dashes, 697

Parentheses, 697
Common Error in Using Parentheses, 698

The Hyphen, 698

The Apostrophe, 699
Common Error in Using Apostrophes, 700

Quotation Marks, 700
Related Punctuation, 700

Block Quotations, 701

Mechanics, 701

Ellipses, 701

Brackets, 702

Italics, 702

Numbers, 703
Special Cases, 703

Abbreviations, 704

Capitalization, 705

This handbook concentrates on style, punctuation, and mechanics. Where appropriate, it defines common errors directly after discussing the correct usage.

Many of the usage recommendations made here are only suggestions. If your organization or professional field has a style guide that makes different recommendations, you should of course follow it.

Also, note that this is a selective handbook. It cannot replace full-length treatments, such as the handbooks often used in composition courses.

Sentence Style

Avoid Sentence Fragments

frag

A *sentence fragment* is an incomplete sentence. Most sentence fragments are caused by one of two problems:

- a missing verb

   FRAGMENT: The pressure loss caused by a worn gasket.

   This example is a fragment because it lacks a verb. (The word *caused* does not function as a verb here; rather, it introduces a phrase that describes the pressure loss.)

   COMPLETE
   SENTENCE: The pressure loss *was caused* by a worn gasket.

   In this revision, *pressure loss* has a verb: *was caused.*

   COMPLETE
   SENTENCE: We identified the pressure loss caused by a worn gasket.

   In this revision, *pressure loss* becomes the object in a new main clause: *We identified the pressure loss.*

   FRAGMENT: A plotting program with clipboard plotting, 3D animation, and FFTs.

   COMPLETE
   SENTENCE: It *is* a plotting program with clipboard plotting, 3D animation, and FFTs.

   COMPLETE
   SENTENCE: A plotting program with clipboard plotting, 3D animation, and FFTs *will be released* today.

- a dependent element used without an independent clause

   FRAGMENT: The article was not accepted for publication. *Because the data could not be verified.*

   Because the data could not be verified is a fragment because it lacks an independent clause: a clause that has a subject and verb and could stand alone as a sentence. It needs more information to be complete.

COMPLETE SENTENCE:	The article was not accepted for publication because the data could not be verified.

In this revision, the dependent element is joined to the independent clause that precedes it.

COMPLETE SENTENCE:	Because the data could not be verified, the article was not accepted for publication.

In this revision, the dependent element is followed by the independent clause.

FRAGMENT:	*Delivering over 150 horsepower.* The two-passenger coupe will cost over $32,000.
COMPLETE SENTENCE:	Delivering over 150 horsepower, the two-passenger coupe will cost over $32,000.
COMPLETE SENTENCE:	The two-passenger coupe will deliver over 150 horsepower and cost over $32,000.

Avoid Comma Splices

cs

A *comma splice* is the error in which two independent clauses are joined, or spliced together, by a comma. Independent clauses can be linked correctly in three ways:

- by a comma and a coordinating conjunction

COMMA SPLICE:	The 909 printer is our most popular model, it offers an unequaled blend of power and versatility.
CORRECT:	The 909 printer is our most popular model, for it offers an unequaled blend of power and versatility.

In this case, a comma and one of the coordinating conjunctions (*and, or, nor, but, for, so,* and *yet*) link the two independent clauses. The coordinating conjunction explicitly states the relationship between the two clauses.

- by a semicolon

COMMA SPLICE:	The 909 printer is our most popular model, it offers an unequaled blend of power and versatility.
CORRECT:	The 909 printer is our most popular model; it offers an unequaled blend of power and versatility.

In this case, a semicolon links the two independent clauses. The semicolon creates a somewhat more distant relationship between the two clauses than the comma-and-coordinating-conjunction link; the link remains implicit.

- by a period or other terminal punctuation

COMMA SPLICE:	The 909 printer is our most popular model, it offers an unequaled blend of power and versatility.

CORRECT: The 909 printer is our most popular model. It offers an unequaled
blend of power and versatility.

In this case, the two independent clauses are separate sentences. Of the three ways to punctuate the two clauses correctly, this punctuation suggests the most distant relationship between them.

Avoid Run-On Sentences

run

A run-on sentence (sometimes called a *fused sentence*) is a comma splice without the comma. In other words, two independent clauses appear without punctuation between them. Any of the three strategies to fix a comma splice fixes a run-on sentence.

RUN-ON   The 909 printer is our most popular model it offers an unequaled
SENTENCE: blend of power and versatility.

CORRECT: The 909 printer is our most popular model, for it offers an unequaled
blend of power and versatility.

CORRECT: The 909 printer is our most popular model; it offers an unequaled
blend of power and versatility.

CORRECT: The 909 printer is our most popular model. It offers an unequaled
blend of power and versatility.

Avoid Ambiguous Pronoun References

ref

Pronouns must refer clearly to the words or phrases they replace. Ambiguous pronoun references can lurk in even the most innocent-looking sentences:

UNCLEAR: Remove the cell cluster from the medium and analyze it. (Analyze what,
the cell cluster or the medium?)

CLEAR: Analyze the cell cluster after removing it from the medium.

CLEAR: Analyze the medium after removing the cell cluster from it.

CLEAR: Remove the cell cluster from the medium. Then analyze the cell cluster.

CLEAR: Remove the cell cluster from the medium. Then analyze the medium.

Ambiguous references can also occur when a relative pronoun such as *which*, or a subordinating conjunction such as *where*, introduces a dependent clause:

UNCLEAR: She decided to evaluate the program, which would take five months.
(What would take five months, the program or the evaluation?)

CLEAR: She decided to evaluate the program, a process that would take five
months. (By replacing *which* with *a process that,* the writer clearly
indicates that the evaluation will take five months.)

CLEAR: She decided to evaluate the five-month program. (By using the
adjective *five-month,* the writer clearly indicates that the program will
take five months.)

UNCLEAR: This procedure will increase the handling of toxic materials outside the
plant, where adequate safety measures can be taken. (Where can
adequate safety measures be taken, inside the plant or outside?)

CLEAR: This procedure will increase the handling of toxic materials outside the plant. Because adequate safety measures can be taken only in the plant, the procedure poses risks.

CLEAR: This procedure will increase the handling of toxic materials outside the plant. Because adequate safety measures can be taken only outside the plant, the procedure will decrease safety risks.

As the last example shows, sometimes the best way to clarify an unclear reference is to split the sentence in two, eliminate the problem, and add clarifying information. Clarity is always a fundamental characteristic of good technical communication. If more words will make your writing clearer, use them.

Ambiguity can also occur at the beginning of a sentence:

UNCLEAR: Allophanate linkages are among the most important structural components of polyurethane elastomers. They act as cross-linking sites. (What act as cross-linking sites, *allophanate linkages* or *polyurethane elastomers*?)

CLEAR: Allophanate linkages, which are among the most important structural components of polyurethane elastomers, act as cross-linking sites. (The writer has changed the second sentence into a clear nonrestrictive modifier and combined it with the first sentence.)

Your job is to ensure that the reader will know exactly which word or phrase the pronoun is replacing by either restructuring the sentence or dividing it in two.

If you begin a sentence with a pronoun that might be unclear to the reader, be sure to follow it immediately with a noun that clarifies the reference.

UNCLEAR: The new parking regulations require that all employees pay for parking permits. These are on the agenda for the next senate meeting. (What are on the agenda, the *regulations* or the *permits*?)

CLEAR: The new parking regulations require that all employees pay for parking permits. These regulations are on the agenda for the next senate meeting.

Compare Items Clearly

compar

When comparing or contrasting items, make sure your sentence clearly communicates the relationship. A simple comparison between two items often causes no problems: "The X3000 has more storage than the X2500." However, don't let your reader confuse a comparison and a simple statement of fact. For example, in the sentence "Trout eat more than minnows," does the writer mean that trout don't restrict their diet to minnows or that trout eat more than minnows eat? If a comparison is intended, a second verb should be used: "Trout eat more than minnows do." And if you are introducing three items, make sure that the reader can tell which two are being compared:

AMBIGUOUS: Trout eat more algae than minnows.

CLEAR: Trout eat more algae than they do minnows.

CLEAR: Trout eat more algae than minnows do.

Beware of comparisons in which different aspects of the two items are compared:

> ILLOGICAL: The resistance of the copper wiring is lower than the tin wiring.
>
> LOGICAL: The resistance of the copper wiring is lower than that of the tin wiring.

In the illogical construction, the writer contrasts *resistance* with *tin wiring* rather than the resistance of copper with the resistance of tin. In the revision, the pronoun *that* substitutes for *resistance* in the second part of the comparison.

Use Adjectives Clearly

adj

In general, adjectives are placed before the nouns that they modify: *the plastic washer.* In technical communication, however, writers often need to use clusters of adjectives. To prevent confusion, use commas to separate coordinate adjectives, and use hyphens to link compound adjectives.

Adjectives that describe different aspects of the same noun are known as coordinate adjectives:

> *portable, programmable* CD player
>
> *adjustable, removable* housings

In these cases, the comma replaces the word *and.*

Note that sometimes an adjective is considered part of the noun it describes: *electric drill.* When one adjective modifies *electric drill,* no comma is required: *a reversible electric drill.* The addition of two or more adjectives, however, creates the traditional coordinate construction: *a two-speed, reversible electric drill.*

The phrase *two-speed* is an example of a compound adjective—one made up of two or more words. Use hyphens to link the elements in compound adjectives that precede nouns:

> a *variable-angle* accessory
>
> increased *cost-of-living* raises

The hyphens in the second example prevent the reader from momentarily misinterpreting *increased* as an adjective modifying *cost* and *living* as a participle modifying *raises.*

But note that hyphens are not used when the adjective follows the noun:

> raises to cover increased *cost of living*

A long string of compound adjectives can be confusing even if hyphens are used appropriately. To ensure clarity in such a case, put the adjectives into a clause or phrase following the noun:

> UNCLEAR: an *operator-initiated, default-prevention* technique
>
> CLEAR: a technique *initiated by the operator to prevent default*

Maintain Number Agreement

Number disagreement commonly takes one of two forms in technical communication: (1) the verb disagrees in number with the subject when a prepositional phrase intervenes or (2) the pronoun disagrees in number with its referent when the latter is a collective noun.

Subject-Verb Disagreement

| agr s/v |

A prepositional phrase does not affect the number of the subject and the verb. The following examples show that the object of the preposition can be plural in a singular sentence, or singular in a plural sentence. (The subjects and verbs are italicized.)

INCORRECT: The *result* of the tests *are* promising.

CORRECT: The *result* of the tests *is* promising.

INCORRECT: The *results* of the test *is* promising.

CORRECT: The *results* of the test *are* promising.

Don't be misled by the fact that the object of the preposition and the verb don't sound natural together, as in *tests is* or *test are*. As long as the subject and verb agree, the sentence is correct.

Pronoun-Antecedent Disagreement

| agr p/a |

The problem of pronoun-antecedent disagreement occurs most often when the antecedent, or referent, is a collective noun—one that can be interpreted as either singular or plural, depending on its usage:

INCORRECT: The *company* is proud to announce a new stock option plan for *their* employees.

CORRECT: The *company* is proud to announce a new stock option plan for *its* employees.

In this example, *company* acts as a single unit; therefore, the singular verb, followed by a singular pronoun, is appropriate. When the individual members of a collective noun are emphasized, however, plural pronouns and verbs are appropriate: *The inspection team have prepared their reports.* Or *The members of the inspection team have prepared their reports.*

Use Tenses Correctly

Two tenses are commonly misused in technical communication: the present tense and the past perfect tense.

| t |

Present Tense

The present tense is used to describe scientific principles and recurring events.

INCORRECT: In 1992, McKay and his coauthors argued that the atmosphere of Mars *was* salmon pink.

CORRECT: In 1992, McKay and his coauthors argued that the atmosphere of Mars *is* salmon pink.

Notice that although the argument was made in the historical past—1992—the point is expressed in the present tense, because the atmosphere of Mars continues to be salmon pink. When the date of the argument is omitted, some writers express the entire sentence in the present tense.

McKay and his coauthors *argue* that the atmosphere of Mars *is* salmon pink.

Past Perfect Tense

The past perfect tense is used to describe two events that occurred in the past, with one occurring before the other.

We *had begun* excavation when the foreman *discovered* the burial remains.

Had begun is the past perfect tense. The excavation began before the burial remains were discovered.

The seminar *had concluded* before I *got* a chance to talk with Dr. Tran.

Punctuation

The Period

Periods are used in the following instances.

1. At the end of sentences that do not ask questions or express strong emotion:

   The lateral stress still needs to be calculated.

2. After some abbreviations:

   M.D.

   U.S.A.

   etc.

   (For a further discussion, see "Abbreviations" later in this appendix.)

3. With decimal fractions:

   4.056

   $6.75

   75.6 percent

The Exclamation Point

The exclamation point is used at the end of a sentence that expresses strong emotion, such as surprise or doubt:

The nuclear plant, which was originally expected to cost $1.6 billion, eventually cost more than $8 billion!

Because technical communication requires objectivity and a calm, understated tone, exclamation points are rarely used.

The Question Mark

The question mark is used at the end of a sentence that asks a direct question:

What did the commission say about effluents?

Do not use a question mark at the end of a sentence that asks an indirect question:

He wanted to know whether the procedure had been approved for use.

When a question mark is used within quotation marks, the quoted material takes no other end punctuation:

"What did the commission say about effluents?" she asked.

The Comma

The comma is the most frequently used punctuation mark, as well as the one about whose usage writers most often disagree. Following are the basic uses of the comma.

- To separate the clauses of a compound sentence (one composed of two or more independent clauses) linked by a coordinating conjunction (*and, or, nor, but, so, for, yet*):

  Both methods are acceptable, but we have found that the Simpson procedure gives better results.

  In many compound sentences, the comma is needed to prevent the reader from mistaking the subject of the second clause for an object of the verb in the first clause:

  The RESET command affects the field access, and the SEARCH command affects the filing arrangement.

  Without the comma, the reader is likely to interpret the coordinating conjunction *and* as a simple conjunction linking *field access* and *SEARCH command*.

- To separate items in a series composed of three or more elements:

  The manager of spare parts is responsible for ordering, stocking, and disbursing all spare parts for the entire plant.

  Despite the presence of the conjunction *and*, most technical-communication style manuals require that a comma follow the second-to-last item. The comma clarifies the separation and prevents misreading. For example, sometimes in technical communication the second-to-last item will be a compound noun containing an *and*.

  CONFUSING: The report will be distributed to Operations, Research and Development and Accounting.

CLEAR: The report will be distributed to Operations, Research and Development, and Accounting.

The series comma is also useful in preventing another kind of misreading. Look at the following beginning of a sentence:

According to Johnson, Smith and Jones . . .

Smith and Jones could be part of a series:

According to Johnson, Smith, and Jones, the new technology will never live up to its potential.

Or *Smith and Jones* could be the subject of a clause:

According to Johnson, Smith and Jones have done outstanding work this quarter.

Using the series comma prevents this ambiguity.

- To separate introductory words, phrases, and clauses from the main clause of the sentence:

However, we will have to calculate the effect of the wind.

To facilitate trade, the government holds a yearly international conference.

Whether the workers like it or not, the managers have decided not to try the flex-time plan.

In each of these three examples, the comma helps the reader understand the sentence. Notice in the following example how the comma actually prevents misreading:

Just as we finished eating, the rats discovered the treadmill.

The comma is optional if the introductory text is brief and cannot be misread.

CORRECT: First, let's take care of the introductions.

CORRECT: First let's take care of the introductions.

- To separate the main clause from a dependent clause:

Although most of the executive council saw nothing wrong with it, the advertising campaign was canceled.

Most PCs use green technology, even though it is relatively expensive.

- To separate nonrestrictive modifiers (parenthetical clarifications) from the rest of the sentence:

Jones, the temporary chairman, called the meeting to order.

- To separate interjections and transitional elements from the rest of the sentence:

Yes, I admit your findings are correct.

Their plans, however, have great potential.

- To separate coordinate adjectives:

The finished product was a sleek, comfortable cruiser.

The heavy, awkward trains are still being used.

The comma here takes the place of the conjunction *and*. If the adjectives are not coordinate—that is, if one of the adjectives modifies the combination of the adjective and the noun—do not use a comma:

They decided to go to the first general meeting.

- To signal that a word or phrase has been omitted from an elliptical expression:

  Smithers is in charge of the accounting; Harlen, the data management; Demarest, the publicity.

  In this example, the commas after *Harlen* and *Demarest* show that the phrase *is in charge of* has been omitted.

- To separate a proper noun from the rest of the sentence in direct address:

  John, have you seen the purchase order from United?

  What I'd like to know, Betty, is why we didn't see this problem coming.

- To introduce most quotations:

  He asked, "What time were they expected?"

- To separate towns, states, and countries:

  Bethlehem, Pennsylvania, is the home of Lehigh University.

  He attended Lehigh University in Bethlehem, Pennsylvania, and the University of California at Berkeley.

  Note the use of the comma after *Pennsylvania*.

- To set off the year in a date:

  August 1, 1999, is the anticipated completion date.

  Note the use of the comma after *1999*. If the month separates the date and the year, the first comma is not used; the numbers are not next to each other:

  The anticipated completion date is 1 August 1999.

- To clarify numbers:

  12,013,104

  (European practice is to reverse the use of commas and periods in writing numbers: periods signify thousands, and commas signify decimals.)

- To separate names from professional or academic titles:

  Harold Clayton, Ph.D.

  Marion Fewick, CLU

  Joyce Carnone, P.E.

Note that the comma also follows the title in a sentence:

Harold Clayton, Ph.D., is the featured speaker.

Common Errors in Using Commas

- The writer does not use a comma between the clauses of a compound sentence:

  INCORRECT: The mixture was prepared from the two premixes and the remaining ingredients were then combined.

  CORRECT: The mixture was prepared from the two premixes, and the remaining ingredients were then combined.

- The writer does not use a comma (or uses just one comma) to set off a nonrestrictive modifier:

  INCORRECT: The phone line, which was installed two weeks ago had to be disconnected.

  CORRECT: The phone line, which was installed two weeks ago, had to be disconnected.

  See the discussion of restrictive and nonrestrictive modifiers in Chapter 14.

- The writer does not use a comma to separate introductory words, phrases, or clauses from the main clause, when misreading can occur:

  INCORRECT: As President Canfield has been a great success.

  CORRECT: As President, Canfield has been a great success.

- The writer does not use a comma (or uses just one comma) to set off an interjection or a transitional element:

  INCORRECT: Our new statistician, however used to work for Konaire, Inc.

  CORRECT: Our new statistician, however, used to work for Konaire, Inc.

- The writer joins two independent clauses with a comma, creating a comma splice (a comma used to "splice together" independent clauses not linked by a coordinating conjunction):

  INCORRECT: All the motors were cleaned and dried after the water had entered, had they not been, additional damage would have occurred.

  CORRECT: All the motors were cleaned and dried after the water had entered; had they not been, additional damage would have occurred.

  CORRECT: All the motors were cleaned and dried after the water had entered. Had they not been, additional damage would have occurred.

For more information, see "Avoid Comma Splices" earlier in this appendix.

- The writer uses unnecessary commas:

  INCORRECT: Another of the many possibilities, is to use a "first in, first out" sequence.

  CORRECT: Another of the many possibilities is to use a "first in, first out" sequence.

In the incorrect sentence, the comma should be removed because it separates the subject (*Another*) from the verb (*is*).

  INCORRECT: The schedules that have to be updated every month are, numbers 14, 16, 21, 22, 27, and 31.

  CORRECT: The schedules that have to be updated every month are numbers 14, 16, 21, 22, 27, and 31.

In the incorrect sentence, the comma separates the verb from its complement: the word or words that complete the meaning of the sentence.

  INCORRECT: New and old employees who use the processed order form, do not completely understand the basis of the system.

  CORRECT: New and old employees who use the processed order form do not completely understand the basis of the system.

In the incorrect sentence, a comma separates the subject and its restrictive modifier from the verb.

  INCORRECT: The company has grown so big, that an informal evaluation procedure is no longer effective.

  CORRECT: The company has grown so big that an informal evaluation procedure is no longer effective.

In the incorrect sentence, the comma separates the predicate adjective *big* from the clause that modifies it.

  INCORRECT: Recent studies, and reports by other firms confirm our experience.

  CORRECT: Recent studies and reports by other firms confirm our experience.

In the incorrect sentence, the comma separates the two elements in the compound subject.

The Semicolon

;

Semicolons are used in the following instances.

- To separate independent clauses not linked by a coordinating conjunction:

  The second edition of the handbook is more up-to-date; however, it is more expensive.

- To separate items in a series that already contains commas:

  The members elected three officers: Jack Resnick, president; Carol Wayshum, vice president; Ahmed Jamoogian, recording secretary.

In this example, the semicolon acts as a "supercomma," keeping the names and titles clear.

Common Error in Using Semicolons

Sometimes writers incorrectly use a semicolon when a colon is called for:

INCORRECT: We still need one ingredient; luck.

CORRECT: We still need one ingredient: luck.

The Colon   Colons are used in the following instances.

> :

- To introduce a word, phrase, or clause that amplifies, illustrates, or explains a general statement:

    The project team lacked one crucial member: a project leader.

    Here is the client's request: we are to provide the preliminary proposal by November 13.

    We found three substances in excessive quantities: potassium, cyanide, and asbestos.

    The week was productive: fourteen projects were completed and another dozen were initiated.

    Note that the text preceding a colon should be able to stand on its own as a main clause:

    INCORRECT: We found: potassium, cyanide, and asbestos.

    CORRECT: We found potassium, cyanide, and asbestos.

- To introduce items in a vertical list, if the sense of the introductory text would be incomplete without the list:

    We found the following:

        potassium
        cyanide
        asbestos

    See Chapter 11 for more information on constructing lists.

- To introduce long or formal quotations:

    The president began: "In the last year . . ."

Common Error in Using Colons

Writers sometimes incorrectly use a colon to separate a verb from its complement:

INCORRECT: The tools we need are: a plane, a level, and a T-square.

CORRECT: The tools we need are a plane, a level, and a T-square.

CORRECT: We need three tools: a plane, a level, and a T-square.

The Dash

—

Dashes are used in the following instances.

- To set off a sudden change in thought or tone:

  The committee found—can you believe this?—that the company bore full responsibility for the accident.

  That's what she said—if I remember correctly.

- To emphasize a parenthetical element:

  The managers' reports—all 10 of them—recommend production cutbacks for the coming year.

  Arlene Kregman—the first woman elected to the board of directors—is the next scheduled speaker.

- To set off an introductory series from its explanation:

  Wetsuits, weight belts, tanks—everything will have to be shipped in.

  When a series *follows* the general statement, a colon replaces the dash:

  Everything will have to be shipped in: wetsuits, weight belts, and tanks.

  With most word-processing programs, you have to use a special character to make a dash; there is no dash key on the keyboard. Another way to make a dash is to use two uninterrupted hyphens (--). Do not add space before or after the hyphens. This form of the dash looks less professional than the special dash character.

Common Error in Using Dashes

Sometimes writers incorrectly use a dash as a "lazy" substitute for other punctuation marks:

INCORRECT: The regulations—which were issued yesterday—had been anticipated for months.

CORRECT: The regulations, which were issued yesterday, had been anticipated for months.

INCORRECT: Many candidates applied—however, only one was chosen.

CORRECT: Many candidates applied; however, only one was chosen.

Parentheses

( )

Parentheses are used in the following instances.

- To set off incidental information:

  Please call me (x3104) when you get the information.

  Galileo (1546–1642) is often considered the father of modern astronomy.

H. W. Fowler's *Modern English Usage* (New York: Oxford University Press, 2nd ed., 1965) was finally revised in 1996.

- To enclose numbers and letters that label items listed in a sentence:

  To transfer a call within the office, (1) place the party on HOLD, (2) press TRANSFER, (3) press the extension number, and (4) hang up.

  Use both a left and a right parenthesis—not just a right parenthesis—in this situation.

Common Error in Using Parentheses

Sometimes writers incorrectly use parentheses instead of brackets to enclose their interruption of a quotation (see the discussion of brackets):

INCORRECT: He said, "The new manager (Farnham) is due in next week."

CORRECT: He said, "The new manager [Farnham] is due in next week."

The Hyphen   Hyphens are used in the following instances.

-

- In general, to form compound adjectives that precede nouns:

  general-purpose register

  meat-eating dinosaur

  chain-driven saw

  Note that hyphens are not used after adverbs that end in *-ly*:

  newly acquired terminal

  Also note that hyphens are not used when the compound adjective follows the noun:

  The Woodchuck saw is chain driven.

  Many organizations have their own preferences about hyphenating compound adjectives. Check to see if your organization has a preference.

- To form some compound nouns:

  vice-president

  editor-in-chief

- To form fractions and compound numbers:

  one-half

  fifty-six

- To attach some prefixes and suffixes:

  post-1945

  president-elect

- To divide a word at the end of a line:

We will meet in the pavil-
ion in one hour.

Whenever possible, avoid such breaks; they slow the reader down. When you do use them, check the dictionary to make sure you have divided the word *between* syllables. Do not assume that your word processor has divided the word correctly.

The Apostrophe

Apostrophes are used in the following instances.

- To indicate the possessive case:

the manager's goals   the employee's credit union

the worker's lounge   Charles's T-square

For joint possession, add the apostrophe and the *s* to only the last noun or proper noun:

Watson and Crick's discovery

For separate possession, add an apostrophe and an *s* to each of the nouns or pronouns:

Newton's and Galileo's ideas

Make sure you do not add an apostrophe or an *s* to possessive pronouns: *his, hers, its, ours, yours, theirs.*

- To indicate the possessive case when a noun modifies a gerund:

We were all looking forward to Bill's joining the company.

In this case, the gerund—*joining*—is modified by the proper noun *Bill.*

- To form contractions:

I've   shouldn't

can't   it's

The apostrophe usually indicates an omitted letter or letters. For example, *can't* is *can(no)t*, *it's* is *it(i)s*.

Some organizations discourage the use of contractions; others have no preference. Find out the policy your organization follows.

- To indicate special plurals:

three 9's

two different JCL's

the why's and how's of the problem

As in the case of contractions, it is a good idea to learn the stylistic preferences of your organization. For plurals of numbers and acronyms, some style guides omit the apostrophe: *9s, JCLs*. Usage varies considerably.

Common Error in Using Apostrophes

Writers sometimes incorrectly use the contraction *it's* in place of the possessive pronoun *its*.

INCORRECT: The company does not feel that the problem is *it's* responsibility.

CORRECT: The company does not feel that the problem is *its* responsibility.

Quotation Marks

" "

Quotation marks are used in the following instances.

- To indicate titles of short works, such as articles, essays, or chapters:

  Smith's essay "Solar Heating Alternatives"

- To call attention to a word or phrase that is being used in an unusual way or in an unusual context:

  A proposal is "wired" if the sponsoring agency has already decided who will be granted the contract.

  Don't use quotation marks as a means of excusing poor word choice:

  The new director has been a real "pain."

- To indicate direct quotation, that is, the words a person has said or written:

  "In the future," he said, "check with me before authorizing any large purchases."

  As Breyer wrote, "Morale *is* productivity."

  Do not use quotation marks to indicate indirect quotation:

  INCORRECT: He said that "third-quarter profits will be up."

  CORRECT: He said that third-quarter profits will be up.

  CORRECT: He said, "Third-quarter profits will be up."

Related Punctuation

Note that if the sentence contains a *tag*—a phrase identifying the speaker or writer—a comma separates it from the quotation:

Wilson replied, "I'll try to fly out there tomorrow."

"I'll try to fly out there tomorrow," Wilson replied.

Informal and brief quotations require no punctuation before the quotation marks:

She said "Why?"

In the United States (unlike most other English-speaking nations), commas and periods at the end of quotations are placed within the quotation marks:

> The project engineer reported, "A new factor has been added."
>
> "A new factor has been added," the project engineer reported.

Question marks, dashes, and exclamation points, however, are placed inside the quotation marks when they apply only to the quotation and outside the quotation marks when they apply to the whole sentence:

> He asked, "Did the shipment come in yet?"
>
> Did he say, "This is the limit"?

When a punctuation mark appears inside a quotation mark at the end of a sentence, do not add another punctuation mark.

INCORRECT:	Did she say, "What time is it?"?
CORRECT:	Did she say, "What time is it?"

Block Quotations   When quotations reach a certain length—generally, more than four lines—switch to a block format. In word-processed manuscript, a block quotation is usually

- indented 10 spaces from the left-hand margin
- typed without quotation marks
- introduced by a complete sentence followed by a colon

(Different organizations observe their own variations on these basic rules.)

> McFarland writes:
>> The extent to which organisms adapt to their environment is still being charted. Many animals, we have recently learned, respond to a dry winter with an automatic birth control chemical that limits the number of young to be born that spring. This prevents mass starvation among the species in that locale.
>
> Hollins concurs. She writes, "Biological adaptation will be a major research area during the next decade."

Mechanics

Ellipses

. . .

Ellipses (three spaced periods) indicate the omission of some material from a quotation. A fourth period with no space before it precedes ellipses when the sentence in the source has ended and you are omitting material that follows:

> "Send the updated report . . . as soon as you can."
>
> Larkin refers to the project as "an attempt . . . to clarify the issue of compulsory arbitration. . . . We do not foresee an end to the legal wrangling . . . but perhaps the report can serve as a definition of the areas of contention."

In the second example, the writer has omitted words after *attempt* and after *wrangling*. In addition, the writer uses a sentence period plus three spaced periods after *arbitration*, which ends the source's sentence, to indicate that the following sentence has been omitted.

Brackets

Brackets are used in the following instances.

- To indicate words added to a quotation:

  "He [Pearson] spoke out against the proposal."

  A better approach would be to shorten the quotation:

  The minutes of the meeting note that Pearson "spoke out against the proposal."

- To indicate parenthetical information within parentheses:

  (For further information, see Charles Houghton's *Civil Engineering Today* [New York: Arch Press, 1997].)

Italics

ital

Italics (or underlining) are used in the following instances.

- For words used as words:

  In this report, the word *operator* will refer to any individual who is actually in charge of the equipment, regardless of that individual's certification.

- To indicate titles of long works (books, manuals, and so on), periodicals and newspapers, long films, long plays, and long musical works:

  See Houghton's *Civil Engineering Today*.

  We subscribe to the *Wall Street Journal*.

  In the second example, note that the word *the* is not italicized or capitalized when the title is used in a sentence.

- To indicate the names of ships, trains, and airplanes:

  The shipment is expected to arrive next week on the *Penguin*.

- To set off foreign expressions that have not become fully assimilated into English:

  The speaker was guilty of *ad hominem* arguments.

  Check a dictionary to determine whether the foreign expression has become assimilated.

- To emphasize words or phrases:

  Do not press the ERASE key.

  If your word processor does not have italic type, indicate italics by underlining:

  Darwin's <u>Origin of Species</u> is still read today.

Numbers

[num]

The use of numbers varies considerably. Therefore, in choosing between words and numerals, consult the guidelines that your organization or research area follows. Many organizations use the following guidelines.

- Use numerals for technical quantities, especially if a unit of measurement is included:

  3 feet   43,219 square miles

  12 grams   36 hectares

- Use numerals for nontechnical quantities of 10 or more:

  300 persons

  12 whales

  35 percent increase

- Use words for nontechnical quantities of fewer than 10:

  three persons

  six whales

- Use both words and numerals

  – For back-to-back numbers:

    six 3-inch screws

    fourteen 12-foot ladders

    3,012 five-piece starter units

    In general, use the numeral for the technical unit. If the nontechnical quantity would be cumbersome in words, use the numeral.

  – For round numbers over 999,999:

    14 million light-years

    $64 billion

  – For numbers in legal contracts or in documents intended for international readers:

    thirty-seven thousand dollars ($37,000)

    five (5) relays

  – For addresses:

    3801 Fifteenth Street

Special Cases

- If a number begins a sentence, use words, not numerals:

  Thirty-seven acres was the size of the lot.

  Many writers would revise the sentence to avoid this problem:

  The lot was 37 acres.

- Don't use both numerals and words in the same sentence to refer to the same unit:

  On Tuesday the attendance was 13; on Wednesday, 8.

- Write out fractions, except if they are linked to technical units:

  two-thirds of the members

  3½ hp

- Write out approximations:

  approximately ten thousand people

  about two million trees

- Use numerals for titles of figures and tables and for page numbers:

  Figure 1

  Table 13

  page 261

- Use numerals for decimals:

  3.14

  1,013.065

  Add a zero before decimals of less than one:

  0.146

  0.006

- Avoid expressing months as numbers, as in *3/7/98*: in the United States, this means March 7, 1998; in many other countries, it means July 3, 1998. Use one of the following forms:

  March 7, 1998

  7 March 1998

- Use numerals for times if A.M. or P.M. are used:

  6:10 A.M.

  six o'clock

Abbreviations

ab

Abbreviations save time and space, but use them carefully; you can never be sure that your readers will understand them. Many companies and professional organizations provide lists of approved abbreviations.

Analyze your audience in determining whether and how to abbreviate. If your readers include nontechnical people unfamiliar with your field, either write out the technical terms or attach a list of abbreviations. If you are new in an organization or are writing for publication for the first time in a certain field, find out what abbreviations are commonly used. If for any reason you are unsure about whether or how to abbreviate, write out the word.

The following are general guidelines about abbreviations.

- You may make up your own abbreviations. For the first reference to the term, write it out and include the abbreviation in parentheses. In subsequent references, use the abbreviation. For long works, you might want to write out the term at the start of major units, such as chapters.

  The heart of the new system is the self-loading cartridge (slc).

  This technique is also useful, of course, in referring to existing abbreviations that your readers might not know:

  The cathode-ray tube (CRT) is your control center.

- Most abbreviations do not take plurals:

  1 lb

  3 qt

- Most abbreviations in scientific writing are not followed by periods:

  lb

  cos

  dc

  If the abbreviation can be confused with another word, however, use a period:

  in.

  Fig.

- Spell out the unit if the number preceding it is spelled out or if no number precedes it:

  How many square meters is the site?

Capitalization

cap

For the most part, the conventions of capitalization in general writing apply in technical communication.

- Capitalize proper nouns, titles, trade names, places, languages, religions, and organizations:

  William Rusham

  Director of Personnel

  Quick-Fix Erasers

  Bethesda, Maryland

Italian

Methodism

Society for Technical Communication

In some organizations, job titles are not capitalized unless they refer to specific persons:

Alfred Loggins, Director of Personnel, is interested in being considered for vice president of marketing.

- Capitalize headings and labels:

A Proposal to Implement the Wilkins Conversion System

Mitosis

Table 3

Section One

The Problem

Rate of Inflation, 1985–1995

Figure 6

Appendix B
Guidelines for Speakers of English as a Second Language

Basic Characteristics of a Sentence, 708

Linking Ideas by Coordination, 709

Linking Ideas by Subordination, 709

Verb Tenses, 710

-ing Forms of Verbs, 711

Infinitives, 711

Helping Verbs and Main Verbs, 712

Agreement of Subject and Verb, 712

Conditions, 713

Articles, 714

Adjectives, 715

Adverbs, 715

Omitted Words, 716

Repeated Words, 717

English is notoriously difficult to master, and no brief guide can answer all your questions about the many eccentricities of the language. But this appendix provides a concise guide to some of the most challenging aspects of English for non-native speakers.

The advice offered here is based on two highly regarded brief books:

Ann Raimes. (1992). *Grammar troublespots: An editing guide for students* (2nd ed.). New York: St. Martin's Press.

Diana Hacker. (1995). *A writer's reference* (3rd ed.). New York: St. Martin's Press.

These two books are excellent investments.

Basic Characteristics of a Sentence

A sentence has five characteristics:

- It starts with an uppercase letter and ends with a period, a question mark, or (rarely) an exclamation point.
- It has a subject, usually a noun. The subject is what the sentence is about.
- It has a verb, which tells the reader about what happens to the subject or what the subject does.
- It has a standard word order. The most common sequence in English is subject-verb-object. But you can add information at different locations. For instance, here is a sentence in the basic subject-verb-object order:

subject   verb   object

We hired a consulting firm.

You can add information to the start of the sentence:

Yesterday we hired a consulting firm.

Or to the end of the sentence:

Yesterday we hired a consulting firm: *Sanderson & Associates*.

Or in the middle:

Yesterday we signed *a nontransferable contract* with a consulting firm: Sanderson & Associates.

In fact, any element of a sentence can be expanded.

- It has an independent clause: an idea that can stand alone. The following is a sentence because its idea can stand alone:

  SENTENCE: The pump failed because of improper maintenance.

The following is also a sentence:

  SENTENCE: The pump failed.

But the following is *not* a sentence because the thought is incomplete:

NOT A SENTENCE: Because of improper maintenance.

A result clause is required to complete this sentence:

SENTENCE: Because of improper maintenance, the pump failed.

Linking Ideas by Coordination

One way to connect ideas is by coordination. Coordination means the two ideas are roughly equal in importance. There are four main ways to coordinate ideas:

- Use a semicolon (;) to coordinate two sentences with similar structures:

  The information for bid was published last week; the proposal is due in less than a month.

- Use a comma and one of the coordinating conjunctions (*and, but, or, nor, so, for,* and *yet*) to coordinate two ideas.

  The information for bid was published last week, but the proposal is due in less than a month.

  In this example, *but* clarifies the relationship between the two clauses: you haven't been given enough time to write the proposal.

- Combine two sentences into one. Here are two separate sentences:

  The bridge was completed last year. The bridge already needs repairs.

  One way to combine them would be as follows:

  The bridge was completed last year and already needs repairs.

  Notice that there is no comma after *year* because the two verbs in the sentence have the same subject.

- Use transitional words and phrases within and between sentences (see Chapter 14 for more information on transitional words and phrases).

  The 486 chip has already been replaced. *As a result,* it's hard to find a 486 in a new computer.

Linking Ideas by Subordination

Two sentences can also be linked by subordination, that is, by deemphasizing one of them. There are two basic ways to use subordination:

- Use a subordinating word or phrase. Start with the two sentences about a bridge:

  The bridge was completed last year. The bridge already needs repairs.

  Then combine them as follows:

  The bridge already needs repairs *although* it was completed last year.

In this example, *although* subordinates the clause, leaving *the bridge already needs repairs* as the independent clause. Note that the order of the sentences could be reversed:

Although it was completed last year, the bridge already needs repairs.

In general, it is easier to read a sentence if the subject (*the bridge*) appears before the pronoun that replaces it (*it*). This way, readers don't have to remember the *it* clause until they find out what the subject of the sentence is. Another way to subordinate would be as follows:

The bridge, *which* was completed last year, already needs repairs.

This version emphasizes the *already needs repairs* portion of the sentence and deemphasizes the *was completed last year* portion by putting it in a *which* clause.

- Combine two sentences into one.

Completed last year, the bridge already needs repairs.

In this version, *completed last year* modifies *the bridge*. The independent clause is *the bridge already needs repairs*.

Verb Tenses

t Verb tenses in English can be complicated, but in general there are four kinds of time relationships that you need to understand.

- Simple past, present, and future

Yesterday we *subscribed* to a new ecology journal.

We *subscribe* to three ecology journals every year. (Meaning: We subscribe to the three journals regularly.)

We *will subscribe* to the new ecology journal next year. (Or: We *are going to subscribe* to the new ecology journal next year.)

- An action in progress at a known time

We *were updating* our directory when the power failure occurred.

We *are updating* our directory now.

We *will be updating* our directory tomorrow when you arrive.

- An action completed before a known time

We *had already started* to write the proposal when we got your call. (The writing began before the call.)

We *have started* to write the proposal.

We *will have started* to write the proposal by the time you arrive. (Meaning: Both events are in the future, with the writing beginning before the arrival.)

- An action in progress until a known time

  We *had been working* on the reorganization when the news of the merger was publicized. (Meaning: The work was in progress when the news of the merger was publicized.)

  We *have been working* on the reorganization for over a year. (Meaning: The work will continue into the future.)

  We *will have been working* on the reorganization for two years by the time it occurs.

-ing Forms of Verbs

-ing   English uses the *-ing* form of verbs in four major ways:

- As part of a verb in a sentence:

  We are *shipping* the materials by UPS.

- To add extra information in a sentence:

  Analyzing the sample, we discovered two anomalies.

  The sample *containing* the anomalies appears on Slide 14.

  We studied the sample, *thinking* it could be important.

- As an adjective:

  the *leaking* pipe

- As a noun:

  Writing is the best way to learn to write.

  The *-ing* form used as a noun is called a *gerund*. In the previous example, the gerund acts as the subject of the sentence. A gerund can also act as a direct object:

  The designer tried *inserting* the graphics by hand.

Infinitives

inf   Infinitive phrases are used in three main ways:

- As a noun:

  The editor's goal for the next year is *to publish the journal on schedule.*

- As an adjective:

  The company requested the right *to subcontract the project.*

- As an adverb:

  We established the schedule ahead of time *to prevent the kind of mistake we made last time.*

Helping Verbs and Main Verbs

| help |

There are a number of helping verbs in English. The following discussion explains four categories of helping verbs.

- Modals. There are nine modal verbs: *can, could, may, might, must, shall, should, will,* and *would*. After a modal verb, use the base form of the verb (the form of the verb used after *to* in the infinitive).

  base form
  ↓
  The system *must meet* all applicable codes.

- Forms of *do*. After a helping verb that is a form of *do—do, does,* and *did—*use the base form of the verb.

  base form
  ↓
  Do we *need* to include the figures for the recovery rate?

- Forms of *have* plus the past participle (usually the *-ed* or *-en* form of the verb). To form one of the perfect tenses (past, present, or future), use a form of *have* plus the past participle of the verb.

  PAST PERFECT: We *had written* the proposal before learning of the new RFP.

  PRESENT PERFECT: We *have written* the proposal according to the instructions in the RFP.

  FUTURE PERFECT: We *will have written* the proposal by the end of the week.

- Forms of *be*. To describe an action in progress, use a form of *be* and the present participle (the *-ing* form of the verb).

  We *are testing* the new graphics tablet.

  The company *is considering* flextime.

  To create the passive voice, use a form of *be* and the past participle.

  The piping *was installed* by the plumbing contractor.

  See the discussion of active and passive voice in Chapter 13.

Agreement of Subject and Verb

| agr s/v |

The subject and the verb in a clause or sentence must agree in number. Five major constructions are important to remember:

- Simple agreement

  The new *valve* is installed according to the manufacturer's specifications.

  The new *valves* are installed according to the manufacturer's specifications.

- Agreement when the clause or sentence contains information between the subject and the verb.

  The *result* of the tests is included in Appendix C.

  The *results* of the test are included in Appendix C.

- Agreement when the clause or sentence contains special pronouns and quantifiers. Pronouns that end in -*body* or -*one*—such as *everyone, everybody, someone, somebody, anyone, anybody, no one,* and *nobody*—are singular. In addition, quantifiers such as *something, each,* and *every* are singular.

  SINGULAR: *Everybody* is invited to the preproposal meeting.

  SINGULAR: *Each* of the members is asked to submit billable hours by the end of the month.

- Agreement when the clause or sentence contains a compound subject. In such cases, the verb must be plural.

  COMPOUND   *The contractor and the subcontractor* want to meet to resolve
  SUBJECT: the difficulties.

- Agreement when a relative pronoun such as *who, that,* or *which* begins a clause. In such cases, make sure the verb agrees in number with the noun that the relative pronoun refers to.

  The numbers that *are* used in the formula *do* not *agree* with *the ones* we were given at the site. (*Numbers* is plural, so *that* is treated as a plural noun.)

  The *number* that *is* used in the formula *does* not *agree* with *the one* we were given at the site. (*Number* is singular, so *that* is treated as a singular noun.)

Conditions

Four main types of condition are used with the word *if* in English:

- Conditions of fact:

  If you *see* "Unrecoverable Application Error," the program *has crashed.*

  If rats *are* allowed to eat as much as they want, they *become* obese.

  For conditions of fact, use the same verb tense in both clauses. In most cases, use the present tense.

- Future prediction:

  If we *win* this contract, we *will have* to add three more engineers.

  For prediction, use the present tense in the *if* clause. Use *will, can, should,* or *might* plus the base form of the verb in the independent clause.

- Present-future speculation:

If I *were* president of the company, I *would be* much more aggressive.

Notice that the present-future speculation usage suggests a condition contrary to fact. For instance, the example above implies that you are not president of the company. Use *were* in the *if* clause. Use *would, could,* and so on plus the base form of the verb in the independent clause.

- Past speculation:

If we *had won* this contract, we *would have* had to add three engineers.

Notice here too that the example implies that the condition is contrary to fact. Use *had* plus the past participle in the *if* clause. Use *would have, could have,* and so on in the independent clause. You should be aware of one other point about *if* conditions. You can restructure the sentence and not use *if* at all:

Had we won this contract, we would have had to add three engineers.

Articles

art

Few aspects of English can be as frustrating to the nonnative speaker of English as the correct usage of the simple words *a, an,* and *the.* Although there are a few rules that you should try to learn, remember that there are many exceptions and special cases. Here are three general guidelines.

- Singular proper nouns—those that name specific persons, places, and things—do not usually take an article:

Taiwan

James Allenby

But plural proper nouns often do take an article:

the United States

the Allenbys

- Countable common nouns take an article:

the microscope

a desk

Countable nouns are persons, places, or things that can be counted: *one microscope, two microscopes; one desk, two desks.*

But uncountable common nouns generally do not take an article:

overtime

equipment

integrity

research

information

How can you be sure if a word is countable or uncountable? Unfortunately, you can't. You have to keep a list.

- Common nouns can be referred to as either specific or nonspecific. The specific form takes *the;* the nonspecific form takes either *a* or *an.*

  Our department received funding for *an experiment. The experiment* will take six months to complete.

  In the first sentence in this example, *experiment* is a nonspecific noun; in the second sentence, *experiment* is a specific noun because it has been identified in the previous sentence.

Adjectives

adj Keep in mind three main points about adjectives in English:

- Adjectives do not take a plural form.

  a complex project

  two complex projects

- Adjectives can be placed before the nouns they modify or later in the sentence.

  The *critical* need is to reduce the drag coefficient.

  The need to reduce the drag coefficient is *critical.*

- Adjectives of one or two syllables take special endings to create the comparative and superlative forms.

Positive	*Comparative*	*Superlative*
big	bigger	biggest
heavy	heavier	heaviest

  Adjectives of three or more syllables take the word *more* for the comparative form and the words *the most* for the superlative form.

Positive	*Comparative*	*Superlative*
qualified	more qualified	the most qualified
feasible	more feasible	the most feasible

Adverbs

adv Like adjectives, adverbs are modifiers, but their placement in the sentence is somewhat more complex. Remember five points about adverbs.

- Adverbs modify verbs.

Management terminated the project *reluctantly.*

- Adverbs also modify adjectives.

The executive summary was *conspicuously* absent.

- Adverbs also modify other adverbs.

The project is going *very* well.

- Adverbs that describe how an action takes place can appear in different locations in the sentence.

Carefully the inspector examined the welds.

The inspector *carefully* examined the welds.

The inspector examined the welds *carefully.*

But the adverb should not be placed between the verb and the direct object.

INCORRECT: The inspector examined *carefully* the welds.

- Adverbs that describe the whole sentence can also be placed in different locations in the sentence.

Apparently, the inspection was successful.

The inspection *apparently* was successful.

The inspection was successful, *apparently.*

Omitted Words

om   Except for imperative sentences, in which the subject *you* is understood (*Get the correct figures*), all sentences in English require a subject.

The company has a policy on conflict of interest.

Do not omit the *there* or *it.*

INCORRECT: Are three reasons for us to pursue this issue.

CORRECT: *There* are three reasons for us to pursue this issue.

INCORRECT: Is important that we seek his advice.

CORRECT: *It* is important that we seek his advice.

See the discussion of expletives in Chapter 14.

Repeated Words

- Do not repeat the subject of a sentence.

    INCORRECT: The company we are buying from *it* does not permit us to change our order.

    CORRECT: The company we are buying from does not permit us to change our order.

- In an adjective clause, do not repeat an object.

    INCORRECT: The technical communicator does not use the same software that we were writing in *it*.

    CORRECT: The technical communicator does not use the same software that we were writing in.

It is not correct because *that* is the object of the preposition *in*.
    In an adjective clause, do not use a second adverb.

    INCORRECT: The lab where we did the testing *there* is an excellent facility.

    CORRECT: The lab where we did the testing is an excellent facility.

Appendix C
Commonly Misused Words

This appendix explains the proper usage of some of the most commonly misused words and phrases in technical communication. A brief list like this one cannot replace a full-length work: I recommend Harry Shaw's *Dictionary of Problem Words and Expressions* (New York: McGraw-Hill, 1987) and Theodore M. Bernstein's *The Careful Writer: A Modern Guide to English Usage* (New York: Atheneum, 1965).

The definitions in this appendix provide brief explanations and/or examples. The examples are enclosed within quotation marks.

accept, except. "We will not *accept* delivery of any items *except* those we have ordered."

adapt, adopt. *Adapt* means to adjust or to modify; *adopt* means to accept. "Management decided to adapt the quality-circle plan rather than adopt it as is."

affect, effect. *Affect* is a verb: "How will the news *affect* him?" *Effect* is most commonly a noun: "What will be the *effect* of the increase in allowable limits?" *Effect* is also (rarely) a verb meaning to bring about or cause to happen: "The new plant is expected to *effect* a change in our marketing strategy."

already, all ready. "The report had *already* been sent to the printer when the writer discovered that it was not *all ready*."

alright, all right. *Alright* is a misspelling of *all right*.

among, between. In general, *among* is used for relationships of more than two items, *between* for only two items. "The collaboration *among* the writer, the illustrator, and the printer," but "the agreement *between* the two companies."

amount, number. *Amount* is used for noncounting items; *number* refers to counting items: "the *amount* of concrete," but "the *number* of bags of concrete."

assure, ensure, insure. *To assure* means to put someone's mind at ease: "let me *assure* you." *To ensure* and *to insure* both mean to make sure: "the new plan will *ensure* [or *insure*] good results." Some writers prefer to use *insure* only when referring to insurance: "to *insure* against fire loss."

can, may, might. *Can* refers to ability: "We *can* produce 300 chips per hour." *May* refers to permission: "*May* I telephone your references?" *Might* refers to possibility: "We *might* see declines in prices this year."

compliment, complement. A *compliment* is a statement of praise: "The owner offered a gracious *compliment* to the architect on his design." *Compliment* is also a verb: "The owner graciously *complimented* the architect." A *complement* is something that fills something up or makes it complete, or something that is an appropriate counterpart: "The design is a perfect *complement* to the landscape." *Complement* is also a verb: "The design *complements* the landscape perfectly."

could of. This is not a correct phrase; it is a corruption of *could've*, the contraction of *could have*: "She *could have* mentioned the abrasion problem in the report."

criteria, criterion. *Criteria*, meaning standards against which something will be measured, is plural; *criterion* is singular.

data, datum. *Data* is plural; *datum* is singular. However, the distinction is fading in popular usage, although not in some scientific and engineering applications. Check to see how it is spelled in your company or field.

discreet, discrete. *Discreet* means careful and prudent: "She is a very *discreet* manager; you can confide in her." *Discrete* means separate or distinct: "The company will soon split into three *discrete* divisions."

effective, efficient. *Effective* means that the item does what is is meant to do; *efficient* also carries the sense of accomplishing the goal without using more resources than necessary. "Air Force One is an *effective* way to fly the president around, but it is not *efficient*; it costs some $40,000 per hour to fly."

farther, further. *Farther* refers to distance: "one mile *farther* down the road." *Further* means greater in quantity, time, or extent: "Are there any *further* questions?"

feedback. Many writers will not use *feedback* to refer to a response by a person: "Let me have your *feedback* by Friday." They limit the term to its original meaning, dealing with electricity, because a human response involves thinking (or should, anyway).

fewer, less. *Fewer* is used for counting items: "*fewer* bags of cement"; *less* is used for noncounting items: "*less* cement." It's the same distinction as between *number* and *amount*.

foreword, forward. A *foreword* is a preface, usually written by someone other than the author, introducing the book. *Forward* refers to being in advance: "The company decided to move *forward* with the project."

i.e., e.g. *I.e.,* Latin for *id est,* means *that is. E.g.,* Latin for *exempli gratia,* means *for example.* Writers often confuse them. That's why I recommend using the English versions. Also, add commas after them: "Use the main entrance, *that is,* the one on Broadway."

imply, infer. The writer or speaker *implies;* the reader or listener *infers.*

input. People who don't like to give their *feedback* also don't like to offer their *input.*

its, it's. *Its* is the possessive pronoun: "The lab rat can't make up *its* mind." *It's* is the contraction of *it is:* "*It's* too late to apply for this year's grant." Why do people mix up these two words? Because they remember learning that possessives take apostrophes—"Bob's computer"—when they use the possessive form of it, they add the apostrophe. But *its* is a possessive pronoun, like *his, hers, theirs, ours,* and *yours,* a word specifically created to fulfill only one function: to indicate possession. It is not the possessive form of another word, and therefore it does not take an apostrophe.

-ize. Many legitimate words end in *-ize,* such as *harmonize* and *sterilize,* but many writers and readers can't stand new ones (such as *prioritize*) when there are perfectly fine words already (such as *rank*).

lay, lie. *Lay* is a transitive verb meaning to place: "*Lay* the equipment on the table." *Lie* is an intransitive verb meaning to recline: "*Lie* down on the couch." The complete conjugation of *lay* is *lay, laid, laid, laying;* of *lie,* it is *lie, lay, lain, lying.*

lead, led. *Lead* is the infinitive verb: "We want to *lead* the industry." *Led* is the past tense: "Last year we *led* the industry."

parameter. This is a mathematical term referring to a constant whose value can vary according to its application. Many writers object to the nonmathematical uses of the term, including such concepts as perimeter, scope, outline, and limit. (You guessed it: the same people who don't provide *input* or *feedback* don't use *parameter* very much either.)

phenomena, phenomenon. *Phenomena* is plural; *phenomenon* is singular.

plain, plane. *Plain* means simple and unadorned: "The new company created a very plain logo." *Plane* has several meanings: an airplane, the act of smoothing a surface, the tool used to smooth a surface, and the flat surface itself.

precede, proceed. *Precede* means to come before: "Should Figure 1 *precede* Figure 2?" *Proceed* means to move forward: "We decided to *proceed* with the project despite the setback."

shall, will. *Shall* suggests a legal obligation, particularly in a formal specification or contract: "The contractor *shall* remove all existing debris." *Will* suggests intent; it does not suggest a legal obligation: "We *will* get in touch with you as soon as possible to schedule the job."

sight, site, cite. *Sight* refers to vision; *site* is a place; *cite* is a verb meaning to document a reference.

than, then. *Than* is a conjunction used in comparisons: "Plan A works better *than* Plan B." *Then* is an adverb referring to time: "First we went to the warehouse. *Then* we went to the plant."

their, there, they're. *Their* is the possessive pronoun: "They brought *their* equipment with them." *There* refers to a place—"We went *there* yesterday"—or in expletive expressions—"*There* are three problems we have to solve." *They're* is the contraction of *they are.*

to, too, two. *To* is used in infinitive verbs ("*to* buy a new microscope") and in expressions referring to direction ("go *to* Detroit"). *Too* means excessively: "The refrigerator is *too* big for the lab." *Two* is the number 2.

viable. This is a fine Latin word meaning able to sustain life: "*viable* cell culture" and "*viable* fetus." Many writers avoid such clichés as *viable alternative* (while they're avoiding *input* and *feedback*).

weather, whether. *Weather* refers to sunshine and temperature. *Whether* refers to alternatives. "The demonstration will be held outdoors *whether* or not the *weather* cooperates."

who's, whose. *Who's* is the contraction of *who is. Whose* is the possessive case of *who:* "*Whose* printer are we using?"

-wise. Don't say, "*RAMwise,* the computer has 16 MB." Instead, say "The computer has 16 MB of RAM."

Xerox. The people at Xerox become unhappy when writers ask for a xerox copy. The correct word is *photocopy; Xerox* is a copyrighted term.

your, you're. *Your* is the possessive pronoun: "Bring *your* calculator to the meeting." *You're* is the contraction of *you are.*

Appendix D
Selected Bibliography

Technical Communication

Allen, O. J., & Deming, L. H. (1994). *Publications management: Essays for professional communicators*. Amityville, NY: Baywood.

Beer, D. F. (1992). *Writing and speaking in the technology professions: A practical guide*. New York: IEEE.

Blicq, R. S., & Moretto, L. A. (1995). *Writing reports to get results: Quick, effective results using the pyramid method*. Piscataway, NJ: IEEE.

Boiarsky, C. R., & Soven, M. K. (1995). *Writings from the workplace: Documents, models, cases*. Needham Heights, MA: Allyn & Bacon.

Brusaw, C. T., Alred, G. J., & Oliu, W. E. (1993). *Handbook of technical writing*. (5th ed.). New York: St. Martin's.

Day, R. A. (1995). *Scientific English: A guide for scientists and other professionals*. (2nd ed.). Phoenix, AZ: Oryx.

Dombrowski, P. M. (Ed.). (1994). *Humanistic aspects of technical communication*. Amityville, NY: Baywood.

Dragga, S., & Gong, G. (1989). *Editing: The design of rhetoric*. Amityville, NY: Baywood.

Hoft, N. L. (1995). *International technical communication: How to export information about high technology*. New York: Wiley.

Markel, M. (1994). *Writing in the technical fields: A step-by-step guide for engineers, scientists, and technicians*. New York: IEEE.

Mathes, J. C., & Stevenson, D. W. (1991). *Designing technical reports: Writing for audiences in organizations*. (2nd ed.). Indianapolis: Bobbs-Merrill.

Pickett, N. A., & Laster, A. A. (1993). *Technical English*. (6th ed.). New York: HarperCollins.

Sides, C. H. (1991). *How to write and present technical information*. (2nd ed.). Phoenix: Oryx.

Also see the following journals:
IEEE Transactions on Professional Communication
Journal of Business and Technical Communication
Journal of Technical Writing and Communication
Technical Communication
Technical Communication Quarterly

Ethics

Beauchamp, T. L., & Bowie, N. E. (1997). *Ethical theory and business*. (5th ed.). Englewood Cliffs, NJ: Prentice Hall.

Brockmann, R. J., & Rook, F. (Eds.). (1989). *Technical communication and ethics*. Washington, DC: Society for Technical Communication.

Galler, B. A. (1995). *Software and intellectual property protection*. Westport, CT: Quorum.

720

Velasquez, M. G. (1992). *Business ethics: Concepts and cases*. (3rd ed.). Englewood Cliffs, NJ: Prentice Hall.

Collaborative Writing

Blyler, N. R., & Thralls, C. (Eds.). (1993). *Professional communication: The social perspective*. Newbury Park, NY: Sage.

Cross, G. A. (1993). *Collaboration and conflict: A contextual exploration of group writing and positive emphasis*. Cresskill, NJ: Hampton.

Ede, L., & Lunsford, A. (1990). *Singular texts/plural authors: Perspectives on collaborative writing*. Carbondale: Southern Illinois University.

Forman, J. (Ed.). (1992). *New visions of collaborative writing*. Portsmouth, NH: Boynton/Cook.

Lay, M. M., & Karis, W. M. (1991). *Collaborative writing in industry: Investigations in theory and practice*. Amityville, NY: Baywood.

Research Techniques

Berkman, R. I. (1994). *Find it fast: How to uncover expert information on any subject*. (3rd ed.). New York: Harper Perennial.

Berkman, R. I. (1995). *Find it online!* New York: Windcrest/McGraw-Hill.

Lavin, M. R. (1991). *Business information: How to find it, how to use it*. (2nd ed.). Phoenix, AZ: Oryx.

Levin, J. (1995). *The federal Internet source*. (2nd ed.). Washington, DC: National Journal Inc. and NetWeek L.L.C.

Malinowsky, H. R. (1994). *Reference sources in science, engineering, medicine, and agriculture*. Phoenix, AZ: Oryx.

Maxwell, B. (1995). *How to access the federal government on the Internet*. Washington, DC: Congressional Quarterly Books.

Mount, E., & Kovacs, B. (1991). *Using science and technology information sources*. Phoenix, AZ: Oryx.

Zimmerman, D. E., & Muraski, M. L. (1994). *The elements of information gathering*. Phoenix, AZ: Oryx.

Usage and General Writing

Burchfield, R. W., & Fowler, H. W. (Eds.). (1996). *The new Fowler's dictionary of modern English usage*. (3rd ed.). New York: Oxford University.

Bush, D. W., & Campbell, C. P. (1994). *How to edit technical documents*. Phoenix, AZ: Oryx.

Corbett, E. P. J. (1990). *Classical rhetoric for the modern student*. (3rd ed.). New York: Oxford University.

Maggio, R. (1991). *The dictionary of bias-free usage: A guide to nondiscriminatory language*. Phoenix, AZ: Oryx.

Partridge, E. (1995). *Usage and abusage: A guide to good English*. New edition. Whitcut, J., ed. New York: Norton.

Strunk, W., & White, E. B. (1995). *The elements of style*. (3rd ed.). Needham Heights, MA: Allyn & Bacon.

Williams, J. (1996). *Style: Ten lessons in clarity and grace*. (5th ed.). Glenview, IL: Scott Foresman.

Style Manuals

American National Standards, Inc. (1979). *American National Standard for the preparation of scientific papers for written or oral presentation*. ANSI Z39.16—1972. New York: American National Standards Institute.

CBE Style Manual Committee, Council of Biology Editors. (1994). *Scientific style and format: The CBE manual for authors, editors, and publishers.* (6th ed.). Chicago: Cambridge University Press.

The Chicago manual of style. (1993). (14th ed.). Chicago: University of Chicago, 1993.

Dodd, J. S. (Ed.). (1986). *The ACS style guide: A manual for authors and editors.* Washington, DC: American Chemical Society.

Li, X., & Crane, N. B. (1996). *Electronic style: A guide to citing electronic information.* (Rev. ed.). Westport, CT: Meckler.

Nagle, J. (1995). *Handbook for preparing engineering documents: From concept to completion.* New York: IEEE.

Pollack, G. (1977). *Handbook for ASM editors.* Washington, DC: American Society for Microbiology.

Publications manual of the American Psychological Association. (1994). (4th ed.). Washington, DC: American Psychological Association.

Rubens, P. (1994). *Science and technical writing: A manual of style.* New York: Henry Holt.

U.S. Government Printing Office style manual. (1988). (Rev. ed.). New York: Outlet.

Also, many private corporations, such as John Deere, DuPont, Ford Motor Company, General Electric, and Westinghouse, have their own style manuals.

Graphics, Design, and Multimedia

Baird, R. N., et al. (1992). *The graphics of communication.* (6th ed.). San Diego: Harcourt Brace.

Campbell, K. S. (1995). *Coherence, continuity, and cohesion: Theoretical foundations for document design.* Hillsdale, NJ: Erlbaum.

Cleveland, W. S. (1994). *Elements of graphing data.* Summit, NJ: Hobart.

Harris, R. L. (1996). *Information graphics: A comprehensive illustrated reference.* Atlanta, GA: Management Graphics.

Horton, W. (1991). *The icon book: Visual symbols for computer systems and documentation.* New York: Wiley.

Horton, W. (1991). *Illustrating computer documentation: The art of presenting information graphically on paper and online.* New York: Wiley.

Horton, W., Taylor, L., Ignacio, A., & Hoft, N. L. (1996). *The web page design cookbook.* New York: Wiley.

Johnson, S. (1996). *Electronic publishing construction kit.* New York: Wiley.

Jones, G. E. (1995). *How to lie with charts.* Alameda, CA: Sybex.

Kelvin, G. V. (1992). *Illustrating for science.* New York: Watson-Guptill.

Kosslyn, S. M. (1994). *Elements of graph design.* New York: W. H. Freeman.

Leonard, D. C., & Dillon, P. M. (1994). *Multimedia technology from a to z.* Phoenix, AZ: Oryx.

Nielsen, J. (1995). *Multimedia and hypertext: The Internet and beyond.* Boston: AP Professional.

Parker, R. C. (1996). *Looking good in print: Deluxe CD-ROM edition.* Chapel Hill, NC: Ventana.

Shushan, R. (1993). *Desktop publishing by design: Aldus PageMaker edition.* (3rd ed.). Redmond, WA: Microsoft.

Tufte, E. R. (1983). *The visual display of quantitative information.* Cheshire, CT: Graphics Press.

Tufte, E. R. (1990). *Envisioning information.* Cheshire, CT: Graphics Press.

Tufte, E. R. (1997). *Visualizing explanations.* Cheshire, CT: Graphics Press.

Vaughan, T. (1996). *Multimedia: Making it work.* (3rd ed.). Berkeley, CA: Osborne Mc-Graw-Hill.

Wheildon, C. (1995). *Type & layout: How typography and design can get your message across—or get in the way.* Berkeley, CA: Strathmoor.

White, J. V. (1990). *Color for the electronic age.* New York: Watson-Guptill.

Williams, R. (1994). *The non-designer's design book.* Berkeley, CA: Peachpit.

Also see the following journals:

Graphic Arts Monthly

Graphics: USA

The Internet

Cady, G. H., & McGregor, P. (1996). *Mastering the Internet.* (2nd ed.). San Francisco: Sybex.

Crumlish, C. (1996). *The Internet for busy people.* Berkeley, CA: Osborne McGraw-Hill.

Ernst, T., & Engst, A. (1995). *Create your own home page.* Indianapolis, IN: Hayden.

Lemay, L. (1995). *Teach yourself Web publishing in a week.* Indianapolis, IN: Sams.

Manger, J. J. (1995). *The essential Internet guide.* New York: McGraw-Hill.

Peal, D. (1995). *Access the Internet!* Alameda, CA: Sybex.

Also see the following journals:

InfoWorld

Internet Week

Internet World

Web Week

Wired

Technical Manuals

Barker, T. (1991). *Perspectives on software documentation: Inquiries and innovations.* Amityville, NY: Baywood.

Bias, R. G., and Mayhew, D. J. (Eds.). (1994). *Cost-justifying usability.* Boston: AP Professional.

Brockmann, R. J. (1990). *Writing better computer documentation: From paper to hypertext. Version 2: The First Quarter Century of Service.* New York: Wiley-Interscience.

Crown, J. (1992). *Effective computer user documentation.* New York: Van Nostrand Reinhold.

Dumas, J. S., & Redish, J. C. (1993). *A practical guide to usability testing.* Norwood, NJ: Ablex.

Forbes, M. (1992). *Writing technical articles, speeches, and manuals.* (2nd ed.). New York: Krieger.

Hackos, J. T. (1994). *Managing your documentation projects.* New York: Wiley.

Haydon, L. M. (1995). *The complete guide to writing and producing technical manuals.* New York: Wiley.

Horton, W. (1994). *Designing and writing online documentation: Hypermedia for self-supporting products.* (2nd ed.). New York: Wiley.

Lanyi, G. (1994). *Managing documentation projects in an imperfect world.* Columbus, OH: Battelle.

Nielsen, J., & Mack, R. L. (Eds.). (1994). *Usability inspection methods.* New York: Wiley.

Price, J., & Korman, H. (1993). *How to communicate technical information: A handbook of software and hardware documentation.* Redwood City, CA: Benjamin/Cummings.

Rubin, J. (1994). *Handbook of usability testing: How to plan, design, and conduct effective tests.* New York: Wiley.

Slatkin, E. (1991). *How to write a manual.* Berkeley, CA: Ten Speed.

Steehouder, M., Jansen, C., van der Poort, P., and Verheijen, R. (Eds.). (1994). *Quality of technical documentation.* Amsterdam: Editions Rodopi B. V.

Velotta, C. (1995). *Practical approaches to usability testing for technical documentation.* Arlington, VA: Society for Technical Communication.

Weiss, E. H. (1992). *How to write usable user documentation.* (2nd ed.). Phoenix, AZ: Oryx.

Whitaker, K. (1995). *A guide to publishing user manuals.* New York: Wiley.

Wieringa, D., Moore, C., & Barnes, V. (1993). *Procedure writing: Principles and practices.* Columbus, OH: Battelle.

Woolever, K. R., & Loeb, H. M. (1994). *Writing for the computer industry.* Englewood Cliffs, NJ: Prentice Hall.

Oral Presentations

Anholt, R. H. (1994). *Dazzle 'em with style: The art of oral scientific presentation.* New York: W. H. Freeman.

D'Arcy, J. (1992). *Technically speaking: Proven ways to make your next presentation a success.* New York: AMACOM.

Smith, T. C. (1991). *Making successful presentations: A self-teaching guide.* (2nd ed.). New York: Wiley.

Proposals

Bowman, J. P., & Branchaw, B. P. (1992). *How to write proposals that produce.* Phoenix, AZ: Oryx.

Freed, R. C., Freed, S., & Romano, J. (1995). *Writing winning business proposals.* New York: McGraw-Hill.

Hill, J. W., & Whalen, T. (1993). *How to create and present successful government proposals.* New York: IEEE.

Miner, L. E., & Griffith, J. (1993). *Proposal planning and writing.* Phoenix, AZ: Oryx.

Stewart, R. D., & Stewart, A. L. (1992). *Proposal preparation.* (2nd ed.). New York: Wiley-Interscience.

Index

abbreviations. *See also* punctuation
    in index, 303
    mechanics of, 704–05
    multicultural definitions for, 124–25
abstract, 158, 160, 167
    descriptive, 274, 275
    informative, 274–76, 283
    in reports, 533
abstract services, 158, 160
accessibility, 14
accessing tools, 316–19
accuracy, 13, 32
    of index, 299, 406
    of information, 145, 166, 169–70, 222
    in letters, 596, 610
    of meetings minutes, 513
    in revision, 406
actions
    documenting in résumé, 568
    in reports
        advocacy of, 522–24
        identification of, 498
    verb usage for, 710–11
active voice, 432, 439–40
adjectives, 688, 692–93, 695, 698. *See also*
        punctuation
    usage, 715, 717
    and verbs, 711
adjustment letter, 612–15
advance organizer, 408
adverbs
    usage, 715–16, 717
    and verbs, 711
agreement, number, 689
agreement, subject/verb, 712–13
analogy, 244–45. *See also* definitions
analytical report, 521–22
APA documentation style, 283–95
apostrophe, 699–700. *See also* punctuation
appearance of document, 15
appendices
    in back matter, 297
    definition placement in, 248
    design, 318
    in proposals, 476–77, 479–81
    in reports, 506, 533
application letter, 4–5, 580–85
appreciation letter, 585, 588
argument
    definition, 195
    development of, 195–98
    guidelines for drafting, 193–95
    introducing and concluding the body,
        224–29
    logical consistency in, 190, 191–92, 214,
        219, 223–24
    organizational patterns for, 198–224, 227

cause and effect, 221–24
        chronological, 199–200
        classification and partition, 211–17
        comparison and contrast, 207–11
        general to specific, 203–06
        more important to less important,
            206–07
        problem-methods-solution, 217–20
        spatial, 200–03
        whole-by-whole or part-by-part, 209–11
    outlining review, 190–93
articles (grammatical), 714–15. *See also*
        journal article
ASCII text, for electronic résumé, 577
attitude
    to time, 119–20
    toward reader/audience, 113
    "you attitude," 596–98, 607
audience. *See also* multicultural audience
    audience profiles, 7–8, 115–18
    categories of readers, 102–10
        expert, 103–05
        general, 108–10
        manager, 104–08
        technician, 104, 106
    characteristics of readers, 7–8, 10, 14,
        110–15
    compared to purpose, 101
    definitions, determination for, 237,
        238–39, 241, 249, 251
    expectations and needs of, 7–8, 10, 14, 51,
        90, 100–125, 196, 199, 213, 222–23,
        269, 311, 348–51, 362–63, 405, 407,
        465, 535–36, 625
    multiple, 118
    for oral presentation, 660, 661–62, 666,
        670, 676
    primary and secondary, 101–02, 137
    testing, 404–05, 650–56
    for Web page, 89–90, 93
authorship, 32, 164, 169, 170. *See also*
        copyright law; references
    documentation of, 284–86, 290, 292

back matter. *See also* front matter
    contents requirements, 295–304
        appendices, 297
        glossary, 296–97, 298, 442
        index, 298–304
        list of symbols, 296–97
    definition, 269
    in manuals, 644–46
    in reports, 533
background
    in executive summary, 279
    for graphics, 354, 358–59
    for report, 190, 206, 497–98

balance, 320–21. *See also* design
bar graph. *See* graphics
benchmark testing, 144
bibliography, 158, 167
    to this book, 721–25
bindings, 315–16. *See also* design
block diagrams, 377–78. *See also* design
block quotations, 701
boxes, 338, 339, 356. *See also* design
brackets, 702. *See also* punctuation
brainstorming, 9, 147, 148, 190
branching, to generate ideas, 151–52, 190
bribery, 124
brochures, 211, 313, 326, 381–82
browsers, 81, 85–86, 91, 95
    URL addresses, 83
bullets, 427, 430. *See also* design; graphics
business customs, cultural patterns for,
        123–24

candor, 25–26. *See also* honesty
capitalization, 705–06. *See also* punctuation
career. *See also* job search
    communication skills affecting, 4–5, 557,
        580
case, possessive, 699–700
case, typographical, 332
catalogs, 211. *See also* brochure
cause-and-effect
    analysis of in report, 522
    patterns for in argument, 221–24
CD-ROM, 10, 154–55
characteristics, required/desired, 209
chartjunk, 348, 349
charts. *See* graphics; *specific charts*
checklist
    for revision, 402–03, 427
    use in graphics, 378–79
chronology, for argument organization,
        199–200
chunking, 321–22, 323
circular definitions, 240. *See also* definitions
claim, within argument, 221
claim letter, 610–12, 613
clarity, of information, 13, 76, 145
classification
    in argument, 211–17
clichés, 443–44, 598–600. *See also* language;
        vocabulary
clip art. *See* graphics
clustering, 151, 190
codes of conduct, 37–40, 131
coherence, 399–423. *See also* given/new
        strategy; headings; lists; paragraphs
collaborative writing
    advantages/disadvantages of, 51–52
    conflict reduction in, 57–59

collaborative writing *(continued)*
  constraints on, 132–34
  description of, 49
  with electronic tools, 74–81
    electronic mail, 4, 68, 74–77
    groupware, 77–80
  and gender, 60–62
  for manuals preparation, 633–37
  and multiculturalism, 62–63
  organizing and maintaining a group, 53–57
  patterns of, 49–51
  styles and templates for, 71–72
collective nouns, 689
colon, 696–97. *See also* punctuation
color. *See also* design
  in graphics, 356–60, 375, 390
  of paper, 314–15, 318, 356
  for safety warnings, 627
columns, 327–30. *See also* design
comma, 691–95. *See also* punctuation
  comma splices, 685–86, 694
communication. *See also* electronic
      communication
  nonverbal, 123–24
  value of in workplace, 4–5, 557, 580
comparing items, 687–88
comparison and contrast, for argument
      organization, 207–11
completion (final) report, 9, 218, 463–64
compound adjectives, 688
compound sentences, 691, 694
comprehensiveness
  of document, 14
  of index, 299, 406
  of information, 145
conciseness, 14–15
conclusions
  of argument, 227–28
  of descriptions, 253–56
  in instruction manuals, 631–33
  in reports, 521, 523, 527–28, 533
confidentiality, 26, 38. *See also* ethics
conflict reduction, 52, 57–59
conjunctions, coordinating, 691, 695
constraints in writing, 131–34
contents, table of, 269, 276–79, 533
contractions, 699
contract law, 34–35, 637
contract theory, 27–28
contrast
  and color, 358–59. *See also* design
  and comparison, 244
coordinate adjectives, 688, 692–93
coordinating conjunctions, 691, 695
coordination, 191–92
copying. *See also* printing
  résumé, 562
  on word processor, 269
copyright law, 29–32, 86–88, 354. *See also*
      legal issues
correlative constructions, 436–37
countable nouns, 714–15
cover letter, 270–71, 272
covers
  for manuals, 638–39
  qualities of, 269, 271
cross-reference
  in document, 318, 319
  in index, 318
  in report, 506

cultural patterns, 119–24. *See also*
      multicultural audience
cutaway drawings, 385, 387

danger information. *See* safety warnings
dangling modifier, 437
dash, 697. *See also* punctuation
database
  online, 183
  use of for research, 154, 183
dates
  cultural patterns for, 121
  in memos, 494
  punctuation, 693
deadlines, 57, 133
decimal system. *See also* numbers
  for manuals, 630–31
  for table of contents, 277
decision matrix, for research analysis, 531–32
definitions
  and descriptions, 250–56
  determination of placement for, 246–48
  drafting of, 248
  and graphics, 242
  role of, 236–37, 249
    analogy, 244–45
    comparison and contrast, 244
    etymology, 245
    exemplification, 242
    history of term, 246
    negation, 245
    partition, 243
    principle of operation, 243
  types of, 238–46
    circular, 240
    extended, 240–46
    parenthetical, 238–39
    sentence, 239–40
  writing situation analysis for, 237–38,
      249–50
demonstrative pronouns, 417–18
descriptions, 248–60. *See also* mechanisms,
      description of; process description
design. *See also* format; graphics
  accessing tools, 316–19
  bindings, 315–16
  boxes, 338, 339, 356
  columns, 327–30
  and cultural sensitivity, 121–22
  design principles, 320–21
    balance and symmetry, 320–21
  document-design elements, 313, 326
  goals, 309–11
  for increased readability, 9–10
  indentation
    hanging, 338, 430
    headings, 337–38, 356
    lists, 426–30
  justification, 335–37
  and learning theory, 321–23
    chunking, 321–22, 323
    filtering, 323, 325
    queuing, 322–23, 324
  margins, 327, 330
    marginal gloss, 327, 339
    notes in, 506
  minimalist, 314
  page count, 314
  page grids, 323, 324–26, 328, 329
    thumbnail sketch, 324, 325

page size, 312–14, 327
paper grades and types, 314–15, 318
planning, 311–12
rules, 338, 356
for safety information, 627
screens, 338–39, 356
titles and headings, 337–38
typography, 323, 330–37
  case, 332
  hyphenation, 337
  justification, 335–37
  line length/spacing, 334–35
  type families, 331–32
  type size, 332–34
  typefaces, 330–31, 337
for Web page, 88–96, 90–92
white space, 326–27
desktop publishing, 309, 324. *See also* word
      processors
deviation bar graphs, 369, 370
diagrams. *See also* design; graphics
  in graphics, 104, 361, 376–78
    block, 377–78
    organization chart, 378
    schematic, 377
  network, 477, 479, 480
dictionary, 158. *See also* thesaurus
directives. *See also* reports
  preparing, 499–500
disabled persons, 449
disclaimer, 34–35
discussion groups, 155–56. *See also*
      electronic mail
document-delivery services, URL address, 159
documentation, 2, 3. *See also* instructions;
      manuals; references
  of sources, 32, 281, 365
    APA/MLA standards, 283–95
documents. *See also* proposals; reports
  cultural patterns for, 121–22
  internal and external, 31
  modular, 118
dot leaders, 364, 365
drafting. *See also* outline; revision
  collaboration in, 50
  definitions, 248
  guidelines for, 193–95
  of index, 300–301
  instructions/manuals, 628–46
  in note taking, 166
  reports, 493
  on word processor, 69–72
draw and paint programs, 324, 391–94
drawings. *See* graphics

e-mail. *See* electronic mail
editor, documentation of, 287, 292
editing, 167
education
  documentation of
    in application letter, 583
    in résumé, 565–67
  of reader/audience, 110–11
electronic communication, 10, 68–69, 118. *See
      also* communication; World Wide Web
  for collaborative writing, 74–81
  with Web pages, 81–88
  on word processor. *See* word processors
electronic mail (e-mail), 4, 68. *See also* World
      Wide Web

advantages of, 74–75, 118, 585
and discussion groups, 155–56
documentation of, 277, 294
and groupware, 77–80
guidelines for, 75–77
as informal report, 492, 494, 499
netiquette for, 76–77, 87
for résumés, 576
and Web pages, 83, 93–94
electronic résumés, 576
ellipses, 701–02. *See also* punctuation
elliptical expressions, 693
emoticons, 75
employee, collaboration by, 51–52
employer
employee obligations to, 25–27
influence of in reports, 523
obligations to public, 27–29
reflecting in meetings minutes, 513
researching for employment, 557–58
Web sites, 558, 560
employment documentation. *See also*
documentation; references
in application letter, 584
in résumé, 567
English. *See also* multicultural audience
Simplified, 449–50, 451, 645–47
environment, corporate responsibility to, 29
ethics. *See also* honesty; legal issues
and accuracy, 13
bribery, 124
in collaborative environment, 58
definition, 23
employee obligations to employer, 25–27,
528
employee obligations to environment, 29
employee obligations to public, 27–29, 198
contract theory, 27–28
due-care theory, 28
strict-liability theory, 28–29
ethical and legal issues, 29–36, 131, 135
codes of conduct, 37–40, 131
contract law, 34–35
copyright law, 29–32, 86–88, 354
fair use of words and graphics, 30–31,
86–88
liability law, 35–37
plagiarism, 30–31, 283
trademark law, 32–33
whistleblowing, 40–42
guidelines for, 42–43
whistleblowing, 40–42
etymology, 245. *See also* definitions
euphemisms, 43, 444
evaluation materials, 54–56
evidence
persuasive, 196–97
use in argument, 196–97, 222
exclamation point, 690–91. *See also*
punctuation
executive summary, 107, 110. *See also*
proposals; reports
in front matter, 277, 279–83
background, 279
findings, 279–82
in reports, 205, 533, 534, 541, 546
exemplification, 242. *See also* definitions
expert reader, 103–05. *See also* audience
expletives, in sentences, 432–33
exploded diagrams, 385, 387

extemporaneous presentations, 660
extended definitions, 240–46. *See also*
definitions

fair use of words and graphics, 30–31, 88.
See also copyright law; legal issues
feasibility report, 200, 207, 209–10, 211,
528–33
criteria establishment for, 530, 531–32
options determination for, 530–33
field report, 502–04
file transfer protocol (FTP), 68, 81, 83, 155,
288, 289, 295. *See also* World Wide
Web
filtering, 323, 325
financial constraints, 133, 135, 312, 346, 353
in feasibility report, 528–29
findings, in reports, 535
flowchart, 355, 379–80, 381
follow-up letters to job search, 585–90
acceptance of job, 588, 589
appreciation, 585, 588
job rejection acknowledgment, 588, 590
footers, design of, 318
footnotes
definitions placement in, 247–48
for tables, 365
formality, 75–76, 437–38
format. *See also* design
constraints on, 122, 134, 136
for electronic mail, 75–76
for outline, 192–93
for reports, 493–99, 540
templates and styles for, 69–71
fonts, 330–31, 337. *See also* design
forms, as informal report, 494, 499
fragments, sentence, 684–85
freewriting, for research ideas, 148–49
front matter. *See also* back matter
contents requirements, 270–83
abstract, 274–76, 283
cover, 271
executive summary, 279–83
headings, 276–77
letter of transmittal, 270–71, 272
list of illustrations, 279
pagination, 277–79
table of contents, 276–79
title page, 271–73
definition, 269, 270, 273
documentation
APA style, 283, 284–89, 290
MLA style, 289–96
in manuals, 637–38
in reports, 533
FTP (file transfer protocol), 68, 81, 83, 155,
288, 289, 295. *See also* World Wide
Web
fused sentences, 686

Gantt chart, 355, 478, 480
gender roles
in collaborative writing, 60–62
cultural patterns for, 119, 125
and nonsexist language, 448–49
and reader's characteristics, 112
general reader. 108–10. *See also* audience
general to specific, for argument
organization, 203–06
generating ideas, 146–52

gerunds, 711
given/new strategy, 400–401
glossary, 110, 124
in back matter, 296–97, 298
definition placement in, 248, 442
in reports, 533
government documents, 160–61
grammar checker, 73. *See also* word
processors
grammatical errors, 15, 449
grammatical parallelism, 405, 415
graphics. *See also* design; text
animation (motion) in, 355, 387–89
appropriateness of, 361–87
in arguments, 200, 203, 206, 207, 214
background in, 354, 358–59
bar graphs, 104, 361, 365–69, 370, 391,
477, 478, 480
deviation, 369, 370
grouped, 368, 370
100-percent, 369, 370
pictographs, 369–71
subdivided, 369, 370
benefits of, 345–47
bullets, 427, 430
charts, 361
chartjunk, 348, 349
CPM, 479
flowcharts, 355, 379–80, 381
Gantt chart, 355, 478, 480
network diagrams, 477, 479, 480
PERT chart, 355, 479
checklists, 378–79, 427
clip art, 354, 369, 386, 627
color use in, 356–60, 375, 390
computer-generated, 354–55, 391–94
paint and draw programs, 324, 391–94
spreadsheets, 391, 392
copyright protection for, 86–87
creation of, 351–55
cultural sensitivity to, 121, 125, 356, 359
in definitions, 239, 242
for definitions, 242
diagrams, 361, 376–78
digital, 86–87, 90, 93–95, 156
documentation of, 283
dot leaders, 364, 365
drawings
cutaway, 385, 387
exploded, 385, 387
line, 384–86
phantom view, 385, 387
and fair use, 30–31, 88
figures, 361
grid lines, 368
for instructions and process description, 35,
378–81, 627, 631, 633, 635, 637, 647
labeling for (captions), 348–50, 375, 383,
413
line graphs, 371–74, 391
ratio, 373–74
stratum (area), 372–73
list of illustrations for, 279
logic boxes, 380
logic trees, 380–81
logical relationships in, 362, 376–78
maps, 386–87
modifying for audiences, 104, 110, 200
for multicultural audience, 121, 356, 359,
388–91

graphics *(continued)*
numerical values, 361–76
on/for Web site, 90, 93–94, 354
for oral presentations, 668–72
patterns, 346, 358
photographs, 361, 381–84
pictographs, 369–71
pie charts, 374–76, 391
placement on page, 330, 348–50
planning, 350–51
recommended readings in, 479
revising, 355–56
for safety information, 627
screen shots, 384, 385
tables, 361, 362–65, 366, 391, 477, 478
and text, 10, 200, 203, 206, 207, 214, 348–50
transparencies/slides, 358–59, 391, 668, 669
graphs. *See* graphics
grid lines, 368
grouped bar graphs, 368, 370
groupthink, 52
groupware. *See* software

hanging indentation, 338, 340, 562–63
headers, design of, 318, 356
headings, 93, 107. *See also* titles
capitalization in, 706
design of, 337–38
in executive summary, 107
in front matter, 276–77
in letters, 596, 600
in manuals, 638
in memos, 494–95, 496–97
revision of, 407
helping verbs, 712
high-tech tools. *See* electronic communication
honesty. *See also* ethics; legal issues
and ethics, 11–13, 25–26, 42–43
with graphics, 347–48, 367, 370, 375, 381–82
of information, 42–43, 145, 563
in oral presentation, 679
whistleblowing, 40–42
hypertext markup language (HTML), 72, 82, 93–94. *See also* World Wide Web
instructions in, 647–50
recommended readings for, 650
hyphen, 698–99. *See also* punctuation

icon. *See also* graphics
use in document, 316, 317
ideas
in argument, 228–29
attachment to, 58, 59
development of, 146–52
linking
by coordination, 709
by subordination, 709–10
no copyright protection for, 31
in paragraphs, 413, 415
parenthetical, 283
if, conditions of usage, 713–14
illustrations, 197. *See also* graphics
imperative mood
for instructions, 631, 646–47
usage, 716
impromptu oral presentations, 660
inadequate sampling, 223

indentation
hanging, 338, 430
headings, 337–38, 356
in letters, 596
lists, 426–30
in résumé, 562–63
index, 158, 159
creation, 269, 406
in back matter, 298–304
cross-references, 299–300
guidelines for, 302–04
for newspapers, 159–60
indirect quotation, 700
individual orientation, compared to group orientation, 121
infinitives, 711
information. *See also* research
accessibility of, 14, 36, 229, 269, 277, 309–11, 348
characteristics of, 144–45, 161–69
constraints on, 132, 135, 194
evaluating, 169–70
given/new, 400–401
government, 160–61
placement in sentence, 430
summarizing, 166–69
information for bid (IFB), 461
Information Mapping®, 451–52
information research center, 152
informational report, 520–21
inquiry letter, 176, 605–08
instructions. *See also* manuals; process description
for children, 35
compared to process descriptions, 248, 628
design, 313, 326
graphics in, 35, 356, 359, 378–81, 627, 631, 633, 635, 637, 647
drafting
conclusion, 631–33
introduction, 628–30
maintenance tips, 631
step-by-step process, 630–31
troubleshooter's guide, 631–33
ethical and legal issues, 35–36
headings and titles for, 407–08
in hypertext, 647–50
for non-native English speakers, 646–47
safety emphasis in, 10, 36, 205, 356, 359, 624–25
safety information preparation, 626–28
and usability testing, 625, 650–56
internet. *See* World Wide Web
internet relay chat (IRC), 288
interview
in job search, 558–60
note taking in, 175
for research, 173–80, 184–85
intranets, 68
introductions, 224–27
invisible writing, 194
italics, 702. *See also* punctuation; typeface

jargon. *See also* language; vocabulary
and idioms, 124
revision of, 441–42
job search
planning, 557–58
portfolio, 558
self-inventory, 557

relocation expectations, 571
7 ways to find a position, 558–61
advertisements, 559–60, 582
placement bureaus, 559
university placement office, 559–60
Web-based résumé, 577
World Wide Web
employer sites, 4, 558, 560, 561
job postings/job boards, 88, 576
URL addresses, 560
job-application materials
application letter, 580–87
follow-up letters, 585–90
acceptance of job, 588, 589
appreciation, 585, 588
job rejection acknowledgement, 588, 590
rejection of job offer, 588, 589
résumé, 4–5
analytical, 564, 571–76
chronological, 564–76
inclusion in proposals, 473–76
of nontraditional student, 573–74, 576, 585
of traditional student, 572–73
résumé preparation
appearance, 562–63
contents requirements, 563–64, 577
electronic, 561, 576–78
length limits, 563–64
preparation of, 561–76
references for, 569–71
skills and abilities documentation, 571–72
templates for, 70–71
journal article, 200, 203, 207, 211, 218, 221
documentation of, 287, 293, 294
skimming, 162
journalistic questions, for exploring a topic, 147–48, 171
justification, typographical, 335–37

key ideas, 166. *See also* ideas
key words, 167, 240
in abstract, 274
in argument, 225
in electronic document, 96, 576
in paragraph, 416–17
precision of, 406
in proposal, 472
in report, 535, 544

lab report, 502
language. *See also* punctuation; vocabulary
adapting to reader/audience, 120–21
inoffensive/nonsexist, 448–49
for interviews, 178
for non-native speakers. *See* multicultural audience
for oral presentations, 664–66
body language, 677–78
lead-in, structure for, 428
leading, typographical. *See* line spacing
learning style, 68–69
and graphics, 345–46
learning theory, 321–23
chunking, 321–22, 323
filtering, 323, 325
queuing, 322–23, 324
legal issues, 12–13, 37, 43, 131, 135. *See also* ethics; honesty

copyright law, 29–32, 86–88, 354
 for electronic mail, 75–76
 employee obligations, 29–36
 and multiculturalism, 124
 and World Wide Web, 86–88
letterhead
 in letters, 75, 600
 in memos, 495
letters. *See also specific letter types*
 appearance of, 15
 compared to e-mail, 75
 cultural patterns for, 122
 design of, 313, 326, 334
 format for, 603, 604
 as informal report, 494, 499
 preparation of
  addresses, 600–601
  clichés avoidance, 598–99, 600
  copy line, 603
  enclosure line, 602–03
  headings, 600
  politeness in, 597, 612
  "you attitude" projection, 596–98, 607
liability law, 35–37
library
 for employment research, 561
 for research, 152–60, 183–84
  online catalogs, 157, 159, 184
Likert scale, 178–79. *See also* questions
line drawings, 384–86
line graphs, 371–74, 391
line length, 334
line spacing. *See also* punctuation
 for executive summary, 279
 for letters, 580
 and line length, 334–35
 for résumé, 562
list of illustrations, 279
 in reports, 533
list of symbols
 in back matter, 296–97
 in reports, 533
list of works cited, 289, 291–95
listening skills, 9, 57, 59–60
lists
 checklist, 378–79, 402–03, 427
 converting paragraphs to, 408–10, 414
 punctuation for, 428–30
  turnovers, 429–30
 reference list, 285–89
 revision of, 408–10
 for sentence revision, 425–30, 431
listserv groups, 155–56, 288, 294
logical consistency
 in argument, 190, 191–92, 214, 219, 223–24
 in graphics, 362, 376–78
 in outline, 191–92
 revision for, 402–03, 405–06, 415
logic boxes, 380
logic trees, 380–81

management. *See also* reports
 division-of-labor vs. integrative model, 49
 project approval requirements by, 136–38, 523
managers
 as audience, 104–08
 executive summaries for, 107, 110, 277, 279–83

manuals, 3, 200. *See also* instructions
 design of, 115, 311, 313, 326, 356
 drafting of, 633–46
 reference, 200
maps, 386–87
marginal gloss, 327, 339
margins
 and design, 327, 330
 notes in, 506
 in résumé, 577–78
matrix, decision, for research analysis, 531–32
meaningless modifiers, 445
mechanisms, descriptions of, 248, 249
media, use of for research, 153–56
meetings minutes preparation, 506–13
memos, 4, 49, 209
 design, 313, 326, 334
 elements of, 206, 494–96
 as informal report, 492, 494–99
 as internal proposal, 461
 nonsexist language for, 448–49
methods, 535–36
minutes, meeting, 506–13
MLA documentation style, 283–95
modifiers. *See* punctuation
modular documents, 118
money. *See* financial constraints
mood (grammar), 631, 646–47, 716
morality. *See* ethics
more important to less important, for argument organization, 206–07
multi-user domain (MUD/MOO), 288
multicultural audience
 and collaborative writing, 62–63
 cultural pattern categories for, 119–24
 date representation, 494
 graphics for, 35, 121, 356, 359, 388–91
 instruction manuals for, 646–47
 non-native speakers
  guidelines for, 708–17
  Information Mapping®, 451–52
  instructions for, 35
  Simplified English for, 449–50, 451, 645–47
  translation for, 118, 450, 646
 people-first approach, 449, 450
 recommended readings in, 63, 390–91
 understanding, 118–25
multiple audience, 118

negation, 245. *See also* definitions
negative constructions, 442–43
negative space (white space), 326–27
Netiquette, 76–77, 87. *See also* electronic mail; World Wide Web
network diagrams, 477, 479, 480. *See also* diagrams; graphics
networks, 79
newsgroup, 288, 294
newsletter design, 313, 314, 326, 339
newspaper
 employment advertisements in, 559–60
 indexes for, 159–60
newspaper article, documentation of, 288, 293
nominalization, 433–34
non-native English speaker. *See* multicultural audience
nonsexist language, 448–49
nonverbal communication, 123–24. *See also* communication

note taking
 for interviews, 175
 paraphrasing, 163–65
 quoting, 165, 176
 for research, 161, 162–63, 171, 175
 summarizing, 166–69
nouns
 collective, 689
 noun strings, 407, 443
 punctuation, 698
 and verbs, 711
numbers. *See also* decimal system
 agreement of, 689
 cultural patterns for, 120–21
 as data, 196–97
 in graphics, 361–76
 in index, 303–04
 in lists, 427
 punctuation of, 690, 693, 695, 697, 698, 699–700, 703–04
 Roman numerals, 277–79, 318, 638, 641

objectivity. *See also* honesty
 in reports, 536
objects, descriptions of, 242, 248, 249
obvious statements, 444–45
ombudsperson, 41–42, 131
online catalogs, 157
oral presentations
 extemporaneous, 660
 in general, 660–62
 humor, 666
 impromptu, 660
 memorized, 660
 preparation for
  audience assessment, 660, 661–64, 666, 670, 676
  equipment, 670
  graphics, 668–72
  language choices, 664–66
  rehearsal, 672–75
  sentence/topic outline, 666–68
 preparation of, 662–75
 presentation of
  body language, 677–78
  nervousness, 675–76, 677
  question/answer format, 678–79
  voice control, 676–77
 scripted, 660, 674–75
order letter, 603–05, 606
organization. *See also* employer
 of oral presentation, 670–72
 of outline, 191, 198–99
 references to, 286, 290, 292
 of report, 209, 498, 535
 of résumé, 562
organizational patterns. *See* argument
organization chart, 378
outline. *See also* argument; drafting
 format selection, 192–93
 grouping of items, 190–91
 logical problems avoidance, 191–92
 for oral presentation, 666–68, 671
 word processor function for, 69
oversimplifying, 224

page design. *See* design
page grids, 323, 324–26, 328, 329
 thumbnail sketch, 324, 325
page size, 312–14, 327. *See also* design

pagination
  and design, 318–19
  front matter, 277–79
  in letters, 600
  page count, 314
paint programs, 324, 391–94
paper. *See also* printing
  color choices for, 314–15, 318, 356
  and design, 314–15, 356
  for documents, 115
  grades and types, 314–15, 318
  for memos, 495
  for résumé, 562, 578
paragraphs. *See also* sentence
  revising, 410–19
    converting to lists, 408–10
    manageable length, 413–15
    structuring, 411–19
  space between, 334
  topic sentence placement, 412, 414, 418
  transitional words and phrases, 415–16
  type
    body, 410–11
    transitional, 411, 413, 415–16, 417
parallel elements, in sentence, 434–35
parallelism
  and correlative constructions, 436–37
  faulty, 434–35
  grammatical, 405, 415
  in lists, 428
paraphrasing, for note taking, 163–65
parentheses, 697–98. *See also* punctuation
parenthetical definitions, 238–39. *See also*
    definitions
part-by-part, organizing comparisons,
    209–11
partition
  in argument, 211–17
    subpartitioning, 217
  in definitions, 243
passive voice, 432, 439–40
people-first approach. *See also* multicultural
    audience
  principles, 449, 450
period, 690. *See also* punctuation
periodical indexes, 159
permissions, 32, 77, 176. *See also*
    documentation
personal information
  about reader/audience, 112–13
  in résumé, 557, 569
personnel constraints, 132–33, 135
persuasion
  evidence for, 196–97
PERT chart, 355, 479
phantom view (drawing), 385–87
photographs, 369–71
pictographs, 369–71
pie charts, 374–76, 391
plagiarism, 30–31, 283
politeness, 134, 597, 612
politics, consideration of in argument, 198
pomposity, 447–48
positive constructions, 442–43
possessive case, 699–700
post-hoc reasoning, 224, 527
precision
  of index, 299, 406
  of revision, 406, 438–44
prepositional phrases, 445

presentation graphics, 668–72
prewriting stage
  collaboration in, 50
primary research. *See* research
printing. *See also* copying; paper
  color, 356
  on computer, 72
  copying, 269, 562
  costs, 315
  from Web page, 85
  laser-quality, 562, 578
  professional, 133
problem solving, 8
  in reports, 524–33
  role of research in, 143–44
problem-methods-solutions, for argument
    organization, 217–20
process description, 248. *See also*
    instructions
  compared to instructions, 205, 248, 628
  definitions of, 248, 249, 251–53
  graphics for, 346, 378–81
product information, 4, 35, 42
professionalism, 15
  of letters, 596
  of proposal, 467–68
  of reader/audience, 111–12
  of résumé, 562
progress report, 9, 463, 481, 507–11
  preparation of, 502–06
progress reports, 4, 9
projects. *See also* document; proposal; report
  small and large, 9, 136–38
pronouns
  ambiguous references, 686–87
  demonstrative, 417
  relative, 435
proposals, 4, 49, 51, 200, 207, 209, 211,
    218. *See also* executive summary;
    reports
  as argument, 465
  budget for, 468, 476, 504
  definition of, 460
  deliverables
    goods-and-services proposal, 464
    research proposal, 463–64
  design of, 313, 326, 334
  ethical problems with, 42
  evaluation techniques, 479–81
    formative/summative, 480
    quantitative/qualitative, 480
  external, 460, 465–66
  format for, 481
  internal, 460–61, 466
    sample, 481–87
  logistics of, 460–62
  and persuasion, 464–68
    plan description, 466–67
    professionalism demonstration, 467–68
  solicited, 461–62
    IFB (information for bid), 461–62
    RFP (request for proposal), 461–62,
        465–66
  structure of, 469–70
    appendices, 476–77, 479–81
    budget, 468, 476, 477
    introduction, 471–72
    proposed program, 472–73
    qualifications and experience, 467,
        473–76, 477

  summary, 470–71
    task schedule, 467, 477–78, 480
  unsolicited, 462, 466
public, corporate responsibility to, 27–29
publisher evaluation, 169, 170
pull quotes, 339, 340
punctuation
  adjectives, 688, 692–93, 695, 698
  apostrophe, 699–700
  block quotation, 701
  colon, 696–97
  comma, 691–95
  comma splices, 685–86, 694
  dash, 697
  exclamation point, 690–91
  hyphen, 698–99
  for lists, 428–30
    turnovers, 429–30
  mechanics
    abbreviations, 704–05
    brackets, 702
    capitalization, 705–06
    ellipses, 701–02
    italics, 702
  modifiers
    and correlative constructions, 436–37
    dangling, 437
    meaningless, 445
    misplaced, 435–37
    restrictive/nonrestrictive, 435, 692, 694
    squinting, 436
  nouns, 693
  of numbers, 690, 693, 695, 697–700,
      703–04
  parentheses, 697–98
  period, 690
  question mark, 691
  quotation marks, 693, 700–701
  semicolon, 695–96
  series comma, 691–92
purchase orders, 603
purpose
  in argument, 226
  and definitions, 238, 241, 250
  determination of, 101, 130–31, 213, 226
    in revision, 406, 438
  of graphics, 347, 351
  of memo, 496
  of proposal, 471
  of report, 525–26, 534
  tactics formulation, 134–36
  of Web page, 90

quality control, demonstrating in proposal,
    467
questionnaires, 176–80, 184
question mark, 691. *See also* punctuation
questions
  answering in report, 526
  appropriate, 57–58
  in argument, 228
  for interviews, 174–75, 176–80, 184
  journalistic, 147–48, 171
  Likert scale, 178–79
  for oral presentation, 678–79
  tag questions and qualifiers, 60–61
queuing, 322–23, 324
quotation
  block, 701
  design, 334

documentation of, 165, 176, 283, 292
pull quotes, 339, 340
punctuation for, 693, 700–701
quoting, in note taking, 165, 176

ratio graphs, 373–74
readability, design technique for, 9–10
reader. *See* audience
"real" subject, 432–33
"real" verb, 433–34
reasoning
  within argument, 221–22, 224
  post-hoc, 224, 527
recommendation report, 522–24
recommendations, in reports, 527–28, 533,
  534, 535
redundancy, 446–47
reference librarians, 152
reference list, 285–89
references. *See also* documentation; research
  documentation of, 32, 43, 281, 283–95,
    297, 365
  list of works cited, 289, 291–95
  in proposal, 473
  reference list, 285–89
  in reports, 533
  in résumé, 569–71
remote log-in. *See* Telnet
reports, 49, 205, 209, 221. *See also* executive
  summary; proposals; *specific report
  types*
  abstracts in, 533
  background for, 190, 206, 497–98
  definition, 492
  as deliverable from proposal, 463
  design, 313, 318, 326, 334
  directives, 499–500
  formal
    body, 533–41
    conclusions, 534, 537
    definition, 520
    introduction, 533, 534–35
    methods, 533–34, 535–36
    problem-solving model for, 524–33
    recommendations, 534, 537–41
    results, 534, 536–37
    sample, 542–53
    types of, 520–24
  format for, 493–99, 540
  informal
    formats for, 492–93
    guidelines for, 492–93
    memos as, 492, 494–99
request for proposal (RFP), 461–62,
  465–66
research
  documentation of, 281, 283–95, 365
    in proposal, 473
    in report, 527
  generating ideas for, 146–52
  primary, 143, 171–80
    experiments, 171–73
    field research, 172–73
    inspections, 171
    interviews, 173–76, 184–85
    letters of inquiry, 176
    questionnaires, 176–80, 184
    strategy development, 180–85
  résumé. *See* job-application materials
  role in problem-solving, 143–44

secondary, 143, 152–70
  abstract services, 158, 160
  basic research tools, 156–60
  carrying out, 182–84
  document-delivery services, 159
  government information, 160–61
  newspaper indexes, 159–60
  online catalogs, 157, 159, 184
  periodical indexes, 159
  reference works, 158–59
  research media, 153–56
  understanding the information, 161–70
research report, 4, 463–64
resources
  constraints on, 312, 353
  documenting in proposal, 468
  shared, 58–59
response to inquiry letter, 607, 609
responsibilities
  collaborative, 51
  documentation of
    in application letter, 584
    in résumé, 567–69
  of reader/audience, 111–12
résumé. *See* job-application materials
revision
  choosing words and phrases
    active/passive voice, 432, 439–40
    clichés, 443–44
    concision requirements, 444–48
    euphemisms, 444
    inoffensive/nonsexist language, 448–49
    jargon, 441–42
    meaningless modifiers, 445
    noun strings, 443
    obvious statements, 444–45
    pompous words, 447–48
    positive/negative constructions, 442–43
    prepositional phrases, 445
    redundancy, 446–47
    wordy phrases, 446
  for coherence, 400, 405–06
  collaboration in, 51
  executive summary, 282
  given/new strategy, 400–401
  graphics, 355–56
  headings, 407–08
  index, 301–02, 406
  for international audience. *See* multicul-
    tural audience
  lists, 408–10
  logical consistency for, 402–03, 405–06
    grammatical parallelism, 405, 415
  for multicultural audience, 125
  paragraphs, 410–19
    body/transitional, 410–11, 415–16
    manageable length, 413–15
    topic sentence, 412, 414
    structuring, 411–19
    support, 412–13
  reports, 493
  sentence. *See* sentence revision
  structured writing requirements, 450–52
  techniques, 401–05
    reader testing, 404–05
    self-revision, 401–03
  titles, 406–07
  Web page, 94–95
  whole document, 405–06
  on word processor, 72–74, 335–36, 403

RFP (request for proposal), 461–62, 465–66
rights, for employees, 23–25
Roman numerals. *See also* numbers
  in front matter, 278–79
  in manuals, 638, 641
  use in document, 318
rules, 338, 356. *See also* design
run-on sentences, 686

safety warnings, 10, 36, 356, 359
  color choices for, 627
  definitions, 626
  emphasizing in manuals, 624–28
  employer obligations for, 27–29
  usability testing for, 651
sales letter, 4, 607–10, 611
sampling, inadequate, 527
screens, 338–39, 356. *See also* design
screen shots, 384, 385
scripted oral presentation, 660, 674–75
secondary research. *See* research
self-confidence, 582, 676
semicolon, 695–96. *See also* punctuation
sentence. *See also* paragraphs
  characteristics of, 708–09
  comparison of items in, 687–88
  definitions within, 239–40
  as form of outline, 192–93
  for multicultural audience, 646–47
  number agreement in, 689
  omitted words in, 716
  outlines, 192–93
  topic sentence, 412, 414, 418, 583, 596
  run-on, 686
sentence fragments, 684–85
sentence length
  cultural patterns for, 120, 125
  modifying for audiences, 104, 107, 109,
    110
  for oral presentation, 669
  revising, 430–32
sentence revision. *See also* punctuation;
  revision
  active/passive voice, 432, 439–40
  expletives, 432–33
  formality/informality levels, 437–38
  information placement, 430
  lists for, 425–30
  modifiers
    and correlative constructions, 436–37
    dangling, 437
    meaningless, 445
    misplaced, 435–37
    restrictive/nonrestrictive, 435
    squinting, 436
  "real" subject identification, 432–33
  "real" verb identification, 433–34
  sentence length, 430–32
sentence style, 684–90
series comma, 691–92
sexual harassment, 23
simplification, 224, 527
Simplified English, 449–50, 451, 645–47
sketching
  to generate research ideas, 150–52
    branching, 151–52, 190
    clustering, 151, 190
skimming, 161–62
software. *See also* word processors
  disclaimers for, 34–35

software *(continued)*
documentation of/for, 289, 295, 311, 314, 384
draw and paint programs, 324, 391–94
for graphics, 354–55, 391–94
grammar checker, 73
groupware, 50
and electronic mail, 77–80
for hypertext documents, 72, 649, 650
for oral presentation, 672
project–management, 68, 479
for revision, 72–74, 335–36, 402–03, 430, 434
for translation, 450
spatial development, for argument organization, 200–03
speaking. *See also* oral presentation
to generate research ideas, 149–50
for communicating reports, 493
specificity, 438–39
spell checker, 72–73. *See also* word processors
spreadsheets, 391, 392
squinting modifiers, 436
statistics, 196
status report, 502–06, 521
stratum graphs, 372–73
structured writing principles, 450–52
style, 69–72, 167, 269, 440
style checker, 73
subject
agreement with verb, 712–13
of argument, 226, 227
in sentence, 432–33, 437, 716–17
subordination, 191–92
summary
executive, 107, 110
of key ideas, 166
in manuals, 641
of notes, 166–69
in oral presentation, 665
in proposal, 470–71
in report, 497, 540–41
symbols. *See* graphics

tab, 318
table of contents, 269, 276–79, 533
tables. *See* graphics
talking. *See* oral presentation; speaking
talking (generating ideas), 149–50
task schedule, in proposal, 467, 477–78, 480
technical communicators, 2–3
technician, as audience, 104, 106
Telnet, 81, 83, 289, 295. *See also* World Wide Web
tenses, 689–90
and conditions, 713–14
of verbs, 710–11
testing
audience/reader, 404–05, 625, 650–56
of questionnaire, 180
usability, 625, 650–56
text, 10, 77, 200. *See also* graphics
ASCII format, 577
boilerplate, 476
copyright protections for, 31
and graphics, 10, 200, 203, 206, 207, 214, 348–50
in instructions, 35

placement on page, 327, 330
text table, 366
textual citation, documentation of, 284–85, 289–91
thesaurus, 73–74, 403. *See also* dictionary; word processors
think-aloud protocols, 655
thumbnail sketches, 324, 325
tick marks, 368
time
attitudes to, 119–20
for collaborative writing, 52
time constraints, 133–34, 135, 311–12, 353
for oral presentations, 662, 664
time management, 59–60
time pattern, in reports, 504, 505
timeliness, of information, 145, 170
titles. *See also* headings
design, 337
of documents, 271–73
front matter requirements, 271–73
for oral presentation, 671
revising, 406–07, 694
tone, 134, 136, 540, 644
topic, choosing a, 145–46
topic outlines, 192–93
topic sentence, 412, 414, 418, 583, 596
Toulmin, Stephen, 221
trademark law, 32–33. *See also* copyright law
transitional words and phrases, 415–16
translation, 118, 450, 646. *See also* multicultural audience
transmittal letter, 269, 270–72
transparencies/slides, 358–59, 391, 668, 669. *See also* graphics
trimming, 26
trip report, 4, 143, 199, 500–501
troubleshooting guide, 631–33
truth. *See* ethics; honesty
turnovers, 429–30, 562–63
typeface, 330–34
bold/normal, 337
for electronic résumé, 577
for index, 303
for safety warnings, 626
for transparencies, 668, 669
typography. *See* design

underlining, 702
university, career-placement office, 558–59
usability testing, 625, 650–56
Usenet groups, 83, 155–56, 579
user-friendliness, 624, 625

verbs
in active/passive voice, 439–40, 568
agreement with subject, 712–13
convincing/communicating, 130–31
helping/main, 712
-ing forms, 711
in résumé, 568
in sentence, 433–34, 695
nominalizing, 433–34
tenses of, 710–11
videoconferencing, 81
visuals. *See* graphics
vocabulary. *See also* language
cultural concerns for, 120, 124

modifying for audiences, 104, 107, 109, 110
Simplified English, 449–50, 451, 645–47
voice
active
cultural patterns for, 125
use in résumé, 568
in oral presentation, 676–77
passive, 432, 439–40

warnings. *See* safety warnings
warranty, express and implied, 34
Web. *See* World Wide Web
whistleblowing, 40–42
white space, 326–27
whiteboards, 81
whole-by-whole organization, in comparisons, 209–11
word processors, 10. *See also* software
and drafting tools, 69–72
hypertext markup language, 72, 82, 93–94
outline feature, 69
templates and styles, 69–71
revision tools, 72–74, 403
grammar checkers, 73
spell checkers, 72–73, 403
thesaurus, 73–74, 403
words. *See also* text; vocabulary
commonly misused, 718–20
omitted, 716
repeated, 717
wordy phrases, 446
work experience. *See* employment documentation
work for hire, 3, 31
World Wide Web. *See also* electronic communication
communication on, 79–80
documentation of, 288, 294
file transfer protocols, 68, 81, 83, 155
graphics on/for, 354
home pages, 90–92
frames, 91
hypertext markup language (HTML), 72, 82, 93–94, 647–50
job postings/job boards, 88, 576
employer sites, 4, 558, 560, 561
Web-based résumé, 577
online discussion groups, 155–56
listserv discussion groups, 155–56, 288, 294
Usenet groups, 83, 155–56, 579
Web conversion tools, 72, 94
Web page creation, 4, 49, 88–96. *See also* design
page design, 92–93
site design, 90–92
Web page usage, 81–88
browsers, 81, 83, 85–86, 91, 95
copyright law, 31, 86–88, 354
for research, 83, 155–59, 170, 182–83, 288
search engines, 84–85, 95–96, 158, 579
Uniform Resource Locator (URL), 83–84
writer's block, 195

"you" attitude, 596–98, 607